Electrochemically Enabled Sustainability

Devices, Materials and Mechanisms for Energy Conversion

Electrochemically Enabled Sustainability

Devices, Materials and Mechanisms for Energy Conversion

Kwong-Yu Chan and Chi-Ying Vanessa Li

CRC Press
Taylor & Francis Group
Boca Raton London New York

CRC Press is an imprint of the
Taylor & Francis Group, an **informa** business

First published in paperback 2024

First published 2014 by CRC Press
2385 NW Executive Center Drive, Suite 320, Boca Raton FL 33431

and by CRC Press
4 Park Square, Milton Park, Abingdon, Oxon, OX14 4RN

First issued in hardback 2019

CRC Press is an imprint of Taylor & Francis Group, LLC

© 2014, 2019, 2024 Taylor & Francis Group, LLC

Reasonable efforts have been made to publish reliable data and information, but the author and publisher cannot assume responsibility for the validity of all materials or the consequences of their use. The authors and publishers have attempted to trace the copyright holders of all material reproduced in this publication and apologize to copyright holders if permission to publish in this form has not been obtained. If any copyright material has not been acknowledged please write and let us know so we may rectify in any future reprint.

Except as permitted under U.S. Copyright Law, no part of this book may be reprinted, reproduced, transmitted, or utilized in any form by any electronic, mechanical, or other means, now known or hereafter invented, including photocopying, microfilming, and recording, or in any information storage or retrieval system, without written permission from the publishers.

For permission to photocopy or use material electronically from this work, access www.copyright.com or contact the Copyright Clearance Center, Inc. (CCC), 222 Rosewood Drive, Danvers, MA 01923, 978-750-8400. For works that are not available on CCC please contact mpkbookspermissions@tandf.co.uk

Trademark notice: Product or corporate names may be trademarks or registered trademarks and are used only for identification and explanation without intent to infringe.

Publisher's Note
The publisher has gone to great lengths to ensure the quality of this reprint but points out that some imperfections in the original copies may be apparent.

Library of Congress Cataloging-in-Publication Data

Electrochemically enabled sustainability : devices, materials, and mechanisms for energy conversion / editors, Kwong-Yu Chan, Chi-Ying Vanessa Li.
pages cm
Summary: "With Contributions from leading researchers in their fields, this book provides an overview of the most important electrochemical power sources in development today. Focusing on materials, design, and performance, the text presents the most recent and innovative technologies employed in battery and fuel cell technology. Topics include acid/alkaline batteries, microbial fuel cells, Li-Air, Li-Sulphur, flow batteries, lithium ion batteries, lead acid batteries, supercapacitors, photoelectrochemistry, and carbon dioxide electroreduction. The book discusses the advantages of these fuel cells over conventional power sources and their future applications"-- Provided by publisher.
Includes bibliographical references and index.
ISBN 978-1-4665-7543-1 (hardback)
1. Electric batteries. 2. Direct energy conversion--Materials. 3. Electrochemistry. I. Chan, Kwong-Yu (Chemical engineer), editor of compilation. II. Li, Chi-Ying Vanessa, editor of compilation.

TK2896.E46 2014
621.31'242--dc23
2014007209

ISBN: 978-1-4665-7543-1 (hbk)
ISBN: 978-1-03-291909-6 (pbk)
ISBN: 978-0-429-07210-9 (ebk)

DOI: 10.1201/b17062

Visit the Taylor & Francis Web site at
http://www.taylorandfrancis.com

and the CRC Press Web site at
http://www.crcpress.com

Contents

Preface vii
Editors ix
Contributors xi

1. Electroreduction of Carbon Dioxide 1
   Daniel A. Lowy and Maria Jitaru

2. Microbial Fuel Cells and Other Bio-Electrochemical
   Conversion Devices 55
   Shaoan Cheng and Weifeng Liu

3. Lithium Batteries: Status and Future 121
   Bruno Scrosati and Jusef Hassoun

4. Hollow Mesoporous Carbon with Hierarchical
   Nanoarchitecture in Electrochemical Energy Storage
   and Conversion 163
   Min-Sik Kim, Dae-Soo Yang, Min Young Song, Jung Ho Kim
   and Jong-Sung Yu

5. Layer-Structured Cathode Materials for Energy Storage 191
   Bohang Song, Man On Lai and Li Lu

6. First-Principles Approach for Cathode Design
   and Characterisation 223
   Dong-Hwa Seo, Inchul Park and Kisuk Kang

7. Advanced Batteries and Improvements in Electrode Materials 255
   Ashok Kumar Shukla, Vedam Ganesh Kumar
   and Musuwathi Krishnamoorthy Ravikumar

8. Lead–Carbon Hybrid Ultracapacitors and Their Applications 321
   Ashok Kumar Shukla, Anjan Banerjee
   and Musuwathi Krishnamoorthy Ravikumar

9. Vanadium Flow Batteries: From Materials to Large-Scale
   Prototypes 347
   Huamin Zhang

10. **Physical Properties of Negative Half-Cell Electrolytes in the Vanadium Redox Flow Battery** 395
Asem Mousa and Maria Skyllas-Kazacos

11. ***pH* Differential Power Sources with Electrochemical Neutralisation**.. 429
Huanqiao Li, Chi-Ying Vanessa Li, Guo-Ming Weng and Kwong-Yu Chan

Index 469

Preface

Electrochemical conversion plays a central role and an indispensable linkage in the road map of sustainable energy technologies. Electricity can be generated by renewable sources of hydro, wind, solar power, and biomass, but cannot be stored on a large scale. Although electricity is the most convertible form of energy, it has to be distributed via a grid. Contrary to this, chemical energy is very portable and storable, but not naturally renewable. Storing electricity in chemical form and converting back to end use via electrochemical technologies can, therefore, accelerate the replacement of fossil fuels by renewable sources. The chapters of this book give ample illustrations of how key technologies for sustainability are enabled by electrochemical conversion, prompting the book's title *Electrochemically Enabled Sustainability*.

Lowy and Jitaru show how electrochemical reduction of carbon dioxide solves collectively the problems of carbon capture, energy storage, and generation of portable fuel. Treating wastes with energy recovery is demonstrated with microbial fuel cells by Cheng and Liu. A large portion of the book is devoted to lithium-type batteries, which are applied to vehicle propulsion and energy storage. The broad range of lithium batteries are clearly presented by Scrosati and Hassoun, and the nano-structuring issues are discussed by Shukla et al. Layered-structure materials for lithium ion cathodes are discussed by Song et al., while Kim et al. demonstrated how very high current can be generated with a hierarchical structure. Zhang illustrates the promise of vanadium redox flow batteries convincingly with prototypes for massive electricity storage. The properties of vanadium electrolytes which critically affect performance are addressed by Mousa and Skyllas-Kazacos. Improved performance in hybrid devices is illustrated by Shukla et al. with an ultra-capacitor/lead acid power source; and by Li et al. in various acid–alkaline power sources. In addition, we have a chapter by Seo et al. on first-principles computational approach applied to the design of cathodes for lithium ion batteries.

A wide range of electrochemical devices, the key materials, and operation mechanisms are presented in the chapters. Hence, we added the book's subtitle, *Devices, Materials, and Mechanisms for Energy Conversion*. Though a wide range of topics are covered, this monograph is by no means comprehensive. We only hope to give timely coverage of critical issues in emerging and conventional technologies. This monograph was proposed by Professor Felix Wu as an important project of the Initiative on Clean Energy and Environment (ICEE) at University of Hong Kong. The project started with a workshop on Electrochemical Conversion at the University of Hong Kong where many contributors met. We are fortunate to have the contributed time from this collection of internationally acclaimed experts. Special thanks are given to Jenkin Tsui who gave tireless help in editing and proofreading. Li Ming Leong of CRC Press has been very helpful in smoothing out the publishing progress.

The creation of this cover design has gone through many versions of artwork by Elise Weinger. The cover design was adapted from the *Tai Chi* symbol. It is not our intention to advocate or to discredit *Taoism*. Similarities are drawn between *Yin-Yang* and the polarity in electrochemical conversion. Dynamic interactions of polarity and ions leads to devices and applications. Connotations of recycle, regeneration, conversion, and harmony build the vision of sustainability.

Kwong-Yu Chan and Chi-Ying Vanessa Li

Editors

Kwong-Yu Chan grew up in Hong Kong where he completed his secondary school education. After obtaining his BSc in chemical engineering at University of Alberta, he worked as an assistant lecturer of Hong Kong Polytechnic and a project engineer in Hong Kong Oxygen & Acetylene Co. He completed his PhD thesis on Monte Carlo Simulation of Electrolytes under the supervision of Professor Keith Gubbins at Cornell University. He was a postdoctoral fellow of Professor Robert Savinell at Case Western Reserve University, where he participated in an Al-Air battery project. In 1988, he joined the Department of Chemistry, University of Hong Kong and was promoted to full professor in 2002.

Professor Chan has fundamental and applied research activities in molecular simulation, fuel cells, materials and electrochemical applications. He has published over 150 papers and is a top 1% cited scientist, according to ISI's *Essential Science Indicators*. He is an active member of the Electrochemical Society and American Institute of Chemical Engineers, and has been a regional editor of *Molecular Simulation* and *Journal of Experimental Nanoscience*. Professor Chan has five inventions on the topics of fuel cells, ozone generation and batteries. He received the Croucher Foundation Senior Research Fellowship (2010), Universitas 21 Fellowship (2010), Salzburg Global Fellowship (2008), and the HKU Outstanding Researcher (1998, 2013). Professor Chan held sabbatical visiting positions at the University of Utah and Tsinghua University, and is an honorary adjunct professor of Beijing University of Chemical Technology.

Chi-Ying Vanessa Li graduated with double degrees in BEng and BCom majoring in engineering science and accounting from the University of Auckland in 2005. She then pursued her PhD studies at the University of New South Wales, focusing on hydrogen storage and anode materials under the supervision of Professor Sammy L.I. Chan. Upon completing her studies in 2009, she joined the Department of Chemistry, University of Hong Kong, as a postdoctoral fellow. She has carried out collaborative research with universities in mainland China and Taiwan.

Dr. Li's current work focuses on electrochemistry and catalysis. Her research interests include anode materials on lithium batteries, flow batteries, and MOFs (metal-organic frameworks) for catalytic applications. She has published more than 20 articles in various peer-reviewed journals. Dr. Li is a member of the Electrochemical Society and American Institute of Chemical Engineers. She is also accredited by Engineers Australia as a chartered professional engineer (CPEng), and Institute of Materials, Mineral and Mining as a chartered scientist (CSci).

Contributors

Anjan Banerjee
Solid State and Structural
 Chemistry Unit
Indian Institute of Science
Bangaluru, India

Kwong-Yu Chan
Department of Chemistry
The University of Hong Kong
Hong Kong

Shaoan Cheng
Department of Energy
 Engineering
Zhejiang University
Hangzhou, China

Jusef Hassoun
Department of Chemistry
University of Rome Sapienza
Rome, Italy

Maria Jitaru
Research Institute for Organic
 Auxiliary Products (ICPAO)
Medias, Romania

Kisuk Kang
Department of Materials Science
 and Engineering
Seoul National University
Seoul, Republic of Korea

Jung Ho Kim
Department of Advanced Materials
 Chemistry
Korea University
Sejong, Republic of Korea

Min-Sik Kim
Department of Advanced Materials
 Chemistry
Korea University
Sejong, Republic of Korea

Vedam Ganesh Kumar
Power Logics Company
Seoul, Republic of Korea

Man On Lai
Department of Mechanical
 Engineering
National University of Singapore
Singapore

Huanqiao Li
Department of Chemistry
The University of Hong Kong
Hong Kong

Chi-Ying Vanessa Li
Department of Chemistry
The University of Hong Kong
Hong Kong

Weifeng Liu
Department of Energy Engineering
Zhejiang University
Hangzhou, China

Daniel A. Lowy
FlexEl, LLC
College Park, Maryland
United States of America

Li Lu
Department of Mechanical
 Engineering
National University of Singapore
Singapore

Asem Mousa
School of Chemical Engineering
University of New South Wales
Sydney, Australia

Inchul Park
Department of Materials Science
and Engineering
Seoul National University
Seoul, Republic of Korea

Musuwathi Krishnamoorthy Ravikumar
Solid State and Structural
Chemistry Unit
Indian Institute of Science
Bangaluru, India

Bruno Scrosati
Italian Institute of Technology
Genova, Italy

Dong-Hwa Seo
Department of Materials Science
and Engineering
Seoul National University
Seoul, Republic of Korea

Ashok Kumar Shukla
Solid State and Structural
Chemistry Unit
Indian Institute of Science
Bangaluru, India

Maria Skyllas-Kazacos
School of Chemical Engineering
University of New South Wales
Sydney, Australia

Bohang Song
Department of Mechanical
Engineering
National University of Singapore
Singapore

Min Young Song
Department of Advanced Materials
Chemistry
Korea University
Sejong, Republic of Korea

Guo-Ming Weng
Department of Chemistry
The University of Hong Kong
Hong Kong

Dae-Soo Yang
Department of Advanced Materials
Chemistry
Korea University
Sejong, Republic of Korea

Jong-Sung Yu
Department of Advanced Materials
Chemistry
Korea University
Sejong, Republic of Korea

Huamin Zhang
Dalian Institute of Chemical
Physics
Chinese Academy of Science
Dalian, China

1

Electroreduction of Carbon Dioxide

Daniel A. Lowy and Maria Jitaru

CONTENTS

1.1 Introduction 2
    1.1.1 Carbon Dioxide Emission: A Significant Threat of Our Time 2
    1.1.2 Capture of Carbon Dioxide from the Atmosphere 5
    1.1.3 Synthesis of Fuels from Carbon Dioxide 8
        1.1.3.1 Problem of Global Significance 8
    1.1.4 Artificial Photosynthesis: A Challenge of Present Days.............. 9
    1.1.5 Development of Cost-Effective Electrochemical Reduction of Carbon Dioxide 11
1.2 Electrochemical Reduction of Carbon Dioxide 12
    1.2.1 Thermodynamics of CO_2 Electroreduction 12
    1.2.2 Kinetics of CO_2 Electroreduction 14
    1.2.3 Mediated Electrocatalytic Reduction of CO_2 15
    1.2.4 Electroreduction of CO_2 in Organic Media 18
1.3 Materials, Components and Operation Parameters 20
    1.3.1 Experimental Methodology for CO_2 Electroreduction.............. 20
    1.3.2 Choice of Cathode Materials with Favourable Electrocatalytic Properties 21
    1.3.3 Copper: A 'Magic' Electrode Material 21
    1.3.4 Use of Gas Diffusion Electrodes 24
    1.3.5 Need for High-Purity Electrolytes 25
    1.3.6 Electrochemical Reactors for CO_2 Reduction 25
    1.3.7 Cell Separators 27
    1.3.8 Electrochemical CO_2 Conversion at High Current Density 28
1.4 Organic Products Electrosynthesised from Carbon Dioxide 29
    1.4.1 Synthesis of Formate Salts/Formic Acid 29
    1.4.2 Preparation of Organic Carbonates from CO_2 31
    1.4.3 Synthesis of Methanol 34
        1.4.3.1 Pyridinium Ions: Potent Catalysts Enabling CO_2 Reduction to Methanol 34
    1.4.4 Synthesis of Ethanol and 2-Propanol 37
    1.4.5 Hydrocarbon Synthesis 37
1.5 Future Directions 39
References 43

1.1 Introduction

As carbon dioxide emission represents a significant threat of our time, this chapter reviews (i) the capture of CO_2 from the atmosphere and (ii) its cost-effective electrochemical conversion to fuels or other useful organic products. This chapter addresses the thermodynamics and kinetics of electrochemical CO_2 reduction, including its mediated electrocatalytic reduction and its electroreduction in organic media. This chapter discusses the experimental methodology: the use of gas diffusion electrodes (GDEs) and of various electrochemical reactors, equipped with different cell separators; it emphasises the importance of choosing suitable cathode materials that offer favourable electrocatalytic properties (such as copper, a kind of 'magic' electrode material for liquid fuel synthesis) and of using high-purity electrolytes. Several sections are dedicated to various products that can be obtained by the electroreduction of CO_2: formate salts and/or formic acid, organic carbonates, methanol, ethanol and 2-propanol, as well as short-chain and long-chain hydrocarbons. A discussion of future directions adopted in CO_2 electroreduction concludes this chapter.

1.1.1 Carbon Dioxide Emission: A Significant Threat of Our Time

Accumulation of emitted carbon dioxide has become a severe environmental problem, as it contributes to the greenhouse effect, which leads to global warming. Over the past 130 years, the average temperature of the near-surface of Earth (land and ocean) rose considerably, by close to +1°C. Limiting us to a more specific example, the 2012 November temperature was +0.67°C above the average November temperatures recorded over the past 133 years [13]. With the rapid increase of global population and the unprecedented industrialisation, the energy consumption of mankind grows explosively [1].

At present, more than 85% of the global energy demand is supported by burning fossil fuels, which will continue to play a primary role in the foreseeable future [4]. The combustion of these fossil fuels releases large amounts of carbon dioxide in the atmosphere, and anthropogenic emissions exert a noticeable influence on global climate in a very short period of time [5]. Since the beginning of the industrial age, around 1750, the concentration of carbon dioxide in the atmosphere has increased from 280 to 390 ppm in 2010 [4]. Over the past 2 years, the concentration of CO_2 in the atmosphere has kept growing. It was 391.6 ppm in 2011, while in November 2012, it reached 392.6 ppm [3]. Less than half of the worldwide carbon dioxide emissions originate from large sources, such as coal-burning power plants and chemical facilities, while a significant part of the remaining emissions come from mobile sources, including buses, cars, planes and ships, where capture per ton of CO_2 would be much more costly [6].

The increase in atmospheric carbon dioxide over the past 40 years is shown in Figure 1.1. Data have been recorded at the Mauna Loa Observatory in Hawaii, and the measurements were made and reported independently by two scientific institutions: Scripps Institution of Oceanography (San Diego, CA) and the National Oceanic and Atmospheric Administration (NOAA, Washington, DC) [3]. A horizontal line indicates the upper safety limit for atmospheric CO_2, equal to 350 ppm. As revealed by the graph, starting early 1988, the CO_2 concentration has exceeded the safety limit. While the average annual increase over the 1992–2001 decade was 1.6 ppm per year, from 2002 to 2011, the annual rise was more severe, reaching 2.07 ppm per year [3].

In automotive applications, the combustion of various liquid fuels yields gases, which are emitted at temperatures higher than the ambient temperature. When mixed with the atmosphere, emitted gases increase the air mass temperature. Among the exhaust gases, carbon dioxide poses the most significant concern, as it is a highly stable molecule and, therefore, very difficult to be reconverted into useful chemicals. This is why CO_2 continues to increase in the atmosphere, being the main contributor to the greenhouse effect [7]. Also, carbon dioxide participates in radiative heat transfer. If no other cooling medium can dissipate this thermal emission, global warming will persist; one can only minimise the rate of global warming, rather than eliminating it. As CO_2 represents the primary anthropogenic greenhouse gas, which is believed to be the leading culprit in climate change [1], controlling global warming requires the reduction of thermal emission and of its constituents.

Carbon dioxide recovered from concentrated sources, including metallurgical reduction furnaces, cement kilns and fossil power plants, represents a

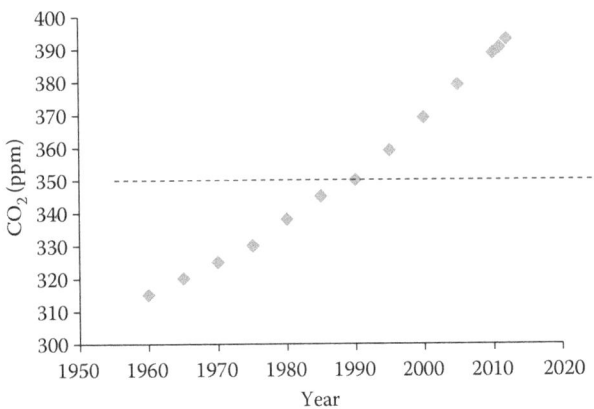

FIGURE 1.1
Increase in atmospheric carbon dioxide starting 1960; graph based on data recorded at Mauna Loa Observatory in Hawaii. The horizontal line indicates the upper safety limit for atmospheric CO_2 (350 ppm). (Adapted from the data collected by the Mauna Loa Observatory and Scripps Institution of Oceanography. See: Earth's CO_2 Home Page. http://co2now.org/ (January 3, 2013).)

valuable feedstock for synthetic fuels provided that the conversion is done by means of renewable energy sources, such as nuclear, solar, wind and hydropower sources [8].

Over the past several years, there has been a growing interest in the capture of CO_2 emissions, and their subsequent permanent immobilisation or chemical conversion to industrially relevant products. A large number of processes have been developed and studied; unfortunately, many of these methods are expensive, since they require either ultra-high-purity CO_2 or are energy intensive. Many purely chemical methods show low product selectivity. Therefore, electrochemical processes are being developed to increase reaction pathway selectivity and to reduce cost, since electrochemistry offers a means for direct control of the surface free energy through the electrode potential [9].

In order to decrease the concentration of CO_2 in the atmosphere, carbon dioxide recovery and utilisation technologies are being developed [10]. Reducing CO_2 in the atmosphere requires (i) the capture of carbon dioxide, followed by (ii) its transportation and storage or (iii) its conversion to useful products, in an affordable manner. As carbon dioxide is the product of the complete oxidation of organic derivatives, its conversion must involve a reduction process.

Six main options are available for the reduction of CO_2: (1) photosynthesis, a natural process, (2) catalytic reduction, (3) photocatalytic reduction (using either semiconductor particles or a photoanode and a photocathode), (4) photoelectrocatalytic (by using either an anode in combination with a photocathode, or conversely, a photoanode in conjunction with a regular cathode), (5) photovoltaic-electrocatalytic (which operates with a combination of a photovoltaic and electrochemical cell) and (6) electrochemical reduction [2,7,11]. Electroreduction of CO_2 shows similarities with photosynthesis, which converts CO_2 to carbohydrates. In both processes, the electrons are supplied by the oxidation of water to molecular oxygen. Also, both conversions proceed under most favourable conditions near to neutral *pH* [12].

A schematic of the main avenues of converting carbon dioxide into useful products is provided in Figure 1.2. The possible pathways towards obtaining various chemicals from CO_2 are shown. Some details on the catalytic reactions shown in Figure 1.2 are the following: CH_4 is obtained via the Sabatier reaction of CO and H_2, urea is synthesised by the reaction of CO_2 and ammonia, while $Na^+C_7H_5O_3$ (sodium salicylate) is obtained by the carboxylation of sodium phenoxide ($C_6H_5O^-Na^+$) with CO_2 [7]. Conversion of CO_2 into fuels and commodity chemicals represent important contemporary energy and environmental challenges. It can be accomplished by either catalytic reductive transformation of CO_2 (see left panel of Figure 1.2) or its electrolytic conversion (see top panel of Figure 1.2) [13].

Another CO_2 conversion process, also contained in Figure 1.2, is the co-electrolysis of CO_2 and steam to produce synthesis gas (referred to as *syn gas*), at 100% efficiency (Equation 1.1). Co-electrolysis is a version of

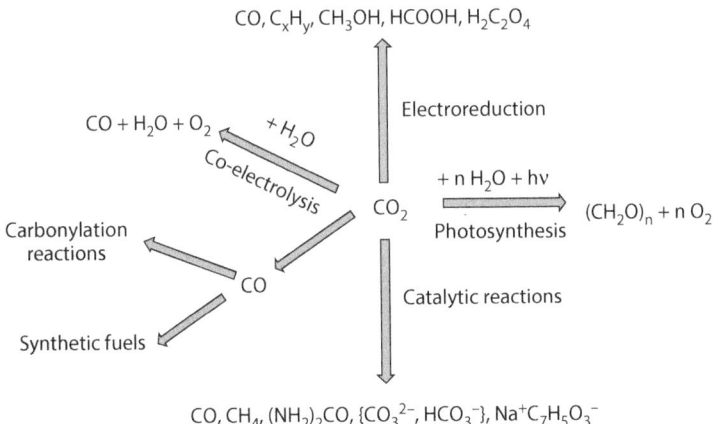

FIGURE 1.2
Main avenues of converting carbon dioxide into various useful chemicals. (Adapted from M. Jitaru, *J. Univ. Chem. Technol. Metallurgy* 42, 2007, 333–344; J. Hartvigsen et al., Carbon dioxide recycling by high temperature co-electrolysis and hydrocarbon synthesis. In N.R. Neelameggham and R.G. Reddy (eds.), *Carbon Dioxide Reduction Metallurgy*, Publication of The Minerals, Metals & Materials Society (TMS), Warrendale, Pennsylvania, 2008, 171–182.)

high-temperature steam electrolysis. This process has been accomplished by using reversible solid oxide fuel cell technology, utilising for the high-temperature process heat generated by a nuclear reactor or a solar dish concentrator. Voltage is applied across a gas-tight and electrically insulating ceramic membrane, typically zirconia doped with trivalent cations (e.g. Y_2O_3). During electrolysis, both water and carbon dioxide are decomposed, O^{2-} ions are transported across the ceramic membrane and a mixture of $\{CO + H_2\}$ is left behind [8].

$$H_2O + CO_2 \rightarrow H_2 + CO + O_2 \tag{1.1}$$

The obtained syn gas may serve as a feedstock for synthetic fuel production, by a subsequent catalytic conversion, for example, in a modular small-scale Fischer–Tropsch reactor [8].

An alternative avenue is to reduce carbon dioxide to its elemental form, by applying energy in a cost-effective manner, a technology widely used by extractive metallurgists for reducing various metal oxides. This process could minimise the amount of CO_2 released in the air, and would result in fuel's improved self-sufficiency.

1.1.2 Capture of Carbon Dioxide from the Atmosphere

Carbon dioxide capture and storage (CSS) have attracted worldwide interest from both fundamental and practical standpoints. There are attempts to develop a technologically and economically advantageous CSS [14,15]. Three basic CO_2 separation and capture avenues have been adopted: (i) pre-combustion capture (the reaction of a primary fuel with oxygen or air to produce H_2), (ii) oxy-fuel combustion (the process needs nearly pure oxygen instead of air for fuel combustion; almost pure CO_2 is generated, which can be stored directly) and (iii) post-combustion capture (i.e. removing CO_2 from flue gas, which consists mainly of N_2 and CO_2, prior to its emission into the atmosphere) [4]. The most mature technology is post-combustion amine absorption, known as *amine scrubbing*, which has been employed on the industrial scale to separate carbon dioxide from natural gas and hydrogen since 1930. It is a robust technology, ready for being tested and used for CO_2 capture from coal-fired power plants [16]. Monoethanolamine (MEA)-based CO_2 capture technology is being applied on the industrial scale, the average cost of captured carbon dioxide being in the range of US$ 23–35 (t of CO_2). Also, a recovery process based on sterically hindered amines has been introduced, and three proprietary solvents (KS-1, KS-2 and KS-3) developed (by the Kansai Electric Power Co., Mitsubishi Heavy Industries, Ltd.). Starting 1999, this technology is applied in a commercial plant, operated by Mitsubishi in Malaysia, to recover 200 tCO_2/day from a flue gas stream. The cost of recovery is of US$ 24/(t of CO_2) [17].

Another approach towards reducing global atmospheric CO_2 attempts to isolate the surplus CO_2 from the biosphere, by permanently fixing it as an insoluble mineral onto the sea bottom. This method is based on the concept that insoluble $CaCO_3$ can be formed by the direct electrolysis of seawater, which yields the hydroxide ions reacting according to Equation 1.2.

$$Ca^{2+}(aq) + 2HCO_3^-(aq) + 2OH^-(aq) \rightarrow CaCO_3(s) + CO_3^{2-}(aq) + 2H_2O\ (l)\ (1.2)$$

Solid calcium carbonate would be disposed by itself, as it can settle directly onto the sea bottom. By this treatment, the concentration of carbonate can be reduced in the seawater, without using any additives or generating secondary wastes; hence, it has potential as a green-oriented method that is able to contribute to solving the problem of global warming [7].

Membranes exhibit a great ability to separate CO_2/H_2 mixtures in pre-combustion capture, and CO_2/N_2 mixtures in post-combustion phase. Membranes are especially designed and manufactured materials, which allow selective gas permeation across them. Membrane processes are continuous, energy efficient passive operations which are mechanically simple and non-polluting. In addition, since membrane separations exploit differences in both solubility and diffusivity, they can provide high selectivity for particular separations. Usually, high-pressure gas streams are preferred in membrane separations [17–23]. A schematic of the membrane separation process is shown in Figure 1.3 (adapted from Ref. [17]), where the gas stream to be separated consists of CO_2 and H_2; molecular hydrogen crosses

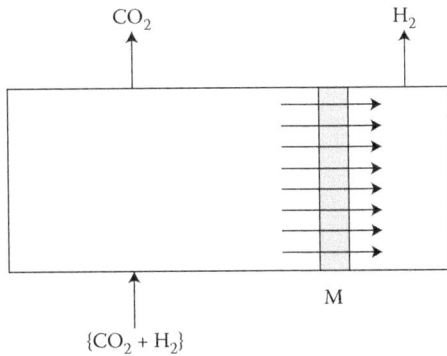

FIGURE 1.3
Schematic of the membrane separation process. (Adapted from K. Thambimuthu, M. Soltanich and J.C. Abandes, (Coordinating Lead Authors) Capture of CO_2. In B. Metz, O. Davidson, H. de Coninck, M. Loos and L. Meyer (eds.), *IPCC Special Report on Carbon Dioxide Capture and Storage.* Cambridge University Press, Cambridge, New York, Sào Paulo, 2005, 110–111, 115.)

the membrane (M), exiting through the chamber to the right, while carbon dioxide is being collected from the chamber to the left.

Some of the available membrane separation processes can already be applied on an industrial scale. Hence, inorganic ceramic membranes (zeolites and their derivatives, e.g. silico aluminophosphates), organic polymer membranes and facilitated transport membranes, which rely on a carrier molecule with high CO_2 affinity to achieve selective CO_2 transport (such as metallic ions or liquid amines), have been used in separating CO_2 from flue gas in post-combustion. As single-stage separation with these membranes is still difficult, new membrane materials are being developed [1]. Typically, the initial separation of carbon dioxide accounts for 60–80% of the total cost of CO_2 sequestration [24,25].

Activated carbons loaded with nickel oxide were useful in carbon dioxide capture. Post-oxidation methods were used for coating activated carbons with NiO, such as the electroless plating of nickel at 573 K, in air stream. Carbon dioxide retained by nickel oxide-loaded activated carbons was monitored by the mass of the CO_2, which was adsorbed at 298 K and 1.0 atm. As post-oxidation time increased, the adsorption capacity of CO_2 was enhanced [26].

Calcium oxide can absorb CO_2 from pre-combustion systems. Natural sources of CaO, such as limestone, $CaCO_3$, produce sorbents that lose reactivity relatively fast. Good CO_2 sorbents were prepared with CaO precursors templated on three natural polysaccharides: chitosan, agar and carrageenan, or three synthetic polymers: poly(acrylic acid), poly(ethylene glycol) and poly(ethylene oxide-*b*-propylene oxide-*b*-ethylene oxide), respectively. Calcium oxide confined onto synthetic polymers exhibited better CO_2 uptake activity and stability than CaO derived from commercial

$CaCO_3$. Given the vast array of available synthetic polymers, one can foresee a promising future for synthetic polymers in the field of CO_2 capture [27].

Alkali-metal-based oxides, such as alkali metal oxides, binary oxides and hydrotalcite-like compounds (layered double hydroxides), are promising adsorbents for CO_2 capture, owing to their relatively high adsorption capacity, low cost and wide availability. The microstructure of these oxides, such as their surface area, porosity, particle size and dispersion, determines the CO_2 adsorption capacity and multi-cycle stability [28].

Several innovative new materials are being researched and tested for the efficient capture of CO_2 from the atmosphere. Olah and co-workers [29] proposed the use of easy-to-prepare solid materials, which are based on fumed silica, impregnated with polyethylenimine (PEI). Such materials were found efficient for capturing CO_2 directly from the air. Despite the very low concentration of the adsorbents, they effectively scrubbed all the CO_2 from the air within a few hours.

Metal-organic frameworks (MOFs) represent a new class of crystalline materials of extremely high porosity that can efficiently capture CO_2. MOFs are composed of metal-containing nodes, which are linked by organic ligands and bridges; these structures are self-assembled by the reaction between various metal ions and organic linkers. This reaction proceeds under mild conditions, and the resulting well-ordered structures are kept together by strong coordination bonds [1]. Among porous materials, MOFs possess the greatest surface area and exhibit the highest CO_2 storage capacity. Hence, MOF-210 has a Langmuir surface area of 10,400 m^2/g and a total CO_2 storage capacity of 2870 mg/g. The volume-specific internal surface area of MOF-210, 2060 m^2/cm^3, is equivalent to the outer surface of nanoparticles (3-nm cubes), and approaches the ultimate adsorption limit for solid materials [30]. Recently, MOF-based membranes have been fabricated and examined for gas separation and CO_2 capture. C-oriented ZIF-69 membranes were synthesised on porous alpha-alumina substrates, to enable for the separation of CO_2/CO gas mixtures [31]. Jeong and co-workers [32] were the first to prepare a MOF membrane (ZIF-8) with controlled microstructure, which showed potential suitability for CO_2 capture in real environments.

1.1.3 Synthesis of Fuels from Carbon Dioxide

1.1.3.1 Problem of Global Significance

Conversion of CO_2 into synthetic fuels and organic precursors addresses three problems of global significance related to fossil resources: (i) shortage of energy, (ii) shortage of carbon resources and (iii) global warming. Reduction of CO_2 to fuels can provide a path for using CO_2 as a chemical feedstock, and by this, it may contribute to mitigate global warming.

Methanol, ethanol or hydrocarbons obtained from CO_2 reduction help in satisfying the increasing demand for energy sources. Electroreduction of CO_2 also enables the preparation of organic molecules containing carbonyl functionality, which can eventually serve as important precursors in organic synthesis [33–35].

Efforts to obtain synthetic hydrocarbon fuels from CO_2 have been motivated by the unparalleled energy density of hydrocarbons, which constitute the backbone of mankind's energy infrastructure [36], and by the need to cope with increasing atmospheric CO_2 released by the burning of fossil fuels. Synthetic fuels originating from CO_2 may become a potential component of a carbon energy cycle, as shown schematically in Equation 1.3, where the fuels can be methane, a longer-chain hydrocarbon, methanol, or C_2–C_3 alcohols [7].

$$CO_2 + energy \rightarrow fuels \rightarrow CO_2 + energy \qquad (1.3)$$

While electricity can be produced in a renewable manner by various methods, the manufacture of dense, portable fuels remains one major challenge. In order to accomplish this goal, one should devise sustainable systems that can produce fuels by converting carbon dioxide and water, utilising naturally available solar, wind or hydro energy [11,37].

Electrochemical methods of converting CO_2 into fuels may play an important role in coping with the problem of anthropogenic carbon dioxide production. Solving the problem of atmospheric CO_2 reduction needs, however, a more generic approach, known as artificial photosynthesis. The next section takes a closer look into this topic.

1.1.4 Artificial Photosynthesis: A Challenge of Present Days

There is one major biological process on Earth that harnesses solar energy for fuel production, and that is photosynthesis. By this process, plants, algae and some bacteria use light energy for reducing atmospheric carbon dioxide to carbohydrates. The reducing agent utilised by plants is hydrogen from water, and as a result, oxygen is also produced [37], while chlorophyll serves as a catalyst [38]. Photosynthesis is a complex multi-stage reaction, which requires 469 kJ/mol of energy, supplied by the sun. Green plants probably began 2500 million years ago, and owing to the released oxygen, the earth's atmosphere reached the present level of 21 vol.% about 580 million years ago, while human presence dates back 'only' 1 million years. Carbohydrates produced by photosynthesis (see right panel in Figure 1.2) represent the energy source for the fossils fuels of coal, oil and natural gas [12]. Therefore, by converting CO_2 into organic compounds, photosynthesis serves as a model paradigm for how to use solar energy for making fuels and how to solve the problem of greenhouse gases. Hence, photosynthesis is a widespread example of CO_2 fixation; it builds organic compounds from carbon dioxide and water using solar energy. Over the past several decades, chemists have studied this naturally occurring carbon fixation process as a model for manufacturing synthetic fuels [14,37,39].

Artificial photosynthesis is defined as the process, which exploits the physics and chemistry underlying natural photosynthesis for harvesting solar energy for technological purposes. Examples include (i) photovoltaic cells, based on inorganic or organic semiconductors, for the production of electricity, (ii) dye-sensitised solar cells, (iii) systems for fuel production based on such devices, (iv) devices for fuel production, based on the use of organic or inorganic photocatalyst and (v) other related systems. A simplified definition of artificial photosynthesis is a process that uses molecular species for converting energy from sunlight in a useful fuel for storage and mobile use [40].

The reduction of carbon dioxide to fuels requires some form of energy to defeat the highly negative free energy of formation of CO_2. Reduction to fuels is more efficient than just capturing CO_2 by pushing it back into the ground [7]. Photosynthetic reactions can be broken down into four steps: (1) harvesting solar energy, (2) light-induced redox reactions, (3) oxidation of water, that is, the use of water as an omnipresent source of electrons and (4) the reductive synthesis of fuels, which can be utilised as a mobile energy source [37,41]. A similar sequence of steps is involved in the electroreduction of carbon dioxide, where one invests electricity instead of solar energy, the redox reactions being induced by the electric field, and the role of chlorophyll is played by an electrocatalyst, while CO_2 serves as the carbon source and water acts as a proton donor.

As anthropogenic carbon dioxide production exceeds the planet's carbon dioxide recycling capability, it causes significant environmental harm [38]. For that reason, the natural carbon cycle should be supplemented by an 'artificial photosynthesis' process, which recycles carbon dioxide into hydrocarbons, using renewable energy sources, hence, reducing reliance on fossil fuels. In artificial photosynthesis, both starting materials (widely available water and CO_2) are renewable, while the required energy for the synthetic carbon cycle can be supplied by any non-fossil-fuel energy source, such as solar, wind, geothermal or nuclear energy. The artificial photosynthesis process begins by capturing CO_2 from natural sources, including the ambient atmosphere, or industrial sources such as smokestacks [38]. It is intended to extract CO_2 directly from the air, using newly developed adsorbent materials. Adsorbed CO_2 can then be converted by chemical or electrochemical reactions into fuels, such as methanol, dimethyl ether, synthetic hydrocarbons or proteins for animal feed. This concept is the basis of the so-called methanol economy [38]. In a broader context, the capture of CO_2 from the atmosphere would enable a closed-loop carbon-neutral fuel cycle. Using renewable or nuclear energy, carbon dioxide and water can be recycled into liquid hydrocarbon fuels in non-biological processes, which constitute the reverse of fuel combustion, as they remove oxygen from CO_2 and H_2O [42].

Several processes have been developed for CO_2 fixation; however, many of them are expensive, since they either require ultra-high-purity CO_2 or are energy intensive. In addition, many of the chemical methods show low product selectivity. Such limitations can be overcome by applying electrochemical procedures, which increase reaction pathway selectivity and reduce costs, as they allow direct control of the surface free energy by setting the electrode potential [9]. Feasible routes towards synthetic fuels are either from direct electroreduction of CO_2 [36] or indirect procedures, where H_2O and CO_2 are electrolysed in solid oxide electrolysis cells to yield syngas (synthesis gas), which contains different amounts of CO and H_2. Then, the syngas is converted to gasoline or diesel fuel by Fischer–Tropsch synthesis [42]. While liquid fuel preparation is in its early stage of development, more mature are the procedures of electrochemical CO_2 reduction to a wide variety of useful products, including carboxylic acids, aldehydes, alcohols and carbon monoxide. Such compounds can be obtained under mild conditions, but at an energy conversion efficiency of only 30–40% [14].

As an overall consideration, any scalable system that uses solar energy for making fuels must be adequately efficient, so that the land areas over which solar energy is collected stays at a socially acceptable size. Also, the cost of such 'light-made' fuels must be economically viable [37].

1.1.5 Development of Cost-Effective Electrochemical Reduction of Carbon Dioxide

A cost-effective electrochemical carbon dioxide reduction would enable a shift to a sustainable energy economy. When coupled to a renewable energy source, the electroreduction of CO_2 could generate carbon-neutral fuels or industrial chemicals, which are conventionally obtained from petroleum. A necessary condition for such a process to take place is the development of active and selective catalysts, which enable the reduction of CO_2 at low overpotential and high current density, and yield the desired products without a significant amount of by-products [43]. Given their close reduction potential values, H^+/H_2O reduction may compete with CO_2 conversion. Therefore, in addition to various reduced carbon species, molecular hydrogen may be discharged. Carbon dioxide reduction proceeds at significant overpotential values, because of the kinetic activation barrier and the large gap between the HOMO (highest occupied molecular orbital) and LUMO (lowest unoccupied molecular orbital) energies. Typically, the electroreduction of CO_2 yields mixtures of reaction products and hydrogen [7]. This finding will be addressed in more detail in the next section on the thermodynamics of carbon dioxide electroreduction.

1.2 Electrochemical Reduction of Carbon Dioxide

1.2.1 Thermodynamics of CO_2 Electroreduction

In aqueous solution, the electrochemical reduction of CO_2 must compete with water electroreduction. Given that hydrogen evolution prevails in acidic solutions, and CO_2 does not exist in basic solutions, and by analogy with photosynthesis, the most favourable conditions for carbon dioxide electroreduction are predicted in close to neutral solutions [44]. Obviously, standard reduction potentials for various CO_2 reductions will vary with *pH* according to the Nernst equation (Equation 1.5) [12]. While the reduction potential for the conversion of CO_2 to CO is low (see Table 1.1), practical CO_2 reduction

TABLE 1.1

Products and Standard Reduction Potentials of Carbon Dioxide Reduction Reactions, at *pH* 6.8, Involving 2–18 Electrons Standard Reduction Potential

Product	Cathode Half-Reaction[a]	Standard Potential [V vs. SHE]
Formic acid	CO_2 (g) + 2H$^+$ (aq) + 2e$^-$ → HCOOH (aq)	−0.020 to −0.114[c]
CO	CO_2 + 2H$^+$ (aq) + 2e$^-$ → CO (g) + H_2O (l)	−0.104
Oxalic acid	2CO_2 (g) + 2H$^+$ (aq) + 2e$^-$ → HOOC–COOH (aq)	−0.088[d]
Formaldehyde	CO_2 + 4H$^+$ (aq) + 4e$^-$ → HCHO (aq) + H_2O (l)	−0.106
Methanol	CO_2 (g) + 6H$^+$ (aq) + 6e$^-$ → CH_3OH (l) + H_2O (l)	−0.030 to +0.031
Glyoxal[b]	2CO_2 (g) + 6H$^+$ (aq) + 6e$^-$ → OHC–CHO (aq) + 2H_2O (l)	−0.160
Methane	CO_2 (g) + 8H$^+$ (aq) + 8e$^-$ → CH_4 (g) + 2H_2O (l)	+0.169
Acetic acid	2CO_2 (g) + 8H$^+$ (aq) + 8e$^-$ → CH_3–COOH (aq) + 2H_2O (l)	−0.260
Glycol aldehyde[b]	2CO_2 (g) + 8H$^+$ (aq) + 8e$^-$ → $HOCH_2$–CHO (aq) + 2H_2O (l)	−0.030
Ethylene glycol[b]	2CO_2 (g) + 10H$^+$ (aq) + 10e$^-$ → $HOCH_2$–CH_2OH (aq) + 2H_2O (l)	+0.200
Acetaldehyde	2CO_2 (g) + 10H$^+$ (aq) + 10e$^-$ → CH_3–CHO (aq) + 3H_2O (l)	+0.050
Ethanol	2CO_2 (g) + 12H$^+$ (aq) + 12e$^-$ → CH_3–CH_2OH (aq) + 3H_2O (l)	+0.090
Ethylene	2CO_2 (g) + 12H$^+$ (aq) + 12e$^-$ → CH_2=CH_2 (g) + 3H_2O (l)	+0.079
Hydroxyacetone[b]	3CO_2 (g) + 14H$^+$ (aq) + 14e$^-$ → HO–CH_2–CO–CH_3 (aq) + 4H_2O	+0.46
Acetone[b]	3CO_2 (g) + 16H$^+$ (aq) + 16e$^-$ → CH_3–CO–CH_3 (aq) + 5H_2O (l)	−0.140
Allyl alcohol	3CO_2 (g) + 16H$^+$ (aq) + 16e$^-$ → CH_2=CH–CH_2OH (aq) + 5H_2O	+0.110
Propionaldehyde	3CO_2 (g) + 16H$^+$ (aq) + 16e$^-$ → CH_3–CH_2–CHO (aq) + 5H_2O (l)	+0.140
1-Propanol	3CO_2 (g) + 18H$^+$ (aq) + 18e$^-$ → CH_3–CH_2–CH_2–OH (aq) + 5H_2O	+0.210

Source: Adapted from M. Jitaru, *J. Univ. Chem. Technol. Metallurgy* 42, 2007, 333–344; K.P. Kuhl et al., *Energy Environ. Sci.* 5, 2012, 7050–7059; M. Gattrell, N. Gupta and A. Co, *J. Electroanal. Chem.* 594, 2006, 1–19.

[a] Each reaction product is shown as a neutral compound.
[b] Marked products were first reported in Ref. [9].
[c] According to Refs. [2] and [44], formate ions, HCOO$^-$, form at −0.225 V vs. SHE.
[d] The standard reduction potential was recalculated for *pH* 6.8.

proceeds at much greater cathode polarisation; at pH 7, CO formation only begins at −0.80 V vs. SHE, and becomes efficient at −1.10 V vs. SHE. Such experiments were conducted on Au electrode, in $KHCO_3$ solution, at room temperature [45]. Hence, a significant overpotential needs to be applied for driving CO_2 electroreduction, and the applied overpotential often causes hydrogen reduction as well. To suppress H_2 evolution, one can use cathodes made of metals with high hydrogen overpotentials (e.g. Hg and Pb) or one can perform CO_2 electroreduction in organic media, as discussed in Section 1.2.4. Table 1.1 lists the most important cathode half-reactions of CO_2, which involve the transfer of 2, 4, 6, 8, 10, 12, 14, 16 or 18 electrons. The standard reduction potential, E_{RHE}, is calculated from the free energy values of each reaction according to Equation 1.4, while the potential values relative to the reversible hydrogen electrode, E_{RHE}, at pH 6.8 are derived from Equation 1.5 (valid at room temperature, 298.15 K).

$$E^o = \frac{\Delta G^o}{nF} \tag{1.4}$$

where $F = 96{,}485.3$ C/mol (Faraday's constant).

$$E_{RHE} = E^o + 0.05916\,pH \tag{1.5}$$

As revealed by data listed in Table 1.1, these reactions proceed at moderate standard potentials. The data in Table 1.1 show that from the thermodynamic standpoint, the most favourable reactions are the reduction of CO_2 to methane, ethylene glycol and propanol. For example, the eight-electron reduction of CO_2 to methane should be possible at +0.169 V vs. RHE. Given that hydrogen evolution becomes thermodynamically possible at 0.00 V vs. RHE, at all negative potential values ($E < 0.00$ V vs. RHE), carbon dioxide reduction will compete with H_2 discharge [46]. Hence, rather than the thermodynamics, the reaction kinetics will determine whether CO_2 reduction proceeds instead of H^+ or water reduction. In order to drive CO_2 reduction selectively, the cathode materials should manifest poor hydrogen discharge activity in the presence of CO_2 [47]. In other words, the selective preparation of CO_2 translates into designing appropriate CO_2 reduction catalysts, based on systematic criteria.

Marković and Ross [46] performed such an analysis for CO_2 reduction to methane, based on an extensive set of density functional theory. They calculated the limiting potential at which each step of the reaction becomes exergonic (or downhill in free energy). Limiting potentials represents a measure of the potential dependence of the electrochemical reaction rate, given that each limiting potential indicates the electrical potential at which one particular elementary step starts to proceed at a considerable rate [46]. Also, the chemical potentials of the bound reduction intermediates can be derived from statistical mechanics treatment of the binding energies of adsorbed species. This approach is able to predict correctly the onset potentials for the formation of various products in CO_2 reduction on the surface of copper cathodes [36,48]. Thermodynamics imposes a limitation on any catalyst to operate at more negative potential values than the equilibrium potential [46].

1.2.2 Kinetics of CO_2 Electroreduction

The CO_2 molecule is linear; its oxygen atoms are weak Lewis bases, while its carbon is electrophilic. Reactions of CO_2 are dominated by the nucleophilic attack at the carbon atom, which results in the bending of the O–C–O bonds [7]. While a wide variety of reduction mechanisms have been proposed, it is unanimously accepted that, whatever the pathway of the reduction process, the first step consists of a single electronation of the CO_2 molecule, to form an anion radical (Equation 1.6) [49,50]:

$$CO_2 + e^- \rightarrow CO_2^{-}\bullet \tag{1.6}$$

This reaction is the rate-determining step. The large energy barrier to the outer-sphere electron transfer arises from the very different geometries of the linear electrically neutral carbon dioxide molecule (the reactant) and the radical anion (the product), which is bent [51]. Therefore, electrochemical reductions require a high overpotential. This thermodynamic barrier can be reduced by protonating the reduction product [52]. In gas phase or in aqueous media, carbon dioxide reacts with hydrated electrons to form the radical anion, which is stabilised by hydration [53–55]. In subsequent reactions, the $CO_2^-\bullet$ species accept protons (from water, which acts as a proton donor) and one or more electrons (from the electrode). A number of reaction products are possible, such as formate ions [10,11], that are generated as shown in Equations 1.7 and 1.8:

$$CO_2^{-}\bullet \text{ (ads)} + H_2O \rightarrow HCOO\bullet \text{ (ads)} + OH^-(aq) \tag{1.7}$$

$$HCOO\bullet \text{ (ads)} + e^- \rightarrow HCOO^-(aq) \tag{1.8}$$

Free protons, present in acidic supporting electrolytes, represent an alternate source of hydrogen ions, which can interact with the adsorbed $CO_2^-\bullet$ anion radicals, as displayed in Equation 1.7a. In both cases, the adsorbed formate radical, $HCOO\bullet$, undergoes a second electronation to formate anions, as in Equation 1.8.

$$CO_2^{-}\bullet + H^+ \rightarrow HCOO\bullet \tag{1.7a}$$

When used in aqueous solution, most flat metallic cathodes yield carbon monoxide and formic acid [56,57]. In nonaqueous media, the radical anion generated according to Equation 1.6 may undergo dimerisation with an electrically neutral CO_2 molecule to form $^-OOC–COO\bullet$ (ads) (Equation 1.9), and then a second electronation to oxalate ions (Equation 1.10) [14,49,50]:

$$CO_2^-\bullet(ads) + CO_2 \rightarrow {}^-OOC - COO\bullet(ads) \quad (1.9)$$

$$^-OOC–COO\bullet(ads) + e^- \rightarrow {}^-OOC–COO^-(aq) \quad (1.10)$$

The dimerisation process to oxalate, described by Equations 1.9 and 1.10, competes with the dimerisation reaction to form $O=\bullet C–O–COO^-$ (Equation 1.9a), and the subsequent reduction/disproportionation, yielding carbon monoxide and carbonate (Equation 1.9b) [58].

$$CO_2^-\bullet(ads) + CO_2(aq) \rightarrow O=\bullet C–O–COO^-(ads) \quad (1.9a)$$

$$O=\bullet C–O–COO^-(ads) + e^- \rightarrow CO(g) + CO_3^{2-}(aq) \quad (1.9b)$$

The above competing mechanism has been reported by Eggins and coworkers [58]. As oxalate formation involves only 2 faraday of charge per mol, yielding a relatively heavy molecule (MW = 90 g/mol), this process appears to be one of the most advantageous from the standpoint of energy efficiency. Technologically, it is also feasible, as the reduction may be conducted in aqueous solution, under ambient conditions [59]. Hence, the electroreduction of CO_2 to formate can be performed as a continuous process, using a filter-press reactor, equipped with lead cathode aqueous solutions. This system only suffers from current density limitation; up to a limit value of 10.5 mA/cm^2, a Faradaic efficiency of 57% could be maintained [60].

Marković and Ross [46] found that an appropriate catalyst for CO_2 electroreduction must enable the effective protonation of adsorbed CO intermediate to adsorbed CHO or COH, which are in a more reduced state than CO. Meanwhile, the catalyst should possess minimal activity for the competing hydrogen evolution reaction. In order to prepare hydrocarbons and alcohols, these two requirements should be met simultaneously. Copper has the ability to perform the hydrogenation of adsorbed CO, a property that is related to its relatively low activity in hydrogen evolution.

1.2.3 Mediated Electrocatalytic Reduction of CO_2

In the most generic sense, electrocatalysis means any mechanism that speeds up a half-cell reaction at the electrode surface. Therefore, electrocatalysis encompasses the factors able to modify the kinetic parameters of the charge transfer processes that occur at the electrode/electrolyte interface. Electrode kinetics is enhanced by materials, which minimise the overpotential at which one particular reaction can proceed. One can evaluate the electrocatalytic activity of various materials by comparing the current generated at a given overpotential. On its turn, mediated electrocatalysis is the variety of effects exerted by a redox system in the homogeneous phase, the redox system being eventually regenerated in a heterogeneous process at the electrode. Mediated electrocatalysis involves an interaction between the mediator, the electrode, and the substrate molecule, for example, carbon dioxide. As a result, the reduction will proceed at a lower overpotential, which translates into a lower specific energy consumption. Also, the selectivity of the process may improve [7]. Using transition metal complexes in CO_2 electroreduction alleviates a great extent problems related to the corrosion and passivation of cathode metals [61].

The reversible homogeneous electrocatalytic effect of pyridinium salts can convert CO_2 to methanol [33,62]. Formic acid can be synthesised by taking advantage of the electrocatalytic effect exerted by aromatic compounds, which contain one or more sp^2 hybridised nitrogen atoms. Such an electron-rich heterocyclic compound is the electron-rich imidazole, which may bind CO_2 more strongly than pyridine, as its six π electrons are distributed over a five-atom ring [33]. This reduction was performed at illuminated iron pyrite semiconductor electrodes, which mimic both the use of solar energy and the most abundant catalyst materials found on Earth. Imidazole was needed only in low concentration (10 mol/dm³), which indicates a heterogeneous electron transfer, diffusion controlled at the electrode surface. The mechanism proposed by Bocarsly, Barton Cole and co-workers [33] is shown in Figure 1.4. By the first electron uptake, imidazole is converted into a radical (Equation 1.11). Then, it loses a hydrogen molecule to form a C-2-based carbene intermediate (Equation 1.12), which has two resonance forms. Next, the carbanion performs a nucleophilic attack on the carbon of CO_2, in the presence of water, which serves as the proton donor (Equation 1.13), and enables its further $2e^-/2H^+$ reduction to formic acid (Equation 1.14).

Imidazolonium cations allowed the synthesis of organic carbamates. For this, the activation of carbon dioxide has been conducted in O_2/CO_2 saturated ionic liquids, via electrochemically generated $O_2^-•$. The electron uptake occurred at a less negative potential than the one needed for the direct cathodic reduction of CO_2. This kind of electrochemical activation has been applied to the C–N bond formation from amines and CO_2 to yield organic carbamates [63].

Trialkylborane additives promote the reduction of carbon dioxide to formate, via bis(diphosphine) Ni(II) and Rh(III) hydride complexes. Late transition metal hydrides, which can be formed by the reaction with molecular H_2, transfer hydride to CO_2 to yield a formate–borane adduct. In order to drive this process, the borane must be of appropriate Lewis acidity; hence, weaker

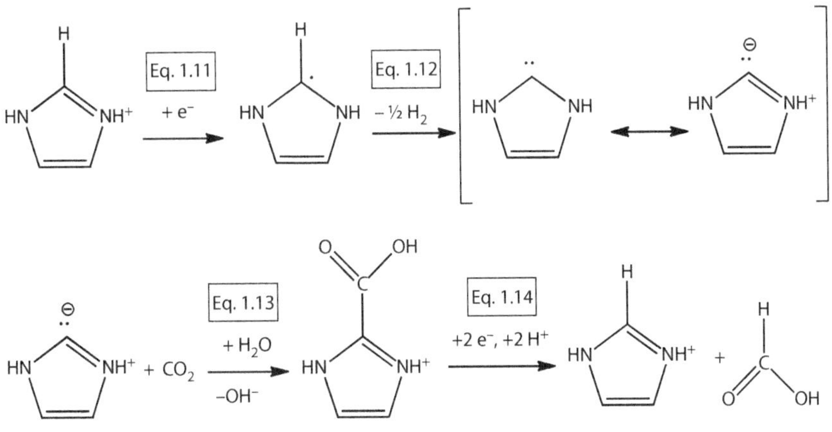

FIGURE 1.4
Mechanism of imidazol-catalysed electrochemical reduction of carbon dioxide to formic acid. For description, refer to details in the text. (Adapted from K.A. Keets et al., *Ind. J. Chem. A*, 51, 2012, 1284–1297.)

acids do not cause a significant hydride transfer enhancement, while stronger acids abstract hydride without CO_2 reduction. The mechanism likely involves a pre-equilibrium hydride transfer, followed by the formation of a stabilising formate–borane adduct [64].

Inorganic redox couples have been shown to be efficient in the mediated CO_2 electroreduction. Hence, Fe^0 porphyrins catalyse the electrochemical reduction of CO_2 to yield CO as the main product. Upon conducting this process in DMF with tetraalkylammonium salts as the supporting electrolyte, the porphyrin is destroyed after only a few catalytic cycles, as the carboxylation and/or hydrogenation of the ring takes places. By contrast, in the presence of a hard electrophile, such as Mg^{2+} ions, the rate of CO production improves spectacularly, while the stability of the catalyst also increases. The first step of the reaction mechanism is the introduction of one molecule of CO_2 into the iron coordination sphere. Next, a second molecule of CO_2 is added, which acts as a Lewis acid, and enables the cleavage of one C–O bond of the first CO_2 molecule, leading by this to CO. This process is accelerated in the presence of Mg^{2+} ions, which serves as a second catalyst. At a low temperature of −40°C, the Mg^{2+} ions facilitate the decomposition of the complex containing two molecules of CO_2, whereas, at room temperature, Mg^{2+} ions triggers the breaking of the bond at the level of the complex, which contains a single molecule of CO_2 in its coordination sphere. The combined action of Fe^0 porphyrins and Mg^{2+} ions offers an example of a bimetallic catalysis, where an electron-rich centre starts the reduction process, while an electron-deficient centre assists the conversion of the bond system [65].

Other examples of bimetallic catalysis include Co^{II}/Co^{I} in cobalt tetra (3-aminophenyl porphyrin) and Ni^{III}/Ni^{II} in [Ni(II)-5,7,12,14-tetramethyl dinaphtho[b,i][1,4,8,11]tetraza[14]annulene]$^{2+}$ complex [66]. When used in THF, iridium dihydride complexes, supported by PCP-type pincer ligands, rapidly insert CO_2 molecules, to yield κ – 2-formate monohydride products. (Pincer ligands are chelating agents that bind tightly to three adjacent coplanar sites, usually on a transition metal in a meridional configuration; anionic ligands with a carbanion as the central donor site and flanking phosphine donors are referred to as PCP pincers.) In acetonitrile–water mixtures, these complexes become efficient and selective catalysts for the electrocatalytic reduction of CO_2 to formate [67]. Also, a homologous series of Ni^{II} complexes, supported by N-heterocyclic carbene-pyridine ligands ((R)bimpy, where R is methyl, ethyl or propyl), have been shown to exhibit high selectivity for reducing carbon dioxide over water under electrocatalytic conditions [68].

1.2.4 Electroreduction of CO_2 in Organic Media

Given the low solubility of carbon dioxide in water, its electroreduction is often conducted in organic solvents, which offer the advantage of greater CO_2 solubility, as listed in Table 1.2 [7,69–72]. At increased external pressure, the CO_2 solubility rises significantly.

The high solubility of CO_2 in solvents with terminal –OH groups is likely to be a result of the strong electron interaction of the high polarity C=O bonds of carbon dioxide with these highly polar hydroxyl groups of the solvent molecules [70]. A schematic of this polar interaction is displayed in Figure 1.5.

TABLE 1.2

Solubility of CO_2 in Water and Various Organic Solvents at 25°C

Solvent	Concentration (mol/L)
Water	0.034
Methanol	0.139
Dimethylsulphoxide (DMSO)	0.138 ± 0.003
Dimethylformamide (DMF)	0.199 ± 0.006
Tetrahydrofuran (THF)	0.205 ± 0.008
Glycerol carbonate	0.250
Acetonitrile (AN)	0.279 ± 0.008
Poly(ethylene glycol), PEG 200[a]	0.343
Poly(ethylene glycol), PEG 300[a]	0.343
Glycerol[a]	0.391

Source: Adapted from M. Jitaru, *J. Univ. Chem. Technol. Metallurgy* 42, 2007, 333–344; C.M. Sánchez-Sánchez et al., *Pure Appl. Chem.* 73, 2001, 1917–1927; O. Achenbrenner and P. Styring, *Energy Environ. Sci.* 3, 2010, 1106–1113; E. Wilhelm and R. Battino, *Chem. Rev.* 73, 1973, 1–9; P. Lühring and A. Schumpe, *J. Chem. Eng. Data* 34, 1989, 250–252.

[a] Solubility in these solvent examined for efficient CO_2 capture.

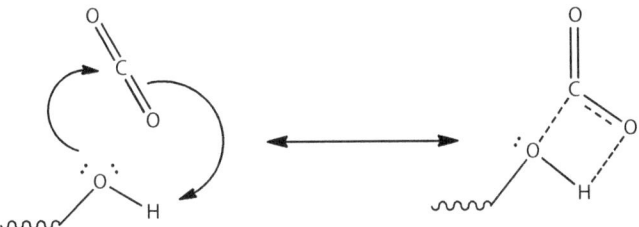

FIGURE 1.5
Polar interaction of a CO_2 molecule with the terminal hydroxyl groups of a solvent. (Adapted from O. Achenbrenner and P. Styring, *Energy Environ. Sci.* 3, 2010, 1106–1113.)

Carbon dioxide solubility was also investigated in binary and ternary mixtures of esters, at high pressure. The goal of the study was to identify an appropriate system for CO_2 capture. Solubility of CO_2 was determined by the constant-volume method in mixtures of {dimethyl carbonate (DMC) + diethyl carbonate}, {DMC + propyl acetate}, {DMC + propylene carbonate} and {DMC + ethylene carbonate}, in the temperature range from 282.0 to 303.0 K. It is found that the solubility of CO_2 in four mixed solvents follows the Henry's law, and the linear compound has a greater ability to dissolve CO_2 than the cyclic compound at the same temperature [73].

Solubility of CO_2 in methanol increases significantly with increasing pressure (Figure 1.6a) and decreasing temperature (Figure 1.6b). Concentrations of CO_2, in mol/dm^3, at different pressure and temperature values were calculated from the data reported by Foster et al. [74].

One should emphasise that the solubility of CO_2 in methanol at temperatures below 0°C is 8–15 times greater than its solubility in water [75]. Therefore, it has proven favourable to conduct the electrochemical reduction of CO_2 in methanol at low temperature values, in a cell equipped with a

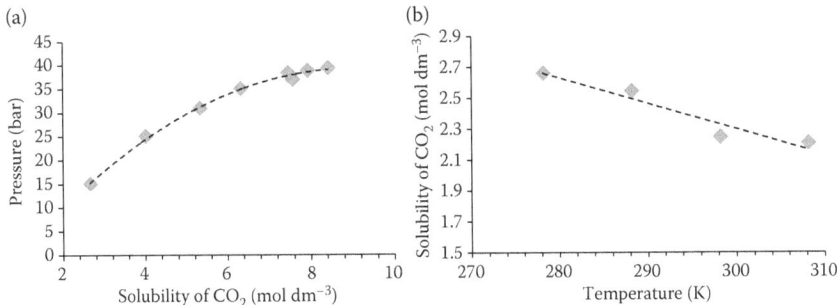

FIGURE 1.6
(a) Dependence of CO_2 solubility in methanol with increasing gas pressure (at a constant temperature of 278.15 K). (b) Temperature dependence of CO_2 solubility in methanol with increasing temperature (at a constant pressure of 15 bar). (Adapted from K. Bezanehtak et al., *J. Chem. Eng. Data* 47, 2002, 161–168.)

copper cathode and a platinum anode. While the catholyte contained benzalkonium chloride in methanol, the anolyte was either 0.1 mol/dm^3 hydrogen carbonate or 80 mol/dm^3 benzalkonium chloride. At 0°C and a cathode potential of −2.2 V vs. SCE, the main products were CO, methane and ethylene, produced at a Faradaic efficiency of 31.8%, 18.9% and 16.9%, respectively. At −15°C, the Faradaic efficiency changed to 24.0% for CO, 39.1% for methane and 4.4% for ethylene [76]. The magnitude of Faradaic efficiency to methane exceeds that reached in an aqueous catholyte (33.3%); however, ethylene can be synthesised at much higher Faradaic efficiency (25.5%) [77].

Reducing carbon dioxide in methanol may contribute to developing a large-scale manufacturing of useful organic products from a readily available, inexpensive raw material: CO_2-saturated methanol from industrial absorbers (the Rectisol process) [78]. Used in coal gasification facilities, Rectisol is a method for removing carbon dioxide and other acid gases during the gasification process. It utilises cold methanol as a solvent for capturing acid gases from a gas stream, by physical absorption under pressure. The major advantage of this process is that it removes hydrosulphuric acid and carbonyl sulphide (COS) to 0.1–1 ppmv (parts per million by volume) in the treated gas, while preserving the H_2 and CO in the synthesis gas. Then, the solvent is regenerated by pressure reduction and stripping with N_2 gas. Typically, the CO_2-rich offgas is discharged in the atmosphere [79]. Carbon dioxide recovered in this process could be converted into useful organic products by electrochemical means.

1.3 Materials, Components and Operation Parameters

1.3.1 Experimental Methodology for CO_2 Electroreduction

The electrochemical conversion of CO_2 into useful products, such as fuels, requires one to supply CO_2 gas to the electrolysis cell by bubbling CO_2 gas continuously in the electrolyte, at a constant flow rate. This can be done at atmospheric pressure (1 bar) or at elevated pressures (30–60 bars). One can perform the electrolysis at high pressure and/or low temperature to increase the otherwise low solubility of CO_2 in the electrolyte solution, or in organic solvents, in which CO_2 solubility is higher than in water [80]. An alternative approach proceeds to the saturation of the electrolyte in CO_2 prior to electrolysis, and keeps the system under continuous stirring over the electroreduction step [81].

Following electrolyte sorption, CO_2 typically undergoes competitive reactions that advance simultaneously on the electrode surface, yielding different products, with a product distribution primarily determined by (i) the identity of the electrode and (ii) the composition of the electrolyte (aqueous or organic, with various possible salts and additives).

1.3.2 Choice of Cathode Materials with Favourable Electrocatalytic Properties

The identity of the cathode materials is essential for the outcome of CO_2 electroreduction. While an earlier electrode classification was based on whether the cathode metal belonged to the *sp*- or the *d*-metal group [50], Hori considered that the performance of various metals is loosely related to the periodic table. For aqueous electrolytes, Hori [82,83] suggested regrouping the electrode metals into two categories: (1) CO formation metals (Cu, Au, Ag, Zn, Pd, Ga, Ni and Pt) and (2) metals that yield formate (Hg, Pb, Zn, In, Sn, Cd and Tl). As discussed in the previous section, copper represents a very special electrode material, enabling the formation of various hydrocarbons, such as methane and ethylene [82].

In nonaqueous electrolytes (such as propylene carbonate), Hg, Tl and Pb yield oxalate, while Cd, Sn and In provide CO [14]. In early experiments, Ito, Ikeda and co-workers [84,85] obtained oxalic acid by the electroreduction of CO_2 on lead electrodes, in organic solvents under high pressure. More recently, Eneau-Innocent et al. reported that lead electrodes, used in tetraethylammonium perchlorate in propylene carbonate, enabled the dimerisation of CO_2 to oxalate [86].

In an aqueous solution, the electrode potentials of CO_2 reduction correlate with the heats of fusion (HoF) of the electrode metals: low-HoF metals (Hg, Tl, Pb, In, Cd and Zn) yield formate, while high-HoF metals (Pt, Pd, Ni, Au, Cu, Ag, Zn, Sn and Ga) form CO [77,87]. The above classification is far from being perfect, and does not cover for all possible scenarios of CO_2 electroreduction. As shown later, in the section on the synthesis of organic carbonate, when used in an ionic liquid, indium cathodes are efficient in the preparation of dimethyl carbonate. Also, copper-based bimetallic electrodes may exhibit an improved catalytic activity in reducing CO_2 to hydrocarbons. Examples include Cu–Ni, Cu–Sn and Cu–Pb alloys. By contrast, for Cu–Ag and Cu–Cd alloy electrodes, the catalytic activity is diluted [81].

1.3.3 Copper: A 'Magic' Electrode Material

Copper and copper alloys exhibit special catalytic effects in the electroreduction of carbon dioxide. They represent unique cathode materials, which can electrocatalytically convert CO_2 and water into hydrocarbons and alcohols, at ambient temperature and atmospheric pressure [2]. So far, copper metal is the only electrode material able to produce significant amounts of hydrocarbons at high reaction rates and over 50% Faradaic yield, over a sustained period of time. Its drawbacks are that a copper electrode can operate only at high overpotential (of almost 1 V), and a mixture of major and minor products are obtained, which contains hydrogen, ethylene and methane [43,47,88]. In these reactions, carbon monoxide appears to be a key intermediate, and its further reduction yields a series of reaction products [2,89]. Copper cathodes can be operated at high current density in aqueous solution yielding useful products, such as methane, ethylene, alcohols (ethanol and 1-propanol) and aldehydes (acetaldehyde and propionaldehyde) [89]. The outstanding selectivity of copper in the electrochemical reduction of CO_2 to hydrocarbons has triggered extensive research towards unveiling the mechanism of the process [36,47,48,90,91]. After performing a rigorous study of CO_2 electroreduction on copper single crystals and copper–gold alloys, Buss-Herman and co-workers [50] concluded that the major products formed at the copper electrodes were CO and CH_4. Methane formed at negative potentials, where adsorbed hydrogen is present at the surface of the copper cathode.

Ethylene was obtained by CO_2 electrolysis at a three-phase interface on copper mesh electrodes, in acidic solution [92–95]. In a 2 mol/dm³ KBr-based aqueous electrolyte of pH 3, at −1.8 V vs. Ag/AgCl, the most abundant product, C_2H_4, obtained at a current efficiency of 63%. The by-products were methane (16.8%), CO (6.2%) and formic acid (2.6%). Importantly, the catalytic activity of the Cu mesh was maintained over the operation, as no catalyst poisoning species were formed [92]. It was found that the modification of copper mesh with Cu^I halide was effective for the concomitant enhancement of ethylene formation and the suppression of hydrogen evolution [93]. The mechanism of CO_2 reduction to ethylene is illustrated schematically in Figure 1.7 [92–96]. The first step consists of the adsorption of the halide (Br⁻) on the surface of the copper electrode, which facilitates the adsorption of CO_2, enabling its subsequent proton-assisted electronation to a chemisorbed carboxylic group. Next, as shown in the second line of Figure 1.7, two neighbouring chemisorbed –COOH groups undergo a dimerisation step, that is, proton-assisted electronation step to form a new C–C bond; this (–CH_2–CO–) intermediate remains adsorbed on the Cu surface, and goes through two consecutive electronation/protonation steps to form the free ethylene.

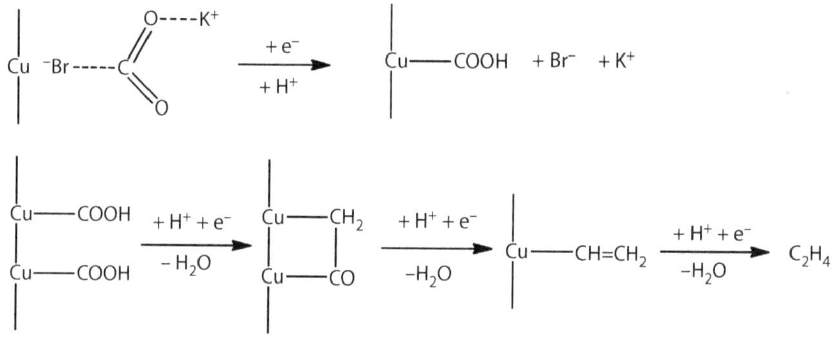

FIGURE 1.7
Schematic of the mechanism of CO_2 reduction to ethylene, in the presence of halides (here: bromine). (Adapted from H. Yano et al., *J. Electroanal. Chem.* 565, 2004, 287–293; K. Ogura, R. Oohara and Y. Kudo, *J. Electrochem. Soc.* 152, 2005, D213–D219.)

In 2007, Jitaru [7] described pathways to methane, ethylene and formic acid on copper electrodes. Thus, the anion radical $CO_2^- \cdot$ formed according to Equation 1.6 may undergo a reaction with another CO_2 molecule to yield carbon monoxide and a carbonate anion radical, according to Equation 1.15.

$$CO_2^- \cdot + CO_2 \rightarrow CO + CO_3^- \cdot \qquad (1.15)$$

Alternatively, the same $CO_2^- \cdot$ species may capture a proton from the supporting electrolyte to be converted into formate ions, as discussed previously (see Equations 1.7, 1.7a and 1.8).

Kyriacou and Schizodimou [97] showed that the rate of carbon dioxide reduction on a cathode made of Cu(88)–Sn(6)–Pb(6) ternary alloy increased in the presence of multivalent metallic cations, of halides, and with increasing acidity of the supporting electrolyte. Hence, when CO_2 reduction was conducted in 1.5 mol/L HCl electrolyte, at the potential of −0.65 V vs. Ag/AgCl, in the presence of various cations, the reaction rate increased with the increasing charge of the cation, in the following order: $Na^+ < Mg^{2+} < Ca^{2+} < Ba^{2+} < Al^{3+} < Zr^{4+} < Nd^{3+} < La^{3+}$. It was found that in an electrolyte containing La^{3+} ions, the reduction rate, at the same potential, was two times greater than in the presence of Na^+ ions. The acceleration effect can be attributed to the participation of the $CO_2^- \cdot$ radical anion in the rate-determining step. Also, in strongly acidic solution, at −0.65 V vs. Ag/AgCl reference, the presence of halides increased the reduction rate in the order $Cl^- < Br^- < I^-$, while decreasing the *pH* below 1 raised the rate by 53%. The main products of CO_2 reduction were CH_3OH, CH_3CHO, HCOOH and CO, and the rate of their formation could be increased at low overpotentials, while the product distribution was controlled by the composition of the electrolyte [97].

Jitaru and co-workers [98] conducted the electroreduction of carbon dioxide on bronze electrodes, having the composition of Sn(85)–Cu(15). They used an aqueous alkaline electrolyte (0.2 M K_2CO_3), and the reaction was performed under CO_2 atmosphere, at 12–25°C. Formate was the main product, formed at a current efficiency of up to 74%, at potentials more negative than −1.6 V SCE [98]. The same research group demonstrated that an electrode consisting of Cu and Cu–Ni nanorods, deposited onto Cu foil and vertically aligned on the foil surface, exhibited high electrocatalytic activity for both oxygen reduction and carbon dioxide electroreduction to formic acid. The Faradaic efficiency was up to 50% [99].

In their milestone paper, Peterson and Nørskov [47] suggest strategies towards improving the activity of the cathode materials relative to Cu. In order to decrease the overpotential relative to Cu, the binding energies of >CO and >CHO to the surface must be decoupled. To accomplish this, one can use the following approaches: (i) *alloy Cu with metals having higher oxygen affinity* (elements with high oxygen affinity may change the potential determining step), (ii) take advantage of *ligand stabilisation* (using homogeneous catalysis by adding a chemical ligand, which can interact with the adsorbed

>CHO that has a planar geometry, but cannot interact with the adsorbed linear >CO; hence, the ligand stabilises the adsorbed >CHO complex, lowering its energy relative to adsorbed >CO); (iii) *tether a ligand* to the electrode surface, such that only the adsorbed planar >CHO can interact with the surface-confined ligand, while the adsorbed linear >C=O cannot; (iv) add a *promoter*, resistant to electrochemical reduction, which can change the relative binding strength of adsorbed species through electronic effects or through structural effects or through a combination of these effects; or (v) to exploit *hydrogen bond formation and solvent effects* (because of the different binding geometries, hydrogen bond formation may affect adsorbed >CHO and >C=O differently; for example, a hydrogen bond acceptor may stabilise adsorbed >CHO preferentially over adsorbed >C=O) [47].

1.3.4 Use of Gas Diffusion Electrodes

The rate of the electrochemical process can be enhanced by using high-throughput GDEs [100–102], which usually consist of Teflon®-bound catalyst particles. GDEs are electrodes offering a solid, liquid and gaseous interface, and an electrically conducting catalyst, which together enable an electrochemical reaction between a liquid and a gaseous phase, which coexist in the pore system of the electrode. Therefore, GDEs are promising for being used in the electrochemical reduction of CO_2, as they alleviate mass transport limitations across the gas/liquid interface and to the surface of the catalyst [100,103].

Typically, GDEs consist of a gas layer (mixture of hydrophobic carbon black and polytetrafluoroethylene (PTFE) dispersion) and a reaction layer (mixture of catalyst powder, hydrophobic and hydrophilic carbon black and PTFE) laminated on a copper mesh current collector [104]. In addition, GDEs allow a good distribution of CO_2 over the surface of the catalyst, and provide high current efficiency for the formation of the desired product, however, at low current density [105,106]. For example, when the working electrode is copper, the purer and smaller the particle size of Cu, the higher the activity and efficiency they show. Reduction of CO_2 to CO and hydrocarbons can be conducted at ~100 times greater rate using a GDE, as compared to Cu foil electrodes [103]. Other effective technological means for improving the electroreduction process include the use of polymer electrolyte membrane cells, which enable the gas-phase electrolysis of CO_2 by enhancing the CO_2 transport [107,108]. Lead-based GDEs have been reported in 2010 [101]. More recently, tin-based GDEs were used [100,109]. Commercially available tin particles were mixed with Nafion® ionomer, and this mixture was used to make an electrode with a gas diffusion layer support, according to a procedure similar to the preparation of electrodes for fuel cells. When the electrolysis is conducted in $NaHCO_3$ solution, under potentiostatic conditions, the exchange current density on the Sn/gas diffusion layer electrode was two orders of magnitude greater than the one recorded on a metallic Sn disc [109].

1.3.5 Need for High-Purity Electrolytes

The purity of the electrolyte is essential, as even trace amounts of impurities (e.g. 5 ppm of lead or iron) interfere with either the surface process at the electrodes or the reactions that occur in the heterogeneous interfacial phase. Therefore, purification of the aqueous solutions by pre-electrolysis is recommended. Neutral and alkaline electrolytes have been used extensively for CO_2 electroreduction, such as aqueous bicarbonates, phosphate, hydroxide and sulphate solutions. Another option is to employ aprotic solvents under ambient pressure or at high pressures. Higher *pH* values and elevated pressures offer the advantage of a greater solubility of CO_2 in the reaction medium, that is, greater concentration of CO_2 [110].

Often, CO_2 reduction is conducted in organic electrolytes, as this offers several advantages over aqueous media: (i) the unwanted hydrogen discharge can be suppressed, (ii) the concentration of water as a reagent can be accurately controlled, (iii) the solubility of CO_2 in organic solvents is much greater than in water, (iv) the electrolysis can be conducted at much more negative polarisation (as organic solvents have a broader potential window than water) and (v) at low water concentration, organic solutions allow for dimerisation reactions, yielding oxalic acid, glyoxylic acid and glycolic acid [14].

1.3.6 Electrochemical Reactors for CO_2 Reduction

As stated by Jaramillo et al. [43], there are no standard experimental methods for investigating the electrochemical reduction of carbon dioxide. It is recommended that the electrochemical reactor should have a working electrode with a large surface area and that one uses a small volume of electrolyte.

For obtaining meaningful results, the experimental setup needs to enable accurate electrochemical measurement of the current and voltage, and it should also allow highly sensitive product identification and quantity of all possible products. Two main types of reactors have proven efficient: (i) batch reactors and (ii) flow reactors, the latter enabling enhanced mass transfer of carbon dioxide, as compared to batch reactors. One can utilise conventional H-cells or one can devise special cells, equipped with parallel electrodes to achieve uniform electric field distribution. Home-designed gas-tight cells were also reported.

In a series of experiments, a flow reactor enabled the reduction to formate of carbon dioxide dissolved in aqueous phosphate buffer, under ambient conditions. The flow reactor enhanced the mass transfer of carbon dioxide, as compared to the batch reactor, and formate was obtained with a maximum current efficiency of 93%. The final formate concentration reached 1.5×10^{-2} mol/dm³ [111].

Electrochemical cells can be built and operated in three different modes [112]:

1. Semi-half discontinuous cells, where one side of the cell separator membrane is in contact with a liquid electrolyte, while the working

electrode is confined to the other side of the membrane, and is in touch with the gas phase, containing high-purity CO_2.
2. Full continuous cells, where the cell is first flushed with a stream of carbon dioxide, and then closed, and voltage applied onto the cell.
3. Semi-half continuous cells, with a design similar to the full continuous cell, but operating in the presence of a continuous flow of CO_2 through the cell.

All three operation modes have proven efficient in a special photoelectrocatalytic cell (PEC) based on nanostructured electrodes, which converted CO_2 to long-chain hydrocarbons (with a chain length >C_5) [112,113]. The concept of PEC originates from Ichikawa [114,115], who made fuels, such as methane, from effluent carbon dioxide, via a contact catalytic process and a photoelectrocatalytic process.

A schematic of the PEC device is displayed in Figure 1.8, drawn after Li [44]. The PEC utilised by Centi and co-workers [113,116,117] had on the cathode side a photocatalyst (nanostructured titania, 1 in Figure 1.8), which photochemically oxidised water to produce molecular oxygen, protons and electrons. Next, the H+ ions crossed the proton-exchange membrane (2 in Figure 1.8), and electrons assisted by protons enabled the

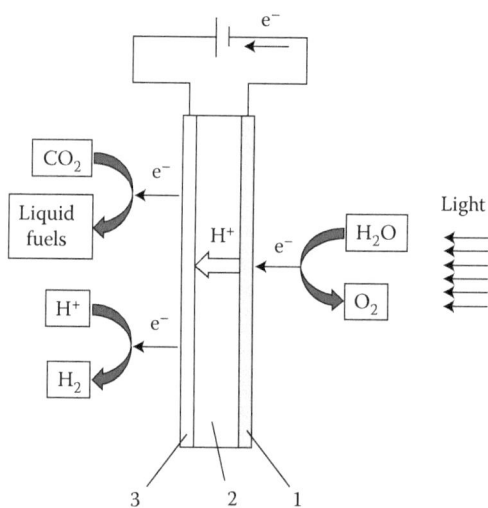

FIGURE 1.8
A schematic of the photoelectrocatalytic (PEC) cell. (Adapted from W. Li, Electrocatalytic reduction of CO_2 to small organic molecule fuels on metal catalysts. In Y. Hu (ed.) *Advances in CO_2 Conversion and Utilization*. ACS Symposium Series, American Chemical Society, Washington, DC, 2010, p. 58.)

reduction of CO_2 to liquid fuels (hydrocarbons or 2-propanol), a process that took place on the cathode side of the cell, that is, on the electrocatalyst confined to the ion exchange membrane (3 in Figure 1.8). Overall, carbon dioxide was converted into fuels by means of solar energy [113]. A competing reaction is the reduction of protons to molecular hydrogen (also shown in Figure 1.8) [44].

1.3.7 Cell Separators

Both divided and undivided cells are being utilised. Cell separators in divided cells may be composite perfluorinated polymer *cation exchange membranes* (e.g. Nafion® 115, Nafion® 117, Nafion® 430 and Nafion 961, manufactured by DuPont, USA) or *non-fluorinated cation exchange membranes* (e.g. CR67-HMR by Ionics, Inc.—General Electric Infrastructure—Water & Process Technologies, USA and Ralex® CM-PAD, reinforced by polyamide or polyester fabrics for improved mechanical properties, manufactured by MEGA in the Czech Republic and distributed worldwide) or an *anion exchange membrane* (e.g. Selemion™ AMV, developed by Asahi Glass Co Engineering, AGC, Japan, Neosepta® AMX by Astom Corporation, USA, and Excellion™ I-200, SnowPure, LLC, USA); the latter allows for the passage of anionic products, such as formate and acetate ions. A typical example of aqueous electrolytes used in a divided cell is alkali bicarbonate; for example, 0.65 mol/dm^3 solution of $NaHCO_3$ may serve as the catholyte, and 1.1 mol/dm^3 $KHCO_3$ as the anolyte. Alternatively, phosphate buffers can act as the electrolyte. Typically, the buffer solutions are set to *pH* 6.8–7.8. It is important to maintain a gas headspace in both compartments of the cell; the headspace is regulated by a mass flow controller. Prior to this, the electroreduced carbon dioxide should be humidified with water by passing the gas through a bubbler. This minimises the evaporation of volatile liquid-phase products. Operating the electrochemical at low temperature, for example, −2°C depresses hydrogen formation significantly, which competes with CO_2 reduction and, therefore, reduces the Faradaic efficiency of the conversion. Based on the results of this work, the proposed electrochemical method appears to be a viable means for removing CO_2 from the atmosphere and converting it into more valuable chemicals [43,81,111,118,119].

A more compact reactor design takes advantage of solid polymer electrolytes (SPEs). Hence, cast Nafion membrane and SPEEK were prepared, and used as cationic SPE; whereas alkali-doped poly (vinyl alcohol), PVA and Amberlyst resin composites were made, and used as anionic SPE. These SPEs were combined with a cathode made of copper, electrodeposited onto porous carbon paper, and an anode consisting of Pt/C on the carbon paper. The main electroreduction products were formic acid, methanol, formaldehyde, carbon monoxide and methane, while undesired hydrogen gas by-product was also detected [120].

A special type of gas diffusion membrane (GDM) was coated with Pt or Pd nanoclusters, and showed enhanced properties with respect to commercial materials for the gas-phase electrocatalytic reduction of CO_2 to fuels [113].

The proper choice of the membrane to be used in CO_2 electroreduction necessitates a circumspect evaluation of membrane properties, so that it complies with the cell operation requirements. The features to be considered include (i) the type of membrane (cationic or anionic), (ii) its electrical resistance (resistance is in the range of 2.5–9.5 Ω/cm^2), (iii) membrane thickness (in the dry state, it is in the range of 0.16–0.75 mm, while in the wet state, it increases by 45–60%), (iv) transport numbers (may exceed 0.98), ion-exchange capacity (1.0–2.8 meq/g), (v) thermal stability (typically 50–100°C), (vi) mechanical properties (burst strength of the membranes may vary in a wide range, from 0.77 to 10.5 kg/cm^2), (vii) changes during swelling (typically, width change is 45–60% and the weight change: 55–70%), and one should also take into account (viii) cost and availability of the selected membrane [121,122].

It has been emphasised in the literature that a combination of GDEs, discussed in Section 1.3.4, and an SPE or an ion-exchange membrane offers an excellent alternative for the electroreduction of carbon dioxide in aqueous media. Such a combination provides an enhanced mass transfer of CO_2 and enables continuous operation [44]. Given that the SPEs do not require the use of a supporting electrolyte, GDE/SPE systems are particularly suitable for the electroreduction of CO_2 in non-conductive organic liquids, and they allow devising and building a compact electrolysis device [123].

1.3.8 Electrochemical CO_2 Conversion at High Current Density

In many instances, products of the electrochemical CO_2 reduction form at low yields. Therefore, in order to generate detectable amounts of products, the electrochemical cells should be operated at a large current density, in the range from 0.5 to 20 mA/cm^2, and for a sufficiently long duration of the electrosynthesis (e.g. 1 h). Because of these relatively large current densities, voltage measurements would suffer significant errors, unless internal voltage drop is compensated. Product separation can be done by gas chromatography (GC), while for identifying the individual components, one can use on-line mass spectrometry (MS), *in situ* Raman spectroscopy and nuclear magnetic resonance (NMR) spectroscopy [2,43].

When employing copper electrodes, they need to be preconditioned prior to each electroreduction experiment. Steps involved are the mechanical polishing of the metal surface and the subsequent electropolishing in concentrated phosphoric acid [43].

Reviewed in the next sections are specific methodologies that enable the conversion of carbon dioxide into formic acid, organic carbonates and liquid fuels: methanol and superior alcohols or hydrocarbons. One section deals with electrochemical reductions performed in nonaqueous media.

1.4 Organic Products Electrosynthesised from Carbon Dioxide

The electrochemical conversion of carbon dioxide yields valuable organic products. Formic acid can be obtained selectively by CO_2 electroreduction; this process being the closest to being implemented as large-scale manufacturing. Formic acid is used in industry for leather production, textile dyeing and finishing, and as coagulant in rubber manufacturing processes. Recently, salts of formic acid (formates) have been used for fighting slippery road conditions in countries with heavy winters and dangerous roads, such as Austria and Switzerland. Formates are more effective than traditional salt treatments in increasing the gripping ability of otherwise slick surfaces treatments; also, formates are more environmentally friendly. Agriculture utilises a very high percentage of formic acid use worldwide, being used as both an antibacterial preservative and pesticide. It is most commonly used as a food additive and is frequently added to animal feed and silage. Also, formic acid is a promising storage material for hydrogen, a future energy carrier.

Electroreducing CO_2 enables the synthesis of organic solvents to hydrocarbons and alcohols that would allow renewable energy sources to be transformed into fuels and chemicals. Over the past three decades, significant work has been done in the field of electrocatalytic and homogeneous reduction of carbon dioxide [124]. To date, however, there are no electrode catalysts available that could accomplish this electroreduction at a low overpotential and reasonable current density.

1.4.1 Synthesis of Formate Salts/Formic Acid

Future energy supplies may use molecular hydrogen as an energy carrier, as hydrogen has a large chemical energy density (39 kWh/kg). The implementation of a hydrogen economy suffers, however, from several unsolved problems, such as the storage of appropriate amounts of hydrogen. A promising hydrogen storage method relies on liquid-phase chemical hydrogen storage materials, such as aqueous sodium borohydride (10.70% hydrogen content by mass of $NaBH_4$), ammonia borane (19.63% hydrogen in $H_3N–BH_3$), hydrazine (9.17% hydrogen in H_2NOH), hydrazine borane (12.93% hydrogen in $H_2NOH–BH_3$) and formic acid (4.39% hydrogen in HCOOH) [125]. The use of formic acid in hydrogen storage would provide reasonable gravimetric and volumetric hydrogen densities, low potential risk and low capital investment, given that it is largely compatible with the current transport infrastructure [99].

Hydrogen can be generated from formic acid–amine adducts at room temperature, and used directly in fuel cells [126,127]. Ruthenium metal carbonyl and hydrido carbonyl complexes exhibit a catalytic activity in the decarboxylation of formic acid. Hence, $[Ru_4(CO)_{12}H_4]$ prepared from $RuCl_3$ and formic acid can decompose formic acid to hydrogen and carbon dioxide [126].

While efficient avenues for releasing hydrogen from formic acid have been demonstrated, the efficient synthesis of formic acid continues to be a challenge [55]. This state of affairs is surprising, if one considers that the electroreduction of carbon dioxide to formic acid has been studied for more than 140 years, and the first report on the process was published by Royer as early as in 1870 [127]. By the beginning of World War I, an additional three papers had been released [128–130]. Nevertheless, the process still needs to be optimised for energy efficiency and selectivity [131,132].

Ruthenium phosphine systems may act as catalysts in the conversion. This process appears to be the most practical application of CO_2 electroreduction, the closest to commercialisation. Laboratory work performed on the 100 A current scale demonstrated that reduction of CO_2 to formate can be accomplished in a trickle-bed continuous electrochemical reactor, under industrially viable conditions; conceptual flow sheets were proposed for two process options, each converting CO_2 at a rate of 100 tons per day [107]. Large-scale electrochemical reduction of CO_2 to formate salts and formic acid on tin or proprietary catalysts can be performed in a flow-through reactor. This process was demonstrated to be feasible from both engineering and economic standpoints [133]. Continuous electrochemical conversion of CO_2 to formate was successfully conducted in a polymer electrolyte membrane cell, where an alkaline ion-exchange membrane was sandwiched between two catalytic electrodes that contained lead and indium for suppressing hydrogen evolution. Given the 80% efficiency of CO_2 conversion to formate, this cell appears to be particularly attractive for large-scale implementation. Performing the electrolysis in a pulsed mode mitigated mass transport limitations, and provided high efficiency [134].

GDEs have been used successfully in the electroreduction of CO_2 to formic acid on tin or lead cathodes, in aqueous solution [57,135,136]. Hence, Kwon and Lee [136] reported the efficient use of nanolayered lead electrodes, prepared by stepwise potential deposition. By this technique, a nanostructured Pb layer forms, which consists of particles and platelets in hexagonal and cubic crystalline form. In electroreductions conducted in aqueous 0.1 mol/dm^3 $KHCO_3$ supporting electrolyte, at 10 mA/cm^2 current density and a temperature of 5°C, cubic lead surface secured the highest Faradaic yield (94.1%), while polycrystalline smooth Pb films enabled only 52.3% yield. The authors suggested that indirect reduction of CO_2 by adsorbed hydrogen atoms (shown in Equations 1.16a, 1.16b and 1.16c) is more likely than direct electroreduction of CO_2 molecules.

$$H^+(aq) + e^- \rightarrow H\bullet\ (ads) \tag{1.16a}$$

$$CO_2 + H\bullet\ (ads) \rightarrow HCOO\bullet\ (ads) \tag{1.16b}$$

$$HCOO\bullet\ (ads) + e^- \rightarrow HCOO^-\ (aq) \tag{1.16c}$$

Subramanian and co-workers [137] performed the two-electron electroreduction of CO_2 to formate ions in an electrochemical membrane reactor, in mildly alkaline $KHCO_3$ buffer solutions (pH 7.5–8.5). Typically, the carbonate concentration was only 0.5 mol/dm^3; therefore, in order to avoid significant ohmic voltage drops through the cell, the conductivity of the electrolyte was augmented by the addition of K_2SO_4. One used a divided filter-press type electrochemical flow reactor, equipped with a composite perfluorinated cation-exchange membrane (Nafion 961 or Nafion 430); the reduction of dissolved CO_2 was conducted at ambient conditions. This flow reactor enhanced the mass transfer of CO_2, as compared to a batch reactor. The anode was titanium mesh, coated with a thin layer of Ir and Ta oxides. The efficiency of various cathode materials (Cd, Pb, Sn and Zn plated stainless-steel woven mesh) for the electroreduction of CO_2 was tested. The optimum values of flow rate, current efficiency and cell voltage were evaluated for the formation of formate in aqueous $KHCO_3$ buffer solutions. By comparing the obtained results, the cathodes were rated as follows: Pb ≈ Sn > Cd ≫ Zn. A maximum current efficiency of 90% was reported when using a lead-coated cathode, for a formate production of 2.35×10^{-1} mol/dm^3, at the flow rate of the catholyte equal to 13 cm^3/min [137].

Tin-based GDEs allowed for the synthesis of formic acid at a high current density of 27 mA/cm^2, a value two orders of magnitude higher than the one enabled by metallic tin disc. The reduction was conducted at –1.6 V vs. NHE, at ambient pressure, with a Faradaic efficiency of 70% [109].

1.4.2 Preparation of Organic Carbonates from CO_2

Organic carbonates represent interest owing to a wide range of applications in the fields of chemical and pharmaceutical industry, including the manufacturing of polymers, electrolyte solvents for electrochemical capacitors and lithium ion batteries, organic solvents for cleaning metal surfaces, fuel additives, and green reagents [138–143]. Lithium ion batteries operate with an organic electrolyte, formed from a salt (such as $LiPF_6$, $LiAsF_6$, $LiBF_4$, $LiClO_4$, etc.), dissolved in a solvent blend of ethylene carbonate and various other linear carbonates, including dimethyl carbonate, diethyl carbonate and ethyl methyl carbonate. Vinylene carbonate is typically used as an additive [142]. Alkyl carbonates enable versatile pathways to organic synthesis in the fields of enzyme chemistry, functional group manipulation, photochemistry, solid-phase synthesis and natural products [139].

Traditional preparation methods of organic carbonates involve the use of carbonyl equivalents, such as the reaction of phosgene with diols or the coupling of haloformates with alcohols or phenols. While both reactions proceed at high yields, they require the use of toxic, expensive and corrosive reagent [140]. Therefore, utilising carbon dioxide has proven a promising alternative to phosgene, owing to its high abundance and environmentally benign nature [141]. Given the ability of electrochemical techniques to activate and reduce CO_2 at room temperature and ambient pressure, the use of these effective and green methods appears as an obvious choice.

In 2008, dimethyl carbonate was obtained by the electrochemical activation of CO_2 in an ionic liquid [144]. Also, cyclic carbonates were successfully synthesised by the electroreduction of CO_2 in the presence of diols in ionic liquids saturated with CO_2. The synthesis was conducted at room temperature, under safe condition, in an undivided cell. No additional electrocatalysts was required, and the ionic liquid could be recycled [145].

Recently, organic carbonates have been prepared under mild conditions by reducing carbon dioxide in the presence of alcohols [146,147]. Hence, organic carbonates were synthesised directly from CO_2 in the presence of alcohols under galvanostatic conditions. The electroreduction was conducted at a copper cathode, at room temperature and normal pressure without catalyst, in a mixture of DMF and acetonitrile. Copper cathodes were more active than Ag, Ni, Ti and stainless steel. Primary and secondary alcohols were converted into the corresponding linear organic carbonates, at moderate and low yields. A schematic of the reactions is given in Equations 1.6, 1.17 through 1.19 [146]. The anion radical generated in the first electronation of CO_2 (Equation 1.6) is rather reactive, and interacts with methanol to form another anion radical (Equation 1.17), which undergoes a second electronation step to form an anion (Equation 1.18), which performs a nucleophilic attack on the methyl iodide added to the system (to serve as a methylating agent), yielding the dimethyl carbonate (Equation 1.19):

$$CO_2 + e^- \rightarrow CO_2^{-} \bullet \quad (1.6)$$

$$CO_2^{-} \bullet + CH_3-OH \rightarrow CH_3-O-CO-OH^{-} \bullet \text{ (ads)} \quad (1.17)$$

$$CH_3-O-CO-OH^{-} \bullet \text{ (ads)} + e^- \rightarrow CH_3-O-CO-O^{-}: + \tfrac{1}{2} H_2(g) \quad (1.18)$$

$$CH_3-O-CO-O^{-}: + CH_3-I \rightarrow CH_3-O-CO-O-CH_3 + I^- \text{ (aq)} \quad (1.19)$$

A long series of alcohols were successfully reacted with CO_2, including benzyl alcohol and its derivatives, as well as cyclohexanol. Meanwhile, tertiary alcohols and phenols have proven inert. Also, cyclic carbonates can be synthesised from CO_2 and diols [146].

Dimethyl carbonate was obtained from CO_2 on an indium cathode, in a high-purity ionic liquid, 1-butyl-3-methylimidazoliumtetrafluoborate ($BMIMBF_4$, >99.99%). Prior to the electroreduction, the ionic liquid was dried carefully, under vacuum at 110°C, as the presence of moisture would narrow the potential window of $BMIMBF_4$. As demonstrated by cyclic voltammetry, the reduction peak occurred at –1.8 V (vs. Ag). The electrolysis experiments were carried out in an undivided cell, magnesium being used as a sacrificial anode. Preceding the electroreduction, the electrolyte was saturated with CO_2, and a constant flow of carbon dioxide through the solution was maintained over the entire duration of the synthesis. Constant-potential electrolysis was conducted under mild conditions, without the need for additional solvents, catalysts, or supporting electrolytes. After CO_2 was activated by passing a certain amount of charge through the system, the $CO_2^-\cdot$ radical anion was converted to dimethyl carbonate by its reaction with methanol, and then the reaction was completed by adding a threefold excess of methyl iodide, a strong methylating agent. The proposed pathway to dimethyl carbonate formation in $BMIMBF_4$ is shown in Figure 1.9, adapted from the work of Liu and co-workers [147].

In the first step, a CO_2 molecule diffuses towards the indium electrode, and gets adsorbed on its surface, where it is activated and the $CO_2^-\cdot$ radical anion forms (Equation 1.20). As $CO_2^-\cdot$ ion pairs with $BMIM^+$, which reacts with methanol to form $CH_3-O-COO^-\cdot$, where the unpaired electron is located on the oxygen atom that originates from methanol (Equation 1.21). Next, molecular hydrogen is released (Equation 1.22), and dimethyl carbonate is formed in a subsequent chemical reaction with methyl iodide (Equation 1.23). This procedure enabled the synthesis of dimethyl carbonate at reasonable Faradaic yields of 73.2–76%, at potentials from –1.7 to –1.9 V (vs. Ag) [68,147].

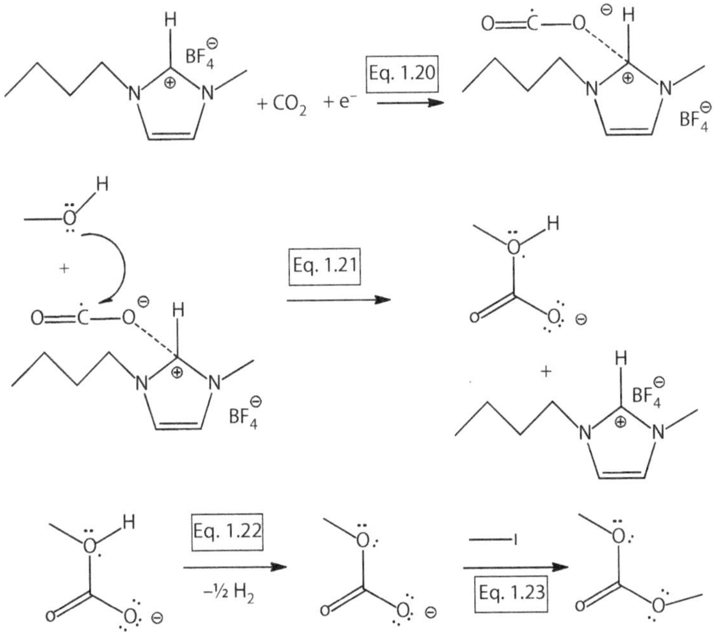

FIGURE 1.9
Multi-step synthesis of dimethyl carbonate, initiated by the $CO_2^-\cdot$ anion radical formed in the electrochemical reduction of CO_2 in an ionic liquid. (Adapted from F. Liu et al., *Int. J. Electrochem. Sci. 7*, 2012, 4381–4387.)

1.4.3 Synthesis of Methanol

Electroreduction of CO_2 in aqueous solutions to form methanol proceeds according to the six-electron transfer reaction shown in Equation 1.24. It is the reverse of the anodic methanol oxidation, which has been extensively investigated in fuel cells.

$$CO_2 \text{ (g)} + 6H^+ \text{ (aq)} + 6e^- \rightarrow CH_3OH \text{ (l)} + H_2O \text{ (l)} \qquad (1.24)$$

The same catalysts and conditions may operate for both reaction directions, that is, either towards the synthesis of methanol or towards its oxidation. Again, CO_2 reduction must be able to compete with the electrochemical reduction of water-yielding hydrogen gas. Thermodynamically, the reduction of CO_2 to CH_3OH is slightly more favourable than the reduction of water. Kinetically, CO_2 reduction can be favoured by electrodes, which act as poor catalysts for water reduction, including metallic Mo, Cu, In, Sn and Sb. Using nearly neutral electrolytes, rather than acidic ones, rendered CO_2 reduction kinetically more favourable [148]. A detailed mechanism of the mediated electroreduction of carbon dioxide to methanol is addressed in the next section.

1.4.3.1 Pyridinium Ions: Potent Catalysts Enabling CO_2 Reduction to Methanol

Since 1994, pyridinium and its substituted derivatives have been identified as effective and stable homogeneous electrocatalysts for the aqueous multiple-electron, multiple-proton reduction of CO_2 to various products, such as formic acid, formaldehyde and methanol. Particularly high Faradaic yields were reached in the reduction of CO_2 to methanol in both electrochemical and photoelectrochemical systems under energetically advantageous conditions [149].

Hence, carbon dioxide can be reduced to methanol on hydrogenated Pd electrodes, in the presence of pyridinium ions, which enable homogeneous catalysis. This reduction takes place at a modest overpotential of ~200 mV. Although the reduction of protons to H_2 competes with methanol formation, methanol Faradaic yields of up to 30% have been reached. Methanol formation proceeds both directly at the electrode surface, and indirectly, by the reduction of pyridinium species to form methanol, pyridine and hydrogen [150].

Bocarsly and co-workers [149,150] reported the mechanistic pathway for the electrochemical conversion of CO_2 to methanol. They described the selective conversion of CO_2 to methanol at a p-GaP semiconductor electrode, catalysed by pyridinium ions, where the reaction was driven by light energy, to yield Faradaic efficiencies near 100%, at potentials well below the standard potential. At illuminated electrodes, cathodic currents of ~20 mA/cm² could be maintained, without an applied bias. At metal electrodes, formic acid and formaldehyde were the intermediate products towards methanol (a 6e⁻ reduced product). Pyridinium radicals played a role in the reduction of both reaction intermediates [149]. The successive steps of the reduction mechanism are displayed in Figure 1.10.

First, the pyridine ring is protonated (Equation 1.25); the base dissociation constant of pyridine is $K_b = 1.49 \times 10^{-9}$. Next, the pyridinium ion undergoes a one-electron transfer to form the pyridinium radical (Equation 1.26). The reduction is coupled to a catalytic reaction of hydrogen generation (Equation 1.27). Hence, steps 2 and 3 constitute an EC sequence (where EC stands for an electrochemical step followed by a chemical step). A competing process, the inner-sphere interaction between the catalyst and the CO_2 substrate, yields a radical pyridinium–CO_2 complex intermediate (Equation 1.28), which cleaves into a hydroxyformyl radical (Equation 1.29). When Pt and Pd electrodes are used, this anion radical can undergo a one-electron

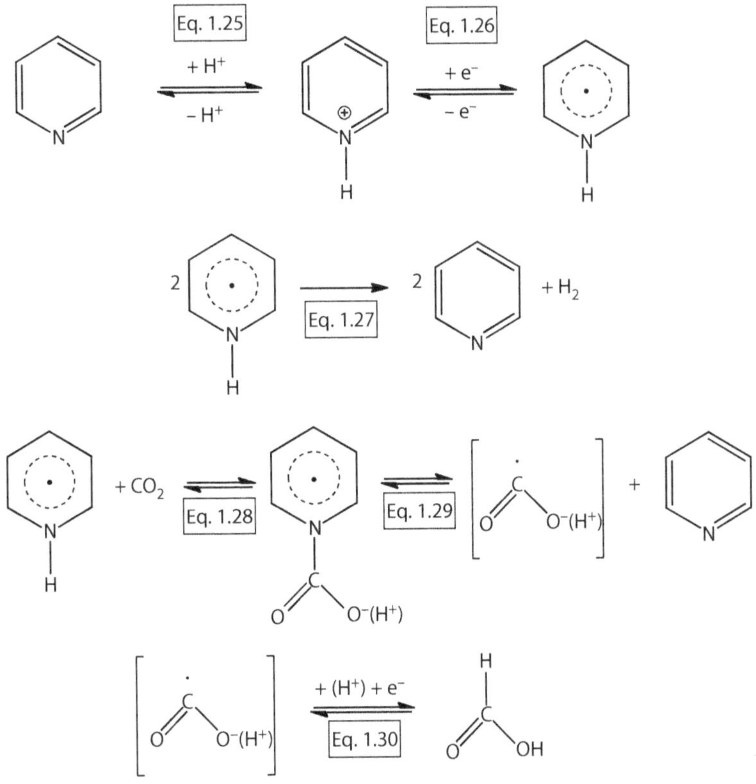

FIGURE 1.10
Inner-sphere-type electron transfer from the pyridinium radical to the carbon dioxide substrate. Refer to the text for details. (Adapted from K.A. Keets et al., *Ind. J. Chem. A*, 51, 2012, 1284–1297; E. Barton Cole et al., *J. Am. Chem. Soc.* 132, 2010, 11539–11551; G. Seshadri, C. Lin and A.B. Bocarsly, *J. Electroanal. Chem.* 372, 1994, 145–150.)

reduction coupled with a protonation, to yield formic acid (Equation 1.30). On p-GaP photoelectrochemical electrodes, the same end product, formic acid, is obtained by the direct reduction of the pyridinium radical–CO_2 complex [149].

As revealed by Figure 1.11, the pyridinium radical can further react with the previously obtained formic acid (Equation 1.31), and the resulting pyridinium–formyl radical adduct may interact with another pyridinium radical, yielding one molecule of formaldehyde and two molecules of pyridine (Equation 1.32). On its turn, formaldehyde may further react with a pyridinium radical to form a more reduced adduct (Equation 1.33), which undergoes a fast reduction with another pyridinium radical to yield methanol (Equation

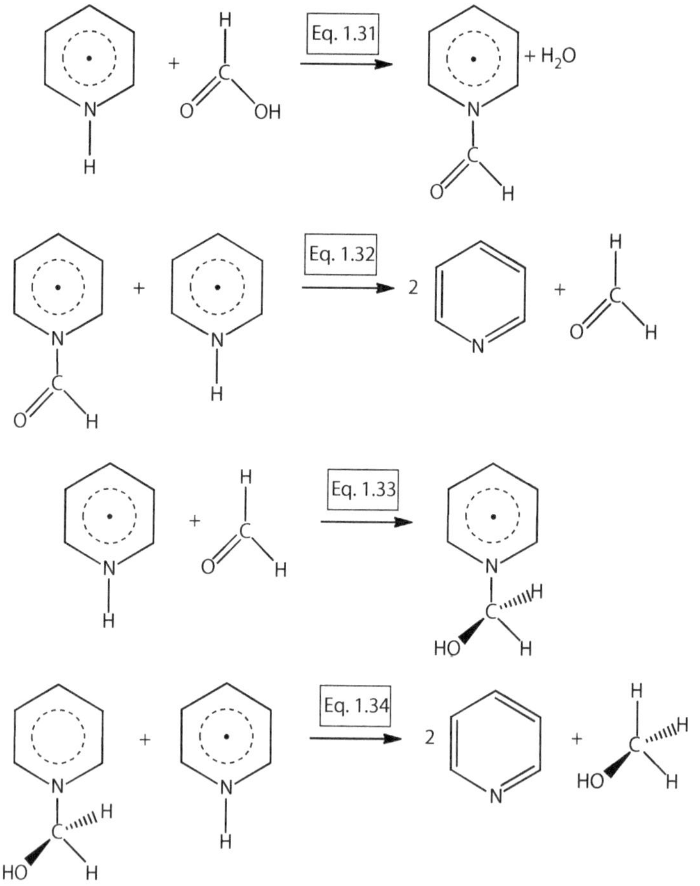

FIGURE 1.11
Mechanism of pyridinium-catalysed electrochemical reduction of formic acid to methanol. Refer to the text for details. (Adapted from K.A. Keets et al., *Ind. J. Chem. A*, 51, 2012, 1284–1297; E. Barton Cole et al., *J. Am. Chem. Soc.* 132, 2010, 11539–11551; G. Seshadri, C. Lin and A.B. Bocarsly, *J. Electroanal. Chem.* 372, 1994, 145–150.)

1.34). Overall, the pyridinium catalyst enables the step-by-step reduction of carbon dioxide to highly reduced species. Hence, pyridinium salts alleviate the need for metal-based multi-electron transfer.

1.4.4 Synthesis of Ethanol and 2-Propanol

Selective electroreduction of CO_2 to ethanol has been reported since the early 1990s. So, Ikeda and co-workers [151] successfully conducted this reaction on mixtures of copper and zinc oxide. They performed the electroreduction potentiostatically, in a w-shaped Pyrex cell with one gas and two liquid chambers, and two lines of gas circulating system. When using a GDE of metal oxides mixed with carbon black (CuO/ZnO = 3/7 + carbon black, 6:5 by mass), the main reduction product was C_2H_5OH, and only small amounts of CO and $HCOO^-$ formed as the by-products, along with comparable amounts of H_2. At a working potential of -1.32 V vs. Ag/AgCl and a partial current density of 4.23 mA/cm^2, ethanol formed with a maximum Faradaic efficiency of 16.7% and a maximum selectivity of 88% [104].

Ethanol and 1-propanol may be formed via the reduction of acetaldehyde and propionaldehyde, respectively, which are intermediates of CO_2 electroreduction on copper cathode, in dilute $KHCO_3$ solution (0.03 mol/dm^3) [89]. Recently, 2-propanol was synthesised by the reduction of CO_2 on iron oxide nanoparticles, deposited on an N-functionalised carbon nanotube (CNT) support (rather than on a pristine or oxidised CNT support). These particles are present as magnetite phase (Fe_3O_4), but their structural anisotropy and size inhomogeneity depend on the preparation method of the carbon surface. High electrocatalytic performance of these materials is a result of the N-functionalisation of the carbon support, as this modification of the support yields small nanoparticles, which are highly populated by reversible chemisorbing sites [152]. Tacconi and co-workers [153] reported the facile electrocatalytic reduction of CO_2 to methanol and 2-propanol on Pt-modified carbon-titanium dioxide cathode nanocomposite matrices ($Pt/C–TiO_2$), in the presence of pyridinium cation, as a solution co-catalyst.

Recently, Peterson and Nørskov [47] compared trends in binding energies for the intermediates identified in CO_2 electrochemical reduction, and presented a 'volcano'-based activity of the experimentally observed variations in transition metal catalysts. They confirmed that copper is the best-known metal electrocatalyst. The authors proposed new strategies to be adopted towards the discovery of potent catalysts that can operate with a reduced overpotential.

1.4.5 Hydrocarbon Synthesis

Reducing carbon dioxide to hydrocarbons is a complex, multi-step process, which involves shared intermediates and multiple reaction patways [90,154,155]. Often, several reactions that are part of the mechanism compete with each other.

Copper electrodes enable the electroreduction of CO_2 to hydrocarbons and alcohols. Intially, CO_2 is reduced to adsorbed >CO, which interacts with the Cu surface in a unique manner, such that further reduction to alcohols and hydrocarbons can take place; by these reactions, methane and even long-chain hydrocarbons may form. The product distribution is affected by cations and anions present in the electrolyte solution, and product selectivity can be improved by the modification of the Cu electrode surface with halides, oxygen or sulphur atoms. Alkanes with chain lengths up to C_{12} can be formed; the amount of charge needed per molecule of alkane (expressed in faraday/mol) increases linearly with the chain length, while the specific gravimetric charge (kAh/kg) decreases with the increase of chain length of the obtained alkane molecule [156]. In this process, the protonation of adsorbed CO was identified as the most important step, which accounts for the overpotential [144]. It was found that halide ions adsorbed onto the copper electrode play a catalytic role; hence, the presence of a Cu–X(ads) catalytic layer facilitates the electron transfer from the electrode to CO_2. When the halide–carbon bond forms, $CuX_{ad}^- - C$, where X_{ads}^- is Br^-, Cl^- or I^-, the electrons flow from the specifically adsorbed halide anion to the vacant orbital of CO_2. In addition, the specifically adsorbed halide anion could suppress the adsorption of protons, leading to an increased hydrogen overvoltage and a greater rate of electrochemical CO_2 reduction [117,118].

To convert CO_2 to long carbon-chain hydrocarbons (>C5) at room temperature and atmospheric pressure, it is possible to use platinum nanoparticles supported on highly conductive and chemically inert carbon black (Vulcan XC-72, Cabot Corporation) and/or carbon cloth as electrodes. A continuous flow cell configuration is required, where the working electrode is in direct contact with gaseous CO_2. This system integrates in a photoelectrochemical device, which utilises solar energy and water to convert CO_2 into fuels [112]. The mechanism of surface reaction proposed by Centi et al. [112] is summarised in a simplified manner by Equations 1.6 and 1.35a through e. Electronation of CO_2 to its anion radical (Equation 1.6) competes with the discharge of protons to molecular hydrogen (Equation 1.35a). Next, the anion radical may undergo a second electronation and protonation to formate ions ($HCOO^-$). Again, the formation of formate ions (Equation 1.35b) competes with the formation of adsorbed CO (Equation 1.35c). In this mechanism, adsorbed carbon monoxide enables the formation of adsorbed methylene, $>CH_2$ (Equation 1.35e), which, on its turn, serves as the key intermediate towards more reduced species: methane (Equation 1.35f) and methanol (Equation 1.35g). Also, adsorbed methylene may undergo a dimerisation to adsorbed ethylene radicals (Equation 1.35 h), which by further reduction yield C_2 products: ethylene (Equation 1.35i) and ethane (Equation 1.35j). Adsorbed methylene groups may also generate C_3, and over C_3 products (Equation 1.35k).

$$CO_2 + e^- \rightarrow CO_2^- \bullet \qquad (1.6)$$

$$H^+ (aq) + e^- \rightarrow \tfrac{1}{2}H_2 (g) \quad (1.35a)$$

$$CO_2^{-}\bullet + e^- + H^+ \rightarrow HCOO^- (des) \quad (1.35b)$$

$$CO_2^{-}\bullet + e^- + 2H^+ \rightarrow -C=O (ads) + H_2O \quad (1.35c)$$

$$-C=O (ads) \rightarrow CO (gas) \quad (1.35d)$$

$$-C=O (ads) + 4H^+ + 4e^- \rightarrow >CH_2 (ads) + H_2O \quad (1.35e)$$

$$>CH_2 (ads) + 2H^+ + 2e^- \rightarrow CH_4 (g) \quad (1.35f)$$

$$>CH_2 (ads) + H_2O \rightarrow CH_3OH \quad (1.35g)$$

$$>CH_2 (ads) + >CH_2 (ads) \rightarrow -CH_2-CH_2- (ads) \quad (1.35h)$$

$$-CH_2-CH_2- (ads) \rightarrow H_2C=CH_2 (g) \quad (1.35i)$$

$$-CH_2-CH_2- (ads) + 2H^+ + 2e^- \rightarrow CH_3-CH_3 (g) \quad (1.35j)$$

$$-CH_2-CH_2- (ads) + >CH_2 (ads) \rightarrow C_3 \text{ and over } C_3 \text{ products} \quad (1.35k)$$

An analysis of the energy balance and economics of CO_2 recycling to hydrocarbon fuels estimated that the full system can feasibly operate at 70% electricity-to-liquid fuel efficiency, and the price of electricity, needed to produce synthetic gasoline, is only 1.0–1.5% of the present sales price of gasoline. In regions where inexpensive renewable electricity is currently available, such as Iceland, fuel production may already be economical. The dominant costs of the process are the electricity cost and the capital cost of the electrochemical cell, where this capital cost is significantly increased, when operating intermittently (on renewable power sources such as solar and wind) [42].

1.5 Future Directions

Conversion of CO_2 to fuels or chemical feedstock is attractive for combating the increasing concentration levels of atmospheric CO_2 and storing energy [134]. The *Green Freedom* concept proposed by Los Alamos National Laboratory scientists aims to remove CO_2 from the air and convert it into gasoline. Air would be blown over a liquid solution of $KHCO_3$, which absorbs the CO_2, and then it would be extracted and subjected to processes that turn CO_2 into fuel: methanol, gasoline or jet fuel [157]. From various available procedures, the 'one-pot' electrochemical synthesis of hydrocarbons from

CO_2 represents an attractive and potentially useful process [90]. Ideally, the conversion should be carried out with electrical energy from a renewable generation system such that 'carbon-neutral' fuels are obtained [134].

Fuel generation from CO_2 on Pt nanoparticle catalysts may be used to produce fuels electrochemically in Mars missions, from ambient CO_2 and water, both present in the atmosphere and soils on Mars [112]. Such fuels made onsite would enable Earth-return propellants and life-support consumables [157,158]. Sridhar and co-workers [158] proposed a new architecture for an *in situ* propellant production plant, which utilises solid oxide electrolysis to accomplish the combined electrolysis of water and carbon dioxide. This system produces methane by means of a Sabatier reactor and oxygen via electrolysis in the optimal oxidiser-to-fuel mixture ratio. It also has the capability to produce additional oxygen for life support needs. Results demonstrate that combined electrolysis enables a competitive system for *in situ* resource utilisation, and is easy to be scaled for sample return from Mars and for possible human missions to the planet [158].

In an effort to reduce the amount of transition metal electrocatalysts needed in CO_2 electroreduction, Perez-Cadenas and co-workers [159] used Ni, Cu or Fe doped carbon xerogels as cathode material. Such doped xerogels exhibited very promising electrocatalytic properties, enabling the conversion of CO_2 to gaseous hydrocarbons C_1–C_4 at atmospheric pressure. The product distribution is determined by the doping metal, as well as by the texture and chemical properties of the carbon xerogel [159]. Carbon-supported iron catalyst is being used in the electrocatalytic reduction of CO_2 as an inexpensive alternative [80].

At present, solar fuels that produce fuels from water and CO_2 present a priority topic of scientific and industrial interest. Research advances on bioroutes, concentrated solar thermal and low-temperature conversion using semiconductors and a photoelectrocatalytic approach are critical in defining challenges, current limits, and identifying priorities on which to focus future research and development. It is essential to produce fuels, which can be transported and stored easily, and can be integrated into the existing energy infrastructure. The role of solar fuels produced from CO_2 in comparison with solar H_2 needs to be addressed. Solar fuels are complementary to solar to electrical energy conversion, but they still need intensified research and development in order to be commercialised [160]. Various methods of achieving light absorption, electron–hole separation and electrochemical reduction of CO_2 are being considered, and the most promising systems for reducing carbon dioxide to CO, formic acid and methanol are identified based on the energy gap matching CO_2 reduction. Different approaches are taken for lowering the overpotentials at which the electroreduction can be conducted, and new catalysts are introduced for achieving high chemical selectivity [161].

Effort is made towards developing large 'artificial trees' that can absorb CO_2 directly from the air. Once CO_2 is 'harvested' from the artificial trees, it can be recycled into synthetic alcohols or synthetic fuels, balancing the hydrocarbon burning of cars and trucks and thus making these into essentially zero-emissions vehicles [162]. Artificial leaves should be developed, able to collect energy in the same way as a natural leaf would. This trend represents a great challenge for the use of renewable energy and a sustainable development.

In order to avoid the problem of intermittency in solar energy, it is necessary to design systems that capture CO_2 directly, and then convert CO_2 into liquid solar fuels that are easy to be stored. An advantageous condition for rendering artificial leaves over natural leaves is that artificial leaves operate at a higher solar energy-to-chemical fuel conversion efficiency, and provide directly fuels, which can be used in power-generating devices. In addition, artificial leaves must be robust, easy to construct and low cost. Devising artificial leaves represent a cutting-edge research area in which various concepts and ideas are under investigation. Product development requires consideration of system engineering, which considerably limits some of the investigated solutions. System design should be included in defining the elements of artificial leaf to be investigated (i.e. catalysts, electrodes, membranes and sensitisers). The main relevant aspects of the cell engineering (mass/charge transport, fluid dynamics and sealing) should also be considered from the very beginnings of the system design [163].

Ongoing fundamental studies may allow a better understanding of the electrochemical conversion of carbon dioxide into various products, and the accrued fundamental knowledge would eventually enable improved manufacturing technologies. It was found that CO_2 reduction to carbon monoxide can be catalysed by Fe^0 complexes, particularly when an *in situ* proton source is available. Iron (0) porphyrin can be prepared electrochemically from Fe^{II} porphyrin, by two consecutive electron uptakes on Hg or glassy carbon electrode. A considerable increase in catalytic activity was accomplished by modifying iron tetraphenylporphyrin through the introduction of phenolic groups in all the ortho and ortho' positions of the tethered phenol rings. Electroreduction experiments were conducted in organic (acetonitrile or DMF) or aqueous-organic media (acetonitrile + water or DMF + water). The modified tetraphenylporphyrin catalyst, which uses iron, one of the most earth-abundant metals, manifests Faradaic yield of CO synthesis above 90% through 50 million turnovers over 4 h of electrolysis at low overpotential (0.465 V), without noticeable degradation. The origin of the enhanced activity appears to be the high local concentration of protons associated with the phenolic hydroxyl substituents [13].

Also, recent results have revealed that metal/metal oxide composite materials are promising catalysts for sustainable fuel synthesis by CO_2 reduction. Thus, in aqueous $NaHCO_3$ solution saturated with CO_2, a Sn electrode covered with a native SnO_x layer exhibited potential-dependent CO_2 reduction activity, unlike an electrode etched to expose fresh Sn^0 surface, prior to being used as a cathode; metallic tin almost exclusively discharged H_2. Next, the Sn-SnO_x catalyst was converted into a composite material, a thin-film catalyst, obtained by the simultaneous electrodeposition of Sn^0 and SnO_x on a Ti electrode. This composite material enabled up to eight-fold higher partial current density and four-fold higher Faradaic efficiency was achieved in CO_2 reduction, as compared to a Sn electrode with a native SnO_x layer on its own [164].

One major problem encountered in the electroreduction of carbon dioxide is the poisoning of the catalyst with adsorbed CO, which forms as an intermediate [90]. To prevent this poisoning, electrochemical activation of CO_2 can be conducted in a tailor-made nanoscale environment, which enables the selective synthesis of formic acid. In the gas phase, methyl thiol can transfer a hydrogen atom to the carbon dioxide anion radicals [55].

Advance has been made on the electrocatalytic side, as well; rhenium patterns were optimised in conjunction with silicon photoelectrodes, such that they provide greater catalytic current and improved stability. Copper complexes capable of absorbing atmospheric CO_2 have been incorporated into an electrocatalytic cycle, and, as it was shown in Section 1.4.3.1 in mechanistic details, metal-free electrocatalysis of CO_2 to methanol was achieved with pyridinium ions [165].

In 2008, Mondal [166] proposed a new mechanistic concept for CO_2 reduction by nanoscale galvanic couples. As uncatalysed electroreduction requires a significant overvoltage, an alternate route has been suggested, which would involve (i) the production of hydrogen from water, by the chemical oxidation of a metal in water, and (ii) the subsequent reduction of the hydronium ion by the released electrons to form hydrogen radicals. Mondal hypothesised that it is essential to prevent the recombination of the obtained hydrogen radicals to H_2 molecules; instead, hydrogen radicals should react directly with CO_2, to yield methanol and other useful products. This scenario may be enabled by a bimetallic catalyst, which would act as a galvanic couple, wherein one metal serves as the electron donor for the production of hydrogen radicals, and the other serves as a catalyst for the reduction of CO_2. Overall, this indirect reduction mechanism may proceed at greater reaction rates. Also, this concept circumvents the need for an external electric field, unlike in the electrochemical process. Mondal emphasised the usefulness of bimetallic electrodes for CO_2 electroreduction on a conceptual basis, without identifying the two metals that need to be combined with each other. Later developments, however, took advantage of bimetallic electrodes in oxidation processes, in the conversion of ethylene, ethanol, propanol and formic acid. There are a few mentions of electroreduction proceeding on such electrodes.

An innovative recent synthetic development is the use of electroreduction of various substrates in the presence of CO_2. Electrocarboxylations have been accomplished on dibromobenzenes [167] and anthrone [168]. When *meta*- and *para*-dibromobenzenes were reduced on glassy carbon and silver electrodes in DMF, in the presence of CO_2, ortho-, meta-, and para monocarboxylation products of methyl bromobenzoate were obtained, while *ortho*-dibromobenzene was converted into a dehalogenated carboxylation product of methyl benzoate. Moderate to good electrocarboxylation yields

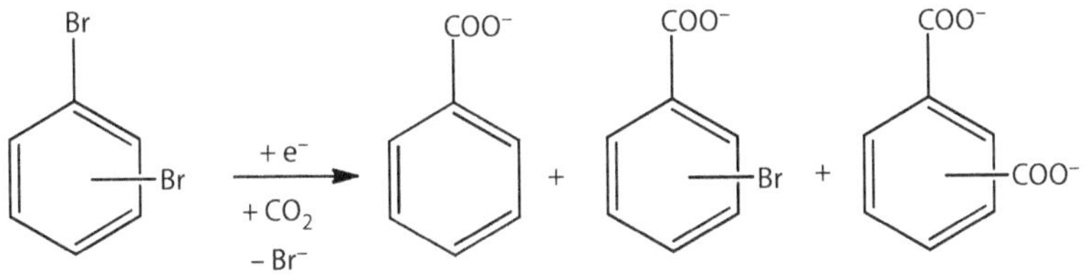

FIGURE 1.12
Electrocarboxylation of *ortho-*, *meta-*, and *para-*dibromobenzenes. (Adapted from Y.C. Lan et al., *J. Electroanal. Chem.* 664, 2012, 33–38.

(49–74%) were accomplished in preparative-scale electrolyses. This process was conducted in the presence of CO_2, in an undivided cell equipped with a magnesium rod, which served as a sacrificial anode. The catalytic ability of Ag was demonstrated in the electroreduction and electrocarboxylation of *ortho-*, *meta-* and *para-*dibromobenzenes (see Figure 1.12) [167].

Electrocarboxylation of anthrone in the presence of CO_2 yielded anthrance-9-carboxylic acid. Under optimised conditions, anthrancene-9-carboxylic acid was obtained in a very high yield of 96.1% [168].

One should conclude by emphasising that in electrochemical reductions, a green reagent is being used—the electron. Because of the presence of other reagents and additives, or as a result of the toxicity of the ionic liquid used as the electrolyte, the electrochemical methodologies are still unable to offer eco-friendly and competitive procedures of organic synthesis [169]. Therefore, future effort needs to be made for devising entirely environmentally friendly electrochemical reduction processes of carbon dioxide.

References

1. J.-R. Li, Y. Ma, M.C. McCarthy, J. Sculley, J. Yu, H.-K. Jeong, P.B. Balbuena and H.-C. Zhou, Carbon dioxide capture-related gas adsorption and separation in metal-organic frameworks, *Coord. Chem. Rev.* 255, 2011, 1791–1823.
2. H. Shibata, *Electrocatalytic CO_2 Reduction—Catalysis Engineering and Reaction Mechanism*, PhD Dissertation, Technische Universiteit Delft, The Netherlands, 2009. ISBN 978-90-6464-341-5.
3. Data collected by the Mauna Loa Observatory and Scripps Institution of Oceanography. See: Earth's CO_2 Home Page. http://co2now.org/ (January 3, 2013).
4. S.A. Rackley, *Carbon Capture and Storage*, Elsevier, Cambridge, MA, 2010.
5. R.E. Hester and R.M. Harrison, *Carbon Capture: Sequestration and Storage*. RSC Publishers, Cambridge, 2010.
6. Research shows feasibility for capturing carbon dioxide directly from air, *Research & Development Magazine*, July 24, 2012.

7. M. Jitaru, Electrochemical carbon dioxide reduction: Fundamental and applied topics (review), *J. Univ. Chem. Technol. Metallurgy* 42, 2007, 333–344.
8. J. Hartvigsen, S. Elangovan, L. Frost, A. Nickens, C. Soots, J. O'Brian and J.S. Herring, Carbon dioxide recycling by high temperature co-electrolysis and hydrocarbon synthesis. In N.R. Neelameggham and R.G. Reddy (eds.), *Carbon Dioxide Reduction Metallurgy*, Publication of The Minerals, Metals & Materials Society (TMS), Warrendale, Pennsylvania, 2008, 171–182.
9. N.S. Spinner, J.A. Vega and W.E. Mustain, Recent progress in the electrochemical conversion and utilization of CO_2, *Catal. Sci. Technol.* 2, 2012, 19–28.
10. M. Aresta (ed.), *Carbon Dioxide Recovery and Utilization*, Kluwer Academic Publishers, Dordrecht-Boston-London, 2003.
11. N.R. Neelameggham, Carbon dioxide reduction technologies: A synopsis of the symposium at TMS 2008, *JOM* 60, 2008, 36–41.
12. M.H. Miles, Electrochemical reduction of carbon dioxide. In N.R. Neelameggham and R.G. Reddy (eds.), *Carbon Dioxide Reduction Metallurgy*, Publication of The Minerals, Metals & Materials Society (TMS), Warrendale, Pennsylvania, 2008, 129–132.
13. C. Costentin, S. Drouet, M. Robert and J.-M. Savéant, A local proton source enhances CO_2 electroreduction to CO by a molecular Fe catalyst, *Science* 338, 2012, 90–94.
14. Y. Hori, Electrochemical CO_2 reduction on metal electrodes. In C. Vayenas et al. (eds.), *Modern Aspects of Electrochemistry*, No. 42, Springer, New York, 2008, 89–189.
15. H. Herzog and D. Colomb, Carbon capture and storage from fossil fuel use, *Encyclopedia of Energy* I, 2004, 277–287.
16. G.T. Rochelle, Amine scrubbing for CO_2 capture, *Science* 325, 2009, 1652–1654.
17. K. Thambimuthu, M. Soltanich and J.C. Abandes, (Coordinating Lead Authors) Capture of CO_2. In B. Metz, O. Davidson, H. de Coninck, M. Loos and L. Meyer (eds.), *IPCC Special Report on Carbon Dioxide Capture and Storage*. Cambridge University Press, Cambridge, New York, Sào Paulo, 2005, 110–111, 115.
18. S. Keskin, T.M. van Heest and D.S. Sholl, Can metal-organic framework materials play a useful role in large-scale carbon dioxide separations? *Chem. Sus. Chem.* 3, 2010, 879–891.
19. J. Zhao, Z. Wang, J.X. Wang and S.C. Wang, Influence of heat-treatment on CO_2 separation performance of novel fixed carrier composite membranes prepared by interfacial polymerization, *J. Membr. Sci.* 283, 2006, 346–356.
20. X. Yu, Z. Wang, Z. Wei, S. Yuan, J. Zhao, J. Wang and S. Wang, Novel tertiary amino containing thin film composite membranes prepared by interfacial polymerization for CO_2 capture, *J. Membr. Sci.* 362, 2010, 265–278.
21. P. Pandey and R.S. Chauhan, Membranes for gas separation, *Prog. Polym. Sci.* 26, 2001, 853–893.
22. M.T. Ho, G. Leamon, G.W. Allinson and D.E. Wiley, Economics of CO_2 and mixed gas geosequestration of flue gas using gas separation membranes, *Ind. Eng. Chem. Res.* 45, 2006, 2546–2552.
23. A.S. Kovvali, H. Chen and K.K. Sirkar, Dendrimer membranes: A CO_2-selective molecular gate, *J. Am. Chem. Soc.* 122, 2000, 7594–7595.
24. E. Favre, Carbon dioxide recovery from post-combustion processes: Can gas permeation membranes compete with absorption? *J. Membr. Sci.* 294, 2007, 50–59.

25. M.T. Ho, G. Allison and D.E. Wiley, Comparison of CO_2 separation options for geo-sequestration: Are membranes competitive? *Desalination* 192, 2006, 288–295.
26. D.I. Jang and S.J. Park, Influence of nickel oxide on carbon dioxide adsorption behaviors of activated carbons, *Fuel* 102, 2012, 439–444.
27. A. Coenen, T.L. Church and A.T. Harris, Biological versus synthetic polymers as templates for calcium oxide for CO_2 capture, *Energy Fuels* 26, 2012, 162–168.
28. S.P. Wang, S.L. Yan, X.B. Ma and J.L. Gong, Recent advances in capture of carbon dioxide using alkali-metal-based oxides, *Energy Environ. Sci.* 4, 2011, 3805–3819.
29. A. Goeppert, M. Czaun, R.B. May, G.K.S. Surya Prakash, G.A. Olah and S.R. Narayanan, Carbon dioxide capture from the air using a polyamine based regenerable solid adsorbent, *J. Am. Chem. Soc.* 133, 2011, 20164–20167.
30. H. Furukawa, N. Ko, Y.B. Go, N. Aratani, S.B. Choi, E. Choi, A.O. Yazaydin, R.Q. Snurr, M. O'Keeffe, J. Kim and O.M. Yaghi, Ultrahigh porosity in metal-organic frameworks, *Science* 329, 2010, 424–428.
31. Y.Y. Liu, E.P. Hu, E.A. Khan and Z.P. Lai, Synthesis and characterization of ZIF-69 membranes and separation for CO_2/CO mixture, *J. Membr. Sci.* 353, 2010, 36–40.
32. M.C. McCarthy, V. Varela-Guerrero, G. Barnett and H.K. Jeong, Synthesis of zeolitic imidazolate framework films and membranes with controlled microstructures, *Langmuir* 26, 2010, 14636–14641.
33. K.A. Keets, E.B. Cole, A.J. Morris, N. Sivasankar, K. Teamey, P.S. Lakkaraju and A.B. Bocarsly, Analysis of pyridinium catalyzed electrochemical and photoelectrochemical reduction of CO_2: Chemistry and economic impact, *Ind. J. Chem. A*, 51, 2012, 1284–1297.
34. T. Yui, Y. Tamaki, K. Sekizawa and O. Ishitani, Photocatalytic reduction of CO_2: From molecules to semiconductors, *Topic Curr. Chem.* 303, 2011, 151–184.
35. C. Finn, S. Schnittger, L.J. Yellowlees and J.B. Love, Molecular approaches to the electrochemical reduction of carbon dioxide, *Chem. Commun.* 48, 2012, 1392–1399.
36. A.A. Peterson, F. Abild-Pedersen, F. Studt, J. Rossmeisl and J.K. Nørskov, How copper catalyzes the electroreduction of carbon dioxide into hydrocarbon fuels, *Energy Environ. Sci.* 3, 2010, 1311–1315.
37. R.J. Cogdell, A.T. Gardiner and L. Cronin, Learning from photosynthesis: How to use solar energy to make fuels, *Phil. Trans. R. Soc. A.* 370, 2012, 3819–3826.
38. G.A. Olah, G.K.S. Prakash and A. Goeppert, Anthropogenic chemical carbon cycle for a sustainable future, *J. Am. Chem. Soc.* 133, 2011, 12881–12898.
39. J.O.'M. Bockris and Sh.U.M. Khan, *Surface Electrochemistry. A Molecular Level Approach*, Plenum Press, New York and London, 1993, 534–541.
40. D. Gust, T.A. Moore and A.L. Moore, Realizing artificial photosynthesis, *Faraday Discuss.* 155, 2012, 9–26.
41. R.J. Coogdell, T.H.P. Brotosudarmo, A.T. Gardiner, P.M. Sanchez and L. Cronin, Artificial photosynthesis—Solar fuels: Current status and future prospects, *Biofuels* 1, 2010, 861–876.
42. C. Graves, S.D. Ebbesen, M. Mogensen and K.S. Lackner, Sustainable hydrocarbon fuels by recycling CO_2 and H_2O with renewable or nuclear energy, *Renewable Sustainable Energy Rev.* 15, 2011, 1–23.
43. K.P. Kuhl, E.R. Cave, D.N. Abram and T.F. Jaramillo, New insights into the electrochemical reduction of carbon dioxide on metallic copper surfaces, *Energy Environ. Sci.* 5, 2012, 7050–7059.

44. W. Li, Electrocatalytic reduction of CO_2 to small organic molecule fuels on metal catalysts. In Y. Hu (ed.) *Advances in CO_2 Conversion and Utilization.* ACS Symposium Series, American Chemical Society, Washington, DC, 2010, p. 58.
45. Y. Hori, A. Murata, R. Takahashi and S. Suzuki, Electrochemical reduction of carbon dioxide to carbon monoxide at a gold electrode in aqueous potassium hydrogen carbonate, *Chem. Commun.* 1987, 728–729.
46. N.M. Marković and P.N. Ross, Surface science studies of modern fuel cell electrocatalysts, *Surf. Sci. Rep.* 45, 2002, 117–229.
47. A.A. Peterson and J.K. Nørskov, Activity descriptors for CO_2 electroreduction to methane on transition-metal catalysts, *J. Phys. Chem. Lett.* 3, 2012, 251–258.
48. W.J. Durand, A.A. Peterson, F. Studt, F. Abild-Pedersen and J.K. Nørskov, Structure effects on the energetics of the electrochemical reduction of CO_2 by copper surfaces, *Surf. Sci.* 605, 2011, 1354–1359.
49. R.P.S. Chaplin and A.A. Wragg, Effects of process conditions and electrode material on reaction pathway for carbon dioxide electroreduction with particular reference to formate formation, *J. Appl. Electrochem.* 33, 2003, 1107–1023.
50. M. Jitaru, D.A. Lowy, B.C. Toma, M. Toma and L. Oniciu, The electrochemical reduction of carbon dioxide on flat metallic electrodes, *J. Appl. Electrochem.* 27, 1997, 875–889.
51. N. Sutin, C. Creutz and E. Fujita, Photo-induced generation of dihydrogen and reduction of carbon dioxide using transition metal complexes, *E. Commnts. Inorg. Chem.* 19, 1997, 67–92.
52. F.R. Keene and B.P. Sullivan, Mechanism of the electrochemical reduction of carbon dioxide catalyzed by transition metal complexes. In B.P. Sullivan, K. Krist and H.E. Guard (eds.), *Electrochemical and Electrocatalytic Reactions of Carbon Dioxide,* Elsevier, New York, Amsterdam, 1993, 118–140.
53. O.P. Balaj, C.K. Siu, I. Balteanu, M.K. Beyer and V.E. Bondybey, Reactions of hydrated electrons $(H_2O)_n^-$ with carbon dioxide and molecular oxygen: Hydration of the CO_2^- and O_2^- ions, *Chem. Eur. J.* 10, 2004, 4822–4830.
54. A. Sanov and R. Mabbs, Photoelectron imaging of negative ions, *Int. Rev. Phys. Chem.* 27, 2008, 53–85.
55. R.F. Höckendorf, C.K. Siu, C. van der Linde, O.P. Balaj and M.K. Beyer, Selective formic acid synthesis from nanoscale electrochemistry, *Angew. Chem. Int. Ed.* 49, 2010, 8257–8259.
56. S. Kaneco, N.-H. Hiei, Y. Xing, H. Katsumata, H. Ohnishi, T. Suzuki and K. Ohta, Electrochemical conversion of carbon dioxide to methane in aqueous $NaHCO_3$ solution at less than 273 K, *Electrochim. Acta* 48, 2002, 51–55.
57. M. Azuma, K. Hashimoto, M. Hiramoto, M. Watanabe and T. Sakata, Electrochemical reduction of carbon dioxide on various metal electrodes in low temperature aqueous $KHCO_3$ media, *J. Electrochem. Soc.* 137, 1990, 1772–1778.
58. B.R. Eggins, C. Ennis, R. McConnell and M. Spence, Improved yields of oxalate, glyoxalate and glycolate from electrochemical reduction of carbon dioxide in methanol, *J. Appl. Electrochem.* 27, 1997, 706–712.
59. Y.B. Vassiliev, V.S. Bagotsky, N.V. Osetrova, A.O. Khazova and N.A. Mayorova, Electroreduction of carbon dioxide. 1. The mechanism and kinetics of electroreduction in aqueous solutions on metals with high and moderate hydrogen overvoltages, *J. Electroanal. Chem.* 189, 1985, 271–294; Electroreduction of carbon dioxide. 2. The mechanism of reduction in Aprotic solvents, ibid. 189, 1985, 295–309.

60. M. Alvarez-Guerra, S. Quintanilla and A. Irabien, Conversion of carbon dioxide into formate using a continuous electrochemical reduction process in a lead cathode, *Chem. Eng. J.* 207, 2012, Special issue, 278–284.
61. A. Rios-Escudero, M. Isaacs, M. Villagrán, J. Zagal and J. Costamagna, Electrochemical reduction of carbon dioxide in the presence of Ni[II]-5,7,12,14-Tetramethyldinaphtho b,i. 1,4,8,11. tetra aza 14. annulene cation, *J. Argentine Chem. Soc.* 92, 2004, 63–71.
62. E.E. Barton, D.M. Rampulla and A.B. Bocarsly, Selective solar-driven reduction of CO_2 to methanol using a catalyzed p-GaP-based photoelectrochemical cell, *J. Am. Chem. Soc.* 130, 2008, 6342–6344.
63. M. Feroci, I. Chiarotto, M. Orsini, G. Sotgiu and A. Inesi, Carbon dioxide as carbon source: Activation via electrogenerated $O_2^{-\bullet}$ in ionic liquids, *Electrochim. Acta* 56, 2011, 5823–5827.
64. A.J.M. Miller, J.A. Labinger and J.E. Bercaw, Trialkylborane-assisted CO_2 reduction by late transition metal hydrides, *Organometallics* 30, 2011, 4308–4314.
65. M. Hammouche, D. Lexa, M. Momenteau and J.M. Savéant, Chemical catalysis of electrochemical reactions—Homogeneous catalysis of the electrochemical reduction of carbon dioxide by iron(0) porphyrins—Role of the addition of magnesium cations, *J. Am. Chem. Soc.* 113, 1991, 8455–8466.
66. M.A. Riqeuelme, M. Isaacs, M. Lucero, E. Trollund and M.J. Aguirre, Electrocatalytic reduction of carbon dioxide at polymeric cobalt tetra(3-amino (phenyl)) porphyrin glassy carbon-modified electrodes, *J. Chil. Chem. Soc.* 48, 2003, 89–92.
67. P. Kang, C. Cheng, Z.F. Chen, C.K. Schauer, T.J. Meyer and M. Brookhart, Selective electrocatalytic reduction of CO_2 to formate by water-stable iridium dihydride Pincer complexes, *J. Am. Chem. Soc.* 134, 2012, 5500–5503.
68. Q.J. Feng and S.Q. Liu, Electrochemical reduction behavior of CO_2 on Cu electrode in ionic liquid ($BMIMBF_4$) during synthesis of dimethyl carbonate, *Asian J. Chem.* 23, 2011, 4823–4826.
69. C.M. Sánchez-Sánchez, V. Mantiel, D.A. Tryk, A. Aldaz and A. Fujishima, Electrochemical approaches to alleviation of the problem of carbon dioxide accumulation, *Pure Appl. Chem.* 73, 2001, 1917–1927.
70. O. Achenbrenner and P. Styring, Comparative study of solvent properties for carbon dioxide absorption, *Energy Environ. Sci.* 3, 2010, 1106–1113.
71. E. Wilhelm and R. Battino, Thermodynamic functions of the solubilities of gases in liquids at 25°C, *Chem. Rev.* 73, 1973, 1–9.
72. P. Lühring and A. Schumpe, Gas solubilities in organic liquids at 293.2 K, *J. Chem. Eng. Data* 34, 1989, 250–252.
73. X. Gui, Z. Tang and W. Fei, Measurement and prediction of the solubility of CO_2 in ester mixture, *Low Carbon Economy* 2, 2011, 26–31.
74. K. Bezanehtak, G.B. Combes, F. Dehghani, N.R. Foster and D.L. Tomasko, Vapor-liquid equilibrium for binary systems of carbon dioxide + methanol, hydrogen + methanol, and hydrogen + carbon dioxide at high pressures, *J. Chem. Eng. Data* 47, 2002, 161–168.
75. T. Mizuno, A. Naitoh and K. Ohta, Electrochemical reduction of CO_2 in methanol at −30°C, *J. Electroanal. Chem.* 391, 1995, 188–201.
76. A. Naitoh, K. Ohta, T. Mizuno, H. Yoshida, M. Sakaia and H. Noda, Electrochemical reduction of carbon dioxide in methanol at low temperature, *Electrochim. Acta* 38, 1993, 2177–2179.

77. Y. Hori, H. Wakebe, T. Tsukamoto and O. Koga, Electrocatalytic process of CO selectivity in electrochemical reduction of CO_2 at metal electrodes in aqueous media, *Electrochim. Acta* 39, 1994, 1833–1839.
78. K. Ohta, M. Kawamoto, T. Mizuno and D.A. Lowy, Electrochemical reduction of carbon dioxide in methanol at ambient temperature and pressure, *J. Appl. Electrochem.* 28, 1998, 717–724.
79. K.W. Crawford and R.A. Orsini, Environmental Assessment: Source Test and Evaluation Report—Rectisol Acid Gas Removal, United States Environmental Protection Agency, EPA-600/S7-84-014, 1984.
80. S. Pérez-Rodriquez, G. García, L. Calvillo, V. Celorrio, E. Pastor and M.J. Lázaro, Carbon-supported Fe catalyst for CO_2 electroreduction to high-added value products: A DEMS study: Effect of the functionalization of the support, *Int. J. Electrochem.* 2011, 2011, Article ID 249804, 1–13.
81. J. Cristophe, Th. Doneux and C. Buess-Herman, Electroreduction of carbon dioxide on copper-based electrodes: activity of copper single crystals and copper-gold alloys, *Electrocatalysis* 3, 2012, 139–146.
82. Y. Hori, A. Murata, R. Takahashi and S. Suzuki, Electrochemical reduction of carbon monoxide to hydrocarbons at various metal-electrodes in aqueous solution, *Chem. Lett.* 8, 1987, 1665–1668.
83. Y. Hori, K. Kikuchi, A. Murata and S. Suzuki, Production of methane and ethylene in electrochemical reduction of carbon dioxide at copper electrode in aqueous hydrogen carbonate solution, *Chem. Lett.* 6, 1986, 897–898.
84. K. Ito, S. Ikeda, T. Iida and A. Nomura, Electrochemical reduction of CO_2 dissolved under high pressure. 3. In non-aqueous electrolytes, *Denki Kagaku*, 50, 1982, 463–469.
85. S. Ikeda, T. Tagaki and K. Ito, Selective formation of formic acid, oxalic acid, and carbon monoxide by electrochemical reduction of carbon dioxide, *B. Chem. Soc. Japan* 60, 1987, 2517–2522.
86. B. Eneau-Innocent, D. Pasquier, F. Ropital, J.M. Leger and K.B. Kokoh, Electroreduction of carbon dioxide at a lead electrode in propylene carbonate: A spectroscopic study, *Appl. Catal. B* 98, 2010, 65–71.
87. K. Hara, A. Tsuneto, A. Kudo and T. Sakata, Electrochemical reduction of CO_2 on Cu electrode under high pressure. Factors that determine the product selectivity, *J. Electrochem. Soc.* 141, 1994, 2097–2103.
88. Y. Hori, K. Kikuchi and S. Suzuki, Production of CO and CH_4 in electrochemical reduction of CO_2 at metal electrodes in aqueous hydrogencarbonate solution, *Chem. Lett.* 11, 1985, 1695–1698.
89. Y. Hori, R. Takahashi, Y. Yoshinami and A. Murata, Electrochemical reduction of CO_2 at a copper electrode, *J. Phys. Chem. B* 101, 1997, 7075–7081
90. M. Gattrell, N. Gupta and A. Co, A review of the aqueous electrochemical reduction of CO_2 to hydrocarbons at copper, *J. Electroanal. Chem.* 594, 2006, 1–19.
91. K.J.P. Shouten, Y. Kwon, C.J. van der Ham, Z. Quin and M.T.M. Coper, A new mechanism for the selectivity of C_1 and C_2 species in the electrochemical reduction of carbon dioxide on copper electrodes, *Chem. Sci.* 2, 2011, 1902–1909.
92. H. Yano, T. Tanaka, M. Nakayama and K. Ogura, Selective electrochemical reduction of CO_2 to ethylene at a three-phase interface on copper(I) halide-confined Cu-mesh electrodes in acidic solutions of potassium halides, *J. Electroanal. Chem.* 565, 2004, 287–293.

93. H. Yano, T. Tanaka, M. Nakayama and K. Ogura, Selective formation of ethylene from CO_2 by catalytic electrolysis at a three-phase interface, *Catal. Today* 98, 2004, 515–521.
94. K. Ogura, Electrochemical and selective conversion of CO_2 to ethylene, *Electrochemistry* 71, 2003, 676–680.
95. K. Ogura, H. Yano and F. Shirai, Catalytic reduction of CO_2 to ethylene by electrolysis at a three-phase interface, *J. Eletrochem. Soc.* 150, 2003, D163–D168.
96. K. Ogura, R. Oohara and Y. Kudo, Reduction of CO_2 to ethylene at three-phase interface. Effects of electrode substrate and catalytic coating, *J. Electrochem. Soc.* 152, 2005, D213–D219.
97. A. Schizodimou and G. Kyriacou, Acceleration of the reduction of carbon dioxide in the presence of multivalent cations, *Electrochim. Acta* 78, 2012, 171–176.
98. M. Jitaru and M. Toma, Electroreduction of carbon dioxide to formate on bronze electrodes, *Studia Univ. Babes-Bolyai. Chemia* 53, 2008, 135–142.
99. M. Jitaru, A.M. Toma, M.C. Tertis and A. Trifoi, Cu–Ni nanostructured electrocatalysts obtained by electrodeposition, *Environ. Eng. Manage. J. (EEMJ)* 8, 2009, 657–661.
100. R.L. Machunda, H.K.K. Ju and J. Lee, Electrocatalytic reduction of CO_2 gas at Sn based gas diffusion electrode, *Curr. Appl. Phys.* 11, 2011, 986–988.
101. R.L. Machunda, J.G. Lee and J. Lee, Microstructural surface changes of electrodeposited Pb on gas diffusion electrode during electroreduction of gas-phase CO_2, *Surf. Interface Anal.* 42, 2010, 564–567.
102. D.T. Whipple, E.C. Finke and P.J.A. Kenis, Microfluidic reactor for the electrochemical reduction of carbon dioxide: The effect of pH, *Solid-State Lett.* 13, 2010, B109–B111.
103. S. Ikeda, K. Ito and H. Noda, Electrochemical reduction of carbon dioxide using gas diffusion electrodes loaded with fine catalysts, nanoscience and nanotechnology. *International Conference on Nanoscience and Nanotechnology-2008*, Shah Alam, Selandor (Malaysia), November 18–21, 2008; *AIP Conf. Proc.* 1136, 2008, 108–113.
104. S. Ikeda, S. Shiozaki, J. Susuki, K. Ito and H. Noda, Electroreduction of CO_2 using Cu/Zn oxides loaded gas diffusion electrodes. In T. Inui, M. Anpo, K. Izui, S. Yanagida and T. Yamaguchi (eds.), *Advances in Chemical Conversions for Mitigating Carbon Dioxide*, Series: Studies in Surface Science and Catalysis, Vol. 114, 1998, 225–230.
105. K.R. Lee, J.H. Lim, J.K. Lee and H.S. Chun, Reduction of carbon dioxide in 3-dimensional gas diffusion electrodes, *Korean J. Chem. Eng.* 16, 1999, 829–836.
106. J. Lee, Y. Kwon, R.L. Machunda and H.J. Lee, Electrocatalytic recycling of CO_2 and small organic molecules, *Chem. Asian J.* 10, 2009, 1516–1523.
107. C. Oloman and H. Li, Electrochemical processing of carbon dioxide, *ChemSusChem* 1, 2008, 385–391.
108. H. Yano, F. Shirai, M. Nakayama and K. Ogura, Electrochemical reduction of CO_2 at three-phase (gas/liquid/solid) and two-phase (liquid/solid) interfaces on Ag electrodes, *J. Electroanal. Chem.* 533, 2002, 113–118.
109. G.K.S. Prakash, F.A. Viva and G.A. Olah, Electrochemical reduction of CO_2 over Sn-Nafion (R) coated electrode for a fuel-cell-like device, *J. Power Sources* 223, 2013, 68–73.
110. K. Ogura, Electrocatalytic reduction of CO_2 to C_2H_4 at a gas/solution/metal interface. In N.R. Neelameggham and R.G. Reddy (eds.), *Carbon Dioxide*

Reduction Metallurgy, Publication of The Minerals, Metals & Materials Society (TMS), Warrendale, Pennsylvania, 2008, 147–160.
111. K. Subramanian, K. Asokan, D. Jeevarathinam and M. Chandrasekaran, Electrochemical membrane reactor for the reduction of carbon dioxide to formate, *J. Appl. Electrochem.* 37, 2007, 255–260.
112. G. Centi, S. Perathoner, G. Winè and M. Gangeri, Electrocatalytic conversion of CO_2 to long carbon-chain hydrocarbons, *Green Chem.* 9, 2007, 671–678.
113. G. Centi, S. Perathoner and Z. Rak, Gas-phase electrocatalytic conversion of CO_2 to fuels over gas diffusion membranes containing Pt or Pd nanoclusters, *Stud. Surf. Catal.* 145, 2003, 283–286.
114. S. Ichikawa, Chemical conversion of carbon dioxide by catalytic hydrogenation and room temperature photoelectrocatalysis, *Energy Conversion Manage.* 31, 1995, 613–616.
115. S. Ichikawa and R. Doi, Hydrogen production from water and conversion of carbon dioxide to useful chemicals by room temperature photoelectrocatalysis, *Catal. Today* 27, 1996, 271–277.
116. G. Centi, S. Perathoner and Z. Rak, Reduction of greenhouse gas emissions by catalytic processes, *Appl. Catal. B: Environmental* 41, 2003, 143–155.
117. G. Centi, S. Perathoner and Z. Rak, Heterogeneous catalytic reactions with CO_2: Status and perspectives, *Stud. Surf. Catal.* 153, 2004, 1–8.
118. K. Ogura and M.D. Salazar-Villalpando, Electrochemical reduction via adsorbed Halide anions, *JOM* 63, 2011, 35–38.
119. K. Ogura, J.R. Ferrell, A.V. Cugini, E.S. Smotkin and M.D. Salazar-Villalpando, CO_2 attraction by specifically adsorbed anions and subsequent accelerated electrochemical reduction, *Electrochim. Acta* 56, 2010, 381–386.
120. L.M. Aeshala, S.U. Rahman and A. Verma, Effect of solid polymer electrolyte on electrochemical reduction of CO_2, *Separation Purification Technol.* 94, 2012, 131–137.
121. Select Commercially Available Ion Exchange Membrane Properties, eet Corporation. http://www.eetcorp.com/lts/membraneproperties.pdf (May 27, 2013).
122. Heterogeneous Ion-Exchange Membranes RALEX, MEGA Corporation. http://www.mega.cz/heterogenous-ion-exchange-membranes-ralex.html (May 27, 2013).
123. J. Jörissen, Ion exchange membranes as solid polymer electrolytes (SPE) in electro-organic syntheses without supporting electrolytes, *Electrochem. Eng.* 41, 1996, 553–562.
124. E.E. Benson, C.P. Kubiak, A.J. Sathrum and J.M. Smieja, Electrocatalytic and homogeneous approaches to conversion of CO_2 to liquid fuels, *Chem. Soc. Rev.* 38, 2009, 89–99.
125. M. Yadav and Q. Xu, Liquid-phase chemical hydrogen storage materials, *Energy Environ. Sci.* 5, 2012, 9698–9725.
126. M. Czaun, A. Goeppert, R. May, R. Haiges, G.K.S. Prakash and G.A. Olah, Hydrogen generation from formic acid decomposition by ruthenium carbonyl complexes. Tetraruthenium Dodecacarbonyl tetrahydride as an active intermediate, *ChemSusChem* 4, 2011, 1241–1248.
127. M.E. Royer, *C. R. Hebd. Seances Acad. Sci.* 70, 1870, 731–732.
128. A. Coehn and S. Jahn, *Ber. Dtsch. Chem. Ges.* 37, 1904, 2836–2842.
129. R. Ehrenfeld, *Ber. Dtsch. Chem. Ges.* 38, 1905, 4138–4143.
130. F. Fischer and O. Prziza, *Ber. Dtsch. Chem. Ges.* 47, 1914, 256–260.

131. W. Leitner, Carbon dioxide as a raw material—The synthesis of formic acid and its derivatives from CO_2, *Angew. Chem.* 107, 1995, 2391–2405; *Angew. Chem. Int. Ed. Engl.* 34, 1995, 2207–2221.
132. J.-M. Savéant, Molecular catalysis of electrochemical reactions. Mechanistic aspects, *Chem. Rev.* 108, 2008, 2348–2378.
133. A.S. Agarwal, Y.M. Zhai, D. Hill and N. Sridhar. The electrochemical reduction of carbon dioxide to formate/formic acid: Engineering and economic feasibility, *ChemSusChem* 4, 2011, 1301–1310.
134. S.R. Narayanan, B. Haines, J. Soler and T.I. Valdez. Electrochemical conversion of carbon dioxide to formate in alkaline polymer electrolyte membrane cells, *J. Electrochem. Soc.* 158, 2011, A167–A173.
135. F. Köleli, T. Röpke and C.H. Hamann, The reduction of CO_2 on polyaniline electrode in a membrane cell, *Synth. Met.* 140, 2004, 65–68.
136. Y. Kwon and J. Lee, Formic acid from carbon dioxide on nanolayered catalyst, *Electrocatalysis* 1, 2010, 108–115.
137. K. Subramanian, G. Gomathi and K. Asokan, Reduction of carbon dioxide to formate in an electrochemical membrane reactor in $KHCO_3$ buffer solutions, In N.R. Neelameggham and R.G. Reddy (eds.), *Carbon Dioxide Reduction Metallurgy*, Publication of The Minerals, Metals & Materials Society (TMS), Warrendale, Pennsylvania, 2008, 187–198.
138. P. Tundo and M. Selva, The chemistry of dimethyl carbonate, *Acc. Chem. Res.* 35, 2002, 706–716.
139. J.P. Parrish, R.N. Salvatore and K.W. Jung, perspectives of Alkyl carbonates in organic synthesis, *Tetrahedron* 56, 2000, 8207–8237.
140. A.G. Shaikh and S. Swaminathan, Organic carbonates, *Chem. Rev.* 96, 1996, 951–976.
141. W.J. Peppel, Preparation and properties of the alkylene carbonates, *Ind. Eng. Chem.* 50, 1958, 767–770.
142. D. Linden and T.B. Reddy (ed.) *Handbook of Batteries*, 3rd edn., McGraw-Hill, New York, 2002, Ch. 35. ISBN 0-07-135978-8.
143. A.B. McEwen, S.F. McDevitt and V.R. Koch, Nonaqueous electrolytes for electrochemical capacitors: Imidazolium cations and inorganic fluorides with organic carbonates, *J. Electrochem. Soc.* 114, 1997, L84–L86.
144. L. Zhang, D.F. Niu, K. Zhang, G.R. Zhang, Y.W. Luo and J.X. Lu, Electrochemical activation of CO_2 in ionic liquid ($BMIMBF_4$): Synthesis of organic carbonates under mild conditions, *Green Chem.* 10, 2008, 202–206.
145. H. Wang, L.X. Wu, Y.-C. Lan, J.Q. Zhao and J.-X. Lu, Electrosynthesis of cyclic carbonates from CO_2 and diols in ionic liquids under mild conditions, *Int. J. Electrochem. Sci.* 6, 2012, 4218–4227.
146. L.X. Wu, H. Wang, L. He, L. Wu, A.J. Zhang, H. Kajiura, Y.M. Li and J.X. Lu, Direct electrosynthesis of organic carbonates from CO_2 with alcohols under mild condition, *Int. J. Electrochem. Sci.* 7, 2012, 5616–5625.
147. F. Liu, S. Liu, Q. Feng, S. Zhuang, J. Zhang and P. Bu, Electrochemical synthesis of dimethyl carbonate with carbon dioxide in 1-butyl-3-methylimidazoliumtetrafluoborate on indium electrode, *Int. J. Electrochem. Sci.* 7, 2012, 4381–4387.
148. M.H. Miles and A.N. Fletcher, Electrochemical studies of carbon dioxide and sodium formate in aqueous solutions, *ACS Symp. Series* 363, 1988, 171–178.
149. E. Barton Cole, P.S. Lakkaraju, D.M. Rampulla, A.J. Morris, E. Abelev and A.B. Bocarsly, Using a one-electron shuttle for the multielectron reduction of CO_2 to

methanol: Kinetic, mechanistic, and structural insights, *J. Am. Chem. Soc.* 132, 2010, 11539–11551.
150. G. Seshadri, C. Lin and A.B. Bocarsly, A new homogeneous electrocatalyst for the reduction of carbon dioxide to methanol at low overpotential, *J. Electroanal. Chem.* 372, 1994, 145–150.
151. S. Ikeda, Y. Tomita, A. Hattori, K. Ito, H. Noda and M. Sakai, Selective ethanol formation by electrochemical reduction of carbon dioxide comprised of the mixtures of copper and zinc oxides, *Denki Kagaku* 61, 1993, 807–809.
152. R. Arrigo, M.E. Schuster, S. Wrabetz, F. Girgsdies, J.P. Tessonnier, G. Centi, S. Perathoner, D.S. Su and R. Schlogl, New insights from microcalorimetry on the FeOx/CNT-based electrocatalysts active in the conversion of CO_2 to fuels, *ChemSusChem* 5, 2012, 577–586.
153. N.R. de Tacconi, W. Chanmanee, B.H. Dennis, F.M. MacDonnell, D.J. Boston and K. Rajeshwar, Electrocatalytic reduction of carbon dioxide using Pt/C–TiO_2 nanocomposite cathode, *Electrochem. Solid State Lett.* 15, 2012, B5–B8.
154. S. Kaneko, K. Iiba, H. Katsumata, S. Suziki and K. Ohta, Electrochemical reduction of high pressure CO_2 at a Cu electrode in cold methanol, *Electrochim. Acta* 51, 2006, 4880–4885.
155. S. Kaneko, Y. Ueno, H. Katsumata, S. Suziki and K. Ohta, Electrochemical reduction of CO_2 in copper particle-suspended methanol, *Chem. Eng. J.* 119, 2006, 107–112.
156. M. Jitaru and D.A. Lowy, Electroreduction of carbon dioxide, In R. Savinell, K. Ota and G. Kreysa (eds.), *Encyclopedia of Applied Electrochemistry: Springer Reference*, Springer-Verlag, Berlin, Heidelberg, 2014 (http://www.springerreference.com).
157. F.J. Martin and W.L. Kubic, Green Freedom, a Concept for Producing Carbon-Neutral Synthetic Fuels and Chemicals, Los Alamos National Laboratory, LA-UR-07-7897 (November 2007).
158. K.R. Sridhar, C.S. Iacomini and J.E. Finn, Combined H_2O/CO_2 solid oxide electrolysis for Mars *in situ* resource utilization, *J. Propulsion Power* 20, 2004, 892–901.
159. A.F. Perez-Cadenas, C. Ros, S. Morales-Torres, M. Perez-Cadenas, P.J. Kooyman, C. Moreno-Castilla and F. Kapteijn, Metal-doped carbon xerogels for the electrocatalytic conversion of CO_2 to hydrocarbons, *Carbon* 56, 2013, 324–331.
160. D. Kaplan, R.S. Baird, H.F. Flynn, F. Howard, J.E. Ratliff, C.R. Baraona, R. Cosmo et al., The 2001 Mars in-situ-propellant-production precursor (MIP) flight demonstration—project objectives and qualification test results, AIAA Space 2000 Conference and Exposition, September 19–21, 2000.
161. G. Centi and S. Perathoner, Toward solar fuels from water and CO_2, *ChemSusChem* 3, 2010, 195–208.
162. B. Kumar, M. Llorente, J. Froehlich, T. Dang, A. Sathrum and C.P. Kubiak, Photochemical and photoelectrochemical reduction of CO_2. In M.A. Johnson and T.J. Martinez (eds.), *Annu. Rev. Phys. Chem.* 63, 2012, 541–545.
163. S. Bensaid, G. Centi, E. Garrone, S. Perathoner and G. Saracco, Toward artificial leaves for solar hydrogen and fuels from carbon dioxide, *ChemSusChem* 5, 2012, 500–521.
164. Y.H. Chen and M.W. Kanan, Tin oxide dependence of the CO_2 reduction efficiency on tin electrodes and enhanced activity for tin/tin oxide thin-film catalysts, *J. Am. Chem. Soc.* 134, 2012, 1986–1989.

165. C.D. Windle and R.N. Perutz, Advances in molecular photocatalytic and electrocatalytic CO_2 reduction, *Coord. Chem. Rev.* 256, 2012, 2562–2570.
166. K. Mondal, CO_2 reduction by nanoscale Galvanic couples. In N.R. Neelameggham and R.G. Reddy (eds.), *Carbon Dioxide Reduction Metallurgy*, Publication of The Minerals, Metals & Materials Society (TMS), Warrendale, Pennsylvania, 2008, 183–186.
167. Y.C. Lan, H. Wang, L.X. Wu, S.F. Zhao, Y.Q. Gu and J.X. Lu, Electroreduction of dibromobenzenes on silver electrode in the presence of CO_2. *J. Electroanal. Chem.* 664, 2012, 33–38.
168. L. Zhang, H. Wang, J.Q. Zhao, B.L. Chen and J.X. Lu, Electrocarboxylation of anthrone to anthracene-9-carboxylic acid in the presence of CO_2, *Chem. Res. Chinese Univ.* 27, 2011, 1027–1030.
169. M. Feroci, M. Orsini, L. Rossi and A. Inesi, The double role of ionic liquids in electroorganic synthesis: Green solvents and precursors of N-heterocyclic carbenes, *Curr. Org. Synth.* 9, 2012, 40–52.

2

Microbial Fuel Cells and Other Bio-Electrochemical Conversion Devices

Shaoan Cheng and Weifeng Liu

CONTENTS

2.1 Introduction 56
2.2 Biocatalysis in Microbial Fuel Cells for Anode 59
    2.2.1 Exo-Cellular Electron Transfer 59
    2.2.2 Microorganism Communities in MFCs 62
    2.2.3 Anodic Electron Transfer Mechanisms 68
        2.2.3.1 Direct Cell-Surface Electron Transfer 68
        2.2.3.2 Direct Electron Transfer via Nanowires 70
        2.2.3.3 Electron Transfer via Exogenous Redox Mediators 70
        2.2.3.4 Electron Transfer via Endogenous Redox Mediators 71
        2.2.3.5 Electron Transfer via Reduced Metabolic Products 72
    2.2.4 Summary: Recent Advances in Understanding the Role of Microorganisms 72
2.3 Microbial Electrochemistry 73
    2.3.1 Voltage Losses (Polarisations) in Microbial Fuel Cells 73
        2.3.1.1 Ohmic Losses 75
        2.3.1.2 Activation Losses 75
        2.3.1.3 Bacterial Metabolic Losses 77
        2.3.1.4 Concentration Losses 77
    2.3.2 Power Generation 78
    2.3.3 Factors Affecting the Performances of Microbial Fuel Cells 82
        2.3.3.1 Microbial Catalytic Activities 83
        2.3.3.2 Anode Performance 84
        2.3.3.3 Cathode Performance 85
        2.3.3.4 Proton Transfer Efficiency 86
        2.3.3.5 Electrolyte Chemistry 87
2.4 Electrode Materials and Scale-Up of Microbial Fuel Cells 89
    2.4.1 Anode Materials 89
    2.4.2 Cathode Materials 91
    2.4.3 Scaling Up of Microbial Fuel Cells 93

2.5　Types of Microbial Fuel Cells 95
　　2.5.1　Microbial Fuel Cells Producing Electricity from Wastewaters ... 96
　　　　2.5.1.1　Air Cathode Microbial Fuel Cells 96
　　　　2.5.1.2　Bio-Cathode Microbial Fuel Cells 98
　　　　2.5.1.3　Continuous Flow Modes and Cell Stacks 99
　　　　2.5.1.4　Microbial Fuel Cells for Removing Specific
　　　　　　　　Pollutants from Wastewaters 99
　　2.5.2　Sediments Microbial Fuel Cells (SMFCs) 100
　　2.5.3　Plant–Microbial Fuel Cells (PMFCs) 102
2.6　Other Microbial–Electrochemical Conversion Devices 104
　　2.6.1　Microbial Electrolysis Cells (MECs) 104
　　2.6.2　Microbial Sensors 106
　　2.6.3　Bacterial Batteries 106
2.7　Outlook 108
References 110

2.1　Introduction

As supplies of fossil fuels dwindle and the global warming crisis increases, there is a strong demand for new energy from renewable sources with minimal negative environmental impact. Significant efforts have been devoted to develop alternative electricity production methods, such as solar, geothermal, wind, nuclear and bioenergy technologies [1,2]. Microbial fuel cells (MFCs) represent a new approach for harvesting bioelectricity from biomass without net carbon emission. Any biodegradable organic matter, including sediments, organic solid wastes, low-strength wastewater, wastewaters containing volatile fatty acids, proteins, recalcitrant cellulose, and even highly concentrated nitrate and/or sulphate, can be used as the substrates for MFCs. The catalysts, electron-producing bacteria, for organic matter oxidation in MFCs are self-sustaining because bacteria can self-replicate during the fuel cell operation. MFCs have many operational and functional advantages, such as high-energy conversion rate (direct conversion of chemical energy stored in substrates into electricity without biogas–electricity conversion), low-energy input, free of aeration and off-gas treatment, and promising potential for widespread applications in locations lack of electrical infrastructures (e.g. rural and polar areas). Besides electricity production, MFCs can also be used for producing biogases, powering devices and acting as bio-sensors (see Section 2.6).

The earliest concept of bioelectricity generation by bacteria was demonstrated by Potter in 1911 [3]. However, very few practical advances were achieved in this field in the following 80 years due to the low output current. In the early 1990s, it was discovered that current density and power output could be greatly enhanced by the addition of electron mediators or electron shuttles, which could transport electrons from inside the bacterial cell to exogenous electrodes [4]. The key breakthrough in MFCs occurred in 1999 when it was demonstrated that an MFC could produce a maximum current of 1.8 mA and a power output of 3.1 mW without addition of electron mediators [5]. In recent years, numerous studies have been carried out on MFCs and rapid advances have been made in this field.

Usually, microorganisms make energy for cell growth via oxidising organic matter, in which electrons are produced and travel through a series of respiration enzymes. The electrons are then released to an electron terminal where the acceptor is reduced. Many electron acceptors such as oxygen, sulphate and nitrate, can diffuse into the cell and accept electrons therein, with the reduced products diffusing out of the cell (Figure 2.1a). However, researchers discovered that some bacteria can transfer electrons outside the cell to an exogenous terminal acceptor, such as iron oxide. Only these bacteria capable of exogenous transfer of electrons can be used in MFCs to produce electricity (Figure 2.1b). We call these bacteria 'exoelectrogens', with 'exo-' for exocellular and 'electrogens' for the ability to directly transfer electrons to an exogenous terminal electron acceptor. They may transfer electrons by various mechanisms, such as through direct contact, through nanowires, using exogenous mediators or endogenous mediators.

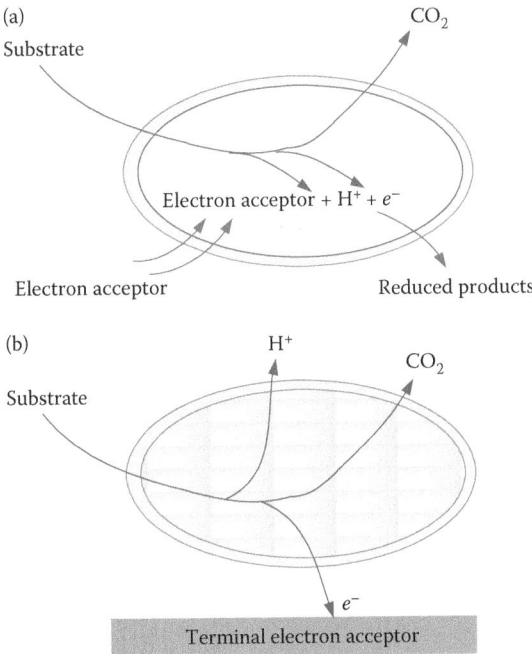

FIGURE 2.1
Schematic diagrams of the electron transfer pathway of (a) common microbes and (b) exoelectrogens.

MFCs are bioelectrochemical systems that directly convert chemical energy available in biomass into electricity using microorganisms as the catalysts. A typical MFC system is shown in Figure 2.2. The structure of an MFC essentially consists of an anode compartment and a cathode compartment separated by a proton exchange membrane. Carbon-based materials and non-corrosive metals are typically used as the electrode materials (Section 2.4). The anode and cathode electrodes are connected by wires and a load, completing the external circuit. Oxygen, a typical electron acceptor, is toxic to the anodic microorganisms and thus inhibits power generation, so it has to be separated from the anode. A proton exchange membrane serves as a barrier for oxygen transfer from cathode chamber to anode chamber to keep the anodic exoelectrogenic bacteria in an anaerobic environment.

At the anode, bio-convertible substrates (electron donors) are oxidised by exoelectrogenic microorganisms, producing electrons and protons. The produced electrons can be transferred to the anode material by electron mediators or shuttles, or by direct membrane-associated electron transfer, or by the so-called nanowires produced by the bacteria [6], or other means yet to be discovered. Electrons donated to the anode pass through an external resistor or other type of electrical device to the cathode, promoted by the electrochemical potential difference between the respiratory enzyme and the electron acceptor at the cathode. Simultaneously with electron transfer, a matching number of protons move from the anode to the cathode through the solution and membrane to maintain electroneutrality. At the cathode, dissolved oxygen (DO) is reduced by the electrons from external circuit and combined with protons to form water. Besides oxygen, other chemical

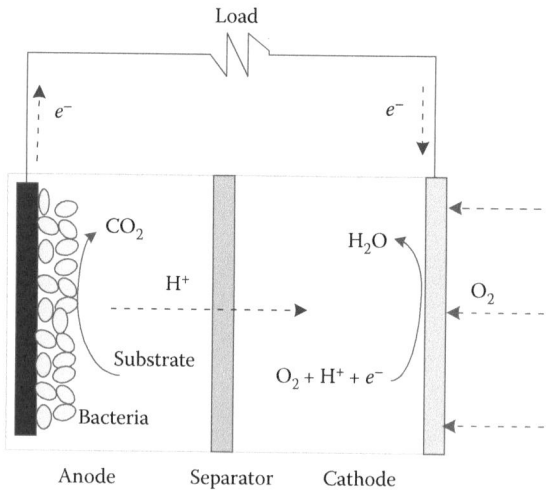

FIGURE 2.2
Schematic diagram of the working principle of an MFC.

oxidisers, such as nitrate, sulphate, ferricyanide or Mn (IV), can also be used as electron acceptors at cathode.

This chapter introduces the principles and applications of MFCs, with emphases on the nature of electricity-producing bacteria, anodic electron transfer mechanisms, power generation of MFCs and efficiency of electrode materials. Different types of MFCs and other microbial–electrochemical conversion devices are also discussed.

2.2 Biocatalysis in Microbial Fuel Cells for Anode

In the anode of an MFC, microorganisms respire and conserve energy using the electrode as the sole electron acceptor, typically under anaerobic conditions. The produced electrons then transfer from the anode to the cathode, resulting in electrical current production. While most of the current microorganism studies focus on the anode in which electrons are transferred from microorganisms to the electrode, in some instances, microbial catalysts can also be found at the cathode, where they accept electrons from the electrode and catalyse the reduction of inorganic ions or the re-oxidation of reduced redox mediators [7]. Since the information on cathodic bacteria and their electron transfer mechanisms is limited [8], here we will primarily focus on the anodic microorganisms and their exo-cellular electron transfer mechanisms.

2.2.1 Exo-Cellular Electron Transfer

Heterotrophic organisms obtain energy of survival from the oxidation of organic compounds or, more precisely, from the free (Gibbs) energy of the oxidation. A large variety of organic compounds, such as lipids, proteins and carbohydrates can be used as carbon and energy sources for bacteria. These organic substrates are chemically converted in the cytoplasm through glycolysis and citric acid cycle, resulting in the reduction of NAD+ (nicotinamide adenine dinucleotide) and FAD (flavin adenine dinucleotide) into NADH and FADH2 (reduced state of NAD+ and FAD) [9]. As shown in Figure 2.3, these electron carriers transfer electrons out of the cell to a terminal electron acceptor (dioxygen or any reducible inorganic compound that can be used) through the cell membrane using a series of membrane mediators. During this process, protons are simultaneously pumped out of the membrane, forming a proton gradient between the inside and outside of cell membrane. It is the energy of proton gradient that facilitates the production of the energy currency of living organisms, ATP (adenosine triphosphate). An electrode serves as the terminal electron acceptor for exoelectrogens in the anode of MFCs.

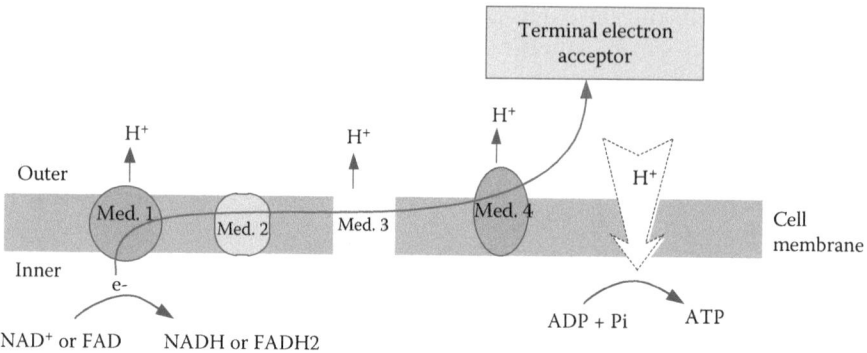

FIGURE 2.3
Schematic representation of bacterial membrane respiration. Note that the number of components of the electron transport chain varies with species.

Electron transfer mechanisms may differ from one organism to another, depending on the nature (soluble or membrane-bonded) of the last redox mediator that contacts the electrode. It is important to recognise that although microorganisms have had billions of years to optimise electron transfer to natural extracellular electron acceptors, such as Fe^{3+} and Mn^{4+} oxides, it is unlikely that they have been evolved to cater for the most effective strategies for electricity production. Thus, there is huge potential for the microorganisms in MFCs to enhance their current-producing capacity, probably via adaptive evolution or genetic engineering [10].

As mentioned above, free electrons are transferred during respiratory via a respiratory chain in which their energy gradually decreases, and are subsequently transferred to an externally terminal electron acceptor. The change of free energy under biological standard conditions ($\Delta G^{0'}$) during this process can be expressed as

$$\Delta G^{0'} = -nF[E^{0'}(\text{acceptor}) - E^{0'}(\text{donor})] \quad (2.1)$$

where $E^{0'}$ denotes the biological standard potential; biological standard conditions mean standard conditions measured at *pH 7*.

Thus, for a given substrate (the electron donor), the more positive the redox potential of a terminal electron acceptor, the higher the energy gain for an organism. In the anodes of MFCs, exoelectrogens (facultative* or obligate

---

* Facultative anaerobes: microorganisms that make adenosine triphosphate (ATP) by aerobic respiration if oxygen is present, but are also capable of growing by fermentation in the absence of oxygen.

anaerobes*) carry out an anaerobic respiration using organic substrates as electron donor and electrode as a transient electron acceptor. For this reason, the growth and exoelectrogenic activity of anodic microorganisms depend strongly on the electrode potential.

In the cathodes, inorganic or organic compounds are used as terminal acceptors. The overall energy change during a whole MFC process ($\Delta G_{total}^{0'}$) can be calculated according to Equation 2.1. The biological standard potentials of some substrates and electron acceptors are listed in Table 2.1. As shown in Table 2.1, compared to other electron acceptors (such as nitrate, sulphate, metal ions (Fe^{3+}) and fumarate), oxygen has the highest energy gain with highest E^0 value (+0.82 V).

Figure 2.4 depicts the energy flux in an MFC. As shown, the microorganisms retain a distinct portion of the total Gibbs free energy for its survival and reproduction ($\Delta G_{biol.}^{0'}$), with the remaining energy converted into electricity ($\Delta G_{elec.}^{0'}$)

$$\Delta G_{elec.}^{0'} = \Delta G_{total}^{0'} - \Delta G_{biol.}^{0'} \quad (2.2)$$

The energy gain for the microorganisms (hence, the loss of electric energy for an MFC) allows the maintenance of the bacterial vitality, a prerequisite for long-term fuel cell operation. However, excess biological energy gain would result in limited electric energy output. Consequently, it is important to find suitable metabolic and electron transfer paths that simultaneously allow sustainable MFC operation and maximum electric energy output [6]. Regulating the metabolic pathways of certain exoelectrogens by genetic

TABLE 2.1

Biological Standard Potentials of Some Biological Electron Donors and Electron Acceptors

Redox Couples	$E°/V$
CO_2/glucose	−0.43 [11]
CO_2/formate	−0.43 [11]
$2H^+/H_2$	−0.42 [11]
CO_2/acetate	−0.28 [11]
SO_4^{2-}/HS^-	−0.22 [11]
Fumarate/succinate	+0.33 [11]
NO_3^-/NO_2^-	+0.43 [11]
MnO_2/Mn^{2+}	+0.60 [12]
Fe^{3+}/Fe^{2+}	+0.77 [11]
$\frac{1}{2}O_2/H_2O$	+0.82 [11]

---

* Obligate anaerobes: Microorganisms that live and grow in the absence of molecular oxygen; they will die in the presence of oxygen.

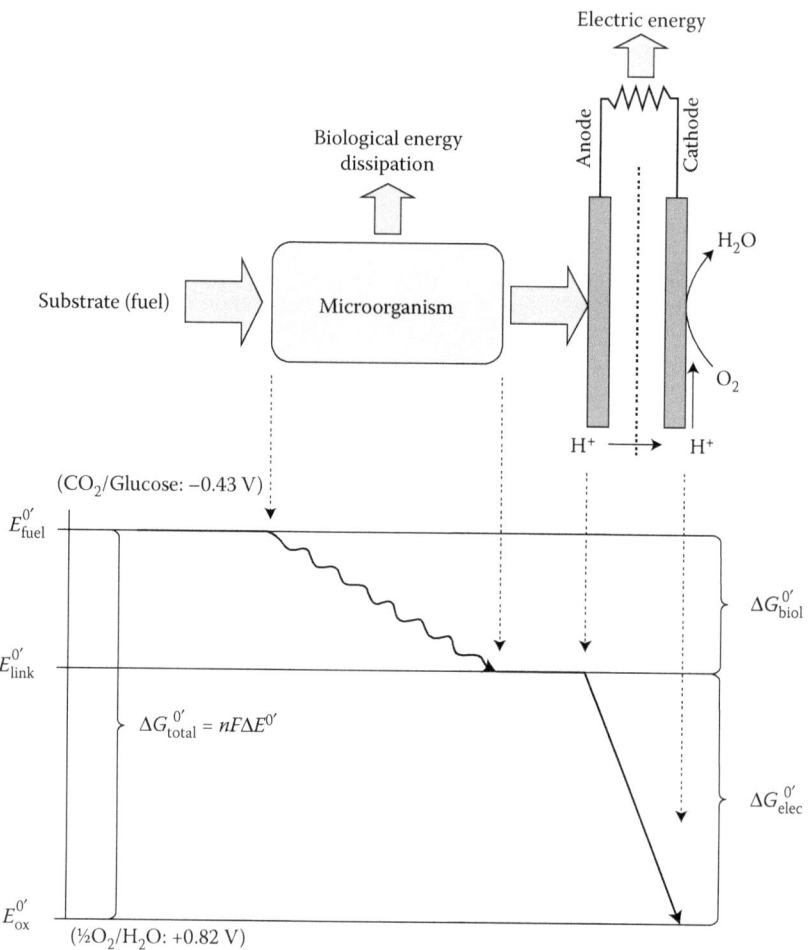

FIGURE 2.4
Schematic diagram of the energy flux in an MFC. (Reprinted with permission from U. Schröder, Anodic electron transfer mechanisms in microbial fuel cells and their energy efficiency, *Phys. Chem. Chem. Phys.* 9, 2007, 2619–2629. Copyright 2007, The Royal Society of Chemistry.)

modification or selecting efficient electron-producing microbes for MFCs might be applicable for this purpose.

2.2.2 Microorganism Communities in MFCs

Numerous bacteria that can generate electricity have been isolated and investigated. Studies on these pure cultures provide us valuable knowledge about the properties of exoelectrogens, as well as the exocellular electron transfer mechanisms. Some examples of exoelectrogenic active bacteria are listed in

Table 2.2 (microorganisms which transfer electrons using exogenous mediators are excluded here, due to the technological and environmental infeasibility of this approach, see Section 2.2.3). Generally, most exoelectrogens belong to the α-, β-, γ- and δ-proteobacteria (Gram negative); while some non-proteobacteria and yeast are also capable of exo-cellular electron transfer. Direct comparison of the electricity-producing ability between different microorganisms is not an easy task, since the actual electron transfer mechanism might not always be known, and the experimental conditions may differ significantly in different studies. Systematic studies using identical fuel cell design are required in future studies, so that a comparison between microbial species can be made. While pure cultures are powerful tools for mechanistics studies, their electricity-producing ability is generally low compared to mixed cultures (Table 2.3). In fact, as demonstrated in Table 2.3, the highest power densities in MFCs are almost always produced by inoculating the anode with a rich and diverse source of bacteria (~1970 mW m^{-2}), such as wastewater or sludge, instead of pure cultures such as *Shewanella putrefaciens* (10 mW m^{-2}) or *Geobacter sulfurreducens* (49 mW m^{-2}). A great diversity of

TABLE 2.2

Examples of Electricity-Producing Microorganisms without Exogenous Mediators in MFCs

Classes	Microorganisms	Substrates	References
γ-Proteobacteria	*Shewanella putrefaciens*	Lactate	[13]
γ-Proteobacteria	*Shewanella oneidensis*	Lactate	[14]
γ-Proteobacteria	*Aeromonas hydrophila*	Acetate	[15]
γ-Proteobacteria	*Escherichia coli*	Glucose	[16]
γ-Proteobacteria	*Pseudomonas aeruginosa*	Glucose	[17]
δ-Proteobacteria	*Desulfuromonas acetoxidans*	Acetate	[18]
δ-Proteobacteria	*Desulfobulbuspropionicus*	Lactate and propionate	[19]
δ-Proteobacteria	*Geobacter metallireducens*	Acetate and butyrate	[20]
δ-Proteobacteria	*Geobacter sulfurreducens*	Acetate	[21]
β-Proteobacteria	*Rhodoferax ferrireducens*	Glucose	[22]
α-Proteobacteria	*Rhodopseudomonaspalustris*	Volatile acids, yeast extract and thiosulphate	[23]
α-Proteobacteria	*Ochrobactrumanthropi* YZ-1	Acetate, lactate, propionate, butyrate, glucose and sucrose	[24]
α-Proteobacteria	*Acidiphilium* sp. 3.2 Sup5	Glucose	[25]
Non-proteobacteria	*Clostridium* sp. EG3	Glucose	[26]
Non-proteobacteria	*Geothrixfermentans*	Acetate, lactate and propionate	[27]
Yeast	*Saccharomyces cerevisiae*	Glucose	[28]
Yeast	*Pichiaanomala*	—	[29]

TABLE 2.3

Examples of Current and Power Generation of MFCs Using Pure Cultures and Mixed Cultures

Microbes	Current Density (mA m^{-2})	Power Density (mW m^{-2})	References
Pure Culture			
Aeromonas hydrophila	120	18	[15]
Desulfuromonas acetoxidans	125	14	[18]
Geobacter metallireducens	320	38	[20]
Geobacter sulfurreducens	65	49	[21]
Rhodoferax ferrireducens	31	33	[22]
Shewanella putrefaciens	312	10	[30]
Mixed Culture (Inoculum)			
Domestic wastewater	8000	1970	[31]
Pre-acclimated bacteria	5000	1015	[32]
Domestic wastewater	3000	483	[33]
Domestic wastewater	8800	1370	[34]
Pre-acclimated bacteria	7000	1220	[35]
Pre-acclimated bacteria	2300	750	[36]

Note: The fuel cell design and experimental conditions are different in each case so that direct comparisons should not be made.

microorganisms has been found in the anode biofilms of MFCs. An example of the phylogenetic tree of the bacterial community from an MFC is illustrated in Figure 2.5.

The nature and the diversity of microorganisms present in MFCs is a function of several factors, including (i) the inoculation source, (ii) the nature of the substrate, (iii) the operating conditions, (iv) system architectures, and (v) electron acceptors at the cathode. A low coulombic efficiency in some MFCs suggests that in addition to exoelectrogenes, some non-electricity-generating microorganisms which are sustained by alternative metabolisms such as fermentation and methanogenesis, are also involved in the biofilms. In most situations proteobacteria (gram negative) dominate the communities' composition in MFCs. Although iron-reducing bacteria (proteobacteria) such as *Shewanella* and *Geobacter* species are thought to be the major contributors to electricity generation in MFCs, community analysis reveals a much greater diversity of bacteria exists in the anodic biofilm.

The ratios between α-, β-, γ- and δ-proteobactria may differ greatly according to the nature of the inoculum. It was shown, for example, that α-proteobacteria comprised 64.5% of the communities present in an MFC fed with artificial wastewater and only 10.8% when fed by river water [37]. Jung and Regan [38] examined the anodic bacteria compositions of two-chamber MFCs inoculated with anaerobic sludge and fed with different substrates (acetate, lactate and glucose). Results showed that all anode communities

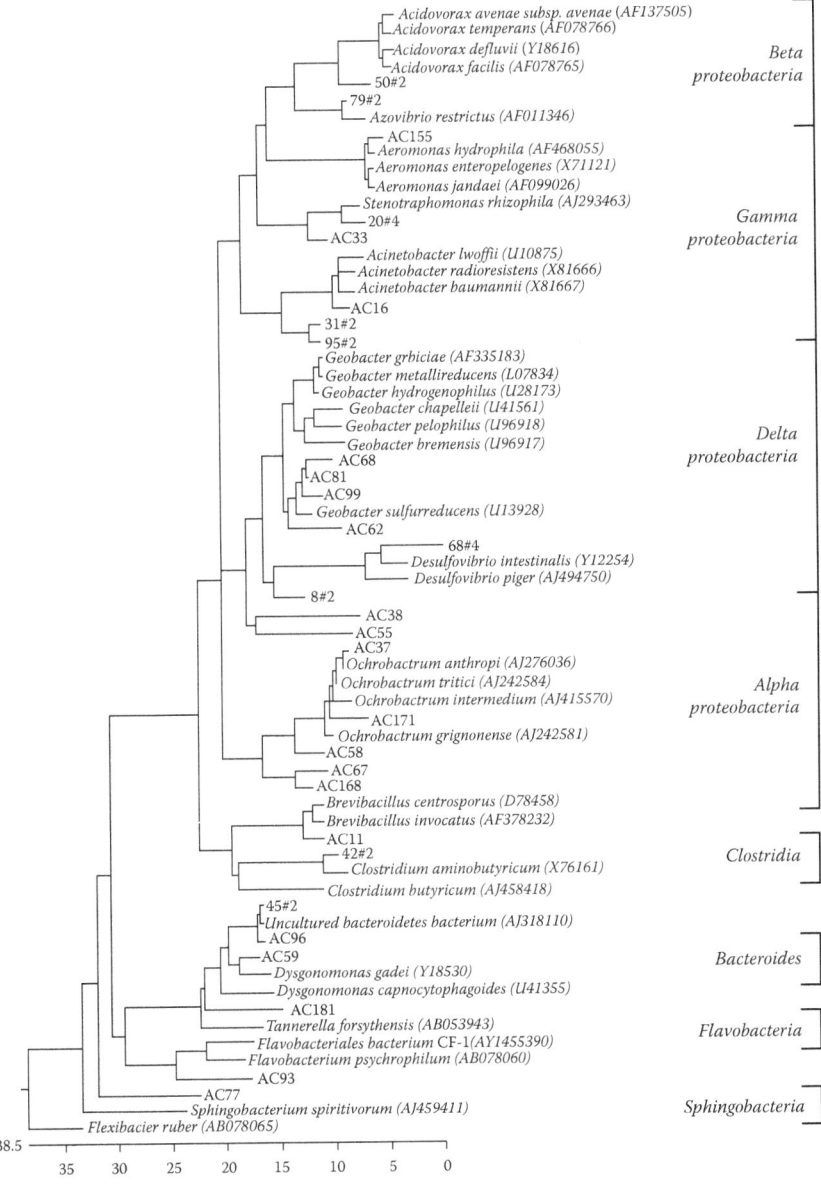

FIGURE 2.5
Example of a phylogenetic tree showing 16S rDNA gene sequences of the bacterial community form an MFC enriched with acetate. (Reprinted with permission from J. Lee et al. Use of acetate for enrichment of electrochemically active microorganisms and their 16S rDNA analyses, *FEMS Microbiol. Lett.* 223, 2003, 185–191. Copyright 2003, Wiley-Blackwell.)

contained genetic sequences closely related to G. *sulfurreducens* (>99% similarity) and an uncultured bacterium in the *Bacteroidetes* class (99% similarity). However, *Firmicutes* were found only in glucose-fed MFCs, presumably serving the role of converting complex carbon into simple molecules and scavenging oxygen.

The chemical reaction occurring at the cathode also affects the community composition of anode biofilms, as gases and some soluble organic or inorganic species tend to be permeable through the membrane to some extent [6]. For example, when oxygen is used as electron acceptor in cathodes some oxygen will diffuse into the anode chamber, influencing the biofilm composition via poisoning anaerobic exoelectrogens, or promoting the growth of aerobic bacteria. This will lead to a decreased electricity generation efficiency of the biofilm. Other species in the catholytes, such as ammonia, nitrate and sulphate can also diffuse into the anode chamber. Bacteria which use these electron acceptors directly will grow and contribute to the diversity of the community, but not directly associated with electricity generation.

With oxygen as the terminal electron acceptor, the dominant bacteria in the anode biofilm varies with the inoculum, substrate and operation conditions. For instance, Phung et al. [39] observed a majority of β-proteobacteria (related to *Leptothrix* spp) in the microbial community of an MFC inoculated with river sediment and fed with river water, while an α-proteobacteria (mainly *Actinobacteria*) dominant community was observed in the same device fed with low concentration of glucose and glutamate. A nearly even distribution among the α-, γ- and δ-proteobacteria was reported by Lee et al. [13] in the anode biofilm of an acetate-fed MFC system that was inoculated with activated sludge. In an ethanol-fed two-chamber MFC [40], the majority (83%) of cloned 16S rRNA gene sequences were related to β-proteobacteria, primarily *Dechloromonas*, *Azoarcus* and *Desulfuromonas*.

Instead of using oxygen as an electron acceptor, ferricyanide was used as the electron acceptor in a glucose-fed MFC [18]. The bacterial community showed great phylogenetic diversity, containing gene sequences derived from bacteria of the taxa *Firmicutes*, γ-, β- and α-proteobacteria. The predominant bacteria were the facultative anaerobes capable of hydrogen production, such as the Gram-negative* *Alcaligenes faecalis* and Gram positive* *Enterococcus gallinarum*. In some systems, a large percentage of the clones are uncharacterised. For example, in an MFC using wastewater as an inoculum, starch as the substrate and DO as the terminal oxidiser, the developed microbial community was found to consist of 36% unidentified clones, 25% β- and 20% α-proteobacteria, and 19% *Cytophaga*, *Flexibacter* and *Bacteroides* groups on the basis of sequences from RFLP-screened 16S rDNA clones [41].

---

* Gram-positive bacteria are those that are stained dark blue or violet by Gram staining. This is in contrast to gram-negative bacteria, which cannot retain the crystal violet stain, instead taking up the counterstain (safranin or fuchsine) and appearing red or pink.

A clear pattern of dominance by δ-proteobacteria emerges in sediment MFCs. A strict anaerobic condition is maintained in these systems through immersing the anode in anoxic sediment. Microbial community analysis, via characterisation of 16S rRNA genes, revealed that there was an enrichment of microorganisms in the *Geobacteraceae* family on anodes harvesting electricity from a diversity of marine and fresh water sediments [42,43]. Typically, Geobacteraceae account for over half of the microorganisms on the energy-harvesting anodes. In contrast, they generally constitute less than 5% of the community on normal solid materials that do not act as an anode. In marine environments, the predominant Geobacteraceae are *Desulfuromonas* species, which prefer marine salinities, whereas in freshwater environments *Geobacter* species predominate [43]. The above results suggest a high correlation of the communities' composition to the operational parameters of the fuel cell. In addition to the communities' composition, their electrochemical activity, thus the anode performance is also influenced by these parameters.

Liu et al. [44] demonstrated that the bio-electrochemical activity of anode biofilm is strongly dependent on the electrode material. It was presumed that the material structure characterisation, rather than the real surface area, might be an important factor for the formation of electrochemically active biofilm. Biofilms form well on graphite rod and polycrystalline carbon rod, showing similar catalytic activity. Carbon fiber veil or carbon-paper-based materials, having a large microbially accessible surface leads to a projected current density ~40% higher than on graphite rod [44]. The *pH* value also played a crucial role for the development and current production of anodic microbial biofilms. It was demonstrated that only a narrow *pH* window (ranging from *pH* 6 to 9) was suitable for growth and operation of biofilms derived from *pH*-neutral wastewater [45]. The maximum current density for anode biofilms growth (821 µA cm^{-2}) was obtained at *pH* 7. However, Yuan et al. [46] examined a maximum power density (1170 ± 58 mW m^{-2}) at *pH* 9.0 in a two-chamber MFC, which is 29% higher than those working at *pH* 7.0.

Operation temperature is another important factor determining the formation and bioelectrocatalytic performance of anodic biofilms. Patil et al. [47] demonstrated that wastewater-derived biofilms can operate within a temperature range from about 0 to 45°C, and the biofilms showed increased bio-electrocatalytic performance at elevated operation temperature. In another study [48] however, anodic biofilm formed at a lower temperature (15°C) was observed to produce a higher maximum power and more constant voltage and yield a much lower anodic resistance than that cultivated at high temperature (25°C). As the internal resistance and Coulombic efficiency varied in different systems, it is difficult at this stage to understand how and why these factors affect the subsequent microbial community. Systematic comparisons of the effect of MFC architecture (i.e. sediment, non-sediment, oxygen, ferricyanide and poised-potential MFCs), substrate and inoculum are required for a better understanding of the community which evolves in MFC systems.

2.2.3 Anodic Electron Transfer Mechanisms

The anodic electron transfer mechanism in MFC is a key issue in understanding how MFCs work. As discussed above, the redox active species at the end of the electron transfer chain 'links' the solid electrode in MFCs anodes, completing the exo-cellular electron transfer (Figure 2.3). These linking species, for example, may be a soluble redox shuttle, an outer membrane redox protein or a pili (nanowire). For an efficient electron transfer, the linking species must fulfill the following requirements [6]:

  i. Be able to physically contact the electrode surface.
  ii. Be electrochemically active, that is, it should possess a low oxidation overpotential at given electrode surfaces.
  iii. The standard potential of the linking species, should be as close to the redox potential of the primary substrate as possible, or must at least be significantly negative to that of the terminal electron acceptor (usually oxygen).

Various mechanisms of electron transfer and linking species have been identified and exploited. Basically there are two major types of linking species: (i) soluble compounds (artificial or self-produced mediators), and (ii) compounds bonded to the microbial cell membrane (membrane-bond proteins or nanowires). Accordingly, electron transfer mechanisms from microorganisms to the electrode can be divided into five primary types, which will be discussed in the following section: (1) direct cell-surface electron transfer, (2) direct electron transfer via nanowires, (3) electron transfer via exogenous redox mediators, (4) endogenous redox mediators and (5) reduced metabolic products.

2.2.3.1 Direct Cell-Surface Electron Transfer

Some microorganisms can directly transfer electrons to the electrode via a physical contact of the cell membrane or a membrane organelle with the anode. No diffusional redox species are involved in this electron transfer process. As illustrated in Figure 2.6a, the direct electron transfer requires the microorganisms to possess (1) membrane-bound protein relays which transfer electrons from the inside of the bacterial cell to its outside, and (2) an outer membrane (OM) redox protein which accepts the electrons and delivers them to an external, solid electron acceptor (a metal oxide or an MFC anode). The most studied OM redox proteins are c-type cytochromes, which are involved in metal-reducing microorganisms such as *Geobacter*, *Rhodoferax* and *Shewanella*. These bacteria often have to rely on solid terminal electron acceptors like iron(III) oxides in their natural environments,

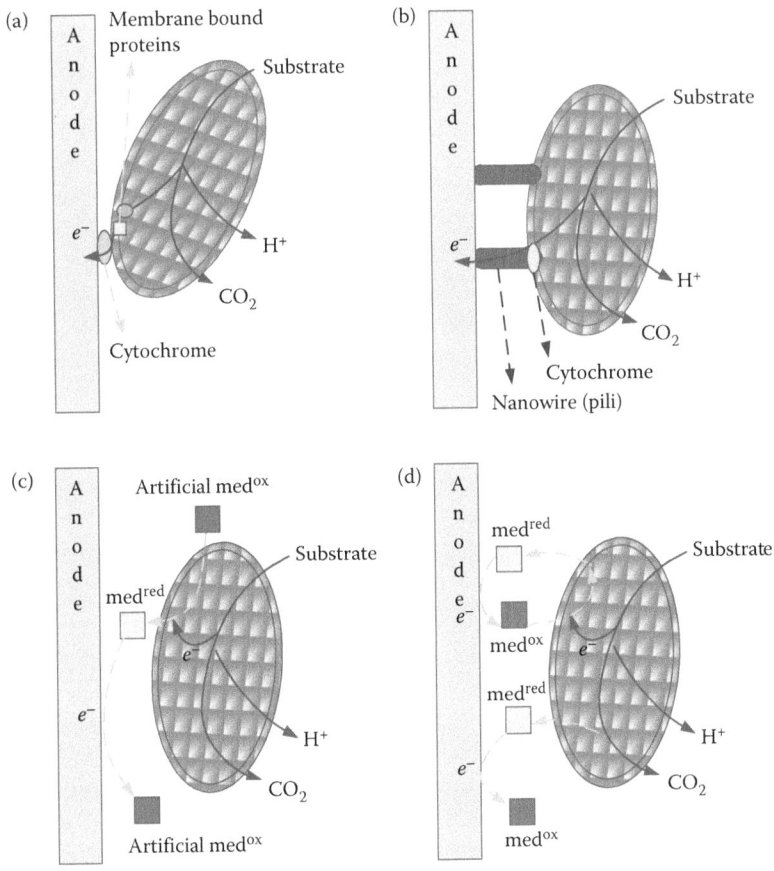

FIGURE 2.6
(See colour insert.) Schematic illustration of electron transfer via (a) direct contact by membrane-bound cytochromes; (b) membrane-bound nanowires; (c) artificial mediator and (d) self-produced redox mediators. The oval represents the microorganism cell.

while in MFCs the anodic electrode resume the role of the solid electron acceptor.

Since the direct electron transfer via outer membrane cytochromes requires the physical contact of the bacterial cell to the fuel cell anode, only bacteria in the first monolayer at the anode surface are electrochemically active and responsible for electricity generation. Consequently, the MFC performance is limited by the maximum cell density in this bacterial monolayer. For example, maximum current densities as low as 0.6 and 6.5 $\mu A\ cm^{-2}$ were achieved for MFCs based on pure *Shewanella putrefaciens* [13] and *Geobacter sulfurreducens* [21], respectively.

2.2.3.2 Direct Electron Transfer via Nanowires

Some microorganisms (e.g. *Geobacter, Rhodoferax* and *Shewanella* strains) can evolve electronically conductive molecular pili (nanowires) that allow the microorganism to reach more distant solid electron acceptors [49]. As shown in Figure 2.6b, the pili are connected to the membrane-bound cytochromes. These cytochromes deliver electrons to the pili, which then transfer the electrons to metal oxides or electrode via physical contact. The conductive nanowires make it possible for cells that are not in direct contact with the anode surface to transfer electrons to the electrode. Thus, thick electroactive biofilms develop on the electrode surface, resulting in higher levels of power production. For instance, Reguera and coworkers [50] reported a 10-fold increase of current production (from 0.8 to 12 mA) upon nanowire formation of *Geobacter sulfurreducens*. Formation of such nanowires also facilitates possible interspecies electron transfer within the biofilm. It has been observed that *Pelotomaculum thermopropionicum* bacteria produce thick conductive nanowires that connect them to methanogens in a biofilm [51].

2.2.3.3 Electron Transfer via Exogenous Redox Mediators

In early MFC studies, in order to achieve appreciable power densities, artificial electron shuttles or mediators are required to add to carry electrons from inside the cell to the electrode. In 1931, Cohen introduced electrochemically active redox products (potassium ferricyanide or benzoquinone) for the first time to facilitate the electron transfer from bacterial cultures to immersed electrodes, thus improving the current generation capacity of anaerobic microbes [52]. Exogenous electron mediators (artificial med^{ox} and med^{red}) are typically capable of crossing cell membranes. As shown in Figure 2.6c, they can diffuse across the cell membrane (med^{ox}) and be reduced within the cell (med^{red}) via accepting electrons from one or more electron carriers. The reduced mediators then transport across the cell membrane to the external of the cell, and deliver their electrons to the terminal electron acceptor (electrode surface). Artificial mediators are important in MFCs that use microorganisms such as *Escherichia coli, Pseudomonas, Proteus,* and *Bacillus* species, which are unable to effectively transfer electrons derived from central metabolism to the outside of the cell [53]. Common electron shuttles include thionine, benzylviologen, 2,6-dichlorophenolindophenol, 2-hydroxy-1,4-naph-thoquinone and various phenazines, phenothiazines, phenoxoazines, iron chelates, and neutral red.

In most instances, MFCs using artificial electron shuttles have an incomplete oxidation of most substrates, producing fermentation products. This results in a low conversion efficiency of fuel to electricity. Continual addition of electron shuttles is also required in practical MFCs operations, and is considered technologically and economically unfeasible. Furthermore, many of the electron shuttles used in previous studies are toxic to humans and could not be responsibly released into the environment. Therefore, this approach has been generally abandoned in current MFCs studies, with the exception in some fundamental research.

2.2.3.4 Electron Transfer via Endogenous Redox Mediators

In 2004, Rabaey and coworkers demonstrated that *Pseudomonas aeruginosa* could transfer electrons from inside the cell to an electrode via self-produced chemical mediators (endogenous), such as pyocyanin and related compounds, and produce electricity in an MFC [17]. Hence, exogenous mediators were not required to participate in these systems. Besides *Pseudomonas aeruginosa*, some *Shewanella* species and *Geothrixferementans* can also produce electron shuttles to facilitate exocellular electron transfer. However, *Geobacter* species, another Fe^{3+}-reducer, do not transfer electrons via this approach. For the applications of MFCs, electron transfer via endogenous redox mediators are of special interest, as this process makes the electron transfer independent of exogenous redox shuttles.

As depicted in Figure 2.6d, a mediator serves as a reversible terminal electron acceptor, transferring electrons from the bacterial cell either to the MFC anode or into aerobic layers of the biofilm, where it becomes re-oxidised and available for subsequent redox processes. One molecule can thus serve for thousands of redox cycles. Consequently, production of small amounts of redox mediators (directly in the anode biofilm) enables the organisms to dispose electrons at adequate rates. Especially in batch systems, these endogenous redox mediators can effectively facilitate the electron transfer, and increase the efficiency of current generation. The presence of these electron mediators facilitates the development of thick electro-active biofilms, in which exoelectrogens can transfer electrons to the electrode even several millimeters away. However, in an open or flow system the shuttles will be lost rapidly from the site of release, leading to a decrease in Coulombic and energy efficiency. Besides, it has been demonstrated that biosynthesising an electron mediator is energetically expensive resulting in additional biological loss [54]. Consequently, microorganisms that transfer electrons via self-produced shuttles are expected to be disadvantaged in open environments. This might explain why species from the Geobacteraceae, which transfer electrons through direct cell-surface contact or nanowires, predominate over other species under Fe^{3+}-reducing conditions in many sedimentary environments.

Another significant factor limiting the effectiveness of electricity production by the microorganisms that produce an electron shuttle is that they could only partially oxidise their organic fuels. For example, the *Psuedomonas* species have a fermentative metabolism in MFCs, leaving most of the electrons initially available in glucose as fermentation products [55]. *Shewanella* species can only incompletely oxidise a limited number of organic acids such as lactate and pyruvate to acetate under anaerobic conditions [13], limiting the efficiency of electricity production (maximum current density of 8 mA m^{-2}).

2.2.3.5 Electron Transfer via Reduced Metabolic Products

When fermentative microorganisms and yeast are used for MFCs operation, the microbial fermentative or photo-heterotrophic processes will produce certain energy-rich reduced metabolites, which can be abiotically oxidised at the anode surface to provide electron. These products might include hydrogen, alcohols, ammonia or formate. However, up to now only a few studies documented this mechanism or directly quantified which reduced products were oxidised at the anode [19,56]. With the low yield of such fermentation products, and the slow electrode reactions, these systems are inherently inefficient. Although it is possible to modify the composition of anodes to increase their reactivity with some metabolic end products, these electrodes tend to foul with oxidation products.

One example of such reduced metabolic products that readily react with electrodes is H_2S, which is produced by sulphate reducers in MFCs [19]. The primary oxidation product in this process is insoluble sulphur (S^0). Reduction of sulphate to sulphide requires eight electrons, but oxidation of sulphide to S^0 releases only two of these eight electrons at the electrode surface. For example, when pyruvate was used as the substrate, the above electrochemical reactions might be expressed as

$$4CH_3COCOO^- + SO_4^{2-} + 4H_2O \xrightarrow{\text{microbes}} 4CH_3COO^- + 4HCO_3^- + H_2S + 2H^+ \tag{2.3}$$

$$H_2S \xrightarrow{\text{Anode}} 2H^+ + S^0 + 2e^- \tag{2.4}$$

Therefore, this is not an efficient method for transferring electrons to an anode. Furthermore, the *Desulfovibrio* species typically used in such systems only partially oxidise their organic electron donors to acetate, which will not react with electrodes, further limiting the efficiency. Symbiosis with exoelectrogens that can utilise acetate or genetically engineering the *Desulfovibrio* species might be useful to improve the electricity generation efficiency.

2.2.4 Summary: Recent Advances in Understanding the Role of Microorganisms

As discussed above, the exocellular electron transfer mechanisms in MFC systems are rather complex. Different microbes may transfer electrons via quite distinct approaches, and one specific microorganism may have more than one electron transfer paths. For example, *Shewanella* spp. have outer membrane cytochromes for direct electron transfer by contact, but they can also produce electrically conductive nanowires. *S. oneidensis* also produces flavins that can function as electron shuttles. MFCs are mostly inoculated with complex natural sources, such as wastewaters, sludges, and river/marine sediments. The anode biofilms in these systems generally consist of a wide diversity of microorganisms, including electricity-producing bacteria as well as those without electrochemical activity (i.e. non-exoelectrogenic bacteria). The presence of non-exoelectrogenic bacteria may disrupt the electrical conductivity of the biofilm and occupy electrode surface, thus reducing the efficiency of ultimate electricity generation. A combination of exoelectrogenic species and non-exoelectrogenic species may facilitate an effective conversion of complex organic matter to electricity. For instance, fermentative microorganisms in an anodic biofilm can convert glucose (unusable for *Geobacteria* species) into acetate, which is subsequently utilised by *Geobacteria* species to produce electricity.

Interspecies electron transfer within anodic biofilms emerges as an interesting subject in MFC community studies. There are now compelling lines of evidence for cell–cell interactions based on nanowires or electron mediators. For example, it is well established that close proximity of fermentative and methanogenic bacteria is advantageous to both microorganisms to facilitate the interspecies hydrogen transfer [57]. However, it has been recently observed that *Pelotomaculum thermopropionicum* can produce thick conductive pili, which connect these fermentative bacteria with a methanogen (*Methanothermobacter thermautotrophicus*) in co-cultures [58]. Much remains to be discovered about the microbiology of MFC biofilms and the mechanisms of interspecies electron transfer within. Such information might be useful for us to optimise MFC design and improve the electricity generation.

2.3 Microbial Electrochemistry

2.3.1 Voltage Losses (Polarisations) in Microbial Fuel Cells

Voltage is the first consideration for the performance of any electrical source. Voltage generation of MFCs is far more complicated to understand or predict than that of a chemical fuel cell. In the biofilm of an MFC, various bacteria may grow symbiotically, setting different potentials. The electrochemical activity of bacteria is sensitive to the operation conditions, such as temperature, pH, and fuel cell configuration. Even the potential produced by even a pure culture is hard to predict, we can only calculate the maximum voltage limits of an MFC based on thermodynamic relationships between the substrates and electron acceptors. Typical anode and cathode potentials in MFCs at both standard conditions and operating conditions are shown in Table 2.4.

The theoretical maximum cell voltage for an acetate-fed air-cathode MFC is 1.1 V (E_{emf}, obtained in terms of Nernst equation). However, the measured MFC voltage is considerably lower than this value. In an open circuit when

TABLE 2.4
Standard Electrode Potentials E^0 and Corresponding Adjusted Values ($E^{0\prime}$) at Typical Conditions in MFCs (E^0 Values from Ref. [59]) and $E^{0\prime}$ Calculated Using the Nernst Equation; All Potentials are Shown against NHE)

Electrodes	Reactions	Conditions	E^0 (V)	$E^{0\prime}$ (V)
Anode	$2H^+ + 2e^- \rightarrow H_2$	$pH = 7, 298$ K	0.000	−0.414
	$2HCO_3^- + 9H^+ + 8e^- \rightarrow CH_3COO^- + 4H_2O$	$HCO_3^- = 5$ mM, $CH_3COO^- = 5$ mM, $pH = 7, 298$ K	0.187	−0.296
	$2HCO_3^- + 9H^+ + 8e^- \rightarrow CH_3COO^- + 4H_2O$	$HCO_3^- = 5$ mM, $CH_3COO^- = 16.9$ mM, $pH = 7, 298$ K	0.187	−0.300
	$6CO_2 + 24H^+ + 24e^- \rightarrow C_6H_{12}O_6 + 6H_2O$	$pH = 7, 298$ K	−0.014	−0.428
Cathode	$O_2 + 4H^+ + 4e^- \rightarrow 2H_2O$	$pO_2 = 0.2, pH = 7, 298$ K	1.229	0.805
	$O_2 + 4H^+ + 4e^- \rightarrow 2H_2O$	$pO_2 = 0.2, pH = 10, 298$ K	1.229	0.627
	$O_2 + 2H^+ + 2e^- \rightarrow H_2O_2$	$pO_2 = 0.2, [H_2O_2] = 5$ mM, $pH = 7, 298$ K	0.695	0.328
	$MnO_2(s) + 4H^+ + 2e^- \rightarrow Mn^{2+} + 2H_2O$	$[Mn^{2+}] = 5$ mM, $pH = 7, 298$ K	1.230	0.470
	$MnO_4^- + 4H^+ + 3e^- \rightarrow MnO_2 + 2H_2O$	$[MnO_4^-] = 10$ mM, $pH = 3.5, 298$ K	1.700	1.385
	$Fe(CN)_6^{3-} + e^- \rightarrow Fe(CN)_6^{4-}$	$[Fe(CN)_6^{3-}] = [Fe(CN)_6^{4-}], 298$ K	0.361	0.361

Source: From B.E. Logan, *Microbial Fuel Cells*, 2008, Copyright Wiley-VCH Verlag GmbH & Co. KGaA. Reprinted with permission.

no current is flowing, for example, the maximum MFC voltage achieved thus far is ~0.80 V [61]. This difference between the measured cell voltage and the cell E_{emf} is caused by various voltage losses as illustrated in Figure 2.7a. These losses can be expressed as the sum of the anode overpotential, the cathode overpotential and the ohmic loss of the system.

$$E_{cell} = E_{emf} - \left(\sum \eta_{anode} + \left| \sum \eta_{cathode} \right| + IR_\Omega \right) \qquad (2.5)$$

where $\Sigma\eta_{anode}$ and $\Sigma\eta_{cathode}$ are the overpotentials of the anode and the cathode, respectively, and IR_Ω is the sum of all ohmic losses which are proportional to the generated current (I) and ohmic resistance of the system (R_Ω). The overpotentials of the electrodes are generally current dependent; for example they are especially evident at low current densities where the voltage decreases rapidly. In an MFC, the overpotential of an electrode generally arises from three losses: (i) activation losses; (ii) bacterial metabolic losses and (iii) mass transport or concentration losses. Typical polarisation curves of an MFC shown in Figure 2.7b is derived from its respective cathode and anode polarisation curve in Figure 2.7c.

2.3.1.1 Ohmic Losses

The ohmic losses (or ohmic polarisation) in an MFC arise from resistance to the ion (including proton) conduction through the membrane (if present) and electrolytes, the resistance to electron flow through the electrodes, and interconnections (i.e. the linear voltage drop region in Figure 2.7b). Reducing the ohmic losses is the most important objective for the optimisation of MFC design. This can be achieved by reducing the electrode spacing, using a membrane and electrode with a low resistivity, good contacts throughout the circuit, and (if practical) increasing solution conductivity to the maximum tolerated by the bacteria [31]. However, it must be noted that reducing ohmic losses does not always lead to increase in power generation. For example, when the electrode spacing is reduced to below 2 cm, power generation of air-cathode MFCs will decrease, even though there is a reduction in ohmic losses. This is due to anode contamination by excessive oxygen permeation from the cathode.

2.3.1.2 Activation Losses

Activation losses (or activation polarisation) correspond to the energy loss from initiating the redox reactions at the electrodes, and during the electron transfer through the cell membrane (i.e. via mediator, nanowire or surface-bond cytochrome) to the anode surface. Activation losses are apparent at low current densities (i.e. the first region in Figure 2.7b and c), and increase steadily as the current density increases. Reducing these losses can be achieved by using improved catalysts at the cathode (e.g. using a Pt catalyst),

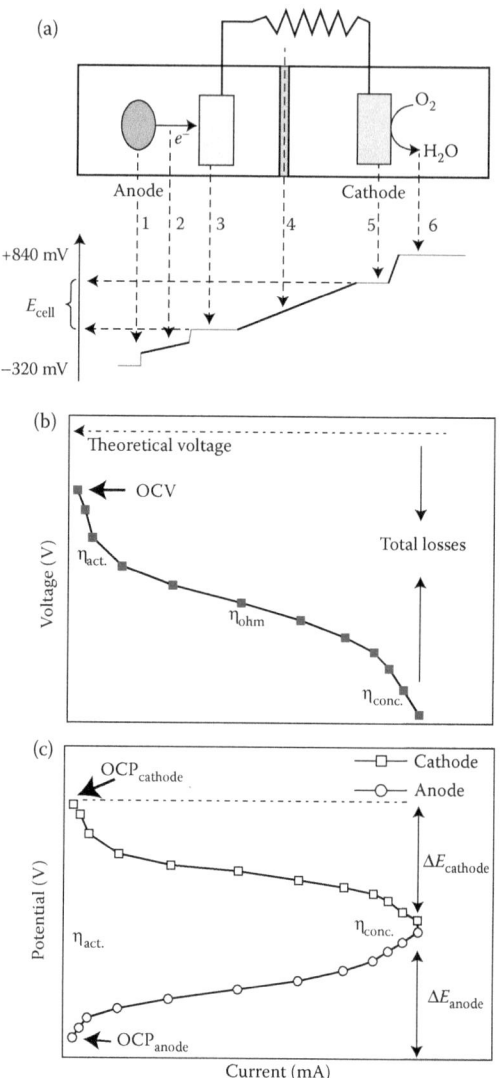

FIGURE 2.7
(a) Potential losses during electron transfer in MFCs at different stages. Stage 1—loss owing to bacterial electron transfer; Stage 2—losses owing to electrolyte resistance; Stage 3—losses due to the anode resistance; Stage 4—losses at the MFC resistance (useful potential difference) and membrane resistance losses; Stage 5—losses due to the cathode resistance; Stage 6—losses owing to electron acceptor reduction. (b) The typical current–voltage polarisation curve of an MFC and (c) the typical current–potential polarisation curves of cathode and anode of an MFC. (Reprinted from *Trends Microbiol.*, 23, K. Rabaey and W. Verstraete, Microbial fuel cells: Novel biotechnology for energy generation, 291–298. With permission from Elsevier; F. Zhao, R.C.T. Slade and J.R. Varcoe, Techniques for the study and development of microbial fuel cells: An electrochemical perspective, *Chem. Soc. Rev.* 38, 2009, 1926–1939. Reproduced by permission of The Royal Society of Chemistry. Copyright 2005, Elsevier and 2009, Royal Society of Chemistry.)

increasing the electrode surface area, and changing the electrode chemical properties (e.g. reduce O-group content and increase positive charges on the anode) to improve the electrochemical activity of biofilms, and facilitate electron transfer between the bacteria and the electrode(s) [31,32].

2.3.1.3 Bacterial Metabolic Losses

Voltage losses due to bacterial metabolism are a consequence of bacteria deriving energy through oxidising substrates. Most bacteria are able to discharge many protons across the membrane per substrate oxidised using the tricarboxylic acid cycle, thus more ATP are obtained than essentially required to transport one electron exogenously. In an MFC, bacteria transport electrons from a substrate at a low potential (e.g. acetate, −0.296 V) to the anode, the final electron acceptor. Hence, the anode potential determines the energy gain for the bacteria. The larger the difference between the redox potential of the substrate and the anode potential, the higher the metabolic energy gain for the bacteria, but the lower the maximum attainable MFC voltage. With an aim to maximise the cell voltage, the potential of the anode should be kept as low (negative) as possible. However, electron transport might be inhibited if the anode potential becomes too low, and more energy will be lost due to substrate fermentation (if possible). More investigation is required to ascertain the impact of a low anode potential on the stability of power generation.

2.3.1.4 Concentration Losses

Concentration losses (or concentration polarisation) take place when the flux of reactants to the electrode or the flux of products from the electrode are insufficient, limiting the current production rate. These losses occur mainly at high current densities (e.g. 3 A m^{-2}), when the diffusion rate of chemical species to or from the electrode surface is limited (i.e. the high current region in Figure 2.7b and c).

At the anode, concentration losses are caused by either insufficient supply of reduced species toward the electrode, or limited discharge of oxidised species from the electrode surface. This will increase the ratio between the oxidised and the reduced species at the electrode surface, which can produce an increased electrode potential (as demonstrated by the OCP_{anode} curve in Figure 2.7c). At the cathode the reverse may occur, resulting in a drop in cathode potential ($OCP_{cathode}$ curve in Figure 2.7c). Moreover, diffusional gradients and thus mass transport limitations in the bulk liquid may occur in poorly mixed systems, which can also limit the substrate flux to the biofilm.

Another important problem affecting the voltage generation is proton flux limitation from the anode or towards the cathode. Proton accumulation in the anodic biofilm will lower the local *pH*, reducing the bacterial electrochemical activity. Limited proton transfer to the cathode will result in elevated *pH* around the cathode, thus limiting the voltage generation, thereby maintain a sufficient buffer capacity of the system for a stable and efficient electricity production.

2.3.2 Power Generation

Maximisation of power output (i.e. the power that is usable) is the ultimate goal of most MFC studies. The power output of an MFC is calculated from the measured cell voltage and the current as

$$P = IE_{cell} \tag{2.6}$$

Normally, the voltage is measured across a fixed external resistor (R_{ext}), while the current is obtained from Ohm's law ($I = E_{cell}/R_{ext}$). Thus, power output of MFC system is usually calculated as

$$P = \frac{E_{cell}^2}{R_{ext}} \tag{2.7}$$

As discussed above, the maximum cell voltage (E_{cell}) that can be measured in MFC is its OCV, which is always lower than the E_{emf} of the cell due to various potential losses. The MFC can be viewed as having two resistors linked in series, an external load (R_{ext}) and an internal resistance (R_{int}). The theoretical maximum power of the total system is therefore

$$P_{t,emf} = \frac{E_{emf}^2}{R_{int} + R_{ext}} \tag{2.8}$$

In MFC studies, our concern is the maximum power that can be used, that is, the power output (P_{max}). The theoretical maximum power output can be expressed as

$$P_{max} = \frac{E_{emf}^2}{(R_{int} + R_{ext})} \times \frac{R_{ext}}{(R_{int} + R_{ext})} = \frac{R_{ext} E_{emf}^2}{(R_{int} + R_{ext})^2} \tag{2.9}$$

Since OCV, other than E_{emf}, is the maximum voltage that can be achieved in an MFC, we therefore generally use the following equation to determine the maximum power output.

$$P_{max} = \frac{R_{ext}(OCV)^2}{(R_{int} + R_{ext})^2} \tag{2.10}$$

As seen from Equation 2.10, P_{max} reaches its maximum value $OCV^2/4R_{int}$, when the internal resistance equals the external resistance. It is clear that a higher maximum power output is achieved when the system has a smaller internal resistance. Therefore, minimising the internal resistance is the key approach to maximise MFC performance.

When we compare the performance of different MFC systems, the power output usually has to be normalised to some characteristic of the fuel cell. For MFCs aiming to produce electricity, it is common to normalise the power output by the projected anode surface area, because the anode is where the biological reaction occurs. The power density (P_{anode}, W m^{-2}) therefore is

$$P_{anode} = \frac{E_{cell}^2}{A_{anode} R_{ext}} \quad (2.11)$$

Note that application of the above normalisation is reasonable only when the power output is limited by the anode performance. In many instances, the cathode reaction is thought to limit overall power generation [64,65], or the surface area of anode material is difficult to measure (e.g. when the anode chamber is filled with large amounts of carbon granules of various surface areas). In such cases it is more useful to normalise the power production by the cathode surface area ($A_{cathode}$).

$$P_{cathode} = \frac{E_{cell}^2}{A_{cathode} R_{ext}} \quad (2.12)$$

For an effective comparison, the projected surface areas of all components, the specific surface area (if known), and the methodology should always be clearly stated when reporting the power density.

The power density normalised to the fuel cell volume is the most useful engineering value to evaluate the performance of an MFC. Using the volumetric power density is convenient to perform engineering calculations for size and costing of fuel cell, and is useful for comparison to chemical fuel cells and other wastewater treatment techniques. The volumetric power density, P_V (W m^{-3}), is expressed as

$$P_V = \frac{E_{cell}^2}{V R_{ext}} \quad (2.13)$$

where V (m^3) is the total fuel cell volume (i.e. the empty bed volume). Using the total fuel cell volume as a basis for the calculation is to follow the convention in environmental engineering calculations. In some cases, however, other volume basis may be employed to normalise the power output for a more reasonable comparison. For example, the anode chamber volume is used as a basis for calculation when comparing between dual-chamber and single-chamber fuel cell, because there is no 'cathode chamber' for an air-cathode MFC. However, if cells are stacked together, the air spacing between the liquid chambers should be included for the overall fuel cell volume. If microorganisms are grown outside the fuel cell, the volume where the microbes grow should also be included in the total fuel cell size.

To opt for high power output MFCs, it is essential to convert as much electrons stored in the substrates as possible into current. The recovery of electrons measured in terms of 'Coulombic efficiency', C_E, is defined as the ratio of electrons actually recovered as current, to the electrons in the initial organic matter in the anode. For a fed-batch MFC system, the Coulombic efficiency over a period of time t_b, C_E is calculated as

$$C_E = \frac{M \int_0^{t_b} I dt}{nFV_{anode}\Delta c} \qquad (2.14)$$

where Δc (mol L^{-1}) is the change of substrate concentration over the batch cycle (which is completed when the cell voltage is lower than threshold, usually <50 mV) during time t_b; M (g mol^{-1}) is the molecular weight of the substrate; F is Faraday's constant; n is the number of electrons released per mole of substrate oxidation; and V_{anode} (L) is the volume of liquid in the anode chamber. For complex substrates, COD* is normally employed to represent the substrate concentration, so C_E becomes

$$C_E = \frac{8 \int_0^{t_b} I dt}{FV_{anode}\Delta COD} \qquad (2.15)$$

where 8 is obtained based on M_{O_2} = 32 g mol^{-1} and n = 4 moles e$^-$ per mole of oxygen. For MFCs that run in a continuous flow mode, the Coulombic efficiency can be expressed on the basis of COD as

$$C_E = \frac{8I}{Fq\Delta COD} \qquad (2.16)$$

where I (A) is the current generated under steady conditions, q (L s^{-1}) is the volumetric influent flow rate, and ΔCOD is the difference in the influent and effluent COD. The Coulombic efficiency depends strongly on the bacterial metabolism at the anode. The maximum C_E can achieve levels of 90% or more with pure cultures and non-fermentable substrate such as acetate. The Coulombic efficiency is reduced when the anodic bacteria utilise alternate electron acceptors, such as those present in the medium (or wastewater),

---

* COD (Chemical oxygen demand, g (O$_2$)/L) indicates the mass of oxygen that is needed to fully oxidize the organic compounds in per litre of solution.

or those diffusing from the catholyte like oxygen, other than from the anode electrode. Furthermore, competitive processes, such as fermentation and/or methanogenesis, by non-electricity-producing bacteria in the biofilm will consume the substrates rather than generating current, thus reducing the Coulombic efficiency.

The energy efficiency of an MFC, η_t, reflects the energy recovered from the system compared to the energy content in the initial substrates. It can be calculated by dividing the combustion heat of the starting substrates over a time interval t by the total power produced, or

$$\eta_t = \frac{\int_0^t E_{cell} I \, dt}{\Delta H m_s} \quad (2.17)$$

where ΔH (J mol^{-1}) is the combustion heat and m_s (mole) is the amount of substrate consumed. ΔH values for specific compounds can be easily obtained, but they are not known for actual wastewaters with a complex composition. Thus, the above calculation is only available for MFCs with the influents of a known composition. Reported energy efficiencies of MFCs range from 2% to 50% for easily biodegradable substrates [65,66]. As a reference, the electric energy efficiency for thermal conversion of methane does not exceed 40%, thus MFCs holds its competitiveness over these systems.

Bacteria in MFCs utilise parts of the energy in the substrates to support growth and reproduction, resulting in an increase in biomass, and a decrease in Coulomic and energy efficiencies. The substrate consumption for bacteria growth is represented by the net cell yield, $Y_{X/C}$ (g biomass COD (g substrate COD)$^{-1}$), which is expressed as

$$Y_{X/C} = \frac{X}{\Delta COD} \quad (2.18)$$

where X (g COD) is the biomass produced over time (either t_b or hydraulic retention time). The growth rate can be measured directly by determining the biomass (g COD) built up on the electrode surface and discharged in the effluent. The cell yield of MFCs is typically lower than that of traditional aerobic processes. This makes MFCs particularly attractive since sludge disposal is highly economically and environmentally unfavorable. Net cell yields ranging from 0.07 and 0.22 g biomass COD (g substrate COD)$^{-1}$ have been observed for MFCs, while the values for typical aerobic wastewater treatment are generally around 0.4 g biomass COD (g substrate COD)$^{-1}$ [62]. A lower cell yield in MFCs is reported since a significant part of the substrate energy is converted into electrical power in these systems, thus leaving limited energy for biomass growth.

The total COD (substrate) removal in MFC arises from three consumption approaches: (i) conversion into electrical current (via the Coulombic efficiency), (ii) conversion into biomass (via the growth yield) and (iii) consumption by other unknown processes. When the electron and energy recovery efficiencies are determined, the fraction of COD that is removed by unknown processes, θ, can be obtained as

$$\theta = 1 - C_E - Y_{X/C} \tag{2.19}$$

2.3.3 Factors Affecting the Performances of Microbial Fuel Cells

Up to date, performances of laboratory MFCs are still much lower than the ideal performance. Electricity generation in an MFC is a combined effect of (i) microbial catabolism, (ii) electron transfer from microbes to the anode (anode performance), (iii) reduction of electron acceptors at the cathode (cathode performance), and (iv) proton transfer from the anode to cathode. Factors affecting each of the above processes may have a great influence on the overall MFC performance, thus appropriate experimental conditions is very important. Many modifications have been carried out to improve each of these processes, leading to higher total power output of MFCs. As listed in Table 2.5, major factors affecting the power generation of MFCs include: microbial catalytic activity, anode and cathode performances, proton transfer efficiency and solution chemistry.

TABLE 2.5

Major Factors Affecting the Power Generation of MFCs

Factors	Methods for Facilitation of Power Generation of MFCs
Microbial catalytic activity	Use mixed cultures
	Inoculated with pre-acclimated bacteria
	Apply appropriate anode potential
	Add surfactants into the anode biofilm
Anode performance	Add redox mediators in anode chamber
	Use anode materials of high conductivity, good microbial compatibility and large surface area
	Anode modification with conductive polymers, metal oxides, etc.
Cathode performance	Increase the concentration of electron acceptors
	Cathode modification with efficient catalysts
	Increase the operation temperature
Proton transfer efficiency	Reduce electrode spacing
	Use suitable membrane or eliminate membrane
	Use suitable buffer
Solution chemistry	Increase the operation temperature
	Employ near neutral *pH*
	Increase the solution conductivity

2.3.3.1 Microbial Catalytic Activities

The anodic microbial catalysts are expected to play a key role for the power density and efficiency of an MFC. The highest power generation of MFCs seems to be produced by those operating with a mixed culture, or a microbial community, rather than those operating with a pure culture. The structure and activity of the bacterial community are sensitive to various environmental conditions, such as solution pH, electrode potential, ionic strength, and temperature. Since these environmental parameters can also affect other processes (e.g. the proton transfer efficiency, and cathode performance), it is now more difficult to conclude a quantified relationship between the microbial community and these parameters. Moreover, the configuration and operating mode of MFCs also appear important for the composition and activity of anodic biofilms.

The inoculum source is another factor affecting the anodic microbial community and activity. Kim et al. [67] suggested that start-up of an MFC is most successful with biofilm harvested from the anode of an existing MFC. In another study, enrichment of the bacteria on the anode of an MFC resulted in increased power output and a change in the bacterial community [17].

Although it has been observed by phylogenetic methods that microbial communities form various patterns, this does not appear to influence microbial activity and MFC performance. Instead, electrochemical techniques that enhance the anodic bacterial growth and electron transfer tend to improve the MFCs performance (e.g. via control of the anode reduction/oxidation potential, or addition of exogenous compounds). The metabolism and electrochemical activity of microbes show strong dependence on the anode potential [49]. It is therefore possible to control the growth and activity of microbial community by adjusting the anode potential to a specific value. The optimum potential for electron transfer for a specific bacterium usually equals to the potential of the respiratory enzyme used as a terminal electron acceptor. Enrichment of microbes respiration at different negative potentials can be achieved via polarising the anode potential to more negative values during biofilm cultivation. Microorganisms unable to survive in these increased negative anode potentials will be out-competed by other bacterial species. MFCs operating with such microbial community are expected to produce high-voltage output and energy efficiency. Wen et al. [68] reported that addition of surfactants (e.g. Tween 80) into MFCs could greatly improve the power output from 0.6 to 5.2 W m^{-2}. It was speculated that adding surfactants into the anode biofilm can increase the permeability of cell membranes, which reduces the electron transfer resistance through the cell membrane, and increases the electron transfer rate and number, consequently enhancing the current and power output. Chemically 'perforating' pores and channels on the bacterial membrane using some cationic reagents, such as chitosan, and ethylenediamine tetraacetic acid, are also reported to significantly improve the electron transfer rate and power density of MFCs

[69]. The enhancements were attributed to the increased mediator excretion, and enhanced direct electron transfer through the bacterial membrane.

2.3.3.2 Anode Performance

Voltage losses at the anode result from overpotential and ohmic losses. The latter arises from resistances to electron and ion flow through the electrodes and interconnections. Other than increasing the microbial catalytic activity as mentioned above, soluble redox mediators have been added to a batch anode to improve the electron transfer to reduce the overpotential losses. Benchmark for a redox mediator is to possess a low potential, favorably in the order of −300 mV or more. These redox mediators (e.g. benzylviologen, phenazines, phenothiazines, phenoxoazines, iron chelates and neutral red) would then enable bacteria to have a sufficiently high interaction rate with the electrode, resulting in a high Coulombic and energetic efficiency.

The anode material can affect both the overpotential and ohmic losses, with influence on the microbial attachment, electron transfer, and/or direct substrate oxidation. The most frequently used anodes are carbon-based materials, such as carbon cloth, graphite felt and carbon brush. The advantages of these materials are their high electric conductivity, good microbial compatibility and large surface area. Generally, material structures with a higher specific surface area are expected to provide more colonised sites for bacteria, resulting in a higher power output. However, one should note that only those surfaces accessible to bacteria are responsible for current generation; those pores with a diameter less than 2 nm, for example, are too small for bacteria to access, thus are not beneficial to electricity production.

Modifying the anodes with appropriate methods can increase the power production in MFCs. Metals and/or metal oxides, and conductive polymers have been used for this purpose. Park and Zeikus [30] reported that current production of an MFC was enhanced from 0.02 mW m^{-2} using a graphite anode to 10.2 mW m^{-2} when Mn^{4+} ions were incorporated into the anode. Coating the anode with polyaniline (PANI) or modified PANI polymers, such as fluorinated PANI and PANI/carbon nanotube composite increased the current density by 10-fold [70,71]. Moreover, treatment of a carbon cloth anode with ammonia gas increased the surface charge of the electrode and improved MFC performance [32]. These studies have shown that anode modification is a practical approach to enhance MFCs power output. Care should be taken to ensure the stability of the modified electrodes. This would be particularly true in MFCs containing microbial communities (e.g. those derived from anaerobic sludge), since they may possess a wide variety of catabolic abilities. Therefore, in assessing MFCs, it is desirable that the durability of any modified electrodes be clarified. Reported anode materials can operate for over six months without decline in performance [72].

2.3.3.3 Cathode Performance

The cathode reaction efficiency depends on a variety of factors, including the concentration and species of the oxidant (electron acceptor), catalyst performance, electrode structure and operational conditions.

Oxygen, hexacyanoferrate ($K_3[Fe(CN)_6]$) and permanganate ($KMnO_4$), which possess high redox potentials, have been used as electron acceptors at the cathode. Although hexacyanoferrate and permanganate may produce a similar or higher cathode potential than oxygen ($OCP_{cathode}$ of 0.70 and 1.38 V vs. ~0.4 V; [73]), oxygen remains as the most commonly used and promising cathodic oxidant, due to its free and clean nature. In many cases, oxygen reduction on the electrode surface takes place slowly, resulting in a high reduction overpotential. This cathode reaction thus becomes the rate-limiting step in an MFC. Hence, improving the cathodic reaction rates is critical for building an MFC with high efficiency and high-power output. Different approaches have been explored for this purpose, primarily including cathode modification with catalysts and optimisation of operational conditions.

Many chemical and biological catalysts have been tested to improve the cathode performance. The use of catalysts on the cathode surface can significantly lower the activation overpotential and promote the kinetics of oxygen reduction at the electrode surface. Various precious metals (e.g. platinum and gold) and non-precious metals (e.g. ferric iron, manganese oxides, iron complexes and cobalt complexes) have been widely employed as catalysts in cathode materials, because of their favorable low overpotential for oxygen reduction. Among them, non-precious metal catalysts are of special interest, since they are less expensive and more abundant than precious metals such as Pt), with an almost similar catalytic ability as the latter. Microorganisms can also be used as catalysts and mediators in the cathode. The advantages of biocathodes over abiotic cathodes are the lower construction and operation costs, and possible additional functions (such as denitrification and desulphurisation).

Increasing the oxidant concentration at the cathode will lead to a higher performance of MFCs. According to the Nernst equation, increasing the oxidant concentration will result in a higher ratio of reactant to product activities, and a higher thermodynamic voltage, although this improvement is small due to the logarithmic nature of the equation. More importantly, the cathodic reaction kinetics can be significantly accelerated with increased oxidant concentration. For instance, the maximum power output of an air-cathode MFC increased by 15.8% (from 0.095 to 0.11 mW) when the catholyte was saturated with pure oxygen (38 mg L^{-1} DO) instead of air (7.9 mg L^{-1} DO) [74]. Generally, the reaction rate (r) increases linearly with the oxidant concentration [O] according to the relationship $r = k[O]^\alpha$, where k is the rate constant and α is the stoichiometric coefficients of the oxidant.

The operating temperature can also affect the kinetics of oxidant reduction, mass and proton transfer, thus determining the cathode performance. Liu et al. [75] reported a 9% increased power output in a single-chamber membrane-less MFC when the operating temperature increased from 20°C to 32°C, mainly due to an increase in the cathodic potential at high current densities (from about −200 mV to −100 mV at 0.36 mA cm^{-2}). However, the operating temperature should not exceed the temperature tolerance of microorganisms. Otherwise, the MFC performance would be adversely affected by deactivation of microorganisms.

2.3.3.4 Proton Transfer Efficiency

In many MFCs, the proton transfer efficiency from the anode to the cathode is the rate-limiting step and a major cause of internal resistance. Although equivalent amounts of protons and electrons are produced at the anode in MFCs, the migration rate of protons to the cathode is much slower than that of the electrons. It arises from the fact that the migration of electrons is forced by the potential difference between the two electrodes, while the migration of protons is caused by diffusion. A proton exchange membrane (PEM), if present, functions as a proton transfer barrier, and further decreases the proton diffusion rate. Since proton transport inside the fuel cell is slower than its production rate in the anode and its consumption rate in the cathode, a *pH* difference between the two electrodes occurs without buffer. For example, in the absence of any buffer solution, Gil et al. [76] detected a *pH* difference of 4.1 (9.5 at cathode and 5.4 at anode) after a 5-h operation with an initial *pH* of 7 in both chambers. Accumulation of protons at the anode will suppress the microbial activity, thus the electricity production, whereas a limited proton concentration at the cathode may reduce the cathodic reduction rate.

Proton-transfer efficiency depends on the distance between the two electrodes, the type of membrane used, and the type and concentration of buffer. At present, there is no standard method to measure the proton transfer efficiency. It can only be qualitatively determined by the *pH* difference between the anode and cathode chambers. According to the diffusion theory, reducing the anode-to-cathode distance facilitates the proton transfer. Should the electrodes be too close to each other, there will be adverse effects on the activities of microorganisms [60].

Membranes used in MFCs include cation-exchange, anion-exchange, bipolar membranes and other types (such as ultrafiltration membrane). Although they are required as a separator for the anode and cathode electrolytes, they also establish a barrier to proton transfer, contributing to the internal resistance. It has been found that the presence of a PEM reduced the maximum power density from 494 to 262 W m^{-2} in a single-chamber air-cathode MFC [65]. Nafion is the most popularly used proton exchange membrane because of its high proton permeability. However, it is also permeable for other cations (e.g. Na^+, K^+, NH_4^+, Ca^{2+}, and Mg^{2+}) at high efficiencies. Since the concentrations of these cationic species are typically 10^5 times higher than the proton concentration ([H$^+$] = 10^{-7} M) in MFCs, they would accumulate in the cathode chamber, causing an increase in *pH*, and a decrease in MFC performance [77]. Now it has been found that other less expensive membranes (such as CMI-7000 and AMI-7001 made by Membrane International Inc.) can give a power output comparable to that of Nafion [78].

The ratio of membrane surface-area-to-system volume is an important factor affecting the maximum power output. Oh and Logan [79] demonstrated that the power generation of MFC decreases with the increase of proton exchange membrane surface area over a relatively large range. For a fixed anode and cathode surface area ($A_{Anode} = A_{Cathode}$ = 22.5 cm^2), the power density normalised to the anode surface area increased with the proton exchange membrane size in the order 45 mW m^{-2} (A_{PEM} = 3.5 cm^2), 68 mW m^{-2} (A_{PEM} = 6.2 cm^2), and 190 mW m^{-2} (A_{PEM} = 30.6 cm^2).

Adding a suitable buffer into the electrolytes helps to maintain a constant *pH* (typically *pH* 7) in both anode and cathode, thereby achieving a suitable proton concentration for the growth of microorganisms and cathodic oxygen reduction. The highly concentrated ions in buffers could increase the solution conductivity, and reduce the internal resistance. It is generally accepted that the presence of buffer is indispensable for a well-performed MFCs. An ideal buffer can maintain constant *pH* without interfering the chemical reactions or microbial physiology, and provide sufficient proton at the cathode for high power densities. Commonly used buffers in MFC studies include phosphate buffers, bicarbonate buffers and some synthetic zwitter ionic buffers, such as MES (2-[N-morpholino] ethane sulphonate), HEPES (4-(2-hydroxyethyl)-1-piperazineethanesulphonic acid), and PIPES (piperazine-N,N-bis [2-ethanesulphonate]). It has been shown that increasing the concentration of the phosphate buffer from 50 to 100 mM will increase the power density from 697 to 1010 W m^{-3}, and a power density of 1550 W m^{-3} was obtained when using a bicarbonate buffer of 200 mM at *pH* 9.0 [80].

2.3.3.5 Electrolyte Chemistry

In addition to the above factors, the electrolyte chemistry, such as the temperature, *pH* and salinity, also shows great influences on the performances of MFCs. MFCs can operate over a wide temperature ranges (4–55°C), depending on the tolerance of the bacteria (psychrophilic, mesophilic or thermophilic microbes). Changes in temperature would influence the structure and activity of the anodic microbial community, the system kinetics (activation energy, mass transfer coefficients, solution conductivity) and thermodynamic (free Gibbs energy, electrode potentials), thereby affecting the overall power output and COD removal efficiency. It has been observed that increasing the system temperature from 4°C to 35°C resulted in an increased power density production from 15.1 to 174.0 mW m^{-3}, and increased COD removal from 58% to 94% for single-chamber MFCs [81]. The MFC start-up time is also affected by the temperature. Cheng et al. [82] detected that the startup time for single-chamber MFCs increased from 50 h at 30°C to 210 h at 15°C. At temperatures below 15°C MFCs could not start up after one month of operation. However, if an MFC was first started up at temperature of 30°C, its effective power output can be achieved even at temperatures as low as 4°C in the subsequent operation.

The electrolyte *pH* can affect both the electrochemical and biochemical processes in an MFC. Conventionally, MFCs are operated with a near neutral electrolyte *pH* to maintain optimal conditions for bacterial growth and electricity generation. However, the *pH* of actual wastewaters covers a broader range (from acidic to alkaline), thus the performance of MFCs under various *pH* need to be understood for a better application prospect. Though anodic bacteria generally require a near neutral *pH* for their optimal growth, an acidic *pH* may give higher proton transfer rates and higher activity of intracellular electron carriers. Zhang et al. [83] observed that as the electrolyte acidity increased from *pH* 7 to *pH* 4 in a two-chamber air-cathode MFC, the voltage outputs gradually decreased by 31.6% from 339 to 232 mV, and the power generation decreased by 59.9% from 237 to 95 mW m^{-2}; however, a faster COD removal was achieved. Deterioration of anodic biofilm was demonstrated to be responsible for the decreased MFC performance under acidic conditions. MFCs operated at *pH* ≤ 4 may cause long term, even irreversible reduction of MFC performances. However, Raghavulu et al. [84] obtained highest power output in a two-chamber air-cathode MFC at *pH* 6 (compared to *pH* 7 and *pH* 8), but highest COD removal at neutral conditions.

On the cathode electrode, oxygen reduction prefers alkaline conditions to obtain a lower reduction overpotential. Both Puig et al. [85] and He et al. [86] reported the highest current and power generation at *pH* 9.0–9.5 in two-chamber air-cathode MFCs. Differences in the above results may arise from the different operating conditions (such as substrates, electrolyte conductivity, temperature) and MFC configurations in these studies.

The salinity of electrolytes has antagonistic effects on the performances of MFCs. Increasing the salinity would result in an increased conductivity, and therefore facilitates proton transfer and reduces the internal resistance. However, high salt concentrations are known to adversely affect the physiology of anaerobic microbial consortia. Liu et al. [75] observed that increasing the solution ionic strength from 100 to 400 mM by adding NaCl increased power output from 720 to 1330 mW m^{-2} for a single-chamber air-cathode MFCs. In another study using a two-chamber air-cathode MFC [87], adding up to 20 g L^{-1} of NaCl reduced the overall internal resistance by 33% and increased the maximum power production by 30%. Further increases in salt concentration, however, reduced the power production due to inhibition of the microbial activity. Therefore, an optimal concentration of 20 g L^{-1} of NaCl should be used for a highest power output.

2.4 Electrode Materials and Scale-Up of Microbial Fuel Cells

Electrodes are one of the most critical factors governing the performance and cost of MFCs. The physicochemical properties of electrodes (e.g. surface area, electric conductivity, and chemical stability) determine the microbial attachment affinity, electron transfer efficiency, electrode ohmic resistance and the surface reaction rate. Electrodes constitute a major cost in the construction of MFCs, and determine whether the MFC is cost-effective and scalable technology, especially for wastewater treatment. The long-term stability of MFC performance is largely dependent on the durability of electrode materials.

Over the past decade, a variety of electrode materials and configurations have been explored. Generally, both the anode and cathode materials should have high electric conductivity, chemical stability, mechanical strength and low cost. Fulfilling these requirements, carbon-based materials and non-corrosive metals are currently most-widely used as the base electrode materials.

2.4.1 Anode Materials

Anode materials in MFCs should have a good biocompatibility, low resistance and large surface area, to facilitate bacterial growth and electron transfer. A large variety of carbonaceous and several metal materials with unique configurations and surface areas have been developed for this purpose. A summary of commonly used anodes and their characteristics are listed in Table 2.6.

Carbonaceous materials are the most widely used anodes in present MFCs studies, due to their high conductivity and good biocompatibility. Commonly used types include carbon paper, carbon cloth, carbon mesh, carbon felt, reticulated vitrified carbon, graphite rod, graphite felt, graphite fiber brush, and graphite or activated carbon (AC) granules (shown in Figure 2.8). Carbon paper, cloth, mesh and felt are generally used as plain electrodes in MFCs, with the unique advantage of reduced electrode spacing, leading to reduced internal resistance and higher volumetric power density. Wang et al. [33] obtained a maximum power density of 483 mW m^{-2} using a carbon cloth anode and brewery wastewater substrate in a single-chamber MFC. While a graphite rod appears to be a reasonable anode material, its performance is limited due to the low porosity and surface area for microorganism attachment. Using graphite felt or foam with a higher specific surface area may overcome this drawback, and results in a much higher power output. For instance, in a two-chamber MFC, graphite foam or graphite felt electrodes, having almost the same geometric surface area as that of graphite rods, produced approximately 2.4 (0.46 mA) or 3 (0.57 mA)-fold more current than the graphite rods (~0.19 mA) [22].

TABLE 2.6

Comparison of the Characteristics of Commonly Used Anodic Base Materials in MFCs.

Anode Materials	Advantages	Limitations	References
Carbon paper	Easy to connect wiring	Lack of durability, brittle	[88]
Carbon cloth	Flexible, large porosity	Thin, expensive	[33]
Carbon felt	Large porosity	Large resistance	[89]
Carbon mesh	Relatively cheap, large porosity	Thin, easy to deform	[32]
Graphite rod	Good electrical conductivity and chemical stability, relatively cheap, and easy to obtain	Low surface area	[90]
Graphite felt	Large porosity, good conductivity and stability	High cost	[22]
Graphite fibre brush	Higher specific surface, easy to produce	Clogging	[91]
Graphite and AC granule	Fill the anode chamber, inexpensive, long durability	Difficult to collect electron, high ohmic resistance	[92,93]
Stainless steel and titanium	Super electrical conductivity	Smooth surface, poor biocompatibility	[94,95]

The graphite fiber brush (graphite fibers wound around one or more conductive corrosion-resistant metal wires) is considered to be the most promising anode material for future large-scale applications, due to its high surface area, low electrode resistance and commercially abundant. Logan et al. [91] reported a power density of up to 2400 mW m^{-2} in a flat-type air-cathode MFC with a graphite brush anode, four times that obtained with carbon paper as the anode (600 mW m^{-2}). Besides carbonaceous materials, some non-corrosive metals, such as stainless steel and titanium, can also be used an anode materials. However, they demonstrate a lower power generation (4–23 mW m^{-2}) than carbon materials, which may be attributed to their smooth surface preventing bacterial adhesion [94].

Modification of electrode materials can alter the electrode physical and chemical properties to facilitate microbial attachment and electron transfer. Recently, improvement of MFC performance has been reported through various anodic modification methods, including physical or chemical treatments and coating electro-active layers, as seen in Table 2.7. While facilitating power generation, these modifications elevated the cost of electrode, as their synthesis involve extensive equipments, multiple steps, high-temperature conditions and/or chemical consumption. Therefore, a simple or effective modification technique is of urgent need. Up to now, very little is known about the mechanism on how electrode surface properties affect the bacterial activity and electron transfer. This information will be helpful to advance new anode designs.

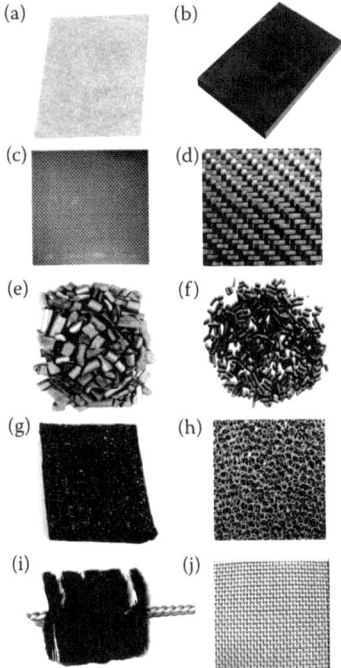

FIGURE 2.8
Photographs of commonly used electrode materials in MFC: (a) carbon paper; (b) graphite plate; (c) carbon cloth; (d) carbon mesh; (e) granular graphite; (f) granular AC; (g) carbon felt; (h) reticulated vitrified carbon; (i) graphite fibre brush; (j) stainless-steel mesh. (Reprinted with permission from *Bioresource Technol*, 102, J. Wei, P. Liang and X. Huang, Recent progress in electrodes for microbial fuel cells, 9335–9344. Copyright 2011, Elsevier.)

2.4.2 Cathode Materials

A basic requirement for cathode material in MFCs is to have a high redox potential and the ability to easily capture protons. Most of the materials mentioned above for anode can also be used for cathode. Presently, commonly used cathode materials include graphite, carbon cloth, carbon paper and stainless-steel mesh. Apparently different to the anodes, a catalyst is usually coupled with a cathode material when oxygen is used as the electron acceptor, due to low oxygen reduction rate at pristine electrode surface.

Pt is the most popular cathode catalyst in MFC studies for its superior catalytic activity for oxygen reduction. For example, Moon et al. [99] obtained a power density of 150 mW m^{-2} using a Pt modified graphite felt cathode in an MFC, while only one third of the value was achieved by pure graphite cathode. However, the high cost of Pt limits its practical application, although the Pt loading was reduced from 0.25–2 mg cm^{-2} to as low as 0.1 mg cm^{-2} [64,100].

Alternatively, many low cost but highly active non-Pt catalysts have been developed in recent years. For example, Morris et al. [100] observed that a

TABLE 2.7

Methods for Anode Modifications and the Improvements in Performance of MFCs

Modification Methods	Base Anode Material	Improvement in Performance	References
Ammonia treatment	Carbon cloth	Increase power density by 20%, reduce start-up time by 50%	[31]
Heat treatment	Carbon brush	Increase power density by 25%	[34]
H_2SO_4 treatment	Carbon brush	Increase power density by 8%	[34]
HNO_3 treatment	Graphite felt	Two fold increase in power density	[96]
Electrochemical oxidation	Graphite felt	Increase current density by 39.5%	[97]
Surface coating with carbon nanotube	Carbon clothes	0.7–30.5-Fold increase in power density	[98]
Surface coating with polyaniline	Graphite felt	1.8-Fold increase in power density	[96]
Surface coating with AQDS or NQ	Graphite disk	0.5–0.7-Fold increase in kinetic activity	[1]
Surface coating with iron oxide	Carbon paper	2.75-Fold increase in the power density	[67]
Surface coating with graphite paste containing Fe_3O_4	Graphite disk	1.1-Fold increase in kinetic activity	[1]
Sb(V) complex in graphite paste	—	0.9-Fold increase in kinetic activity	[10]

double-chamber MFC using PbO_2 as the cathode catalyst could produce 2–4 times more power than that of Pt catalyst, implying that PbO_2 could replace Pt as an inexpensive cathode catalyst in scaling-up MFCs. Other examples of non-Pt catalysts are CoTMPP and iron phthalocyanine (FePc)-based catalysts, MnPc, β-MnO_2, Co-OMS-2, MnO_x and Co/Fe/N/CNT (Table 2.8). Non-Pt catalysts are more suitable to be applied in large-scale MFCs due to their low cost and high catalytic activity. For aerobic biocathodes with oxygen as the terminal electron acceptor, electron mediators, such as iron and manganese, are first reduced by the cathode (abiotically) and then reoxidised by bacteria. Materials used for the anodes are also adaptable for bio-cathodes, such as graphite plate, carbon felt, granular graphite and graphite fiber brush, as well as stainless-steel mesh. However, it remains uncertain which material is more suitable for bio-cathode and the interaction mechanisms between the bacteria and electrode remains unclear.

A catalyst is not always essential for a cathode material. When ferricyanide ($K_3[Fe(CN)_6]$) or permanganate ($KMnO_4$) is used as the terminal oxidant, plain graphite or carbon electrode without any catalysts is sufficient to produce a high power output, due to their high redox potential and small reaction overpotential compared to oxygen. In air-cathode or aqueous

TABLE 2.8
Non-Pt Catalysts Used in the Cathode of MFCs

Catalysts	Cathode Materials	OCP (V)	P_{max} (mW m^{-2})	References
CoTMPP	Carbon cloth	—	369	[64]
PbO$_2$	Ti sheets	—	78	[100]
FePc	Carbon paper	0.319	634	[101]
FePcVC	Carbon paper	0.289	530	[101]
MnPc	Carbon paper	0.285	353	[101]
β-MnO$_2$	Carbon cloth	0.565	172	[35]
MnO$_x$	Carbon cloth	0.714	161	[102]
Co-OMS-2	Carbon cloth	0.147	180	[92]
Co/Fe/N/CNT	Carbon cloth	0.473	751	[103]
Microbes	Carbon felt	0.400	117 (Wm^{-3})	[104]

air-cathodes, carbon materials with high specific area have been tested to reduce the cathodic reaction overpotential without utilising a catalyst. Zhang et al. [35] developed a cathode with AC cold-pressed around a Ni mesh current collector. This air-cathode produced a maximum power density of 1220 mW m^{-2} (from cathode projected surface area), which was higher than a Pt-catalysed carbon cloth cathode (1060 mW m^{-2}). Common carbon materials can get reasonable oxygen reduction activity through some treatments, such as nitric acid treatment, Fe^{3+} doping and thermal treatment. Duteanu et al. [105] employed HNO$_3^-$-treated carbon powder as the cathodic catalyst in an air-cathode MFC, and obtained a current density (1115 mA m^{-2}, at 5.6 mV) greater than those of a Pt-supported carbon cathode. The development of Pt catalyst-free cathodes will lead to a significant reduction in the cost of MFCs, making it more feasible for large-scale applications.

2.4.3 Scaling Up of Microbial Fuel Cells

Scale-up is the next step ahead for MFCs commercialisation, especially in the field of wastewater treatment. During the past decades, tremendous laboratory work has been conducted on mL-scale MFCs [106]. Impressive progresses have been made in terms of increasing power output, lowering system cost and deeper understanding of the bacterial community and electron transfer mechanisms, but there is little information available on the scaling-up of MFCs. Although the maximum power densities of MFCs at laboratory-scale have reached 1.55 kW m^{-3} (normalised to the fuel cell volume) [80] under optimal conditions, the power density decrease significantly (typically less than 35 W m^{-3}) when the fuel cell size was increased from mL-scale to litre-scale [107], indicating that scaling factors have not been well understood or controlled. It has been suggested that MFCs should be able to produce at least 400 W m^{-3} to be competitive with traditional anaerobic digestion [108]. In addition to high power output, simple configuration, low-cost and long-term stability are also essential for the successful application of MFCs at large scale.

Among the various types of MFCs that have been developed, air-cathode MFCs is more promising for practical applications due to their simple configuration, sustainable operation and relatively high power output. A continuous operating mode is considered more suitable for practical application than a fed-batch mode. Up to now, no reports have been published on pilot-scale tests using MFCs, except some sediment MFCs for powering remote devices in seawater applications [109]. However, several studies [82,107,110] have been conducted to detect the factors affecting power output in litre-scale MFCs (fuel cell volume >1 L), and such information will be valuable for guiding the construction and operation of real field fuel cell. In most cases, the high internal resistance, which arises from electrode overpotentials, membrane resistance and electrolyte resistance is limiting the overall power output of a larger MFC.

To minimise internal resistance, electrode overpotentials should be reduced; this can be achieved by increasing the surface area or surface reactivity of electrode materials. Increasing the surface area of the anode can significantly increase power density in smaller fuel cells (28 mL) [91], but it did not appear so important in a larger MFCs [82]. In contrast, the surface area of the cathode is the limiting factor for power production in large-scale MFCs. Cheng and Logan [82] demonstrated that in a 1.6 L air-cathode MFC, doubling the cathode size can increase power by 62% with domestic wastewater, but power was increased by only 12% with doubled anode size. The volumetric power density of MFC was linearly related to the specific surface area of the cathode, and independent of the fuel cell size or configuration. From an economic view, platinum catalysts in cathode should be replaced by non-precious metal-based catalysts or bacteria catalyst (bio-cathodes) to facilitate cathodic oxygen reduction.

Changing the electrolyte composition can improve power densities in both larger and smaller systems. Increasing the ionic strength of larger MFC (520 mL) from 100 to 300 mM, for example, increased power by 25% [60]. In a 20 L stacked, two-chamber MFC, cathode performance was improved by decreasing the *pH*, aerating the catholyte with pure oxygen instead of air, and increasing the flow rate, resulting in a power density increase to 144 W/m^3 [107]. Balance *pH* by complete liquid loop over cathode and anode [111], or acidifying the catholyte of two-chamber MFCs [110] also increased power generation. However, possible cathode contamination by microbes and organics from anode effluent, additional demand of acidating agents, and the need of a membrane are factors which should not be overlooked.

Reducing electrode spacing can proportionally decrease the electrolyte resistance, thus enhancing the performance of MFCs. It can also increase the ratio of the electrode surface area/volume and in turn maximise the volumetric power density. However, possible short circuit and increased oxygen diffusion to the anode would decrease power output when the electrodes get too close. So a separator which prevents electrode contact and oxygen diffusion is needed to keep the electrodes spacing and the internal resistance small. The separator can inhibit proton transfer and lead to *pH* gradients between the electrode chambers, thus increase the internal resistance. These offsetting needs to reduce oxygen transport and facilitate proton transport makes it difficult to design separators. Low-cost cloths have been advised to replace expensive membranes as an effective separator for large-scale applications. A cloth-electrode assembly configuration has proven to reduce the anode–cathode spacing to 0.6 mm and greatly enhances the power generation [112]. In addition, separator deformation and pollution by biofilm growth may occur over time, which will adversely affect the separator performance, so long-term stability should be considered in the design of separators.

Cost reduction is another challenge for the scaling up of MFCs. This can be achieved by highly efficient but less-expensive anode, cathode and separator materials. With a simple configuration and effective current collection, electrodes that contain current collectors are now considered suitable for practical applications. Among the various anode materials, one of the most promising is the graphite fibre brush, which is made by incorporating graphite fibres into a non-corrosive metal core (certain stainless steels or titanium). Metals such as tungsten and stainless steel can also be used in brush forms. Another promising anode material is AC granule, especially when it is linked to a metal mesh current collector. Anode chamber stacked with AC has a high specific surface area (area per mass) for bacterial growth and electricity production. AC is also an excellent cathode material for either chemical or biological cathodes. Even though AC is relatively poor in oxygen reduction compared to Pt-catalysed carbon cloth materials, this is compensated by its large surface area, thereby producing a high volumetric power density [35]. However, much remains to be known about the distribution of microbes, and proton and electron transfer mechanisms inside the AC stack in both anode and cathode, and the fuel cell configuration needs to be optimised for a better performance.

As discussed above, large-scale applications of MFCs would require a high power output (achieved by effective electrodes, low electrode spacing, high solution conductivity and constant solution *pH*), simple fuel cell configuration, low construction cost and long-term stability. With continual efforts put into the search for cost-effective materials and optimal designs, MFCs have potential to become commercially available in the near future.

2.5 Types of Microbial Fuel Cells

MFCs can produce electricity from various organic sources, such as wastewaters, slugs and soils. According to the substrate resources, MFCs are categorised into three types: (i) MFCs producing electricity from wastewaters,

(ii) sediment microbial fuel cells (SMFCs) and (iii) plant microbial fuel cells (PMFCs).

2.5.1 Microbial Fuel Cells Producing Electricity from Wastewaters

The most important application of MFCs is to recover electricity from aqueous organic matter (either synthetic or natural wastewaters). A large variety of fuel cell configurations, electrode materials, catholytes and operating modes were developed for a high electricity production and biomass removal. In this section, we will introduce some typical MFCs primarily based on the catholytes used and the operational modes.

Several compounds with high redox potentials, including oxygen, ferricyanide and permanganate and some high-valence ions (e.g. Fe^{2+}, Cu^{2+}, NO_3^-, NO_4^{2-}) have been used as the cathodic electron acceptors. MFCs using oxygen in the air as the terminal oxidant are referred to as 'air-cathode MFCs', which is the most studied fuel cell type to date for their feasibility in large-scale application.

2.5.1.1 Air Cathode Microbial Fuel Cells

One of the most popular air-cathode MFC configuration is a single-chamber tubular fuel cell, which was originally developed at the Department of Civil and Environmental Engineering, the Pennsylvania State University [65]. The photograph for this fuel cell configuration is shown in Figure 2.9a. Simply,

---

FIGURE 2.9 (continued)
(See colour insert.) Configurations of MFCs: (a) single-chamber air-cathode tube-type system; (b) dual-chamber air-cathode tube-type system with a membrane placed in the middle of the chamber; (c) single-chamber air-cathode bottle reactor MFC; (d) dual-chamber flat-plate fuel cell with a CEM placed between the two electrodes and wastewater, with air flowing on opposite sides of the electrodes; (e) dual-chamber H-type system showing anode and cathode chambers equipped for gas aeration; (f) single-chamber air-cathode fuel cell operating in an a continuous up-flow mode; (g) parallel stacked MFC containing six individual cells. (Reprinted with permissions from *Water Res.*, 39, B.E. Logan et al., Electricity generation from cysteine in a microbial fuel cell, 2005, 942–952; *Bioresource Technol.*, 102, A.E. Tugtas, P. Cavdar and B. Calli, Continuous flow membrane-less air cathode microbial fuel cell with spunbonded olefin diffusion layer, 2011, 10425–10430; H. Liu and B.E. Logan, Electricity generation using an air-cathode single chamber microbial fuel cell in the presence and absence of a proton exchange membrane, *Environ. Sci. Technol.*, 38, 2004, 4040–4046; J.R. Kim et al., Power generation using different cation, anion, and ultrafiltration membranes in microbial fuel cells, *Environ. Sci. Technol.*, 41, 2007, 1004–1009; B. Logan et al., Graphite fiber brush anodes for increased power production in air-cathode microbial fuel cells, *Environ. Sci. Technol.*, 41, 2007, 3341–3346; P. Aelterman et al., Continuous electricity generation at high voltages and currents using stacked microbial fuel cells, *Environ. Sci. Technol.*, 40, 2006, 3388–3394; B. Min and B.E. Logan, Continuous electricity generation from domestic wastewater and organic substrates in a flat plate microbial fuel cell, *Environ. Sci. Technol.*, 38, 2004, 5809–5814. Copyright (2005, 2011), Elsevier and (2004, 2006, 2007), The American Chemical Society.)

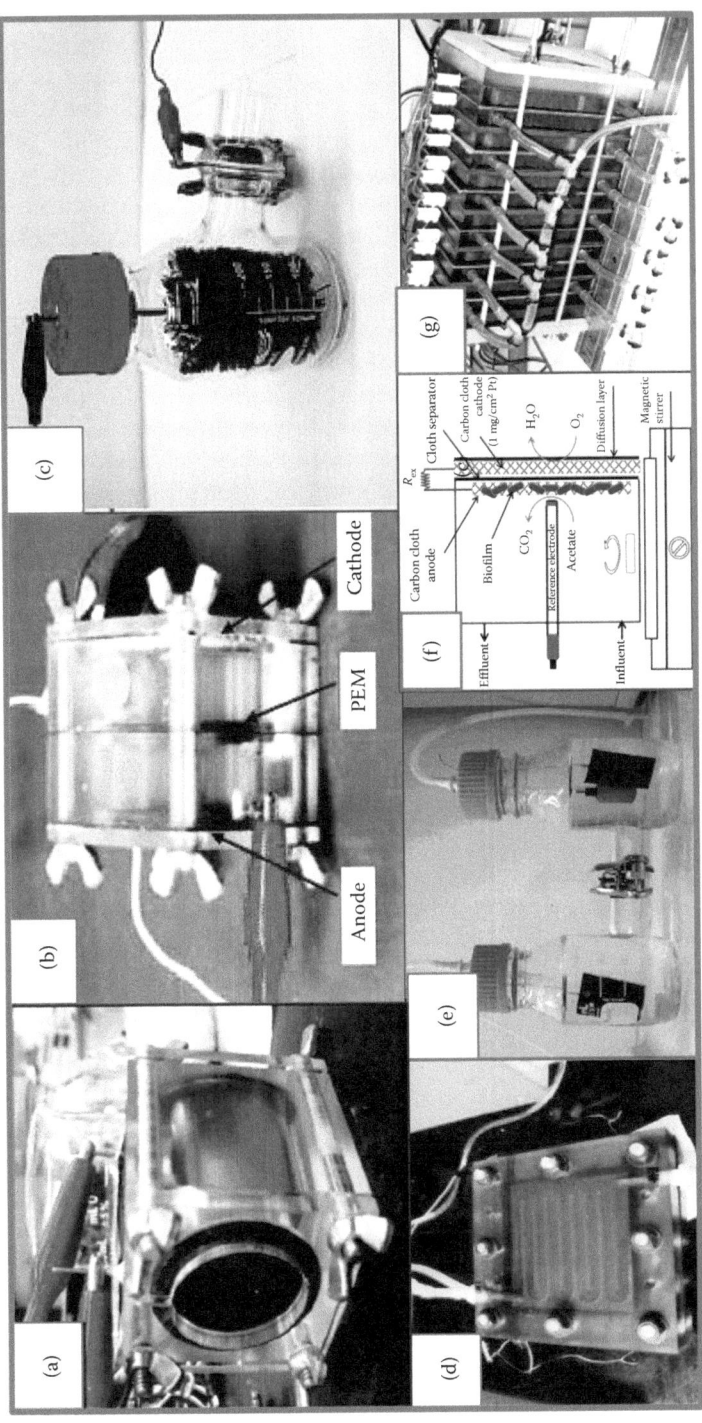

FIGURE 2.9 (continued)

two electrodes are placed on the opposite ends of a cube chamber. The anode (carbon paper, cloth or mesh, or graphite brush) is totally immersed in the electrolyte, and one side of the cathode is faced to the same electrolyte with the other side exposed to open air. The cathode usually consists of a diffusion layer (exposed to air), a conductive supporting material, and a catalyst/binder layer (exposed to water). A cation exchange membrane (CEM) was coated on the internal side of the cathode to prevent oxygen diffusion to the anode. However, CEM was demonstrated in later studies to adversely affect the cathode potential. The simple design makes this fuel cell convenient to examine various factors that affect power production, such as the effects of substrate type, electrode spacing, solution ionic strength, operating temperature, cathode catalysts, anode material and so forth.

Other configurations for air-cathode MFCs include [60] (i) dual-chamber fuel cell, designed to study the effect of different membranes on internal resistance (Figure 2.9b); (ii) bottle fuel cell, for pure/mix culture study and cultivation (Figure 2.9c); (iii) flat plate fuel cell, for a higher MFC performance (Figure 2.9d); (iv) dual-chamber H-type fuel cell for obtaining different electrolyte conditions for the anode and cathode chambers (Figure 2.9e); and other types designed for a specific purpose. With the simple design and cost saving of single-chamber air-cathode MFCs, they have been modified to construct large-size MFCs [82,107,110], which provide valuable knowledge for large-scale applications of MFCs in both wastewater treatment and power generation.

Air can also be aerated into the cathode to provide dissolved oxygen as the terminal oxidant. Such fuel cells are called 'aqueous air-cathode MFCs'. A CEM or an ultrafiltration membrane is necessary to keep the anode from oxygen, giving a two-chamber configuration. However, the performances of these systems are generally low compared to 'air-cathode MFCs', limited by the low DO concentration and high internal resistance. The two-chamber configuration is also adaptable when ferricyanide or permanganate is used as the final electron acceptor. Although MFCs using ferricyanide and permanganate can produce a higher power output than that using oxygen [79] due to the higher redox potential and low system resistance, but they are toxic to the environment and have to be chemically regenerated. Thus, their application is only restricted to fundamental laboratory studies.

2.5.1.2 Bio-Cathode Microbial Fuel Cells

Recently, a new type of MFCs called 'bio-cathode MFCs' rapidly emerges as an attractive research topic. This type of MFCs employs bacteria as the cathodic catalysts to catalyse the reduction of oxygen (aerobic microbes) or other compounds with a high redox potential, such as nitrate and sulphate (anaerobic microbes). Bio-cathode MFCs are generally constructed in a dual-chamber form, with the anode biofilm separated from oxygen or other oxidants by a membrane, thus to ensure a high-power generation and

Coulombic efficiency. The major advantages of these systems are their relatively low cost, good stability and multiple functions for wastewater treatment and biosynthesis. Bio-cathode MFCs show great promise for large-scale applications due to their low-cost and poisoning-resistant nature, but so far, these systems have required the use of DO rather than air, and the fuel cell architecture has not been optimised.

2.5.1.3 Continuous Flow Modes and Cell Stacks

While most MFCs studied on the laboratory scale are in fed-batch mode, a continuous flow mode is considered more suitable for practical applications such as wastewater treatment. Various flow modes, such as up-flow, down-flow and over-flow modes, have been examined for their availability for continuous power production. For example, Tugtas et al. [36] investigated the power production performance of a membrane-less air-cathode MFC fed with acetate in a continuous up-flow mode (Figure 2.9f). A maximum power density of 750 mW m^{-2} was obtained initially, which gradually decreased to 280 mW m^{-2} after 53 days operation as a result of a gradual decrease in the cathode potential. Individual MFCs stacked together in a series are expected to obtain higher overall voltages. However, the fluctuations of biological systems, power generation and cell voltage may be adversely affected when individual cells are joined together. Aelterman et al. [113] examined power generation in a six-cell stack MFCs with acetate as the substrate and ferricyanide as the catholyte (Figure 2.9 g). The connection of these 6 MFC units in a series enabled increased voltages to 2.02 V at a current of 41.0 mA, while retaining high power output at 228 W m^{-3}. The Coulombic efficiency for the stack was only 12% and voltage reversal phenomenon occurred over time. In fact, voltage reversal remains the biggest obstacle for the increase of voltage in stack systems yet to be resolved.

2.5.1.4 Microbial Fuel Cells for Removing Specific Pollutants from Wastewaters

MFCs using certain microbes have a special ability to remove nitrogen and sulphates, as required urgently for most wastewater treatment. In most wastewaters, nitrogen exists in the reduced forms of ammonia (NH_3 or NH_4^+) and organic nitrogen. To remove reduced nitrogen, organic nitrogen is ammonified and ammonia is first oxidised to nitrite and/or nitrate by aerobic nitrifying bacteria, then NO_2^- and/or NO_3^- are reduced to N_2 by anaerobic denitrifying bacteria. In MFCs, simultaneous nitrification and denitrification can be achieved through an optimised design. With pre-enriched nitrified biofilm on the air cathode of a single-chamber MFC [116], an ammonia removal efficiency of up to 97%, and a maximum power density of 900 ± 25 mW m^{-2} were achieved. The maximum total nitrogen removal efficiency reached 94%.

Many studies have revealed that the exoelectrogenic, sulphur-oxidising and sulphate-reducing bacteria in MFCs can harvest electricity during sulphide oxidation [117,118]. By employing a pure culture (*Pseudomonas* sp. C27) to start up a dual-chamber MFC, Lee et al. [117] obtained a maximum power density of 29.3 mW m^{-2} using sulphide as the sole electron donor. Based on the high diversity and activity of anodic microorganisms, MFCs are capable of using a wide range of refractory organic pollutants as the substrate to produce electricity. Such examples include antibiotic compounds [119], pentachlorophenol [120], various dyes [121], pyridine [122], pharmaceutical wastes [123] and so forth. MFCs perform superbly in removing pollutants and generating electricity from wastewaters simultaneously, especially in the presence of acetate or glucose. For example, Wen et al. [119] demonstrated that 91% of the ceftriaxone sodium was biodegraded within 24 h in a single chamber air-cathode MFC. The maximum power density reached 113 W m^{-3} for a 50 mg L^{-1} ceftriaxone sodium + 1000 mg L^{-1} glucose mixed substrate, while the values were only 19 and 11 W m^{-3}, respectively when glucose and ceftriaxone sodium were used as the sole fuel. High valence heavy metal contaminants, such as U(VI), Cr(VI), Cu(II), V(V) and Hg^{2+}, can accept electrons and thus be reduced at the cathode of MFCs. Using a two-chamber glucose-fed MFC, Zhang et al. [124] observed that 68 ± 3.1% of V(V) and 75 ± 1.9% of Cr(VI) (each with initial concentration of 250 mg L^{-1}) could be reduced after a 240 h operation, with a maximum power density of 970 ± 20.6 mW m^{-2}. Chromium was reduced to Cr(III) and deposited on the cathode surface, while most of vanadium can be precipitated from the exhausted catholyte by adjusting the *pH*.

2.5.2 Sediments Microbial Fuel Cells (SMFCs)

Reimers et al. [125] first developed an MFC that can produce electricity from the organic matter in marine sediments. Following this work, several studies have been conducted to produce electricity via harnessing bacterial processes in the sediments of marines, fresh/salt water lakes, riverine sites, septic tanks or other solid sources. These systems are referred to as benthic unattended generators, benthic microbial fuels, or SMFCs. SMFCs can be used to power electronic devices in remote and inaccessible locations, such as the bottom of the seafloor, where it would be expensive and technically difficult to routinely employ traditional batteries. Typically, SMFC consists of an anode embedded in anaerobic sediments connected to a cathode suspended in the overlying aerobic water (Figure 2.10). In the anode, organic matter and other reductive compounds rich in the sediments are utilised by anaerobic bacteria to produce electrons and protons, which are then consumed by the oxygen dissolved in the overlying water at the cathode. The high salinity of the seawater provides good ion conductivity between the electrodes.

Molecular analysis of the microbial community on anode surfaces revealed an obvious predominance of bacteria in the family Geobacteraceae

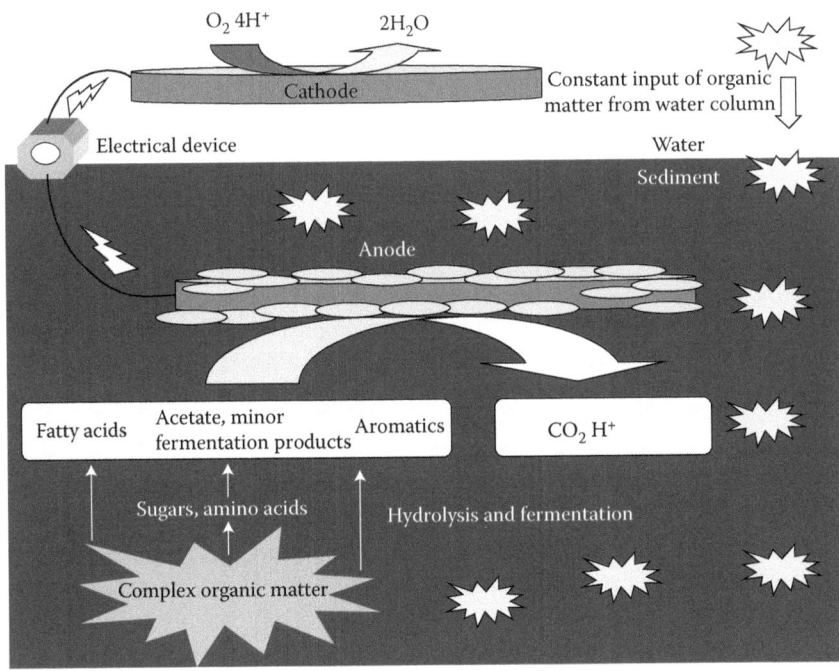

FIGURE 2.10
(**See colour insert.**) Illustration of the mechanism for electricity generation in SMFCs. (Reprinted with permission from *Curr. Opin. Biotechnol*, 17, D.R. Lovley, Microbial fuel cells: Novel microbial physiologies and engineering approaches, 327–332. Copyright 2006, Elsevier.)

in most SMFCs. In marine sediments, *Desulfuromonas* species, which prefer marine salinities, were most abundant on the anodes; whereas *Geobacter* species, which prefer fresh water, predominated on electrodes in freshwater sediments or swine waste. Geobacteraceae are capable of completely oxidising acetate and other simple organic compounds to carbon dioxide to conserve energy and support growth, while simultaneously transport electrons to an electrode. They cannot, however, effectively utilise complex compounds. Instead, some symbiotic microorganisms in the sediments, such as methanogenes and fermentative bacteria, would break down these complex organic assemblages, to produce fermentation products (such as acetate), aromatic compounds and fatty acids which are readily available for the Geobacteraceae. Besides organic carbon sources, inorganic sulphides are also important substrates for SMFCs. In some sulphide-rich sediments, sulphate-reducing bacteria (e.g. *Desulfobulbus* spp and *Desulfocapsa* spp) were found to predominate on the anodes [126]. Sulfides most likely first react with the electrode to form S^0, which is then oxidised by the bacteria to sulphate with the anode serving as the electron acceptor [126].

Structure and active material of the electrodes determine the overall power output of an SMFC to a great extent. Power densities reported for SMFCs are usually normalised to the anode surface area, since the fuel cell volumes are undetectable in most cases. The first SMFC which was built with Pt anode and Pt cathode provided a power density of 15 mW m^{-2} [125]. SMFCs built with graphite disks electrodes generally produced a maximum power density of 18–32 mW m^{-2}, depending on the geochemical characteristics of the site [42]. Other electrode materials include carbon cloth, stainless steel and carbon fibre brush.

Modification of the electrode can effectively improve the system power output. Lowy and Tender [10] modified glassy carbon graphite anodes through adsorption of anthraquinone-1,6-disulphonic acid (AQDS) or incorporating with Sb(V) complex, or through oxidation. The modifications resulted in a 1.9–218 times greater anodic kinetic activities coupled with an 8–60 mV higher anode potential than the plain graphite. The highest kinetic activity (218 times) was obtained by the graphite anode modified with both oxidation and AQDS. The improved performance of the modified anodes may reside in a greater contact surface area to the substrate, which provides more efficient electrocatalysis reaction.

Scott et al. [127] reported that modification of carbon cathode with Fe-Cotetramethoxyphenylporphyrin (FeCoTMPP) improved the maximum power densities of an SMFC from 30 to 62 mW m^{-2}. He et al. [128] observed that application of a rotating cathode in a river SMFC increased the oxygen accessibility to the cathode, and therefore improved the cathode reaction rate, resulting in a higher power production (49 mW m^{-2}) compared to a non-rotating cathode system (29 mW m^{-2}). The performance of SMFCs can also be promoted by pretreatment or incorporating biodegradable fuels into the sediment, since the property or concentration of organic matter in sediments may affect the bacterial metabolism. Improved maximum power density from 2 mW m^{-2} (control) to 84 ± 10 mW m^{-2}, and 83 ± 3 mW m^{-2} were observed after incorporating Chintin 80 and Cellulose, respectively into the sediments of a laboratory SMFC [129].

The long-term stability should be considered when evaluating the performance of an SMFC. The system performance may be restrained after a period of operation, due to an insufficient supply of substrate or electrode poisoning by sulphur deposition or other reasons. For example, in a SMFC field test [130], a maximum sustained power output of 34 mW m^{-2} was obtained over the first 26 days of operation, but reduced to less than 6 mW m^{-2} in the next 72 days.

2.5.3 Plant–Microbial Fuel Cells (PMFCs)

PMFCs are newly emerged MFCs, in which electricity can be generated by microorganisms that use root exudates of aquatic plants as fuel. A typical configuration of a PMFC is illustrated in Figure 2.11. The anode is submerged in a support matrix near the plant roots to facilitate substrates uptake by the

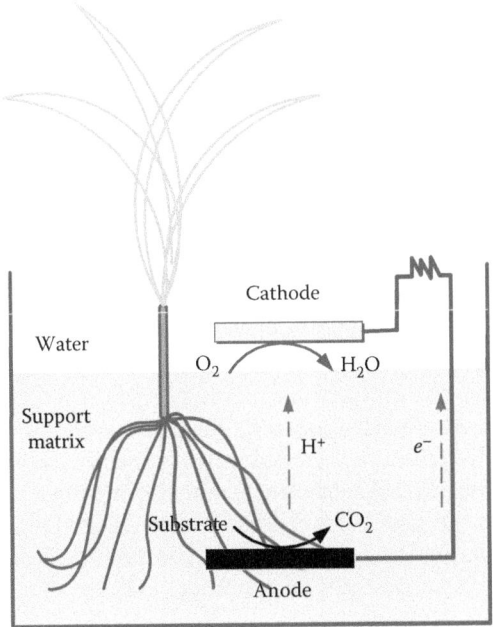

FIGURE 2.11
Schematic illustration of the electricity production mechanism in plant MFCs.

microorganisms. The electricity-producing microbes oxidise the substrates into CO_2, while producing electrons and protons, which are then transferred to the cathode respectively via the external circuit and aqueous environment. The cathode is usually located in overlying water to use oxygen from air, or in the rhizosphere to use oxygen released from the roots as the electron acceptor. A membrane is generally unnecessary for PMFCs, since the oxygen concentration declines over the depth of water and sediment, naturally forming an anoxic environment for the anodic bacteria.

Various aquatic plants, such as reed manna grass (*Glyceria maxima*), rice-plant, *Arundodonax* and *Anisogrammaanomala*, have been used for PMFCs. Water-soluble root exudates of these plants, including carbohydrates, carboxylic acids and amino acids serve as the most important electron donors for electricity production. In addition, soil (if present) can also act as an electron source via chemical or anaerobic respiration processes, such as chemical oxidation of humic acids, iron(II), and sulphur compounds, and microbial oxidation of sulphur and ammonia by specific microbes.

To date, the power generated by PMFCs keeps relatively low. Latest studies show that PMFCs with freshwater plants can generate a maximum power density ranging from several to dozens of mW m^{-2}, whereas that with the seawater plant *Spartinaanglica* can generate a maximum power density of 222 mW m^{-2} [131]. Challenges for the performance improvement include reducing the system internal resistance and improving reaction activities on both electrodes. The support matrix (e.g. flooded soil, vermiculite or graphite granules) is a major contributor to the internal resistance, which interferes with the migration of H^+ between the electrodes and the diffusion of the root exudates to the anode surface. Treatment of anode and cathode materials is promising to increase power generation. Anode modification with certain functional groups would facilitate electron transfer from the bacteria to the electrode surface, thus reducing the anode overpotential [97]. Chemical or biocatalysts, such as platinum, phthalocyanine and microbes, developed on the cathode surface could reduce the cathodic activation energy, thus increasing the overall power output.

A major advantage of PMFCs is sustainability, as the carbon substrate is continuously supplied by plant roots. Since root exudates are originally generated through the photosynthesis process, PMFCs would ultimately convert solar energy into electrical power, thereby reducing methane emissions. As much as 50% less methane emission was observed after submerging an anode in rice field soil [132], with the electrochemically active microbes and methanotrophs in the anode. Studies on wastewater MFCs have shown that ammonium can be oxidised in the anode chamber [133] and nitrate can be reduced to nitrogen in the cathode chamber [134], leading to the removal of nitrogen pollutants from wastewaters. Therefore, PMFCs could be an effective tool to accelerate the removal of nitrate and ammonium in wetland areas.

2.6 Other Microbial–Electrochemical Conversion Devices

2.6.1 Microbial Electrolysis Cells (MECs)

With minor modifications, MFCs can be used to produce hydrogen instead of electricity. In an MFC, the anode potential can approach the theoretical limit of ~-0.3 V for a variety of substrates. If we wish to generate hydrogen at the cathode, a cathode potential of -0.414 V (pH 7, 298 K) must be overcome, which is more negative than the anode potential. Therefore, the overall reaction is not spontaneous. Adding at least 0.114 V [$-0.3 - (-0.414)$] to the circuit can help to decrease the cathode potential and overcome the thermodynamic barrier. In this mode, hydrogen production can be achieved at the cathode, via combining protons and electrons generated from the anodic reaction. MFCs used for hydrogen production is called 'microbial electrolysis cell (MEC)', or alternatively biocatalysed electrolysis systems, or bioelectrochemically assisted microbial reactors (BEAMRs). Various overpotentials at the cathode, practically an external voltage much higher than 0.114 V, generally ~0.25 V or more, must be applied to the circuit to obtain a reasonable current density and hydrogen production rate. Despite this, MECs still remain as a promising technique for hydrogen production over the traditional water electrolysis, which generally needs a voltage input about 1.8–2.0 V.

A typical two-chamber MEC structure is shown in Figure 2.12. MECs function in a similar mode as an air-cathode MFC, except that hydrogen is generated at the cathode instead of water, and an external power source is added. A membrane (either cation or anion exchange membrane, or gaseous diffusion membrane) is essential for most MEC to prevent hydrogen from diffusing back to the anode. In some cases, single-chamber MECs without membrane may also be used to minimise energy losses [135]. Indicators to evaluate the performance of MEC generally include: hydrogen yield, hydrogen recovery efficiency and energy recovery efficiency. A detailed calculation for these parameters can be referred to Logan's paper [60].

During MEC operation, hydrogen may be lost from the cathode chamber, decreasing the hydrogen yield and recovery efficiency. The loss in hydrogen can occur through several processes: (i) diffusion to the anode chamber through the membrane; (ii) abiotic conversion of hydrogen into methane; and (iii) degradation of hydrogen by some cathodic microbes. Thus, it is a challenging task to reduce hydrogen loss to achieve high MEC performance. Another challenge for the application of MECs is the requirement of an external energy supply to increase hydrogen production. The theoretical minimal applied voltage for H_2 production in MFC is 0.114 V. However, in practical MFCs this value increases to 1.8–2.0 V due to the various energy losses inside the MEC system. To achieve a positive energy balance from an MEC, the applied voltage needs to be lower than ~0.6 V [136].

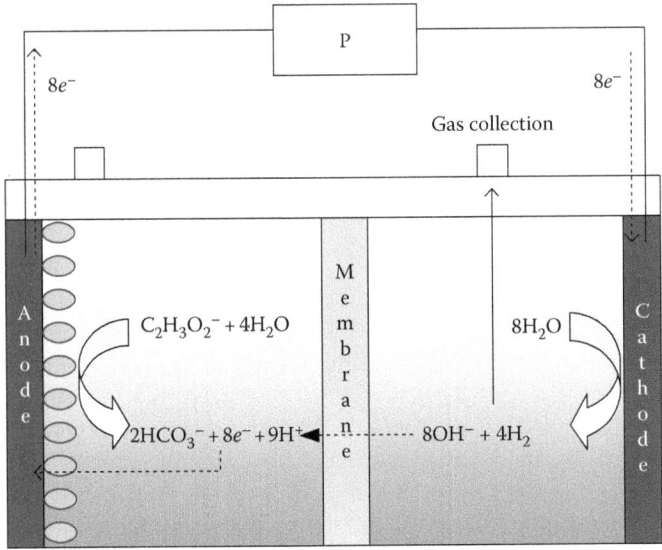

FIGURE 2.12
Schematic illustration of a typical two-chamber MEC.

A high H_2 production rate and high H_2 purity would make the process more sustainable and economical. Relatively high H_2 recovery efficiencies, ranging from 60% to 96% can be obtained for diverse number of donor substrates (e.g. cellulose, glucose, butyrate, lactate, propionate, ethanol or acetate) [61,135]. A high H_2 recovery efficiency of 96% was obtained in a single-chamber MEC using acetate as the substrate at an applied voltage of 0.8 V [135]. This will maximise the desired energy output and minimise the BOD of the effluent. In future applications, MECs may combine with other techniques, such as dark fermentation, to achieve a greater Bio-H_2 yield than an MEC alone. The fermentation process can degrade complex biomass into simple organic products that can be utilised by MECs. Such combinations will make it possible to effectively produce H_2 from complex organic compounds from a variety of wastewaters, such as animal wastes or farming residues.

2.6.2 Microbial Sensors

The MFC technology can be used as a sensor for pollutant analysis and *in situ* process monitoring and control. As mentioned earlier in Section 2.2.2, the electrochemical activity of microorganisms in the anode of MFCs is sensitive to a variety of electrolyte parameters, any changes would affect the microbial activity, thus the electrical current output and Coulombic yield. This makes it possible for MFCs to act as a sensor to detect these parameters. Several types of MFC-based biosensors have been developed for monitoring the biochemical oxygen demand (BOD) [137], microbial activity [138], solution *pH* [139] and toxicity [140] in wastewaters, and showed good stability, accuracy, and wide detection range. A major advantage of MFC-based sensors over other bio-sensors is that they do not require a transducer to translate the signal given by the organisms into an electrical signal. Therefore, MFC-based biosensors are suitable for online detections for the simple structure of high stability [139].

2.6.3 Bacterial Batteries

It has been observed that a number of different bacteria are capable of inhibiting the corrosion of different metals and alloys in corrosive environments. For instance, pitting of Al 2024, rusting of mild steel and tarnishing of brass in artificial seawater can be avoided in the presence of *Shewanella algae*, *Shewanellaana* and *Bacillus subtilis* [141,142]. However, the corrosion protection mechanisms differ for different bacteria. The corrosion potential (E_{corr}) could be shifted in a negative direction by one type of bacteria (e.g. *Shewanella* strains), but also in the positive direction by another bacteria (e.g. *B. subtilis*). These observations provide a basis for the concept of a 'bacterial battery', in which one or both electrode(s) contains certain bacteria, leading to a larger cell voltage than the same battery without microbes.

The concept of bacterial battery was first evaluated by Kuş et al. [143], in whose work *Shewanella oneidensis* MR-1 was added in the growth medium of a galvanic cell containing pure Al 2024 and Cu electrodes (Figure 2.13). As shown in Figure 2.14a, the power output of the cell without bacteria reached its maxima in the first day and dropped drastically in the following days. Instead, the cell voltage and power of the cell with bacteria continuously increased for up to 90 days (Figure 2.14b), and remained almost constant for 200 days. Analysis of the electrochemical impedance spectra (EIS) suggested that a biofilm formed on the Cu surface became thicker and less porous with time. How the bacteria interact with the copper electrode and why they increased the power output still remain unclear. It was presumed that bacterial battery could generate certain oxides to increase the E_{corr} of copper and, thus, the overall power density, as evident by two reduction peaks in the cyclic voltammetry curves of the coated copper [144].

Recently, a membrane was used to separate the anode and cathode compartments, and *S. oneidensis* MR-1 and other bacteria are added in both or only one of the electrode chambers [145]. It was observed that the electrochemical behavior of the copper electrode was significantly affected by the exposure condition. A second time constant was observed in the impedance spectra of copper that was partially immersed in the electrolyte, this can be attributed to the formation of a porous biofilm in the air/liquid/metal interface. Complete immersion of the electrode or deaeration of the electrolyte resulted in one-time-constant spectra. The corrosion potential increased with time for the Cu electrode partially immersed in the electrolyte, while it decreased for the other two exposure conditions. The mechanism on how the exposure condition affects electrode performance in bacterial batteries remains unclear [145].

While bacterial batteries hold a great promise for promoting the power output and long-term stability of galvanic cells, it has to be realised that,

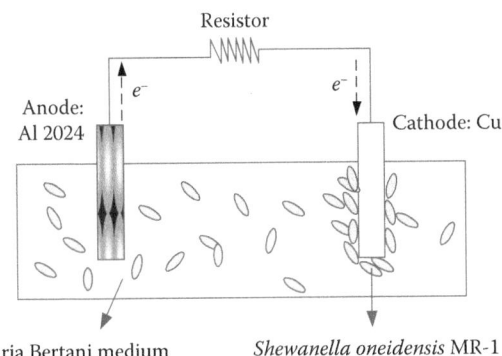

FIGURE 2.13
Schematic diagram of the bacterial battery developed by Kuş et al. (Reprinted with permission from *Corros. Sci.*, 47, E. Kuş et al., The concept of the bacterial battery, 1063–1069, Copyright 2005, Elsevier.)

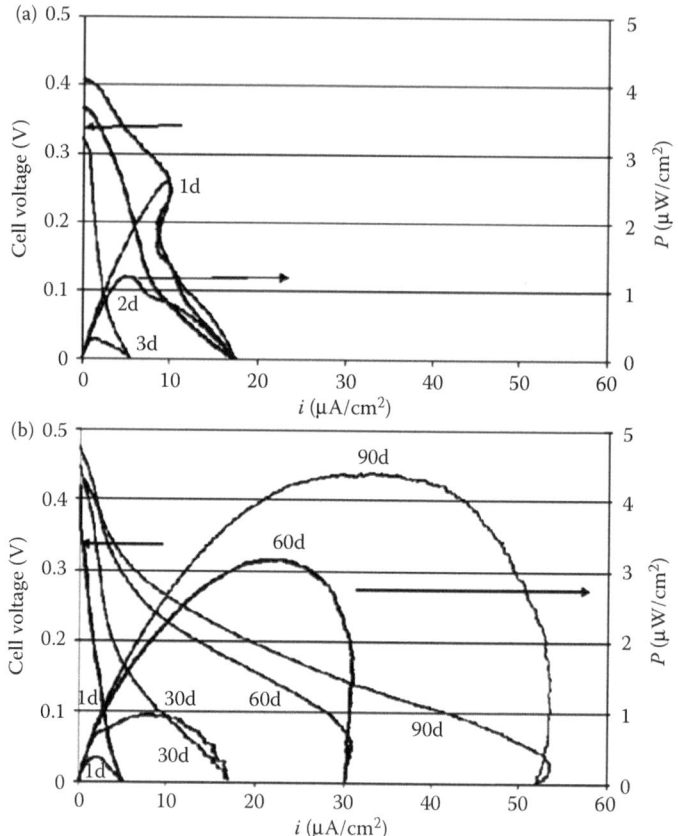

FIGURE 2.14
i–V and i–P curves for the bacterial batteries containing (a) sterile lysogeny broth (LB) medium, and (b) LB + *Shewanella oneidensis* MR-1. (Reprinted with permission from *Corros. Sci*, 47, E. Kuş et al, The concept of the bacterial battery, 1063–1069, Copyright 2005, Elsevier.)

similar to the MFCs, the power output is still low. Increasing the power output level as well as a better understanding of the interaction mechanisms between the bacteria and electrode are needed in future studies.

2.7 Outlook

One of the most attractive applications of MFCs is to produce power from wastewaters at industrial locations, sewage treatment plants or other sites with high quality and reliable influent. The organic compounds in wastewaters can provide sustainable fuels for MFCs. In contrast to conventional wastewater treatment techniques that consume large amounts of electric energy for aeration, MFCs require less power input and can produce energy in the form of electricity or hydrogen. Although the power level produced by MFCs is relatively low at present, advances in technology are promising in achieving a reasonable power level to run a treatment plant [60].

Wastewater aeration and sludge treatment contribute a major operating cost for aerobic systems. Since MFCs are passively aerated, they can yield 50–90% less sludge than a traditional aerobic process, which drastically reduces the energy and cost inputs [108]. Besides, MFCs can operate well at a wide temperature range (4–60°C) over a wide substrate COD loadings. With the low electricity recovery rates of <50% at present, the capital cost of MFCs remains high, limiting their application potentials. Integrating MFCs with traditional anaerobic digestion, up-flow biofilter circuit (UBFC), seems promising for a more effective waste removal and energy production [108]. As MFC technology is continuously improved, MFCs may function more efficiently in a wider range of conditions, either solely or combined with other techniques.

Continuous flow, single-compartment MFCs, and membrane-less MFCs are favored for wastewater treatment due to their feasibility to scale-up [106]. Up to date, tremendous work has been conducted on treating various wastewaters (such as swine, brewery, potato-processing, food-processing and urban wastewaters) by MFCs at a laboratory scale. MFCs have generally been accepted with a benchmark performance of COD removal (20–80%), Coulombic efficiency (10–95%) and electricity generation (0.5–8 A m^{-2}), which vary depending on different wastewater sources, operating conditions and MFC configurations [60]. Large-scale tests will be needed in the future for better understanding of MFC performances at real field applications.

Sediment microbial fuel cells are an excellent precedent of MFC applied to power devices in remote marine and estuarine locations [109]. MFCs hold great promise to produce biofuels from renewable biomass sources. The electricity produced by MFCs can be used for the electrolysis of water to produce hydrogen. Or, more efficiently, hydrogen could be produced directly from biomass sources using the MEC process (see Section 2.6.1).

In addition to applications in electricity and biogas generation, pollutant removal, and microbial sensors, alternative applications of MFCs have also emerged. MFCs have been proposed as a perfect energy supply candidate for autonomous robots by self-feeding different fuels such as sugar, fruit, dead insects, grass and weed [146]. For instance, the robot 'EcoBot-II' solely powers itself by MFCs to perform some behaviours including motion, sensing, computing and communication [147]. MFCs have also been suggested as an application in a spaceship since they can supply electricity by degrading wastes generated on board. A miniature MFC is suggested to be implanted into a human body to power an implantable medical device [146], where MFCs can utilise the nutrients supplied by the human body as a sustainable fuel, to operate the medical devices in the long run.

A major development bottleneck for MFCs is the low process performance when compared to its competitors (e.g. methanogenic anaerobic digesters). In particular, the power output in MFCs systems is relatively low (current maximum 1.55 kW m^{-3}) [80]. The construction of large-scale, highly efficient MFCs is considered to be technically difficult. The high-cost materials used in current MFCs makes it economically unattractive; and further understanding of exoelectrogenic bacteria and their function mechanisms is needed. However, we can see that numerous studies have been performed on MFC technologies in recent years, and many promising improvements for MFCs performance have yielded. To achieve widespread application of MFCs, some recommendations are outlined as below:

i. Larger-scale MFCs to be constructed;
ii. Increase the power output of MFCs;
iii. More low cost and effective electrode materials to be developed;
iv. Reduce the construction cost of MFCs;
v. Better understanding of microorganisms and electron transfer mechanism; and
vi. More MFCs-based systems to be developed, such as hydrogen-production MFCs, sediment MFCs, plant MFCs, and MFCs-based biosensors.

We can expect that with the concerted efforts of more researchers, advances in MFCs will progress more rapidly in the future.

References

1. D.A. Lowy, L.M. Tender, J.G. Zeikus, D.H. Park and D.R. Lovley, Harvesting energy from the marine sediment water interface II: Kinetic activity of anode materials, *Biosens. Bioelectron.* 21, 2006, 2058–2063.
2. J. Wei, P. Liang and X. Huang, Recent progress in electrodes for microbial fuel cells, *Bioresource Technol.* 102, 2011, 9335–9344.
3. M.C. Potter, Electrical effects accompanying the decomposition of organic compounds, *Proc. R. Soc. London, Ser. B* 84, 1911, 260–276.
4. R. Allen and H.P. Bennetto, Microbial fuel-cells: Electricity production from carbohydrates, *Appl. Biochem. Biotechnol.* 39–40, 1993, 27–40.
5. B.H. Kim, D.H. Park, P.K. Shin, I.S. Chang and H.J. Kim, Mediator-less biofuel cell, U.S. Patent 5976719, 1999.
6. U. Schröder, Anodic electron transfer mechanisms in microbial fuel cells and their energy efficiency, *Phys. Chem. Chem. Phys.* 9, 2007, 2619–2629.
7. D.R. Lovley, Powering microbes with electricity: Direct electron transfer from electrodes to microbes, *Environ. Microbiol. Rep.* 3, 2008, 27–35.

8. K.B. Gregory and D.R. Lovley, Remediation and recovery of uranium from contaminated subsurface environments with electrodes, *Environ. Sci. Technol.* 39, 2005, 8943–8947.
9. O. Schaetzle, F. Barriere and K. Baronian, Bacteria and yeasts as catalysts in microbial fuel cells: Electron transfer from micro-organisms to electrodes for green electricity, *Energy Environ. Sci.* 1, 2008, 607–620.
10. D.A. Lowy and L.M. Tender, Harvesting energy from the marine sediment–water interface: III. Kinetic activity of quinone- and antimony-based anode materials, *J. Power Sources* 185, 2008, 70–75.
11. R.K. Thauer, K. Jungermann and R.K. Decke, Energy conservation in chemotrophic anaerobic bacteria, *Bacteriol. Rev.* 41, 1977, 100.
12. Z. He and L.T. Angenent, Application of bacterial biocathodes in microbial fuel cells, *Electroanalysis* 18, 2006, 2009–2015.
13. H.J. Kim, H.S. Park, M.S. Hyun, I.S. Chang, M. Kim and B.H. Kim, A mediator-less microbial fuel cell using a metal reducing bacterium, *Shewanella putrefaciens*, *Enzyme Microb. Technol.* 30, 2002, 145–152.
14. B.R. Ringeisen, E. Henderson, P.K. Wu, J. Pietron, R. Ray, B. Little, J.C. Biffinger and J.M. Jones-Meehan, High power density from a miniature microbial fuel cell using *Shewanella oneidensis* DSP10, *Environ. Sci. Technol.* 40, 2006, 2629–2634.
15. C.A. Pham, S.J. Jung, N.T. Phung, J. Lee, I.S. Chang, B.H. Kim, H. Yi and J. Chun, A novel electrochemically active and Fe(III)-reducing bacterium phylogenetically related to *Aeromonas hydrophila*, isolated from a microbial fuel cell, *FEMS Microbiol. Lett.* 223, 2003, 129–134.
16. Y. Qiao, C.M. Li, S.-J. Bao, Z. Lu and Y. Hong, Direct electrochemistry and electrocatalytic mechanism of evolved *Escherichia coli* cells in microbial fuel cells, *Chem. Commun.*, 11, 2008, 1290–1292.
17. K. Rabaey, N. Boon, S.D. Siciliano, M. Verhaege and W. Verstraete, Biofuel cells select for microbial Consortia that self-mediate electron transfer, *Appl. Environ. Microbiol.* 70, 2004, 5373–5382.
18. D.R. Bond, D.E. Holmes, L.M. Tender and D.R. Lovley, Electrode-reducing microorganisms that harvest energy from marine sediments, *Science* 295, 2002, 483–485.
19. D.E. Holmes, D.R. Bond and D.R. Lovley, Electron transfer by *Desulfobulbus propionicus* to Fe(III) and graphite electrodes, *Appl. Environ. Microbiol.* 70, 2004, 1234–1237.
20. B. Min, S. Cheng and B.E. Logan, Electricity generation using membrane and salt bridge microbial fuel cells, *Water Res.* 39, 2005, 1675–1686.
21. D.R. Bond and D.R. Lovley, Electricity production by *Geobacter sulfurreducens* attached to electrodes, *Appl. Environ. Microbiol.* 69, 2003, 1548–1555.
22. S.K. Chaudhuri and D.R. Lovley, Electricity generation by direct oxidation of glucose in mediatorless microbial fuel cells, *Nat. Biotech.* 21, 2003, 1229–1232.
23. D. Xing, Y. Zuo, S. Cheng, J.M. Regan and B.E. Logan, Electricity generation by *Rhodopseudomonas palustris* DX-1, *Environ. Sci. Technol.* 42, 2008, 4146–4151.
24. Y. Zuo, D. Xing, J.M. Regan and B.E. Logan, Isolation of the exoelectrogenic bacterium *Ochrobactrum anthropi* YZ-1 by using a U-tube microbial fuel cell, *Appl. Environ. Microbiol.* 74, 2008, 3130–3137.
25. A. Borole, H. O'Neill, C. Tsouris and S. Cesar, A microbial fuel cell operating at low pH using the acidophile *Acidiphilium cryptum*, *Biotechnol. Lett.* 30, 2008, 1367–1372.

26. H.S. Park, B.H. Kim, H.S. Kim, H.J. Kim, G.T. Kim, M. Kim, I.S. Chang, Y.K. Park and H.I. Chang, A novel electrochemically active and Fe(III)-reducing bacterium phylogenetically related to *Clostridium butyricum* isolated from a microbial fuel cell, *Anaerobe* 7, 2001, 297–306.
27. D.R. Bond and D.R. Lovley, Evidence for involvement of an electron shuttle in electricity generation by *Geothrix fermentans*, *Appl. Environ. Microbiol.* 71, 2005, 2186–2189.
28. A.L. Walker and C.W. Walker Jr, Biological fuel cell and an application as a reserve power source, *J. Power Sources* 160, 2006, 123–129.
29. D. Prasad, S. Arun, M. Murugesan, S. Padmanaban, R.S. Satyanarayanan, S. Berchmans and V. Yegnaraman, Direct electron transfer with yeast cells and construction of a mediatorless microbial fuel cell, *Biosens. Bioelectron.* 22, 2007, 2604–2610.
30. D.P. Park and J.Z. Zeikus, Impact of electrode composition on electricity generation in a single-compartment fuel cell using *Shewanella putrefaciens*, *Appl. Microbiol. Biotechnol.* 59, 2002, 58–61.
31. S. Cheng and B.E. Logan, Ammonia treatment of carbon cloth anodes to enhance power generation of microbial fuel cells, *Electrochem. Commun.* 9, 2007, 492–496.
32. X. Wang, S. Cheng, Y. Feng, M.D. Merrill, T. Saito and B.E. Logan, Use of carbon mesh anodes and the effect of different pretreatment methods on power production in microbial fuel cells, *Environ. Sci. Technol.* 43, 2009, 6870–6874.
33. X. Wang, Y.J. Feng and H. Lee, Electricity production from beer brewery wastewater using single chamber microbial fuel cell, *Water Sci. Technol.* 57, 2008, 1117–1121.
34. Y. Feng, Q. Yang, X. Wang and B.E. Logan, Treatment of carbon fiber brush anodes for improving power generation in air–cathode microbial fuel cells, *J. Power Sources* 195, 2010, 1841–1844.
35. F. Zhang, S. Cheng, D. Pant, G.V. Bogaert and B.E. Logan, Power generation using an activated carbon and metal mesh cathode in a microbial fuel cell, *Electrochem. Commun.* 11, 2009, 2177–2179.
36. A.E. Tugtas, P. Cavdar and B. Calli, Continuous flow membrane-less air cathode microbial fuel cell with spunbonded olefin diffusion layer, *Bioresource Technol.* 102, 2011, 10425–10430.
37. J. Lee, N.T. Phung, I.S. Chang, B.H. Kim and H.C. Sung, Use of acetate for enrichment of electrochemically active microorganisms and their 16S rDNA analyses, *FEMS Microbiol. Lett.* 223, 2003, 185–191.
38. S. Jung and J. Regan, Comparison of anode bacterial communities and performance in microbial fuel cells with different electron donors, *Appl. Microbiol. Biotechnol.* 77, 2007, 393–402.
39. N.T. Phung, J. Lee, K.H. Kang, I.S. Chang, G.M. Gadd and B.H. Kim, Analysis of microbial diversity in oligotrophic microbial fuel cells using 16S rDNA sequences, *FEMS Microbiol. Lett.* 233, 2004, 77–82.
40. J.R. Kim, S.H. Jung, J.M. Regan and B.E. Logan, Electricity generation and microbial community analysis of alcohol powered microbial fuel cells, *Bioresource Technol.* 98, 2007, 2568–2577.
41. B.H. Kim, H.S. Park, H.J. Kim, G.T. Kim, I.S. Chang, J. Lee and N.T. Phung, Enrichment of microbial community generating electricity using a fuel-cell-type electrochemical cell, *Appl. Biochem. Biotechnol.* 63, 2004, 672–681.

42. L.M. Tender, C.E. Reimers, H.A. Stecher, D.E. Holmes, D.R. Bond, D.A. Lowy, K. Pilobello, S.J. Fertig and D.R. Lovley, Harnessing microbially generated power on the seafloor, *Nat. Biotech.* 20, 2002, 821–825.
43. D.E. Holmes, D.R. Bond, R.A. O'Neil, C.E. Reimers, L.R. Tender and D.R. Lovley, Microbial communities associated with electrodes harvesting electricity from a variety of aquatic sediments, *Microb. Ecol.* 48, 2004, 178–190.
44. Y. Liu, F. Harnisch, K. Fricke, U. Schroder, V. Climent and J.M. Feliu, The study of electrochemically active microbial biofilms on different carbon-based anode materials in microbial fuel cells, *Biosens. Bioelectron.* 25, 2010, 2167–2171.
45. S.A. Patil, F. Harnisch, C. Koch, T. Hübschmann, I. Fetzer, A.A. Carmona-Martínez, S. Müller and U. Schröder, Electroactive mixed culture derived biofilms in microbial bioelectrochemical systems: The role of *pH* on biofilm formation, performance and composition, *Bioresource Technol.* 102, 2011, 9683–9690.
46. Y. Yuan, B. Zhao, S. Zhou, S. Zhong and L. Zhuang, Electrocatalytic activity of anodic biofilm responses to *pH* changes in microbial fuel cells, *Bioresource Technol.* 102, 2011, 6887–6891.
47. S.A. Patil, F. Harnisch, B. Kapadnis and U. Schröder, Electroactive mixed culture biofilms in microbial bioelectrochemical systems: The role of temperature for biofilm formation and performance, *Biosens. Bioelectron.* 26, 2010, 803–808.
48. L. Liu, O. Tsyganova, D.-J. Lee, A. Su, J.-S. Chang, A. Wang and N. Ren, Anodic biofilm in single-chamber microbial fuel cells cultivated under different temperatures, *Int. J. Hydrogen Energy* 37, 2012, 15792–15800.
49. D.R. Lovley, Microbial fuel cells: Novel microbial physiologies and engineering approaches, *Curr. Opin. Biotechnol.* 17, 2006, 327–332.
50. G. Reguera, K.P. Nevin, J.S. Nicoll, S.F. Covalla, T.L. Woodard and D.R. Lovley, Biofilm and nanowire production leads to increased current in *Geobacter sulfurreducens* fuel cells, *Appl. Environ. Microbiol.* 72, 2006, 7345–7348.
51. Y.A. Gorby, S. Yanina, J.S. McLean, K.M. Rosso, D. Moyles, A. Dohnalkova, T.J. Beveridge et al., Electrically conductive bacterial nanowires produced by *Shewanella oneidensis* strain MR-1 and other microorganisms, *Proc. Natl. Acad. Sci.* 103, 2006, 11358–11363.
52. B. Cohen, The bacterial culture as an electrical half-cell, *J. Bacteriol.* 21, 1931, 18.
53. D.R. Lovley, Bug juice: Harvesting electricity with microorganisms, *Nat. Rev. Micro.* 4, 2006, 497–508.
54. K. Rabaey, J. Rodriguez, L.L. Blackall, J. Keller, P. Gross, D. Batstone, W. Verstraete and K.H. Nealson, Microbial ecology meets electrochemistry: Electricity-driven and driving communities, *ISME J* 1, 2007, 9–18.
55. K. Rabaey, N. Boon, M. Hofte and W. Verstraete, Microbial phenazine production enhances electron transfer in biofuel cells, *Environ. Sci. Technol.* 39, 2005, 3401–3408.
56. M. Rosenbaum, U. Schröder and F. Scholz, Investigation of the electrocatalytic oxidation of formate and ethanol at platinum black under microbial fuel cell conditions, *J. Solid State Electrochem.* 10, 2006, 872–878.
57. B.E. Logan and J.M. Regan, Electricity-producing bacterial communities in microbial fuel cells, *Trends Microbiol.* 14, 2006, 512–518.
58. S.I. Ishii, T. Kosaka, K. Hori, Y. Hotta and K. Watanabe, Coaggregation facilitates interspecies hydrogen transfer between *Pelotomaculum thermopropionicum* and *Methanothermobacter thermautotrophicus*, *Appl. Environ. Microbiol.* 71, 2005, 7838–7845.

59. A.J. Bard, R. Parsons and J. Jordan, *Standard Potentials in Aqueous Solution*, Marcel Dekker Inc, New York, 1985.
60. B.E. Logan, *Microbial Fuel Cells*, John Wiley & Sons, Inc., Hoboken, 2008.
61. H. Liu, S. Grot and B.E. Logan, Electrochemically assisted microbial production of hydrogen from acetate, *Environ. Sci. Technol.* 39, 2005, 4317–4320.
62. K. Rabaey and W. Verstraete, Microbial fuel cells: Novel biotechnology for energy generation, *Trends Microbiol.* 23, 2005, 291–298.
63. F. Zhao, R.C.T. Slade and J.R. Varcoe, Techniques for the study and development of microbial fuel cells: An electrochemical perspective, *Chem. Soc. Rev.* 38, 2009, 1926–1939.
64. S. Cheng, H. Liu and B.E. Logan, Power densities using different cathode catalysts (Pt and CoTMPP) and polymer binders (Nafion and PTFE) in single chamber microbial fuel cells, *Environ. Sci. Technol.* 40, 2005, 364–369.
65. H. Liu and B.E. Logan, Electricity generation using an air-cathode single chamber microbial fuel cell in the presence and absence of a proton exchange membrane, *Environ. Sci. Technol.* 38, 2004, 4040–4046.
66. K. Rabaey, G. Lissens, S.D. Siciliano and W. Verstraete, A microbial fuel cell capable of converting glucose to electricity at high rate and efficiency, *Biotechnol. Lett.* 25, 2003, 1531–1535.
67. J.R. Kim, B. Min and B.E. Logan, Evaluation of procedures to acclimate a microbial fuel cell for electricity production, *Appl. Microbiol. Biotechnol.* 68, 2005, 23–30.
68. Q. Wen, F. Kong, F. Ma, Y. Ren and Z. Pan, Improved performance of air-cathode microbial fuel cell through additional Tween 80, *J. Power Sources* 196, 2011, 899–904.
69. J. Liu, Y. Qiao, Z.S. Lu, H. Song and C.M. Li, Enhance electron transfer and performance of microbial fuel cells by perforating the cell membrane, *Electrochem. Commun.* 15, 2012, 50–53.
70. J. Niessen, U. Schröder, M. Rosenbaum and F. Scholz, Fluorinated polyanilines as superior materials for electrocatalytic anodes in bacterial fuel cells, *Electrochem. Commun.* 6, 2004, 571–575.
71. Y. Qiao, C.M. Li, S.-J. Bao and Q.-L. Bao, Carbon nanotube/polyaniline composite as anode material for microbial fuel cells, *J. Power Sources* 170, 2007, 79–84.
72. A.P. Borole, D. Aaron, C.Y. Hamilton and C. Tsouris, Understanding long-term changes in microbial fuel cell performance using electrochemical impedance spectroscopy, *Environ. Sci. Technol.* 44, 2010, 2740–2745.
73. S. You, Q. Zhao, J. Zhang, J. Jiang and S. Zhao, A microbial fuel cell using permanganate as the cathodic electron acceptor, *J. Power Sources* 162, 2006, 1409–1415.
74. S. Oh, B. Min and B.E. Logan, Cathode performance as a factor in electricity generation in microbial fuel cells, *Environ. Sci. Technol.* 38, 2004, 4900–4904.
75. H. Liu, S. Cheng and B.E. Logan, Power generation in fed-batch microbial fuel cells as a function of ionic strength, temperature, and reactor configuration, *Environ. Sci. Technol.* 39, 2005, 5488–5493.
76. G.-C. Gil, I.-S. Chang, B.H. Kim, M. Kim, J.-K. Jang, H.S. Park and H.J. Kim, Operational parameters affecting the performannce of a mediator-less microbial fuel cell, *Biosens. Bioelectron.* 18, 2003, 327–334.
77. R.A. Rozendal, H.V.M. Hamelers and C.J.N. Buisman, Effects of membrane cation transport on *pH* and microbial fuel cell performance, *Environ. Sci. Technol.* 40, 2006, 5206–5211.

78. J.R. Kim, S. Cheng, S.-E. Oh and B.E. Logan, Power generation using different cation, anion, and ultrafiltration membranes in microbial fuel cells, *Environ. Sci. Technol.* 41, 2007, 1004–1009.
79. S.-E. Oh and B. Logan, Proton exchange membrane and electrode surface areas as factors that affect power generation in microbial fuel cells, *Appl. Microbiol. Biotechnol.* 70, 2006, 162–169.
80. Y. Fan, H. Hu and H. Liu, Sustainable power generation in microbial fuel cells using bicarbonate buffer and proton transfer mechanisms, *Environ. Sci. Technol.* 41, 2007, 8154–8158.
81. A. Larrosa-Guerrero, K. Scott, I.M. Head, F. Mateo, A. Ginesta and C. Godinez, Effect of temperature on the performance of microbial fuel cells, *Fuel* 89, 2010, 3985–3994.
82. S. Cheng and B.E. Logan, Increasing power generation for scaling up single-chamber air cathode microbial fuel cells, *Bioresource Technol.* 102, 2011, 4468–4473.
83. L. Zhang, C. Li, L. Ding, K. Xu and H. Ren, Influences of initial pH on performance and anodic microbes of fed-batch microbial fuel cells, *J. Chem. Technol. Biotechnol.* 86, 2011, 1226–1232.
84. S.V. Raghavulu, S.V. Mohan, R.K. Goud and P.N. Sarma, Effect of anodic *pH* microenvironment on microbial fuel cell (MFC) performance in concurrence with aerated and ferricyanide catholytes, *Electrochem. Commun.* 11, 2009, 371–375.
85. S. Puig, M. Serra, M. Coma, M. Cabré, M.D. Balaguer and J. Colprim, Effect of *pH* on nutrient dynamics and electricity production using microbial fuel cells, *Bioresource Technol.* 101, 2010, 9594–9599.
86. Z. He, Y. Huang, A.K. Manohar and F. Mansfeld, Effect of electrolyte *pH* on the rate of the anodic and cathodic reactions in an air-cathode microbial fuel cell, *Bioelectrochemistry* 74, 2008, 78–82.
87. O. Lefebvre, Z. Tan, S. Kharkwal and H.Y. Ng, Effect of increasing anodic NaCl concentration on microbial fuel cell performance, *Bioresource Technol.* 112, 2012, 336–340.
88. M. Sun, F. Zhang, Z.-H. Tong, G.-P. Sheng, Y.-Z. Chen, Y. Zhao, Y.-P. Chen et al., A gold-sputtered carbon paper as an anode for improved electricity generation from a microbial fuel cell inoculated with *Shewanella oneidensis* MR-1, *Biosens. Bioelectron.* 26, 2010, 338–343.
89. Z. Lv, D. Xie, X. Yue, C. Feng and C. Wei, Ruthenium oxide-coated carbon felt electrode: A highly active anode for microbial fuel cell applications, *J. Power Sources* 210, 2012, 26–31.
90. S. You, Q. Zhao, J. Zhang, J. Jiang, C. Wan, M. Du and S. Zhao, A graphite-granule membrane-less tubular air-cathode microbial fuel cell for power generation under continuously operational conditions, *J. Power Sources* 173, 2007, 172–177.
91. B. Logan, S. Cheng, V. Watson and G. Estadt, Graphite fiber brush anodes for increased power production in air-cathode microbial fuel cells, *Environ. Sci. Technol.* 41, 2007, 3341–3346.
92. F.X. Li, Y. Sharma, Y.L. Lei, B.K. and Q.X. Zhou, Microbial fuel cells: The effects of configurations, electrolyte solutions, and electrode materials on power generation, *Appl. Biochem. Biotechnol.* 160, 2010, 168–181.
93. K. Rabaey, P. Clauwaert, P. Aelterman and W. Verstraete, Tubular microbial fuel cells for efficient electricity generation, *Environ. Sci. Technol.* 39, 2005, 8077–8082.

94. C. Dumas, A. Mollica, D. Feron, R. Basseguy, L. Etcheverry and A. Bergel, Marine microbial fuel cell: Use of stainless steel electrodes as anode and cathode materials, *Electrochim. Acta* 53, 2007, 468–473.
95. A. ter Heijne, H.V.M. Hamelers, M. Saakes and C.J.N. Buisman, Performance of non-porous graphite and titanium-based anodes in microbial fuel cells, *Electrochim. Acta* 53, 2008, 5697–5703.
96. K. Scott, G.A. Rimbu, K.P. Katuri, K.K. Prasad and I.M. Head, Application of modified carbon anodes in microbial fuel cells, *Process Saf. Environ. Prot.* 85, 2007, 481–488.
97. X. Tang, K. Guo, H. Li, Z. Du and J. Tian, Electrochemical treatment of graphite to enhance electron transfer from bacteria to electrodes, *Bioresource Technol.* 102, 2011, 3558–3560.
98. H.-Y. Tsai, C.-C. Wu, C.-Y. Lee and E.P. Shih, Microbial fuel cell performance of multiwall carbon nanotubes on carbon cloth as electrodes, *J. Power Sources* 194, 2009, 199–205.
99. H. Moon, I.S. Chang and B.H. Kim, Continuous electricity production from artificial wastewater using a mediator-less microbial fuel cell, *Bioresource Technol.* 97, 2006, 621–627.
100. J.M. Morris, S. Jin, J. Wang, C. Zhu and M.A. Urynowicz, Lead dioxide as an alternative catalyst to platinum in microbial fuel cells, *Electrochem. Commun.* 9, 2007, 1730–1734.
101. E.H. Yu, S. Cheng, K. Scott and B. Logan, Microbial fuel cell performance with non-Pt cathode catalysts, *J. Power Sources* 171, 2007, 275–281.
102. I. Roche, K. Katuri and K. Scott, A microbial fuel cell using manganese oxide oxygen reduction catalysts, *J. Appl. Electrochem.* 40, 2010, 13–21.
103. L. Deng, M. Zhou, C. Liu, L. Liu, C. Liu and S. Dong, Development of high performance of Co/Fe/N/CNT nanocatalyst for oxygen reduction in microbial fuel cells, *Talanta* 81, 2010, 444–448.
104. P. Clauwaert, D. van der Ha, N. Boon, K. Verbeken, M. Verhaege, K. Rabaey and W. Verstraete, Open air biocathode enables effective electricity generation with microbial fuel cells, *Environ. Sci. Technol.* 41, 2007, 7564–7569.
105. N. Duteanu, B. Erable, S.M. Senthil Kumar, M.M. Ghangrekar and K. Scott, Effect of chemically modified Vulcan XC-72R on the performance of air-breathing cathode in a single-chamber microbial fuel cell, *Bioresource Technol.* 101, 2010, 5250–5255.
106. B. Logan, Scaling up microbial fuel cells and other bioelectrochemical systems, *Appl. Microbiol. Biotechnol.* 85, 2010, 1665–1671.
107. A. Dekker, A.T. Heijne, M. Saakes, H.V.M. Hamelers and C.J.N. Buisman, Analysis and improvement of a scaled-up and stacked microbial fuel cell, *Environ. Sci. Technol.* 43, 2009, 9038–9042.
108. T.H. Pham, K. Rabaey, P. Aelterman, P. Clauwaert, L. De Schamphelaire, N. Boon and W. Verstraete, Microbial fuel cells in relation to conventional anaerobic digestion technology, *Eng. Life Sci.* 6, 2006, 285–292.
109. L.M. Tender, S.A. Gray, E. Groveman, D.A. Lowy, P. Kauffman, J. Melhado, R.C. Tyce, D. Flynn, R. Petrecca and J. Dobarro, The first demonstration of a microbial fuel cell as a viable power supply: Powering a meteorological buoy, *J. Power Sources* 179, 2008, 571–575.
110. F. Zhang, K.S. Jacobson, P. Torres and Z. He, Effects of anolyte recirculation rates and catholytes on electricity generation in a litre-scale upflow microbial fuel cell, *Energy Environ. Sci.* 3, 2010, 1347–1352.

111. P. Clauwaert, S. Mulenga, P. Aelterman and W. Verstraete, Litre-scale microbial fuel cells operated in a complete loop, *Appl. Microbiol. Biotechnol.* 83, 2009, 241–247.
112. Y. Fan, S.-K. Han and H. Liu, Improved performance of CEA microbial fuel cells with increased reactor size, *Energy Environ. Sci.* 5, 2012, 8273–8280.
113. P. Aelterman, K. Rabaey, H.T. Pham, N. Boon and W. Verstraete, Continuous electricity generation at high voltages and currents using stacked microbial fuel cells, *Environ. Sci. Technol.* 40, 2006, 3388–3394.
114. B.E. Logan, C. Murano, K. Scott, N.D. Gray and I.M. Head, Electricity generation from cysteine in a microbial fuel cell, *Water Res.* 39, 2005, 942–952.
115. B. Min and B.E. Logan, Continuous electricity generation from domestic wastewater and organic substrates in a flat plate microbial fuel cell, *Environ. Sci. Technol.* 38, 2004, 5809–5814.
116. H. Yan, T. Saito and J.M. Regan, Nitrogen removal in a single-chamber microbial fuel cell with nitrifying biofilm enriched at the air cathode, *Water Res.* 46, 2012, 2215–2224.
117. C.-Y. Lee, K.-L. Ho, D.-J. Lee, A. Su and J.-S. Chang, Electricity harvest from wastewaters using microbial fuel cell with sulfide as sole electron donor, *Int. J. Hydrogen Energy* 37, 2012, 15787–15791.
118. M. Sun, Z.-H. Tong, G.-P. Sheng, Y.-Z. Chen, F. Zhang, Z.-X. Mu, H.-L. Wang et al., Microbial communities involved in electricity generation from sulfide oxidation in a microbial fuel cell, *Biosens. Bioelectron.* 26, 2010, 470–476.
119. Q. Wen, F. Kong, H. Zheng, J. Yin, D. Cao, Y. Ren and G. Wang, Simultaneous processes of electricity generation and ceftriaxone sodium degradation in an air-cathode single chamber microbial fuel cell, *J. Power Sources* 196, 2011, 2567–2572.
120. L. Huang, L. Gan, Q. Zhao, B.E. Logan, H. Lu and G. Chen, Degradation of pentachlorophenol with the presence of fermentable and non-fermentable co-substrates in a microbial fuel cell, *Bioresource Technol.* 102, 2011, 8762–8768.
121. S. Bakhshian, H.-R. Kariminia and R. Roshandel, Bioelectricity generation enhancement in a dual chamber microbial fuel cell under cathodic enzyme catalyzed dye decolorization, *Bioresource Technol.* 102, 2011, 6761–6765.
122. W.-J. Hu, C.-G. Niu, Y. Wang, G.-M. Zeng and Z. Wu, Nitrogenous heterocyclic compounds degradation in the microbial fuel cells, *Process Saf. Environ.* 89, 2011, 133–140.
123. G. Velvizhi and S. Venkata Mohan, Electrogenic activity and electron losses under increasing organic load of recalcitrant pharmaceutical wastewater, *Int. J. Hydrogen Energy* 37, 2012, 5969–5978.
124. B. Zhang, C. Feng, J. Ni, J. Zhang and W. Huang, Simultaneous reduction of vanadium (V) and chromium (VI) with enhanced energy recovery based on microbial fuel cell technology, *J. Power Sources* 204, 2012, 34–39.
125. C.E. Reimers, L.M. Tender, S. Fertig and W. Wang, Harvesting energy from the marine sediment—water interface, *Environ. Sci. Technol.* 35, 2001, 192–195.
126. N. Ryckelynck, H. Stecher and C. Reimers, Understanding the anodic mechanism of a seafloor fuel cell: Interactions between geochemistry and microbial activity, *Biogeochemistry* 76, 2005, 113–139.
127. K. Scott, I. Cotlarciuc, I. Head, K.P. Katuri, D. Hall, J.B. Lakeman and D. Browning, Fuel cell power generation from marine sediments: Investigation of cathode materials, *J. Chem. Technol. Biotechnol.* 83, 2008, 1244–1254.

128. Z. He, H. Shao and L.T. Angenent, Increased power production from a sediment microbial fuel cell with a rotating cathode, *Biosens. Bioelectron.* 22, 2007, 3252–3255.
129. F. Rezaei, T.L. Richard, R.A. Brennan and B.E. Logan, Substrate-enhanced microbial fuel cells for improved remote power generation from sediment-based systems, *Environ. Sci. Technol.* 41, 2007, 4053–4058.
130. C.E. Reimers, P. Girguis, H.A. Stecher, L.M. Tender, N. Ryckelynck and P. Whaling, Microbial fuel cell energy from an ocean cold seep, *Geobiology* 4, 2006, 123–136.
131. M. Helder, D.P.B.T.B. Strik, H.V.M. Hamelers, A.J. Kuhn, C. Blok and C.J.N. Buisman, Concurrent bio-electricity and biomass production in three plant-microbial fuel cells using *Spartina anglica*, *Arundinella anomala* and *Arundo donax*, *Bioresource Technol.* 101, 2010, 3541–3547.
132. A.C. de Rosa, Diversity and Function of the Microbial Community on Anodes of Sediment Microbial Fuel Cells fueled by Root Exudates, PhD thesis, Marburg (Germany): University of Marburg, 2011.
133. Z. He, J. Kan, Y. Wang, Y. Huang, F. Mansfeld and K.H. Nealson, Electricity production coupled to ammonium in a microbial fuel cell, *Environ. Sci. Technol.* 43, 2009, 3391–3397.
134. B. Ma, S. Zhang, L. Zhang, P. Yi, J. Wang, S. Wang and Y. Peng, The feasibility of using a two-stage autotrophic nitrogen removal process to treat sewage, *Bioresource Technol.* 102, 2011, 8331–8334.
135. D. Call and B.E. Logan, Hydrogen production in a single chamber microbial electrolysis cell lacking a membrane, *Environ. Sci. Technol.* 42, 2008, 3401–3406.
136. H.-S. Lee and B.E. Rittmann, Characterization of energy losses in an upflow single-chamber microbial electrolysis cell, *Int. J. Hydrogen Energy* 35, 2010, 920–927.
137. A. Kumlanghan, J. Liu, P. Thavarungkul, P. Kanatharana and B. Mattiasson, Microbial fuel cell-based biosensor for fast analysis of biodegradable organic matter, *Biosens. Bioelectron.* 22, 2007, 2939–2944.
138. Y. Zhang and I. Angelidaki, Submersible microbial fuel cell sensor for monitoring microbial activity and BOD in groundwater: Focusing on impact of anodic biofilm on sensor applicability, *Biotechnol. Bioeng.* 108, 2011, 2339–2347.
139. Z. Liu, J. Liu, S. Zhang, X.-H. Xing and Z. Su, Microbial fuel cell based biosensor for *in situ* monitoring of anaerobic digestion process, *Bioresource Technol.* 102, 2011, 10221–10229.
140. N.E. Stein, H.V.M. Hamelers and C.N.J. Buisman, Influence of membrane type, current and potential on the response to chemical toxicants of a microbial fuel cell based biosensor, *Sens. Actuators B* 163, 2012, 1–7.
141. A. Nagiub and F. Mansfeld, Evaluation of microbiologically influenced corrosion inhibition (MICI) with EIS and ENA, *Electrochim. Acta* 47, 2002, 2319–2333.
142. D. Örnek, T.K. Wood, C.H. Hsu and F. Mansfeld, Corrosion control using regenerative biofilms (CCURB) on brass in different media, *Corros. Sci.* 44, 2002, 2291–2302.
143. E. Kuş, R. Abboud, R. Popa, K.H. Nealson and F. Mansfeld, The concept of the bacterial battery, *Corros. Sci.* 47, 2005, 1063–1069.
144. E. Kuş, K. Nealson and F. Mansfeld, The effect of different exposure conditions on the biofilm/copper interface, *Corros. Sci.* 49, 2007, 3421–3427.

145. E. Kuş, K. Nealson and F. Mansfeld, The bacterial battery and the effect of different exposure conditions on biofilm properties, *Electrochim. Acta* 54, 2008, 47–52.
146. Z. Du, H. Li and T. Gu, A state of the art review on microbial fuel cells: A promising technology for wastewater treatment and bioenergy, *Biotechnol. Adv.* 25, 2007, 464–482.
147. C. Melhuish, I. Ieropoulos, J. Greenman and I. Horsfield, Energetically autonomous robots: Food for thought, *Auton. Robot* 21, 2006, 187–198.

3

Lithium Batteries: Status and Future

Bruno Scrosati and Jusef Hassoun

CONTENTS

3.1 Introduction	121
3.2 Evolution of the Lithium Battery Chemistry	124
3.3 Advanced Anodes	129
3.4 Advanced Electrolytes	133
3.5 Advanced Cathodes	138
3.6 Super Energy Density Batteries: Lithium–Sulphur and Lithium–Air	141
3.6.1 Lithium–Sulphur Battery	142
3.6.2 Lithium–Air Battery	145
3.6.2.1 Aqueous Electrolyte	147
3.6.2.2 Non-Aqueous Electrolyte	148
3.6.2.3 Electrolyte Issue	149
3.6.2.4 Lithium Metal Anode Issue	155
3.7 Conclusions and Outlook	156
References	157

3.1 Introduction

Our present energy policy, still mainly based on burning of fossil fuels, inevitably poses a serious concern due to the CO_2-related global warming (see Figure 3.1).

Accordingly, efforts aimed to assure an efficient use of renewable energy sources, as well as the replacement of internal combustion engines with electric motors for the development of sustainable vehicles, such as hybrid electric vehicles (HEVs) and ultimately, full electric vehicles (EVs), are in progress worldwide (see Figure 3.2). Another advantage of the electric engine over the combustion engine is in its higher fuel efficiency (see Figure 3.3).

The exploitation of alternative and green energy sources, such as solar, wind and geothermal, requires the side support of energy storage systems that can compensate their intermittent characteristic. It is now generally accepted that among various possible choices, the most suitable are

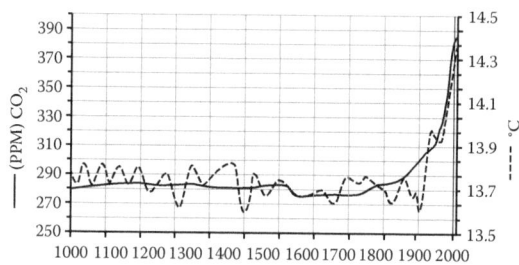

FIGURE 3.1
The 12-year period from 1995 to 2006 was the warmest on record, indicating the immediacy and seriousness of the climate change phenomenon. Computer simulations predict that the average world temperature may rise by approximately 4°C before 2100, posing a potential lethal threat to many living creatures.

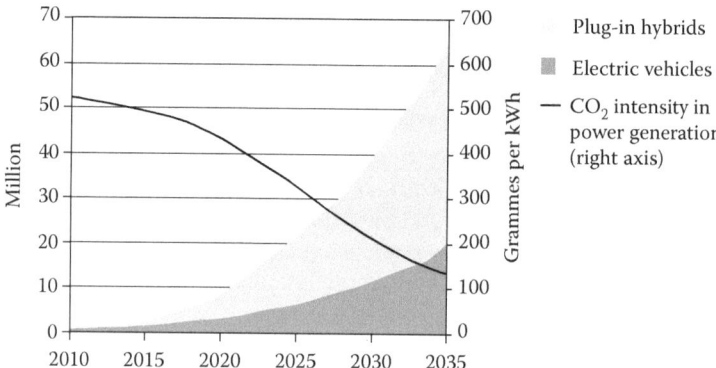

FIGURE 3.2
Previsions tend to indicate that in the year 2035 hybrid and/or electric vehicles will meet a share of 35% of the automobile market, with an associated, considerable reduction of CO_2 emissions.

	Fuel efficiency	On-board efficiency	Well to wheel
ICE	88	16	14
HEV (new prius)	88	37	32
FCHV	58	38	22
		50	29
FCHV prospect	70	60	42

FIGURE 3.3
The replacement of internal combustion engine (ICE) vehicles with electric vehicles results in considerable increase in fuel efficiency.

electrochemical batteries, which offer a portable mean of delivering stored chemical energy as electrical energy with high conversion efficiency and without any gaseous emission.

Particularly, rechargeable batteries, benefiting from high specific energy (high voltage combined with high specific capacity), high rate capability, high safety and low cost, offer the most promising option to support renewable energy plants, as well as to efficiently powered HEVs or EVs. A promising candidate is Li-ion batteries, which today exceed other competing technologies by a factor of at least 2.5 [1,2] thanks to its high energy density, that is, 150 Wh kg^{-1} and 650 Wh L^{-1} (see Figure 3.4).

These unique features make lithium-ion batteries the power source of choice for the portable electronic market (including popular products such as cellular phone, laptop computers, mp3s, as well as power tools). Their annual production amounts to several billions of units globally (see Figure 3.5).

The new trend in the field is that of moving from consumer electronics to electromobility by using lithium batteries for powering the electric engine of hybrid and/or electric sustainable vehicles. However, the present Li-ion batteries, though commercially realised, are not yet at a technological level sufficient to meet the requirements of these vehicles, of which production is still limited to a few demonstration prototypes, mainly of the hybrid type (see Figure 3.6). An effective evolution of the HEV and EV market requires the availability of batteries having lower cost, better safety and especially higher energy density than those presently available. This important step forward can be achieved only by moving to totally new chemistries [3].

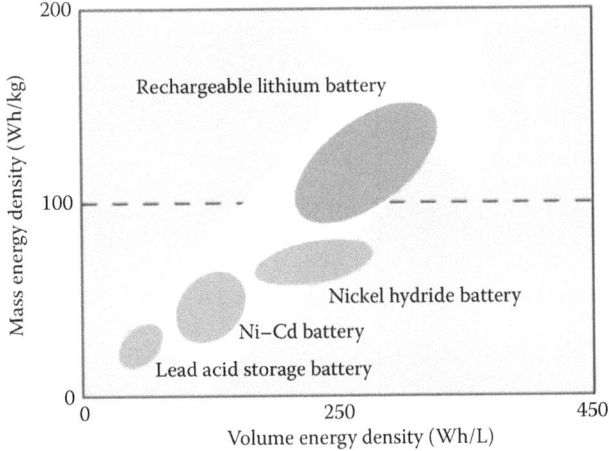

FIGURE 3.4
Mass versus volume energy density of lithium batteries versus most common batteries.

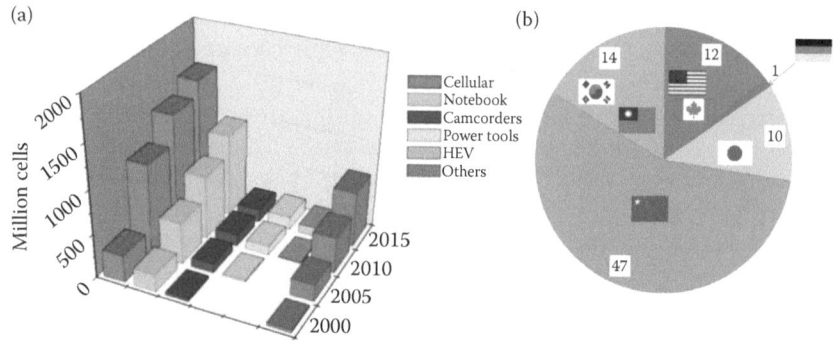

FIGURE 3.5
(See colour insert.) (a) Lithium batteries are fabricated at a rate of several billion units per year since they are the chosen power sources for popular portable electronic devices, in particular, cellular phones and notebooks. (b) Most of the production is concentrated in Asia, with only a limited share in the United States and Europe.

FIGURE 3.6
The production of electric vehicles is still limited to a few demonstration prototypes, mainly of the hybrid type.

3.2 Evolution of the Lithium Battery Chemistry

As a general classification, we may assume that there are two main types of batteries, that is, lithium batteries and lithium-ion batteries. The former use a lithium metal anode, while the conventional configuration of the latter consists of a graphite anode, a lithium cobalt oxide cathode and a liquid, organic carbonate electrolyte [4]. There are various alternative configurations, dependent on the type of the three components: anode, cathode and electrolyte (see Figure 3.7).

Lithium metal is a very appealing electrode material since its electrochemical process, Li $\rightleftarrows$ Li$^+$ + e$^-$, occurs at a very high potential of −3.01 V, associated with a high electrochemical equivalent, namely 3.86 Ah g^{-1} or 2.06 Ah L^{-1}; hence, the use of lithium as anode is expected to give high energy density energy battery systems. However, lithium metal is a very reactive material and decomposes any electrolyte media into which it comes into contact. The decomposition products form a passivation layer on the electrode surface. This layer is not uniform; hence, uneven lithium deposition may occur during charge, giving rise to a cumulative phenomena resulting in the growth of dendrites. Such dendrites may eventually cross the electrolyte to reach the cathode, leading to short circuiting of the cell with associated severe safety hazards (see Figure 3.8).

For this reason, battery manufacturers are reluctant to fabricate batteries based on lithium metal. The only way to safely use lithium as an electrode is by coupling it with a stable electrolyte. The most common examples are solvent-free, polymer membranes formed by the combination of a poly(ethylene oxide) (PEO) matrix and a lithium salt, LiX [5,6]. The excess of negative charge on the oxygen in the PEO chains coordinates by coulombic attraction of the Li$^+$ ions, thus separating them from the anions. By this process, the lithium salt is 'dissolved' in the PEO matrix, analogous to the process of salt dissolution in liquid solvents [5]. The main difference is that while the ions can move with their solvation shell in liquids, this is not possible in the PEO complexes due to the large size and encumbrance of the chains. Therefore, ion transport in the polymer electrolytes requires flexibility of the PEO chains so

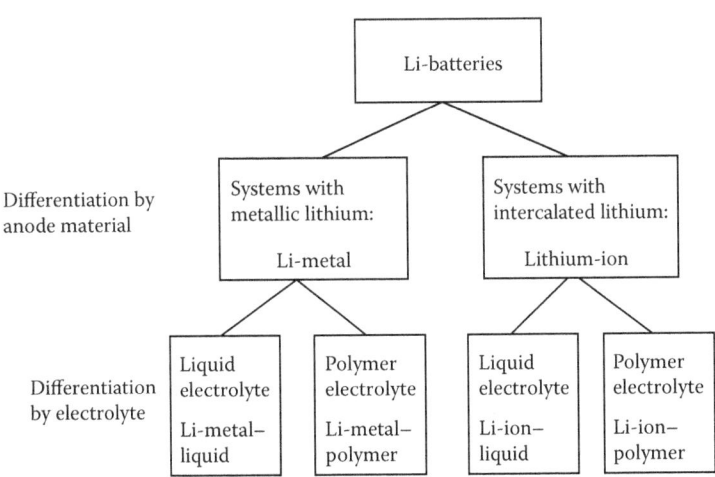

FIGURE 3.7
Lithium and lithium-ion battery systems.

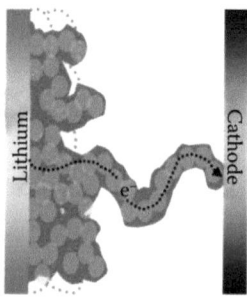 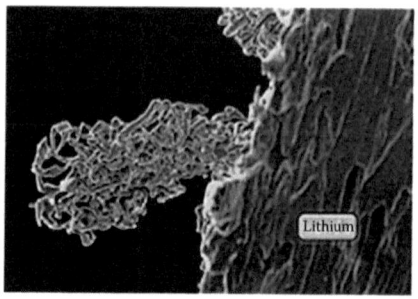

FIGURE 3.8
Owing to the uneven surface, lithium deposition on passivated lithium may give rise to the formation of dendrites that, by crossing the electrolyte, may reach the cathode, resulting in an electronic shunt with elevated heat associated with the current flow. This in turn can cause the electrolyte to ignite and eventually, the system to explode.

that they can unfold to free the Li$^+$ ions to move from one chain loop to the other, as shown in the schematic diagram in Figure 3.9.

The PEO–LiX polymer membranes, due to their fully solid nature, are stable in contact with lithium metal that can then be cycled several hundred times without dendrite formation. In addition, the PEO–LiX electrolytes are of low cost and easily manufactured for the fabrication of plastic-like batteries (see Figure 3.10).

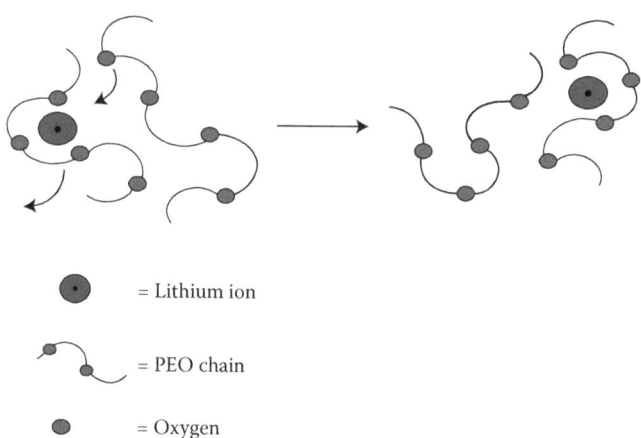

FIGURE 3.9
Schematic diagram of the lithium ion transport mechanism in a polymer electrolyte formed by a complex between a lithium salt and a coordinating PEO polymer. The excess of negative charge on the oxygen in the PEO chains coordinates by coulombic attraction of the Li$^+$ ions, thus separating them from the anions. Because of the large size and encumbrance of the chains, the Li$^+$ ion transport requires flexibility of the chains so that they can unfold to free the Li$^+$ ions to move from one chain loop to the other.

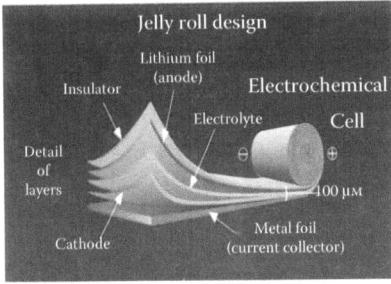
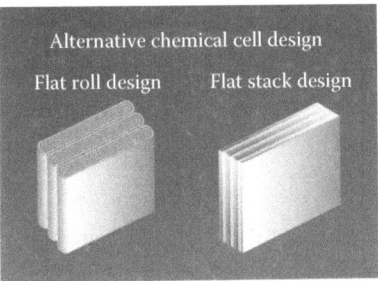

FIGURE 3.10
Concept of the all-solid battery using a dry polymer electrolyte and metallic lithium anode.

These polymer electrolytes were exploited in the late 1990s for the fabrication of large-sized, laminated battery modules based on cells formed by a lithium foil anode and a vanadium oxide cathode, developed jointly by Hydro Québec in Canada and 3M company in the United States [7,8]. The battery module had very good performance in terms of energy density (155 Wh kg^{-1}) and cycle life (600 cycles at 80% depth of discharge (DOD)), and it was proposed as a power source for EVs, a very futuristic concept back in 1996. However, despite this and other successful demonstration projects, the lithium polymer battery project was abandoned and only very recently reconsidered for use in an EV produced in France [9].

In the lithium-ion design, the lithium metal is replaced by graphite, acting as a 'lithium sink' negative electrode to accept and release back lithium ions provided by a 'lithium source' transition metal oxide, for example, lithium cobalt oxide positive electrode and a liquid organic solution, $LiPF_6$ in carbonate solvents, and ethylene carbonate–dimethyl carbonate (EC–DMC) electrolyte [10]. When charging, lithium ions, removed from the $LiCoO_2$ positive electrode, travel through the electrolyte and reach the negative electrode where they are intercalated into the graphite structure. To compensate the transfer of the ionic charges, electrons are also exchanged between the two electrodes (see Figure 3.11).

The choice of graphite as anode is rather surprising since its electrochemical process evolves outside the stability window of most common electrolytes (see Figure 3.12). The electrolyte decomposes at the graphite surface, upon the formation of an electronically blocking but ionically conducting layer, generally named the solid electrolyte interface (SEI) [11]. SEI blocks the decomposition, yet allows ongoing electrochemical process. Therefore, graphite is thermodynamically unstable but kinetically protected. However, this kinetic stabilisation involves the decomposition of the electrolyte, and is in turn accompanied by gas evolution, a phenomenon that, if not properly controlled, can give rise to safety problems.

Although they are commercially established products, lithium-ion batteries still suffer from some practical problems. The most significant limitation

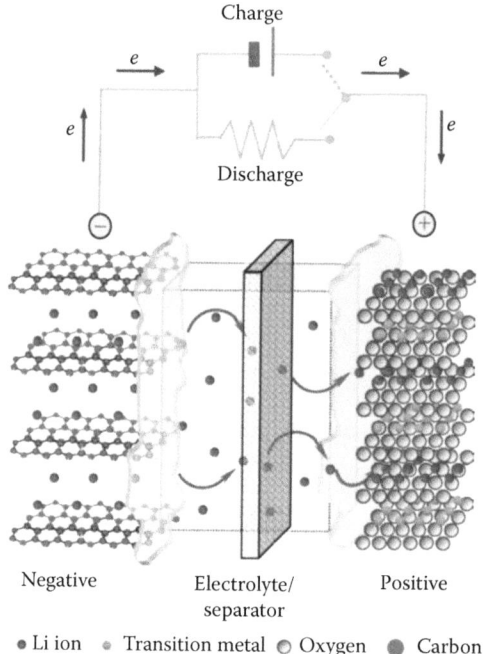

FIGURE 3.11
(See colour insert.) Concept and schematic of a typical lithium-ion battery. (Reprinted, *Encyclopedia of Electrochemical Power Sources*, Vol. 5, K. Xu, Secondary batteries—Lithium recharagable systems: Electrolytes: Overview, in: G. Jurgen (Ed.), pp. 51–70, Copyright 2009, with permission from Elsevier.)

is the safety risks involved. At extreme operating conditions, such as at the upper limit of the charge process, oxygen may be released from the layered $LiCoO_2$ cathode ($Li_2CoO_2 \rightarrow xLiCoO_2 + (1-x)/3Co_3O_4 + (1-x)/3O_3$) $x \leq 1$, which, in the event of local overheating, may react with the flammable organic liquid electrolyte, giving rise to thermal runaway effects, and possibly explosions (see Figure 3.13). Various fire incidents have in fact been reported during battery manufacturing and/or for battery-operated devices. Recently, a Boeing 787 Dreamliner was forced to land due to an onboard lithium battery problem. Another issue is its comparative high cost, since the most common cathode material is based on an expensive element such as cobalt.

Finally, the intercalation chemistry that drives lithium-ion batteries can allow only a maximum exchange of two electrons per mole. This undoubtedly limits the overall battery capacity and consequently restricts the energy density to a maximum of 150–200 Wh kg^{-1}. To meet the basic requirements of batteries for electric transportation, which include safety, low cost and high energy density, new electrode and electrolyte materials have to be developed.

Lithium Batteries

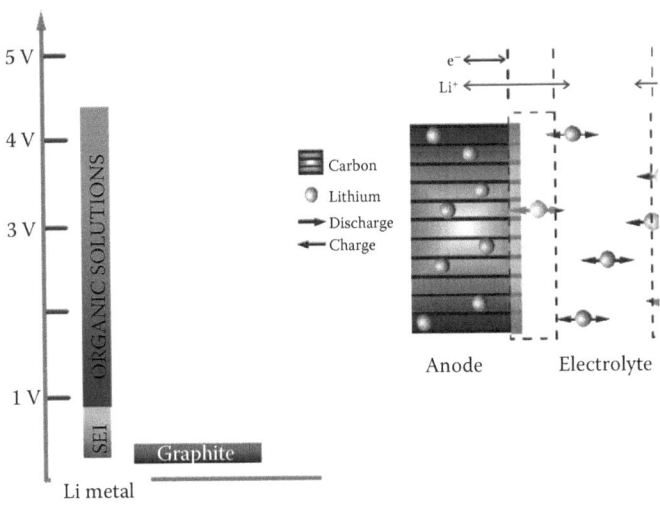

FIGURE 3.12
The electrochemical process of graphite, $x\text{Li}^+ + y\text{C} + xe^- \rightleftarrows \text{Li}_x\text{C}_y$, evolves at about 0.050 V vs. Li, and hence outside the stability window of the most common electrolytes (bottom-left vertical domain indicated by SEI in the figure). Therefore, graphite is theoretically thermodynamically unstable in these electrolytes but kinetically operating in virtue of its surface passivating layers that blocks electron transfer (hence continuous electrolyte decomposition) but not ion transfer (hence allowing the ongoing of the electrochemical process).

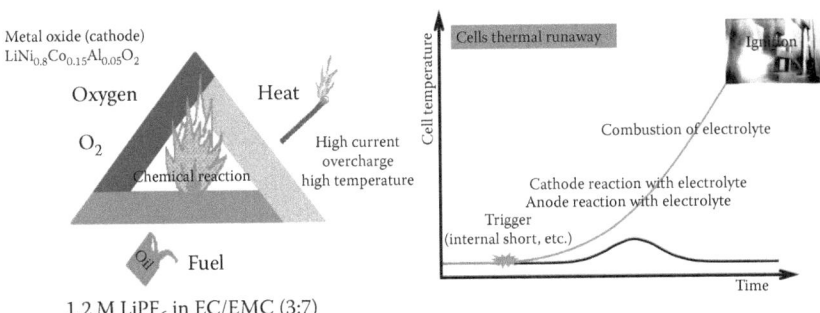

FIGURE 3.13
Under extreme conditions, such as overcharge, oxygen can be released from the LiCoO_2 cathode and, in the event of local overheating, may react with the flammable organic liquid electrolyte, giving rise to a thermal runaway effect that may eventually result in the combustion of the electrolyte. (Courtesy of Professor J. Yamaki, Kyushu University, Fukuoka, Japan.)

3.3 Advanced Anodes

One example of anodes alternative to graphite is lithium titanium oxide, $\text{Li}_4\text{Ti}_5\text{O}_{12}$ (LTO). This material has a defective spinel-framework structure and is characterised by a two-phase electrochemical process with a flat voltage profile at over 1.5 V [12]. The theoretical specific capacity of LTO is lower, and the voltage is higher, than those of conventional graphite. Although this results in lower specific energy, the interest in LTO remains high because of its specific properties, including negligible changes in the lattice structure upon accepting and releasing lithium ions, and an electrochemical process evolving within the stability domain of the electrolyte (see Figure 3.14). For all the above reasons, LTO is a promising anode material and as such, currently used for the development of alternative lithium-ion batteries. The pros and cons of LTO versus conventional graphite are summarised in Table 3.1.

However, breakthrough in the field requires the development of materials having more inherent advantages than LTO over graphite. Appealing examples are lithium metal alloys, for example, lithium–silicon (Li–Si), and lithium–tin (Li–Sn), due to their specific capacity, which largely exceeds that of lithium–graphite, that is, ~4000 mAh g^{-1} for Li–Si and 993 mAh g^{-1} for Li–Sn, versus 370 mAh g^{-1} for Li–C. The potentiality of these materials has been known for some time [13], but their use has been, until recently, prevented by a serious issue associated with the large volume changes experienced during the lithium alloying–de-alloying electrochemical process (see Figure 3.15). This expansion–contraction process induces cracks and, eventually, pulverisation of the electrode, leading it to fail in a few cycles.

The advent of nanotechnology has helped to address this issue [14]. In fact, by reducing the metal particle size to a nanoscale level, the volume change may be controlled and the lithium diffusion length greatly reduced, thus improving the performance of the electrode in terms of life and rate capability. Nanostructured

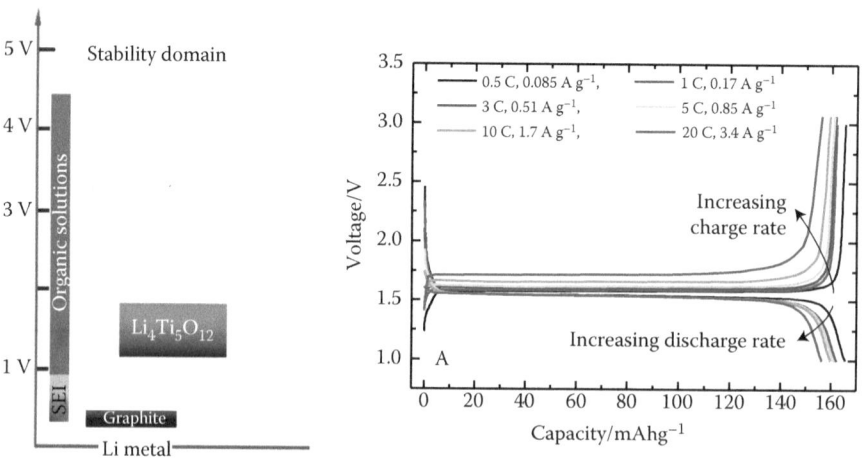

FIGURE 3.14
The electrochemical process of LTO, $Li_4Ti_5O_{12} \leftrightarrows Li^+ + Li_3Ti_5O_{12} + e^-$, occurs at 1.5 V, and hence fully within the stability domain of the electrolyte. There is a general trend of decreasing capacity with increasing charge/discharge rate.

TABLE 3.1
(See colour insert.) Summary of Characteristics and Properties of Anode Materials Designed for Advanced Lithium-Ion Batteries

Material	Theoretical capacity (mAhg^{-1})	Advantages	Issues	Structures
$Li_4Ti_5O_{12}$	170	Negligible volume expansion, low cost, stable electrochemical operation and high thermal stability in the charged–discharged state, already in large-scale production	Low specific capacity, high voltage	
Li_xSn_y	990	High capacity, low cost	Large volume changes. Nanostructures required	
Li_xSi_y	2000	Very high capacity, low cost	Large volume changes. Nanostructures required production announced for 2012 (Panasonic)	
$Sn_xCo_yC_z$	500	High capacity, commercially tested	Cycle life to be demonstrated	
C, MCMB	370	Field tested operation	Operation outside electrolyte stability	

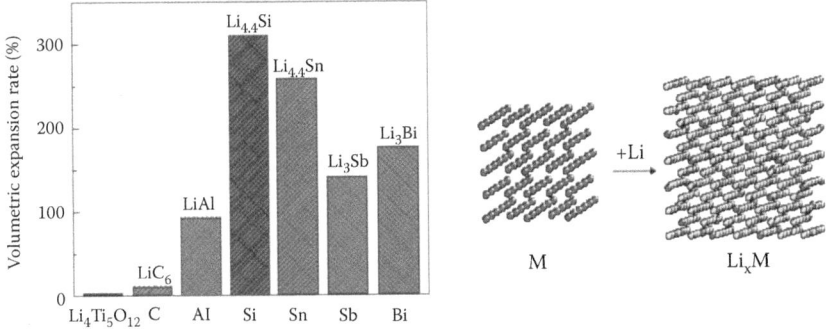

FIGURE 3.15
Upon the electrochemical process, xLi + M (where M can be Al, Si, Sn, Sb, Bi) alloys experience a variation in volume that can exceed 300%.

electrodes, however, are not practically feasible since their large surface area reflects in a high reactivity and in the reduction of tap density, with associated increase in safety hazard and decrease in volumetric energy density. All these aspects are discussed in detail in recent and comprehensive reviews [1,2].

Breakthrough in this field has been achieved through the development of cleverly designed carbon–metal composites. In an oversimplified way, these composites may be described as formed by small-size, metal particles dispersed within a carbon matrix. The carbon matrix, while maintaining in its core the nano-sized metal particle configuration that helps in containing the volume stress, supplies an overall compact structure that assures stability and provides high tap density.

It is now widely assumed that the nanocomposite approach is the most promising in developing electrode materials for practical purposes. The validity of this assumption and of the nanocomposite concept is supported by two convincing examples. One is given by the tin–carbon (Sn–C) composite, whose morphology is shown by the TEM pictures reported in Figure 3.16 [15]. The

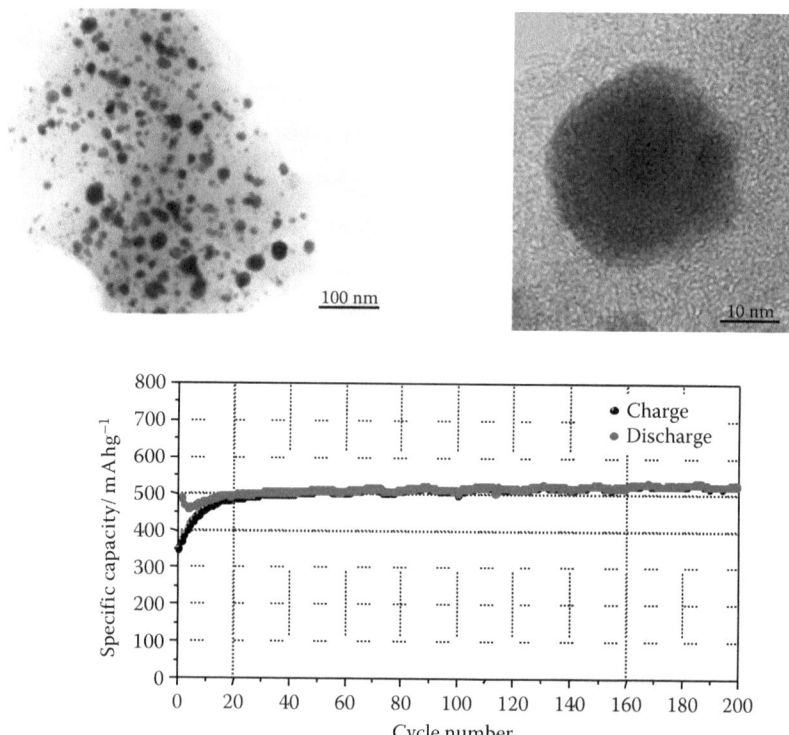

FIGURE 3.16
TEM and HRTM images of a Sn–C nanocomposite and its response in a lithium cell showing a stable and reversible behaviour for over 200 cycles at 0.8C. (G. Derrien et al., Nanostructured Sn–C composite as an advanced anode material in high-performance lithium-ion batteries. *Adv. Mater.* 2007, 19, 2336–2340. Copyright 2007, John Wiley and Sons, Reprinted with permission.)

images clearly show that the tin particles are kept at a nano-size dimension (about 10–30 nm) and that they are evenly dispersed within the carbon matrix. The carbon matrix has a twofold, critical role, in which it provides enough free volume to accommodate the tin's expansion–contraction, assuring the cycling stability, and at the same time, acts as a protecting shell assuring the safe handling of the electrode powder [16].

The above two beneficial roles are clearly demonstrated in Figure 3.16, which shows how the composite can be efficiently cycled in a lithium cell with a charge (lithium–tin alloying process: $SnC + 4.4Li \rightarrow Li_{4.4}Sn + C$)–discharge (lithium–tin de-alloying process: $Li_{4.4}Sn + C \rightarrow SnC + 4.4Li$) columbic efficiency approaching 100%. The long and stable cycle life, combined with the high efficiency, are very convincing evidences that the volume stresses are indeed successfully buffered.

The interest in Li–Si alloys is even greater than that in Li–Sn alloys, and it is understandable to consider that Li–Si offers the highest theoretical specific capacity so far known for lithium battery electrode materials [17]. The volume stress issue, already discussed for tin, also holds for silicon, possibly even at a greater extent. Indeed, the volume change expected for silicon upon full alloying with lithium to form $Li_{4.4}Si$ is of the order of 300% while that for tin at the same $Li_{4.4}Sn$ composition is about 250% (see Figure 3.15). As in the case of tin, the breakthrough in the use of silicon came with the development of appropriate carbon composites [18].

The results reported confirm that tin or silicon composite configuration is the correct approach in bringing the related lithium alloy electrodes to the commercial stage. In fact, their use is now seriously considered by battery manufacturers. For example, the Sn–C–Co ternary composite is presently used by a Japanese manufacturer to produce a battery under the commercial name of Nexelion. The characteristics and properties of metal alloy anodes are summarised in Table 3.1.

3.4 Advanced Electrolytes

The electrolyte in use in most common lithium-ion batteries is a $LiPF_6$–carbonate solvent solution embedded in a porous separator. This is possibly the most critical component impeding the progress of the lithium battery technology. The major issues are (i) the relatively narrow stability domain, which prevents the use of high-voltage cathodes; (ii) the high vapour pressure and the flammability, which affects safety; and (iii) potential hazards to health and the environment. There is, therefore, the need to discard the use of unsafe, liquid organic solutions and employ more inert electrolyte systems.

A possible candidate is solvent-free lithium conducting membranes. The benefits are substantial: the usage of a solid configuration gives increased cell reliability, while at the same time offers modularity in design and ease of handling. Much work has been done on polymer electrolytes for lithium batteries, and the interested reader may find details in a series of excellent reviews, in which the main achievements in the field are thoroughly discussed and evaluated [1,2]. Here, we will comment briefly on the most promising options. A very appealing possibility is obviously one that uses a fully solid, solvent-free membrane. Although various polymer systems of this kind have been proposed, interests today lie on membranes based on homopolymers, such as PEO, hosting a lithium salt, LiX, for example, lithium trifluoromethanesulphonate, $LiCF_3SO_3$. By combining chemical inertness with reasonably good lithium ion transport, the PEO–LiX complexes, in principle, meet the key requirements for an efficient electrolyte separator. In addition, these membranes allow the use of metal lithium anodes, with important reflections in terms of specific energy.

The structure of the PEO-based membranes can be described in terms of polymer chains folding around the lithium ions [5,6] (see Figure 3.17) [19]. The ion transport in the polymer electrolytes requires flexibility of the PEO chains to allow Li^+ ions movement from one chain to the other (see Figure 3.9), a situation that occurs at temperatures above 70°C. This is the main constraint that has so far prevented the wide use of these membranes. Many studies have been carried out with the aim of solving this problem. Some progress has been achieved by dispersing selected nano-size ceramic additives in the polymer bulk [20], widening the useful temperature range. However, successful operation of PEO–LiX membranes at ambient and subambient temperatures is yet to be demonstrated.

The high temperature of operation, however, does not totally rule out these membranes from practical applications in lithium batteries, especially if these are directed to the automobile sector where temperature may not be a critical factor. Accordingly, relevant projects have been launched in recent years for the development of lithium polymer batteries designed for EV application, for example, the above-mentioned electric car developed in France.

The intrinsic value of the polymer battery concept continues to attract industrial interest and some compromise approaches have been adopted while in expectation of a breakthrough in solvent-free membranes. The most common

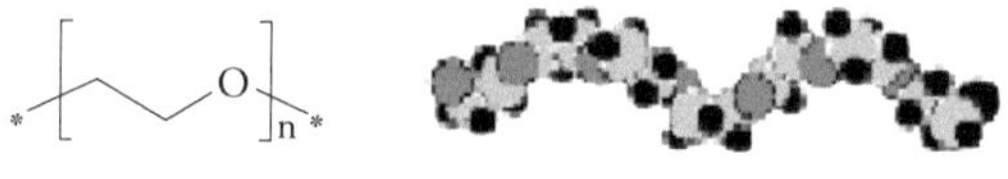

FIGURE 3.17
Polyethylene oxide formula: the light grey atoms are carbon, the dark grey are oxygen and the black are hydrogen. (Reprinted from *Electrochim. Acta*, 45, Y.G. Andreev and P.G. Bruce, Polymer electrolyte structure and its implications, 1417–1423, Copyright 2000, with permission from Elsevier.)

involves the use of solid–liquid hybrid membranes formed by trapping typical liquid lithium-ion solutions (e.g. $LiPF_6$–carbonate solvent) in a polymer matrix, for example, poly(acrylo nitrile) (PAN), or poly(vinylidene fluoride) (PVdF), to form polymer, gel-type electrolytes (GPEs) [21] (see Figure 3.18). The main feature of GPEs is its high conductivity, nearly matching that of a pristine liquid solution; on the other hand, issues typically associated with the presence of liquid, that is, safety and reliability, cannot be resolved completely. Nevertheless, GPEs are presently used by various battery manufacturers for the fabrication of the so-called 'lithium-ion polymer batteries', LiPBs.

Another emerging class of electrolytes are those based on ionic liquids, ILs. These are low-temperature molten salts with important specific properties [22] (see Figure 3.19).

FIGURE 3.18
Gel-type electrolytes (GPEs) are hybrid systems that may be described as formed by a liquid component contained within a polymer network.

Cation	Anion	Tm/°C
Na⁺	Cl⁻	801
Cs⁺	Cl⁻	645
(Pr)₄N	Cl⁻	241
	Cl⁻	87
	NO₃⁻	38
	BF₄⁻	15
	F₃C-SO₂-N-SO₂-CF₃	−14

FIGURE 3.19
Ionic liquids (ILs) are molten salts. NaCl is an example of an ionic compound, but its melting point is too high to be of practical interest. The melting temperature progressively decreases by varying either the cation or the anion of the salts, to reach low values, even below ambient.

Typically, ILs are formed by the combination of a weakly interacting, large cation, for example, of the imidazole type, and a flexible anion, for example, N,N-bis(trifluoromethanesulphonyl) imide, TFSI [22,23]. ILs are not volatile, not flammable, highly conductive, environmentally compatible and can safely operate over a wide temperature range (see Figure 3.20).

This unique combination of favourable properties makes ILs very appealing materials as stable and safe electrolyte media in lithium batteries. These electrolytes are prepared by combining lithium salts, usually with compatible anions, for example, lithium N,N-bis(trifluoromethanesulphonyl) imide, LiTFSI, with the selected IL [24] (see Figure 3.21). The addition of lithium salt generally results in an increase in viscosity, with related reduction in the

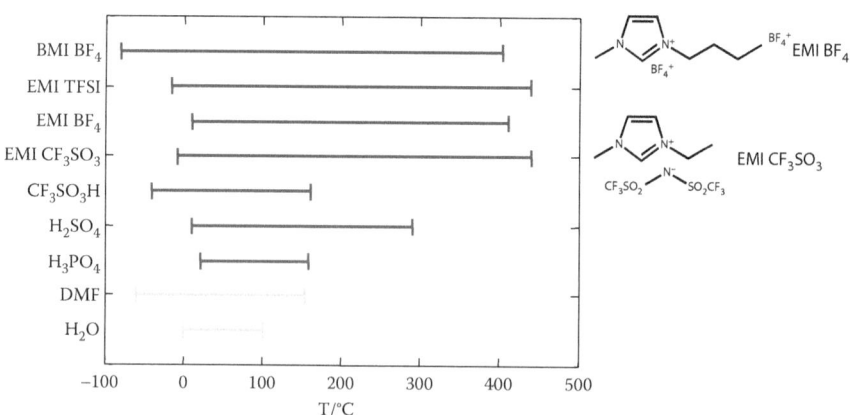

FIGURE 3.20
Temperature range of stability of imidazolium-based ILs (first four lines starting from the top of the figure) in comparison with that of several other liquid electrolyte solution (following five lines). BMI, 1-butyl-3-methylimidazolium; EMI, 1-ethyl-3-methylimidazolium; DMF, dimethylformamide.

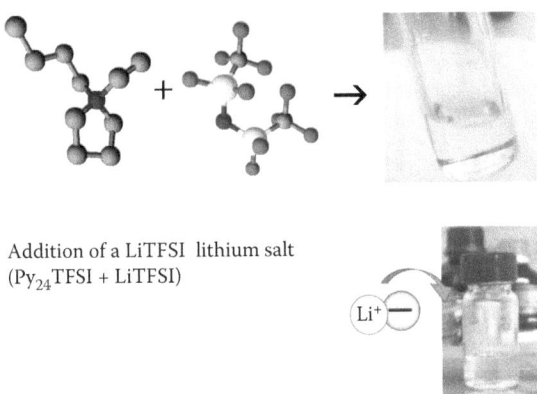

FIGURE 3.21
IL-based solutions to be used as electrolytes in lithium batteries are formed by adding a lithium salt, for example, LiTFSI in a selected IL, for example, a pyrrolidinium-based one, $Py_{24}TFSI$. (Adapted from B. Scrosati, J. Hassoun and Y.-K. Sun, Lithium-ion batteries. A look into the future, *Energy Environ. Sci.* 4, 2011, 3287–3295. Reproduced by permission of The Royal Society of Chemistry.)

overall conductivity, but maintained at levels above 10^{-3} S cm^{-1} (see Figure 3.22), making it still of interest for battery applications.

Many laboratories worldwide are engaged in the investigation of ILs with the aim of establishing their effective potentials as lithium battery electrolytes [23,25–28]. The results, however, are in part contradictory, especially in defining the electrochemical stability of lithium conducting, IL-based solutions. Most commonly, it is believed that these solutions have a limited cathodic stability due to the tendency of imidazolium-based cations to be reduced by electrochemical deprotonation at around 1.5 V vs. Li. This issue

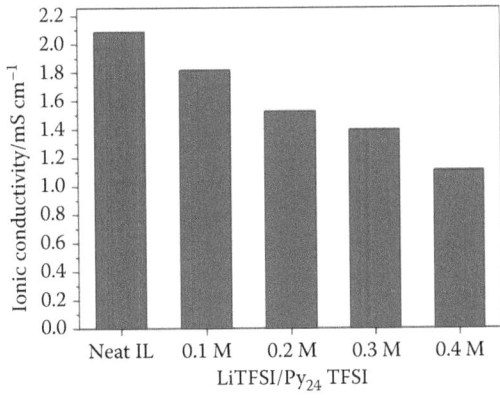

FIGURE 3.22
Ionic conductivity of neat $Py_{24}TFSI$ IL and of its 0.1, 0.2, 0.3, 0.4 M LiTFSI/$Py_{24}TFSI$ solutions at room temperature. The conductivity remains above 10^{-3} S cm^{-1} at all compositions.

TABLE 3.2

(**See colour insert.**) Summary of Characteristics and Properties for Electrolyte Materials Designed for Lithium-Ion Batteries

Material	Advantages	Issues
Liquid organic solutions	High conductivity over a wide temperature range. Liquid state.	Narrow electrochemical stability window. Safety.
Gel-type membranes	High conductivity in a wide temperature range	Solid–liquid hybrid configuration; liquid retention
PEO-LiX, membrane	Solvent-free, fully solid configuration; low cost	Temperature-dependent conductivity; low lithium ion transfer number and limited oxidation stability
Ionic liquids	Non-flammable. High thermal stability. High conductivity	Scarce compatibility with low-voltage anodes

apparently prevents the use of IL-based solutions with common low-voltage anode materials, such as lithium metal, graphite or even $Li_4Ti_5O_{12}$. Many studies are underway to circumvent this problem by developing ILs based on cations more resistant to reduction than those of the imidazolium family. Promising results have been obtained by shifting to aliphatic quaternary ammonium cations having no acidic protons and thus, expected to have a stability domain extending to low voltages [24]. A good example is the IL formed by combining *N-n*-butyl-*N*-ethyl-pyrrolidinium cation with *N,N*-bis(trifluoromethanesulphonyl) imide anion and having LiTFSI as the dissolved lithium salt [24].

The characteristics and properties of advanced electrolytes are summarised in Table 3.2.

3.5 Advanced Cathodes

In the search of new cathodes, particular attention is addressed to materials of the olivine family and, in particular, on lithium iron phosphate, $LiFePO_4$ [29]. This interest is motivated by the many appealing features of this compound, which include good capacity (170 mAh g^{-1}), a two-phase electrochemical process that evolves with a flat 3.5 V vs. Li voltage (see Figure 3.23),

Lithium Batteries

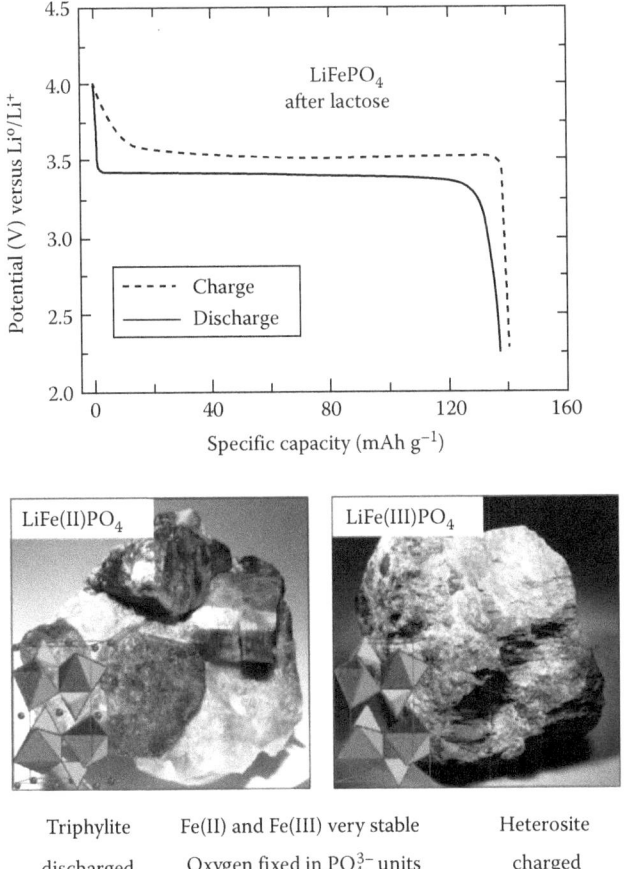

FIGURE 3.23
(See colour insert.) The electrochemical process of LiFePO$_4$ involves two phases, triphylite–heterosite equilibrium resulting in a flat charge–discharge voltage profile evolving around 3.5 V.

and, most significantly, cost, which in principle is much lower than that of LiCoO$_2$. In addition, LiFePO$_4$ is a chemically more stable material, since the PO$_4$ group has stronger covalent bonds than CoO$_2$ (see Figure 3.24), which results in better operational safety [30].

On the other hand, LiFePO$_4$ suffers from a very high intrinsic resistance, which requires special material preparation by sophisticated coating processes (see Figure 3.25) to obtain an enhancement of the electronic conductivity and, consequently, acceptable electrode performance. Low-temperature synthesis of tailor-made LiFePO$_4$ powders via the use of basic media and microwave procedures [31] is very promising in this respect.

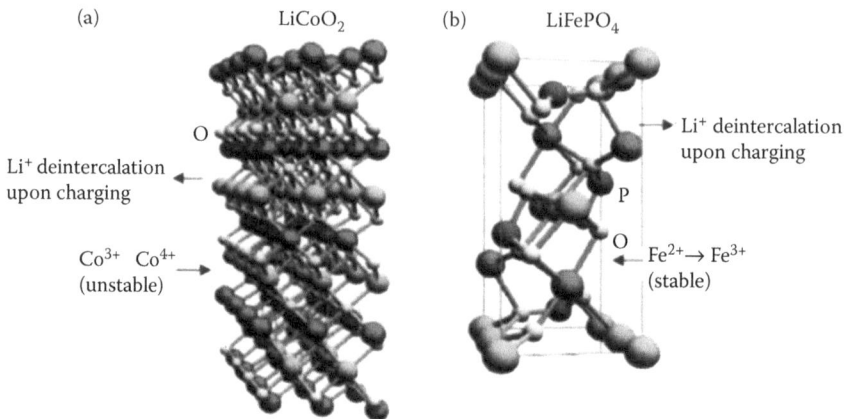

FIGURE 3.24
(**See colour insert.**) Lithium iron phosphate (b) is structurally stronger than lithium cobalt oxide (a), and hence is much less prone to oxygen release during battery operation. (K.E. Aifantis and S.A. Hackney, Current and potential applications of secondary Li batteries, *High Energy Density Lithium Batteries: Material, Engineering, Applications*, 2010, pp. 81–101. Copyright 2010, Wiley-VCH Verlag GmbH & Co. KgaA. Reprinted with permission.)

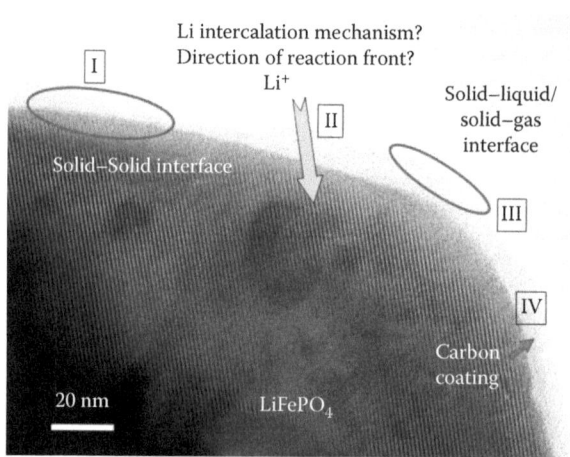

FIGURE 3.25
Thin surface carbon coating (IV) is needed to enhance the overall electronic conductivity of the electrode. The carbon coating results in the inner part in the formation of carbon-electrode solid–solid interface (I) and in the outer part of solid–liquid or solid–gas interface (III). The lithium ion intercalation occurs across the carbon layer into the electrode structure (II). (From 'Report of the Basic Energy Sciences Workshop on Electrical Energy Storage', DOE, July 2007.)

Lithium Batteries

TABLE 3.3

(**See colour insert.**) Characteristics and Properties for Cathode Materials Designed for Lithium-Ion Batteries as Alternatives to the Common $LiCoO_2$

Material	Theoretical capacity (mAhg^{-1})	Advantages	Issues	Structure
$LiNi_{0.5}Mn_{1.5}O_4$	145	High voltage, good specific capacity	Operation within the limits of the electrolyte stability domain	
$LiNi_{1/3}Co_{1/3}Mn_{1/3}O_2$	290	High capacity, high rate	Structure retention upon cycling	
$LiFePO_4$	170	Basic low cost, intrinsic safety and environmental compatibility	Low electronic conductivity, low tap density and needs C coating	
$LiCoO_2$	140	Field-tested operation	High cost	

Other innovative cathode materials attracting interest in lithium-ion battery technology are based on spinel-type or layered structured compounds, such as lithium nickel manganese oxide, $LiNi_{0.5}Mn_{1.5}O_4$ [32], and nickel cobalt manganese oxide, $LiNi_{1/3}Co_{1/3}Mn_{1/3}O_2$ [33]. The former material is characterised by a two-phase electrochemical process reflecting in a flat voltage profile evolving around 4.5 V vs. Li. The theoretical specific capacity is 146 mAh g^{-1}. $LiNi_{1/3}Co_{1/3}Mn_{1/3}O_2$ operates via a typical lithium insertion/de-insertion electrochemical process, characterised by a sloping voltage averaging around 4 V vs. Li. The theoretical specific capacity largely exceeds that of $LiCoO_2$, that is, 290 mAh g^{-1} vs. 140 mAh g^{-1}.

The properties and characteristics of the novel cathode materials are summarised in Table 3.3.

Although these advanced materials allow some innovation in lithium battery technology, their capacity is still not at a level sufficient to produce a significant increase in energy density. Therefore, new electrodes have to be developed to make lithium batteries applicable in the automobile industry.

3.6 Super Energy Density Batteries: Lithium–Sulphur and Lithium–Air

To further advance the science and technology of lithium batteries, new avenues must be opened. Changes in the chemical structures, as described in the previous section, are not sufficient. Improvements in environmental sustainability and energy content are mandatory; these can only be obtained by complete removal of the lithium battery concept, so far mostly based on insertion chemistry. Although the insertion electrodes are based on sustainable 3D metals, such as Ti (TiO_2, $Li_4Ti_5O_{12}$) or Fe ($LiFePO_4$), they are produced from ores, and thus their extraction and manipulation require constantly increasing the input of energy. This expensive fabrication process poses some questions about the long-term viability of lithium batteries.

In addition, insertion reactions are confined to a maximum of one electron transfer per atom of transition metal, thus limiting the specific energy. Significant increases in performance require radical changes in the fundamental electrochemical process, such as a passage from insertion to conversion chemistry which may basically assure operation, implying from two to six electron transfers [3]. Simply, the need is to pass from the classic $xLi + MX_y \rightleftarrows Li_xMX_y$ insertion process to a new $xLi + MX_y \rightleftarrows Li_xX_v + M$ conversion process.

The pioneering work of Tarascon and co-workers introduced a search for high-capacity, conversion electrode materials [34]. A reaction pathway enlisting complete electrochemical reduction of metal oxides, sulphides, nitrides, phosphides and fluorides into a composite consisting of nano-sized particles dispersed in an amorphous Li_mX (X = O, S, N, P) was demonstrated. However, these conversion electrodes suffered from dramatic hysteresis in voltage between charge and discharge, leading to poor energy and voltage efficiency. Following this work, substantial improvement was obtained by using a metal hydride, for example, MgH_2, as the conversion electrode [35]. This electrode shows a reversible process $2Li + MgH_2 \rightleftarrows Mg + 2LiH$, delivering a practical capacity as high as 1480 mAh g^{-1} at an average voltage of 0.5 V vs. Li, combined with a very low charge–discharge polarisation, both properties making it suitable for anode application in practical batteries.

3.6.1 Lithium–Sulphur Battery

The pursuit of high capacity has been addressed also to alternative cathodes, such as sulphur or air–oxygen. A most promising candidate for high-energy systems is the lithium/sulphur (Li/S) battery based on the electrochemical reaction: $16 Li + S_8 \rightleftarrows 8Li_2S$, which, assuming complete conversion, has an energy density of 2500 Wh kg^{-1} and 2800 Wh L^{-1}, in terms of weight and volume, respectively (see Figure 3.26). However, the discharge process involves the progressive formation of various polysulphides according to the sequence: $Li_2S_8 \rightarrow Li_2S_6 \rightarrow Li_2S_4 \rightarrow Li_2S_2 \rightarrow Li_2S$, which, especially in the initial phase, are highly soluble in the electrolyte [36]. This high solubility results in a loss of active mass leading to low utilisation of the sulphur cathode and severe capacity decay upon cycling. The dissolved polysulphide anions, by migration through the electrolyte, may reach the lithium metal anode where they react to form insoluble products on its surface, contributing to the degradation of battery performance [37].

Another issue of the Li/S battery is the low electronic conductivity of S, Li_2S and the intermediate Li–S products, which severely affects the rate

Lithium Batteries

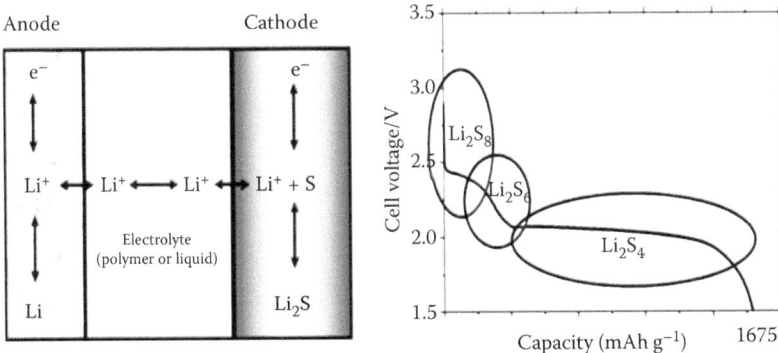

FIGURE 3.26
Schematic diagram of the most conventional Li/S batteries and its typical discharge voltage profile. The circled areas correspond to the progressive formation of the polysulphides during discharge. (Adapted from B. Scrosati, J. Hassoun and Y.-K. Sun, Lithium-ion batteries. A look into the future, *Energy Environ. Sci.* 4, 2011, 3287–3295. Reproduced by permission of The Royal Society of Chemistry.)

capability of the battery. A final problem, often overlooked, is in the use of lithium metal as the preferred anode, which is known to cause serious safety risks due to the uneven deposition upon charge, which may result in the cell being short-circuited and eventually lead to overheating or explosion.

Although investigated by many workers for several decades [38–41], the practical development of the lithium/sulphur battery has been hindered for a long time. There has been renewed interest in this system in the recent years, thanks to a series of important technological breakthroughs. Various strategies to address the solubility issue have been successfully developed. These include, among others, the design of modified organic liquid electrolytes [42] and the use of IL-based electrolytes [43]. An even more effective action to block the dissolution of the reaction products is expected to be provided by the use of completely solvent-free, solid-state, lithium conducting membranes, such as those formed by PEO–lithium salt complexes (see Section 3.4). These membranes have been effectively tested as separators in lithium–sulphur cells by Jeong et al. [44] and more recently by Hassoun and Scrosati [45]. The results are quite encouraging, demonstrating that full capacity can be obtained by solid-state, PEO-based polymer Li/S–C batteries (see Figure 3.27).

An apparent issue is that these batteries have to be operated at around 70–90°C, that is, in the temperature range where the conductivity of the solid-state membranes is sufficiently high [6]. However, this would not be a severe limitation if the battery is addressed to applications where moderately high-temperature operations can be tolerated, such as in the EV area.

Important progress has also been achieved in improving the rate capability of the sulphur electrode. The pioneering work of Nazar and co-workers demonstrated that this goal may be achieved by innovative nanostructures. These authors showed that by fabricating cathodes based on a close interaction

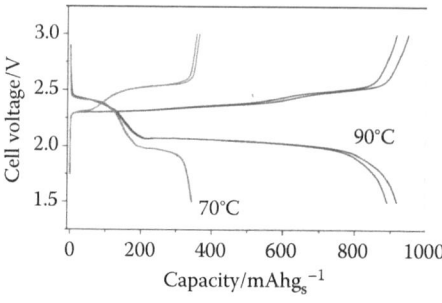

FIGURE 3.27
Charge–discharge voltage profiles of a Li/S battery using a PEO-based solid electrolyte. (J. Hassoun and B. Scrosati, Moving to a solid-state configuration: A valid approach to making lithium-sulfur batteries viable for practical applications, *Adv. Mater.* 2010. 22, 5198–5201. Copyright 2010. John Wiley and Sons. Reprinted with permission.)

of nanostructured sulphur and mesoporous carbon, high reversible capacity and good rate can be obtained [46] (see Figure 3.28). This result confirmed that optimisation in the fabrication of the cathode, for example, replacing simple sulphur–carbon mixtures with sulphur–carbon composites, is a successful approach for improving the conductivity of the electrode, as well as for controlling the solubility of the discharge products [47,48]. Accordingly, various types of nanocomposite S–C electrode structures have been developed. Relevant results have been obtained with a homogeneous dispersion of the sulphur particles in hard carbon spherules, HCS–S [49,50] (see Figure 3.29).

Another step forward in this technology was obtained by changing the cathode from the common sulphur–carbon composite to a lithium sulphide–carbon composite [51]. In this way, the cathode becomes the lithium ion source, allowing replacement of the reactive lithium metal with other more reliable lithium-accepting material, thus obtaining a metal-free, lithium-ion sulphur battery. This concept was first demonstrated by using a tin–carbon anode [52,53] and

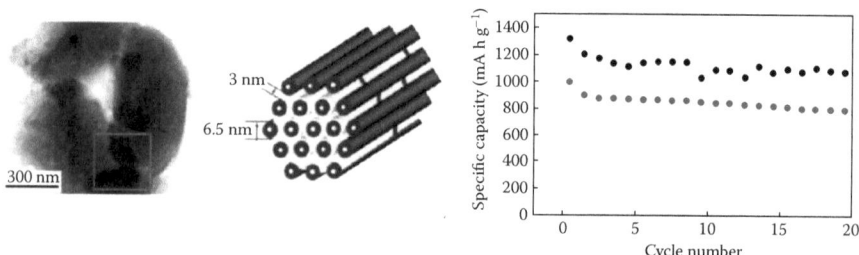

FIGURE 3.28
Advanced, nanostructured sulphur and mesoporous carbon cathode. (Reprinted with permission from *Nat. Mater*, A highly ordered nanostructured carbon-sulphur cathode for lithium-sulphur batteries, X. Ji, K.T. Lee and L.F. Nazar, 8, 2009, 500–506. Copyright 2009, Nature Publishing Group.)

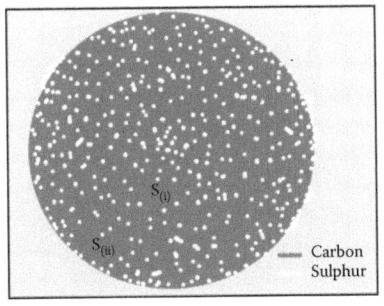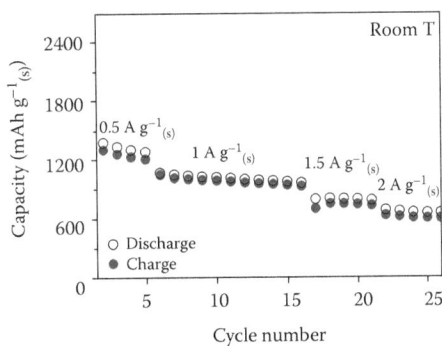

FIGURE 3.29
The Li/S battery using a cathode formed by homogeneously dispersing sulphur particles in a hard carbon spherule (HCS) matrix can cycle with a very high rate capability. S(i) represents internal sulphur while S(ii) represents surface sulphur. (Reprinted from *J. Power Sources*, 202, J. Hassoun et al., A contribution to the progress of high energy batteries: A metal-free, lithium-ion, silicon–sulfur battery, 308–313. Copyright 2012, with permission from Elsevier.)

a gel-type electrolyte giving a battery of the type $SnC/GPE/Li_2S-C$, characterised by the following electrochemical process: $SnC + xLi_2S \rightleftarrows 2Li_xSn + S + C$, where x can extend up to the value of 4.4. The battery is capable of delivering about 800 mAh $g_{Li_2S}^{-1}$ for over 30 cycles (see Figure 3.30).

The lithium metal-free sulphur battery concept was confirmed by using a lithiated silicon–carbon anode, an HCS–S (hard carbon spherules–sulphur composite) cathode (cfr. Figure 3.29) and a $LiCF_3SO_3$TEGDME liquid electrolyte (see Figure 3.31) [50]. The electrochemical process involves the transfer of lithium ions from the anode to the cathode: $Li_xSi-C + S \rightleftarrows Li_xS + xSiC + HCS$ (HCS: hard carbon spherules). The battery delivers a capacity of 500 mAh g^{-1}_S at an average voltage of 1.8 V (see Figure 3.31).

These results consistently contributed to improvements in the technology of lithium/sulphur battery; however, some residual issues still remain to prevent full practical application of this high-energy battery system. Most of the recent works rely on conventional liquid organic carbonate solution as the preferred electrolyte; although the sulphur and/or the lithium sulphide are shielded into a carbon matrix, either mesoporous [46] or spherical [50], thus in principle preventing the direct contact with the electrolyte, it is not yet fully established whether the solubility of the polysulphides is effectively blocked. Another issue still to be solved is in the relatively modest power capability of these batteries.

3.6.2 Lithium–Air Battery

Oxygen as a cathode material is even more promising than sulphur since in principle its use may lead to the development of lithium–air batteries having a theoretical energy density approaching the value provided by gasoline. It is

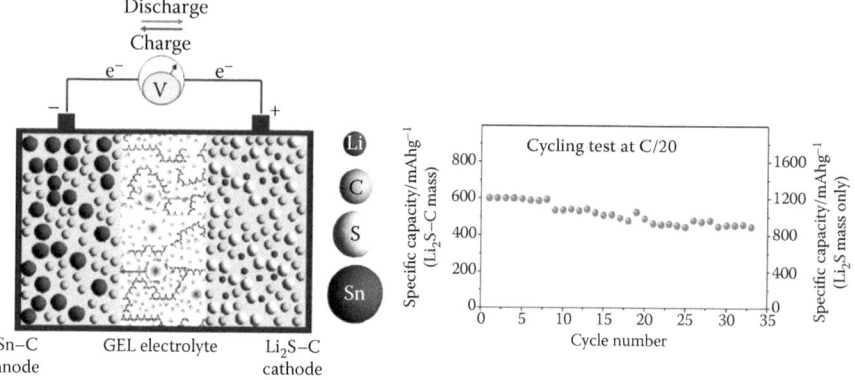

FIGURE 3.30
(**See colour insert.**) Scheme of a SnC/Li$_2$S lithium-ion sulphur battery and its cycling response at a rate of 38 mA cm^{-2} g^{-1} (C/20) in the range of 0.2–4 V. (J. Hassoun and B. Scrosati, A high-performance polymer tin sulfur lithium ion battery, *Angew. Chem. Int. Ed.* 2010, 49, 2371–2374. Copyright 2010, Wiley-VCH Verlag GmbH & Co. KgaA. Reprinted with permission.)

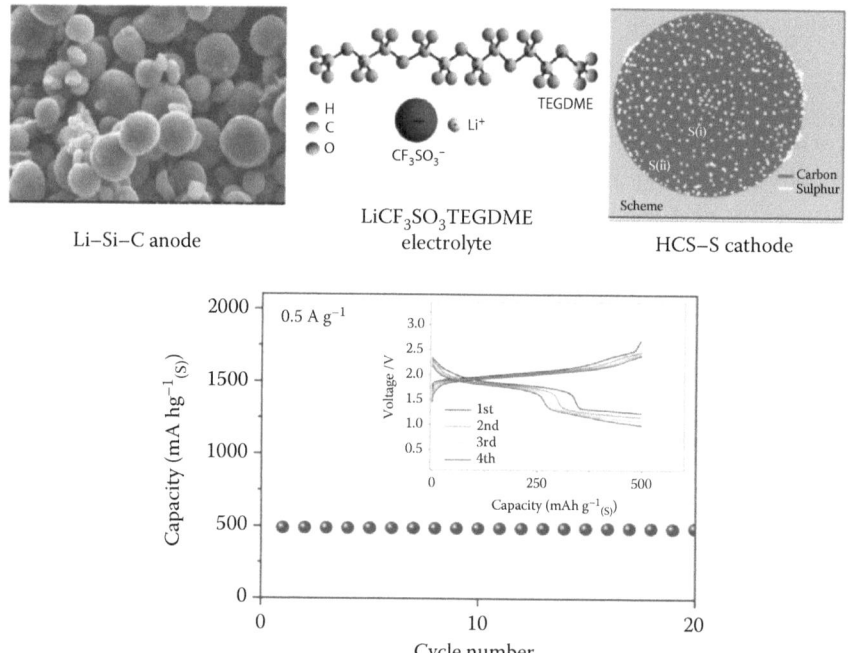

FIGURE 3.31
(**See colour insert.**) Scheme of a LiSiC/S lithium-ion sulphur battery and its cycling response. The two plateaus in discharge are associated with the subsequent formation of polysulphides (cfr. Figure 3.26). (Reprinted from *J. Power Sources*, 202, J. Hassoun et al., A contribution to the progress of high energy batteries: A metal-free, lithium-ion, silicon–sulfur battery, 308–313, 2012. Copyright 2012, with permission from Elsevier.)

this astonishing energy value, much greater than that of any known battery, that has triggered worldwide interest in lithium–air batteries [54]. Accordingly, R&D of lithium–air batteries is heavily funded in various countries involving a large number of academic and industrial laboratories worldwide.

3.6.2.1 Aqueous Electrolyte

There are mainly two approaches, basically varying in the type of electrolyte used, that are presently explored in the development of lithium–air batteries. One uses an aqueous electrolyte in combination with a lithium metal electrode protected by a water-repulsive Li^+ conductive glass–ceramic of the Nasicon type [55–58] or by an anion exchange membrane [55] (see schematic in Figure 3.32). Here, the overall electrochemical process is $O_2 + 4Li + 2H_2O \rightleftarrows 4LiOH$, which has an estimated energy density of the order of 5000 Wh kg^{-1}. This lithium–air aqueous battery operates preferentially in the primary mode. The remaining issues to be solved before achieving full practical application are in the mechanical instability of the protecting film, its high interfacial resistance and the solubility of the reaction products.

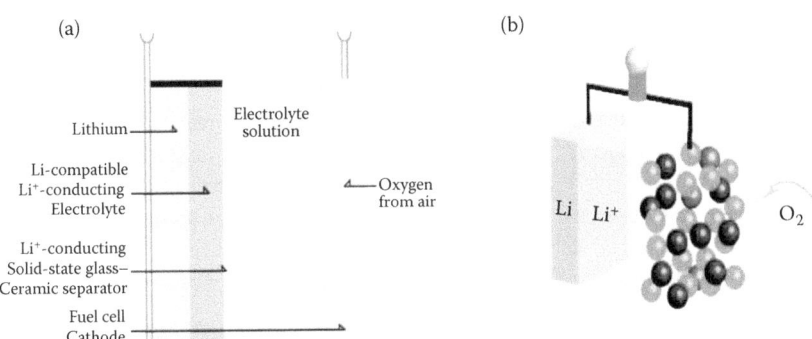

Lithium–air battery with protected lithium metal anode and/or protected cathode (aqueous electrolyte)
$2Li + \frac{1}{2} O_2 + H_2O \rightarrow 2LiOH$
Theor. energy density: 5800 Wh/kg

Lithium–air battery with unprotected lithium metal anode (nonaqueous electrolyte)
$Li + \frac{1}{2} O_2 \rightarrow \frac{1}{2} Li_2O_2$
Theor. energy density: 11,420 Wh/kg

Present lithium ion technology (C–LiCoO$_2$):
Theor. energy density: 480 Wh/kg

FIGURE 3.32
The two types of lithium–air batteries under development. (a) aqueous electrolyte, protected anode design; (b) non-aqueous electrolyte, unprotected lithium design. (Adapted from B. Scrosati, J. Hassoun and Y.-K. Sun, Lithium-ion batteries. A look into the future, *Energy Environ. Sci.* 4, 2011, 3287–3295. Reproduced by permission of The Royal Society of Chemistry.)

3.6.2.2 Non-Aqueous Electrolyte

The second approach considers the use of non-aqueous electrolytes [59–61]. The scheme of the battery, also shown in Figure 3.33, comprises of a lithium metal anode, a separator embedded in a non-aqueous electrolyte, a carbon-supported air electrode (with or without catalyst) and a gas diffusion layer. This is the most appealing and studied framework since its basic process $4Li + O_2 \rightleftarrows 2Li_2O$ theoretically provides the highest energy density. However, initial efforts in this system did not provide convincing evidence of its practicality since the reported performance was limited to few charge–discharge cycles and to a low rate capability.

This poor result was due to a series of severe issues that affected the implementation of the lithium–air battery in its early stage. These are mainly

1. Instability of the electrolyte in the cell environment
2. Low kinetics of the oxygen electrode in the non-aqueous electrolyte
3. Reactivity of the lithium metal anode

All these issues are very critical for the proper operation of lithium–oxygen cells. The electrolyte instability severely affects the cycle life. Indeed, the poor cycling performance observed in the first prototypes was due to this factor. With the formation of lithium peroxide: $2Li + O_2 \rightleftarrows Li_2O_2$, the overall oxygen electrochemical process proceeds via a sequence of intermediates steps: $O_2 + e^- \rightarrow \dot{O}_2^- + 2Li \rightarrow Li_2O_2$, including a radical anion $\dot{O}_2^-$ that readily decomposes most electrolytes [62], in particular the organic carbonate solutions commonly used in lithium-ion batteries (see Figure 3.33).

Therefore, in the early development of lithium–air technology, electrolyte decomposition rather than the desired oxygen reduction process was predominant, which explains their poor cycle life [63]. Accordingly, the choice of a stable electrolyte is one of the challenges in lithium–air batteries.

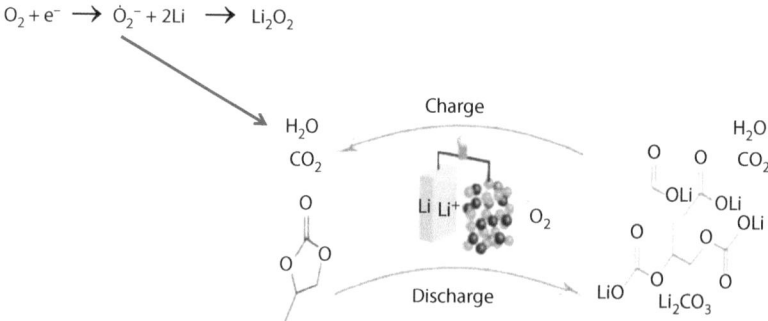

FIGURE 3.33
The oxygen reduction process involves a sequence of intermediates steps, including a radical anion $\dot{O}_2^-$ that readily decomposes the organic carbonate solutions used as electrolytes in the early development of the lithium–air cells.

3.6.2.3 Electrolyte Issue

Much attention has been focused to address the electrolyte issue and thus, to identify a stable medium. Valid examples are di-methoxyethane, DME-based [64] and IL-based solutions [65]; however, their use may be limited by over vapour pressure (in the DME case) or by cost (in the IL case). Also, polymer electrolytes, for example, those based on the combination of PEO and a lithium salt, for example, lithium triflate, $LiCF_3SO_3$, have been considered as suitable lithium–air battery media [62]. In this case, the limitation comes from the thermal dependence of the ionic conductivity that reaches acceptable values only at temperatures above 70°C. Promising alternatives are end-capped glymes, such as tetra (ethylene) glycol dimethyl ether (TEGDME), which, due to their high solvating power and low sensitivity to the oxygen reduction products, are expected to be stable and efficient electrolyte media. It can be assumed that this stability is associated with the chemical inertia of the ether groups, which is much higher than that of the carboxyl groups of conventional organic carbonate solutions (see Figure 3.34).

In glyme electrolytes, such as TEGDME-$LiCF_3SO_3$, the resistance of the ether linkage is well known (e.g. Grignard reagents). Hence, these electrolytes are expected to be nucleophilically stable. Another advantage of the glyme-based electrolyte is the stability of the triple contact between the carbon, the electrolyte and the surrounding O_2. There is some controversy on the effective stability of TGEDME-based electrolytes. Some authors have in fact expressed some concern [66,67] while many others have safely used the tetra-glymes in lithium–air battery studies [61,68–72]. Indeed, it has been recently demonstrated that lithium–air batteries based on these electrolytes can be safely and effectively cycled for hundreds of times [72] (see Figure 3.35).

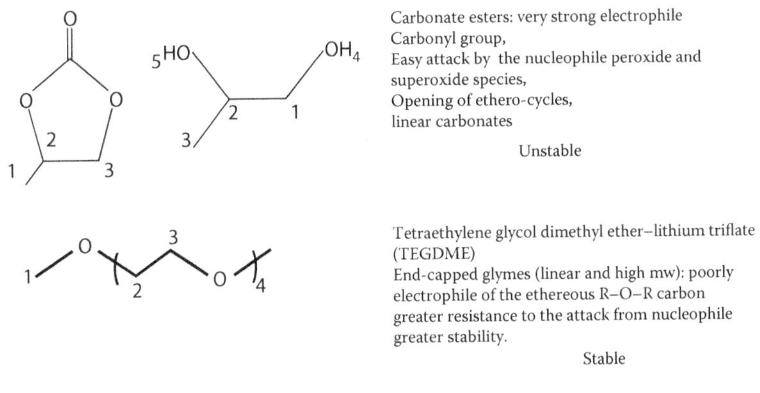

Carbonate esters: very strong electrophile
Carbonyl group,
Easy attack by the nucleophile peroxide and superoxide species,
Opening of ethero-cycles,
linear carbonates

Unstable

Tetraethylene glycol dimethyl ether–lithium triflate (TEGDME)
End-capped glymes (linear and high mw): poorly electrophile of the ethereous R–O–R carbon
greater resistance to the attack from nucleophile
greater stability.

Stable

FIGURE 3.34
Electrolytes for Li–air batteries. Solutions based on end-capped glymes, such as tetra (ethylene) glycol dimethyl ether, which, due to their high solvating power and low sensitivity to the oxygen reduction products, are expected to be stable electrolyte media in lithium–air cells.

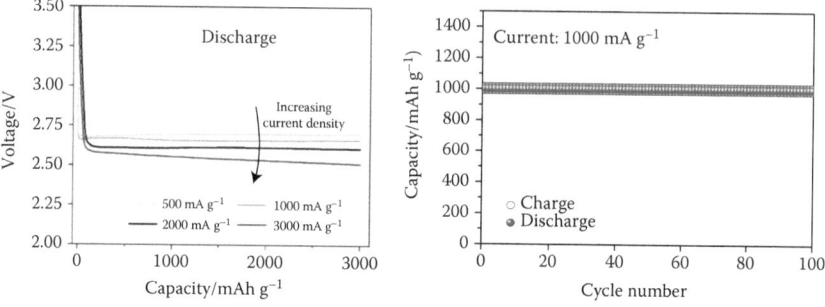

FIGURE 3.35
Discharge curves at various rates and cycle response at 1000 mAh g^{-1} of a Li/O$_2$ battery based on a LiCF$_3$SO$_3$–TGEDME electrolyte. (Reprinted with permission from Nat. Chem. An improved high-performance lithium–air battery, H.-G. Jung et al., 4, 2012, 579–585. Copyright 2012 Nature Publishing Group.)

Another issue affecting the development of non-aqueous lithium–air batteries is in the low kinetics of the oxygen electrode, resulting in a still unsuitable rate capability. Considering that a full automotive power of 100 kW is required for EVs, the Li/O$_2$ battery discharge rates have to be raised from the 10 mA cm^{-2} level to assure reasonable values of electrode area [73]. It is therefore of high importance to improve the rate of the Li/O$_2$ battery. In this respect, the use of catalysts to reduce the oxygen electrode overvoltage is an approach currently pursued and novel electrocatalysts for smoothing out the oxygen reaction are extensively studied worldwide. Although the real effectiveness of the catalytic activity is still under debate [74,75], recent studies confirmed that the oxygen reduction reaction (ORR) is rather insensitive to the presence of catalysts, such as transition metal oxides [60,76] or even noble metals [77–79], since the reaction is satisfactorily promoted by the carbon support alone. This is not the case for the oxygen evolution reaction (OER), which appears to be catalytically sensitive [78].

Many materials have been proposed for use as OER catalysts, including a series of manganese, iron, nickel, cobalt and copper oxides [79,80]. However, the activity of the catalysts may be compromised by various negative effects intrinsic to the battery medium, such as the precipitation on the catalyst surface of Li$_2$O$_2$ formed during discharge and/or dissolution of the catalyst nanoparticles into the electrolyte. In addition, in view of the complex reaction mechanism, it cannot be taken for granted that both the reduction and the re-oxidation process can be influenced by the same catalyst. Indeed, recent work by Gasteiger and co-workers has shown that a 75% efficiency of the battery processes can only be achieved with the use of bifunctional Pt/Au catalyst (see Figure 3.36) [79]. On the other hand, high cost may limit the practical use of this material and more realistic options have to be identified.

The investigation of possible catalysts to promote the kinetics of the oxygen electrode has been undertaken in the recent years. Bruce and co-workers

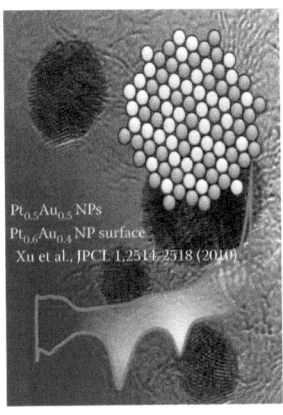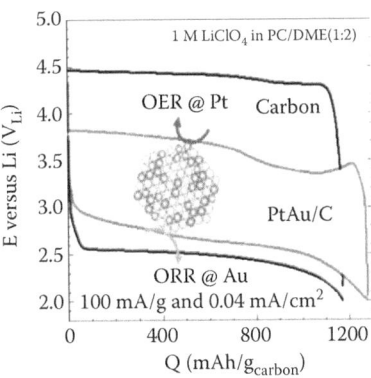

FIGURE 3.36
(See colour insert.) Li–O$_2$ cell discharge/charge profiles of carbon (black, 85 mA g$_{carbon}^{-1}$) and PtAu/C (red, 100 mA g$_{carbon}^{-1}$) in the third cycle at 0.04 mA cm$^{-2}_{electrode}$. (Reprinted with permission from Y.-C. Lu et al., Platinum–gold nanoparticles: A highly active bifunctional electrocatalyst for rechargeable lithium–air batteries, *J. Am. Chem. Soc.* 132, 2010, 12170–12171. Copyright 2010. American Chemical Society.)

[76] have examined a series of metal oxides. Among these, the most promising is α-MnO$_2$ (see Figure 3.37) [80]; in fact, this material is currently used as the preferred catalyst for OER in lithium–oxygen cells.

Additional studies have been conducted on various classes of alternative catalysts materials. Nazar and co-workers have recently reported that nanocrystalline Co$_3$O$_4$ grown on reduced graphene oxide promoted a significant

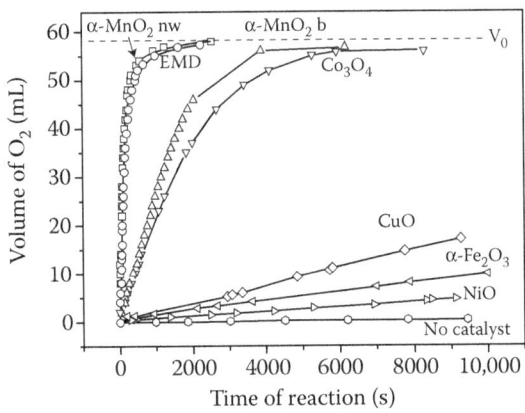

FIGURE 3.37
Time evolution of volume of O$_2$ (as product of the OER process: H$_2$O$_2$ ⇆ H$_2$O + $\frac{1}{2}$O$_2$) at 25°C and 1 atm. Clearly, α-MnO$_2$ is the most effective catalyst. (Reprinted with permission from V. Giordani et al., H$_2$O$_2$ decomposition reaction as selecting tool for catalysts in Li –O2 cells, *Electrochem. Solid-State Lett.* 13, 2010, A180–A183. Copyright 2010. The Electrochemical Society.)

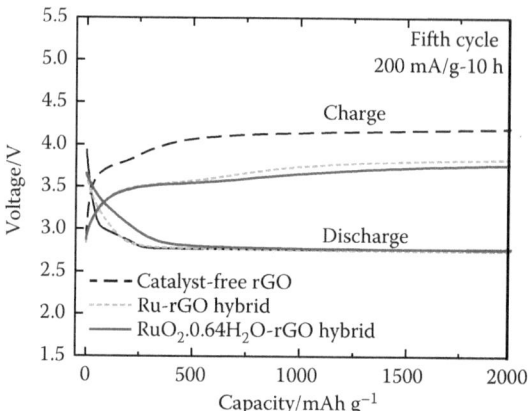

FIGURE 3.38
Comparison between the discharge–charge cycles of Li–air cells using graphene oxide (rGO) only with those using ruthenium-loaded graphene oxide (Ru–rGO hybrid) and ruthenium oxide-loaded graphene oxide (RuO$_2$ · 0.64H$_2$O–rGO hybrid). (Reprinted with permission from H.-G. Jung et al., Ruthenium-based electrocatalysts supported on reduced graphene oxide for lithium-air batteries, *ACS Nano* 7, 2013, 3532–3539. Copyright 2013. American Chemical Society.)

reduction in overpotential for the OER reaction [81]. Another material, metallic mesoporous pyrochlore, which showed promising catalytic activity, was also recently proposed by Nazar and co-workers [82]. Bruce and co-workers demonstrated that replacing carbon with nanoporous gold as oxygen electrode support consistently increases the rate of the OER process, and maintained high capacity delivery for over 100 cycles [83]. Indeed, the nature and the morphology of the support play a key role in promoting the catalyst activity [84,85]. Thus, the use of a two-dimensional graphene support has been proven to be quite effective in enhancing catalytic activity compared to conventional carbon supports, such as Vulcan or carbon black.

The favourable behaviour of the graphene supports may be attributed to a series of positive features, such as (i) a high dispersion and low aggregation of noble metal catalysts resulting from an enhanced interaction between the functionalised graphene surface and the noble metals [86]; and (ii) a large surface area of the graphene support [87]. It has been recently demonstrated that ruthenium-based materials loaded on reduced graphene oxide have a significant catalytic activity in promoting OER in non-aqueous lithium–oxygen cells [88]. Figure 3.38 compares discharge–charge cycles of Li–air cells using ruthenium-loaded graphene oxide (Ru–rGO hybrid), and ruthenium oxide-loaded graphene oxide (RuO$_2$ · 0.64H$_2$O–rGO hybrid), with those using graphene oxide (rGO) only. The role of ruthenium-based catalysts in decreasing the OER overvoltage is clearly demonstrated.

In summary, a comparison of the discharge voltage and capacity retention capability of OER catalysts for Li–O$_2$ cell is displayed in Table 3.4.

Lithium Batteries

TABLE 3.4
List of Possible Catalysts for Li/O₂ Batteries

Catalyst	Current Density (mM g_{carbon}^{-1})	Discharge Voltage (V)	Capacity of First Cycle (mAh g_{carbon}^{-1})	Capacity of (n^{th} Cycle) (mAh g_{carbon}^{-1})	Capacity Retention per Cycle (%)	References
Pt	70	2.55	470	60 (10)	80	[76]
La$_{0.8}$Sr$_{0.2}$MnO$_3$	70	2.6	750	40 (10)	72	[76]
Fe$_2$O$_3$	70	2.6	2700	75 (10)	67	[76]
Fe$_2$O$_3$–carbon added	70	2.6	2500	75 (10)	68	[76]
NiO	70	2.6	1600	600 (10)	90	[76]
Fe$_3$O$_4$	70	2.6	1200	800 (10)	96	[76]
Co$_3$O$_4$	70	2.6	2000	1300 (10)	95	[76]
CuO	70	2.6	900	600 (10)	96	[76]
CoFe$_2$O$_4$	70	2.6	1200	800 (10)	96	[76]
Pt nanoparticles	100	2.6	775	925 (3)	109	[76]
Carbon	85	2.65	1000	—	—	[79]
Au/C	100	2.7	1340	—	—	[79]
PtAu/C	100	2.7	1410	1300 (3)	96	[79]
Nanoporous Au	5000	2.6	3000	3000 (10)	100	[83]
α-MnO$_2$ wires	70	2.65	2950	1625 (10)	94	[60]
β-MnO$_2$ wires	70	—	2375	100 (10)	70	[60]
γ-MnO$_2$	70	—	1900	800 (10)	91	[60]
λ-MnO$_2$	70	—	1875	100 (10)	72	[60]
α-MnO$_2$ bulk	70	—	1500	550 (10)	89	[60]
EMD	70	—	1400	800 (10)	94	[60]
Mn$_3$O$_4$ commercial	70	—	1050	450 (10)	91	[60]
Mn$_2$O$_3$ bulk	70	—	1050	100 (10)	77	[60]
β-MnO$_2$ bulk	70	—	1000	375 (10)	90	[60]

continued

TABLE 3.4 (Continued)
List of Possible Catalysts for Li/O$_2$ Batteries

Catalyst	Current Density (mM g$_{carbon}^{-1}$)	Discharge Voltage (V)	Capacity of First Cycle (mAh g$_{carbon}^{-1}$)	Capacity of (nth Cycle) (mAh g$_{carbon}^{-1}$)	Capacity Retention per Cycle (%)	References
Carbon in LiPF$_6$/TEGDME	70	2.6	2800	700 (4)	63	[82]
Mesoporous Pyrochlore	70	2.6	7000	9000 (4)	109	[82]
Nanocrystalline expanded pyrochlore	70	2.6	5500	5000 (4)	97	[82]
44 wt% Ru-rGo (reduced graphene oxide)	200[a]	2.7	2000	2000 (30)	100	[88]

Source: Adapted from A. Débart et al. *Angew. Chem. Int. Ed.* 47, 2008, 4521–4524; A. Débart et al. *J. Power Sources* 174, 2007, 1177–1182; Y.-C. Lu et al. *J. Am. Chem. Soc.* 132, 2010, 12170–12171; S.H. Oh et al. *Nat. Chem.* 4, 2012, 1004–1010; Z. Peng et al. *Science* 337, 2012, 563–566; H.-G. Jung et al. *ACS Nano* 7, 2013, 3532–3539.

[a] Capacity normalised by the total weight of oxygen electrode (Ru+ rGo).

3.6.2.4 Lithium Metal Anode Issue

Lithium anode still remains the bottleneck on the practical development of lithium–air batteries. Lithium metal is very reactive (see Figure 3.8 and related discussion), and even traces of water must be avoided to prevent side processes that can severely affect the reliability, safety and cycle life of the battery. Coverage of the lithium metal electrode using protective lithium-ion conducting films has been proposed to address this issue [89]; however, even with such films, complete safety is not assured since they are usually fragile and may easily deteriorate or even break during battery operation.

A more promising approach is to replace the lithium metal with an alternative, non-lithium anode, for example, by a lithiated silicon–carbon composite, in order to operate a cell of the type $Li_xSiC/electrolyte/O_2,C$ [90]. Figure 3.39 illustrates the voltage profiles of the Li_xSiC electrode in comparison with that of the oxygen electrode in a $TEGDME–LiCF_3SO_3$ electrolyte, indicating that the two electrodes can be combined into a complete $Li_xSi–O_2$ battery operating around 3 V, according to the process: $xO_2 + 2Li_xSi \rightleftarrows xLi_2O_2 + 2Si$.

Indeed, this lithium-metal-free, lithium-ion, silicon–oxygen battery, although in a preliminary stage [90], gives a promising response, and can operate with a flat discharge plateau at around 2.6 V. The electrochemical behaviour shown in Figure 3.40 suggests that Li–Si alloys can be a suitable candidate to improve the safety of lithium–air battery.

Obviously, the replacement of lithium metal with a lithium metal alloy, where the activity of lithium is necessarily lower than unity, entails a penalty in terms of both voltage and capacity; however, this drawback appears to be favourably counterbalanced by enhancement in safety.

An alternative approach for improving the safety of lithium–air batteries is to use a solid electrolyte, for example, the solvent-free PEO–lithium

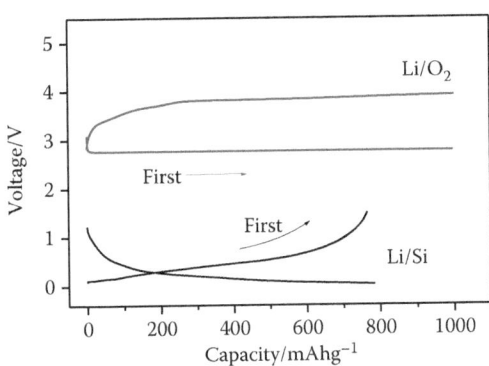

FIGURE 3.39
Voltage profiles of the carbon–oxygen electrode (upper part) and the lithiated-silicon–carbon electrode (lower part) in lithium cells. (Reprinted with permission from J. Hassoun et al. A metal-free, lithium-ion oxygen battery: A step forward to safety in lithium-air batteries, *Nano Lett.* 12, 2012, 5775–5779. Copyright 2012. American Chemical Society.)

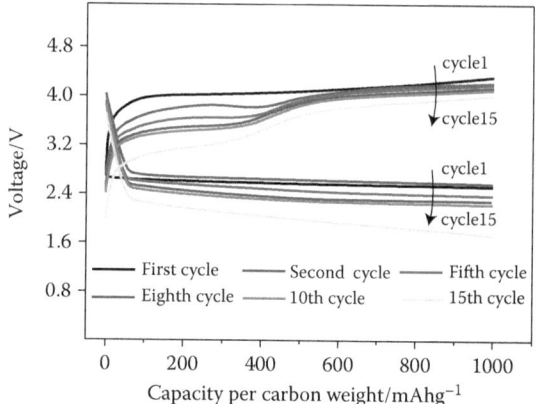

FIGURE 3.40
Voltage profiles and initial cycling behaviour of the $Li_xSiC/TEGDME-LiCF_3SO_3/O_2C$ cell. (Reprinted with permission from J. Hassoun et al. A metal-free, lithium-ion oxygen battery: A step forward to safety in lithium-air batteries, *Nano Lett.* 12, 2012, 5775–5779. Copyright 2012. American Chemical Society.)

salt (see Section 3.4). It has been demonstrated that lithium metal is stable in these electrolytes where it can be safely cycled [6]. In addition, these electrolytes are particularly suitable for use as separators in lithium–air batteries, since the well-documented resistance of the PEO ether linkage combined with the stability towards nucleophiles of most lithium salts (e.g. $LiCF_3SO_3$) is expected to provide a stable medium. Indeed, the application of PEO-based electrolytes for safe and efficient lithium–oxygen cells has been reported [62]. The validity of the solid-state Li–air battery was also proven by a system consisting of a lithium metal anode, two polymer electrolyte membranes on each side of a Li-ion conducting glass ceramic and a composite carbon–glass ceramic O_2 cathode [91–93]. Although some progress has been achieved, the field of solid-state lithium air batteries is still subject to improvement and deserves further attention.

3.7 Conclusions and Outlook

Important progress has been achieved in the recent years on the performance of high-energy batteries. However, further work is essential to assure the practical implementation of lithium–sulphur and lithium–air batteries. While for the former, there is still need to develop the most favourable structure to assure a complete control of the polysulphide solubility, for the latter, much more issues have to be addressed before practical application is viable. Lithium–air is undoubtedly one of the most promising energy storage systems and therefore justifies attention. Considerable funding needs to be devoted to its investigation and development. The road, however, is still long for making lithium–air batteries feasible for vehicle applications; so far, most of the studies have been limited to oxygen-fueled systems, while it is obvious that air cathodes are more cost effective and practically feasible. On the other hand, in open systems, air scrubbing apparatus and/or cathode-protective designs that employ water-blocking membranes would be required. Obviously, the addition of these ancillary parts would lower the attractiveness of the battery by increasing its cost, and greatly reducing the value of the practical energy density.

References

1. B. Scrosati and J. Garche, Lithium batteries: Status, prospects and future, *J. Power Sources* 195, 2010, 2419–2430.
2. B. Scrosati, J. Hassoun and Y.-K. Sun, Lithium-ion batteries. A look into the future, *Energy Environ. Sci.* 4, 2011, 3287–3295.
3. M. Armand and J.M. Tarascon, Building better batteries, *Nature* 451, 2008, 652–657.
4. W. van Schalkwijk and B. Scrosati (Ed.), *Advances in Lithium-Ion Batteries*, Kluwer Academic/Plenum, Boston, USA, 2002.
5. F.M. Gray, *Solid Polymer Electrolytes*, Wiley-VCH, Weinheim, Germany, 1991.
6. F.M. Gray and M. Armand, Polymer electrolytes, in: T. Osaka, M. Datta (Eds.), *Energy System for Electronics*, Gordon and Breach Science Publishers, Tokyo, Japan, 2000, pp. 351–406.
7. G. Kobe, Battery of the future, *Automotive Industries* 176, 1996, 9.
8. C. Donnelly, L. Christensen and D. Kuller, *Electric & Hybrid Vehicle Technology*, SAE, Warrendale, PA, 1996.
9. Groupe Bolloré, *Blue Car.* http://www.bluecar.fr, 2013.
10. K. Xu, Secondary batteries—Lithium recharagable systems: Electrolytes: Overview, in: G. Jürgen (Ed.), *Encyclopedia of Electrochemical Power Sources*, Elsevier, Amsterdam, Netherlands, 2009, Vol. 5, pp. 51–70.
11. E. Peled, D. Golodnitsky, G. Ardel and V. Eshkenazy, The sei model—Application to lithium-polymer electrolyte batteries, *Electrochim. Acta* 40, 1995, 2197–2204.
12. T. Ohzuku, A. Ueda and N. Yamamoto, Zero-strain insertion material of Li[Li$_{1/3}$Ti$_{5/3}$]O$_4$ for rechargeable lithium cells, *J. Electrochem. Soc.* 142, 1995, 1431–1435.
13. M. Winter and J.O. Besenhard, Electrochemical lithiation of tin and tin-based intermetallics and composites, *Electrochim. Acta* 45, 1999, 31–50.
14. A.S. Arico, P. Bruce, B. Scrosati, J.-M. Tarascon and W. van Schalkwijk, Nanostructured materials for advanced energy conversion and storage devices, *Nat. Mater.* 4, 2005, 366–377.
15. G. Derrien, J. Hassoun, S. Panero and B. Scrosati, Nanostructured Sn–C composite as an advanced anode material in high-performance lithium-ion batteries, *Adv. Mater.* 19, 2007, 2336–2340.

16. J. Hassoun, G. Derrien, S. Panero and B. Scrosati, A nanostructured Sn–C composite lithium battery electrode with unique stability and high electrochemical performance, *Adv. Mater.* 20, 2008, 3169–3175.
17. H. Kim, B. Han, J. Choo and J. Cho, Three-dimensional porous silicon particles for use in high-performance lithium secondary batteries, *Angew. Chem. Int. Ed.* 47, 2008, 10151–10154.
18. H. Inoue, *International Meeting on Lithium Batteries, IMLB 2006*, June 18–23, 2006, Abstr. #228.
19. Y.G. Andreev and P.G. Bruce, Polymer electrolyte structure and its implications, *Electrochim. Acta* 45, 2000, 1417–1423.
20. F. Croce, G.B. Appetecchi, L. Persi and B. Scrosati, Nanocomposite polymer electrolytes for lithium batteries, *Nature* 394, 1998, 456–458.
21. J.M. Tarascon, A.S. Gozdz, C. Schmutz, F. Shokoohi and P.C. Warren, Performance of Bellcore's plastic rechargeable Li-ion batteries, *Solid State Ionics* 86–88, Part 1, 1996, 49–54.
22. A. Fernicola, B. Scrosati and H. Ohno, Potentialities of ionic liquids as new electrolyte media in advanced electrochemical devices, *Ionics* 12, 2006, 95–102.
23. M. Armand, F. Endres, D.R. MacFarlane, H. Ohno and B. Scrosati, Ionic-liquid materials for the electrochemical challenges of the future, *Nat. Mater.* 8, 2009, 621–629.
24. A. Fernicola, F. Croce, B. Scrosati, T. Watanabe and H. Ohno, LiTFSI-BEPyTFSI as an improved ionic liquid electrolyte for rechargeable lithium batteries, *J. Power Sources* 174, 2007, 342–348.
25. H. Matsumoto, H. Sakaebe, K. Tatsumi, M. Kikuta, E. Ishiko and M. Kono, Fast cycling of Li/LiCoO$_2$ cell with low-viscosity ionic liquids based on bis(fluorosulfonyl)imide [FSI]–, *J. Power Sources* 160, 2006, 1308–1313.
26. S. Seki, Y. Kobayashi, H. Miyashiro, Y. Ohno, A. Usami, Y. Mita, N. Kihira, M. Watanabe and N. Terada, Lithium secondary batteries using modified-imidazolium room-temperature ionic liquid, *J. Phys. Chem. B* 110, 2006, 10228–10230.
27. J.-H. Shin, W.A. Henderson and S. Passerini, PEO-based polymer electrolytes with ionic liquids and their use in lithium metal-polymer electrolyte batteries, *J. Electrochem. Soc.* 152, 2005, A978–A983.
28. B. Garcia, S. Lavallée, G. Perron, C. Michot and M. Armand, Room temperature molten salts as lithium battery electrolyte, *Electrochim. Acta* 49, 2004, 4583–4588.
29. A.K. Padhi, K.S. Nanjundaswamy and J.B. Goodenough, Phospho-olivines as positive-electrode materials for rechargeable lithium batteries, *J. Electrochem. Soc.* 144, 1997, 1188–1194.
30. K.E. Aifantis and S.A. Hackney, Current and potential applications of secondary Li batteries, in: K.E. Aifantis, S.A. Hackney and R.V. Kumar (Eds.), *High Energy Density Lithium Batteries: Material, Engineering, Applications*, Wiley-VCH Verlag GmbH & Co. KGaA, Weinheim, Germany, 2010, pp. 81–101.
31. S. Beninati, L. Damen and M. Mastragostino, MW-assisted synthesis of LiFePO$_4$ for high power applications, *J. Power Sources* 180, 2008, 875–879.
32. J. Hassoun, P. Reale and B. Scrosati, Recent advances in liquid and polymer lithium-ion batteries, *J. Mater. Chem.* 17, 2007, 3668–3677.
33. P. Reale, D. Privitera, S. Panero and B. Scrosati, An investigation on the effect of Li$^+$/Ni^{2+} cation mixing on electrochemical performances and analysis of the

electron conductivity properties of $LiCo_{0.33}Mn_{0.33}Ni_{0.33}O_2$, *Solid State Ionics* 178, 2007, 1390–1397.

34. P. Poizot, S. Laruelle, S. Grugeon, L. Dupont and J.M. Tarascon, Nano-sized transition-metal oxides as negative-electrode materials for lithium-ion batteries, *Nature* 407, 2000, 496–499.
35. Y. Oumellal, A. Rougier, G.A. Nazri, J.M. Tarascon and L. Aymard, Metal hydrides for lithium-ion batteries, *Nat. Mater.* 7, 2008, 916–921.
36. H.-J. Ahn, K.-W. Kim, J.-H. Ahn and G. Cheruvally, Secondary batteries—Lithium recharagable systems: Lithium-sulfur, in: G. Jürgen (Ed.), *Encyclopedia of Electrochemical Power Sources*, Elsevier, Amsterdam, Netherlands, 2009, Vol. 5, pp. 155–161.
37. S.-E. Cheon, K.-S. Ko, J.-H. Cho, S.-W. Kim, E.-Y. Chin and H.-T. Kim, Rechargeable lithium sulfur battery: II. Rate capability and cycle characteristics, *J. Electrochem. Soc.* 150, 2003, A800–A805.
38. R.D. Rauh, K.M. Abraham, G.F. Pearson, J.K. Surprenant and S.B. Brummer, A lithium/dissolved sulfur battery with an organic electrolyte, *J. Electrochem. Soc.* 126, 1979, 523–527.
39. H. Yamin and E. Peled, Electrochemistry of a nonaqueous lithium/sulfur cell, *J. Power Sources* 9, 1983, 281–287.
40. D. Peramunage and S. Licht, A solid sulfur cathode for aqueous batteries, *Science* 261, 1993, 1029–1032.
41. J. Shim, K.A. Striebel and E.J. Cairns, The lithium/sulfur rechargeable cell: Effects of electrode composition and solvent on cell performance, *J. Electrochem. Soc.* 149, 2002, A1321–A1325.
42. J.H. Shin and E.J. Cairns, Characterization of N-methyl-N-butylpyrrolidinium bis(trifluoromethanesulfonyl)imide-LiTFSI-tetra(ethylene glycol) dimethyl ether mixtures as a Li metal cell electrolyte, *J. Electrochem. Soc.* 155, 2008, A368–A373.
43. L.X. Yuan, J.K. Feng, X.P. Ai, Y.L. Cao, S.L. Chen and H.X. Yang, Improved dischargeability and reversibility of sulfur cathode in a novel ionic liquid electrolyte, *Electrochem. Commun.* 8, 2006, 610–614.
44. S.S. Jeong, Y.T. Lim, Y.J. Choi, G.B. Cho, K.W. Kim, H.J. Ahn and K.K. Cho, Electrochemical properties of lithium sulfur cells using PEO polymer electrolytes prepared under three different mixing conditions, *J. Power Sources* 174, 2007, 745–750.
45. J. Hassoun and B. Scrosati, Moving to a solid-state configuration: A valid approach to making lithium-sulfur batteries viable for practical applications, *Adv. Mater.* 22, 2010, 5198–5201.
46. X. Ji, K.T. Lee and L.F. Nazar, A highly ordered nanostructured carbon-sulphur cathode for lithium-sulphur batteries, *Nat. Mater.* 8, 2009, 500–506.
47. X. Ji and L.F. Nazar, Advances in Li-S batteries, *J. Mater. Chem.* 20, 2010, 9821–9826.
48. D. Aurbach, E. Pollak, R. Elazari, G. Salitra, C.S. Kelley and J. Affinito, On the surface chemical aspects of very high energy density, rechargeable Li–sulfur batteries, *J. Electrochem. Soc.* 156, 2009, A694–A702.
49. N. Jayaprakash, J. Shen, S.S. Moganty, A. Corona and L.A. Archer, Porous hollow carbon@sulfur composites for high-power lithium–sulfur batteries, *Angew. Chem. Int. Ed.* 50, 2011, 5904–5908.
50. J. Hassoun, J. Kim, D.-J. Lee, H.-G. Jung, S.-M. Lee, Y.-K. Sun and B. Scrosati, A contribution to the progress of high energy batteries: A metal-free, lithium-ion, silicon–sulfur battery, *J. Power Sources* 202, 2012, 308–313.

51. T. Takeuchi, H. Sakaebe, H. Kageyama, H. Senoh, T. Sakai and K. Tatsumi, Preparation of electrochemically active lithium sulfide–carbon composites using spark-plasma-sintering process, *J. Power Sources* 195, 2010, 2928–2934.
52. J. Hassoun and B. Scrosati, A high-performance polymer tin sulfur lithium ion battery, *Angew. Chem. Int. Ed.* 49, 2010, 2371–2374.
53. J. Hassoun, Y.-K. Sun and B. Scrosati, Rechargeable lithium sulfide electrode for a polymer tin/sulfur lithium-ion battery, *J. Power Sources* 196, 2011, 343–348.
54. P.G. Bruce, S.A. Freunberger, L.J. Hardwick and J.-M. Tarascon, Li-O_2 and Li-S batteries with high energy storage, *Nat. Mater.* 11, 2012, 19–29.
55. S.J. Visco, E. Nimon and L.C.D. Jonghe, Secondary batteries—Metal-air systems: Lithium-air, in: G. Jürgen (Ed.), *Encyclopedia of Electrochemical Power Sources*, Elsevier, Amsterdam, Netherlands, 2009, Vol. 4, pp. 376–383.
56. K. Nakajima, T. Kato, Y. Inda and B. Hoffman, *Symposium on Energy Storage beyond Lithium Ion; Materials Perspective*, October 7–8, 2010.
57. O. Crowther, B. Meyer, M. Morgan and M. Salomon, Primary Li-air cell development, *J. Power Sources* 196, 2011, 1498–1502.
58. T. Zhang, N. Imanishi, S. Hasegawa, A. Hirano, J. Xie, Y. Takeda, O. Yamamoto and N. Sammes, Li/polymer electrolyte/water stable lithium-conducting glass ceramics composite for lithium–air secondary batteries with an aqueous electrolyte, *J. Electrochem. Soc.* 155, 2008, A965–A969.
59. K.M. Abraham and Z. Jiang, A polymer electrolyte-based rechargeable lithium/oxygen battery, *J. Electrochem. Soc.* 143, 1996, 1–5.
60. A. Débart, A.J. Paterson, J. Bao and P.G. Bruce, α-MnO_2 Nanowires: A catalyst for the O_2 electrode in rechargeable lithium batteries, *Angew. Chem. Int. Ed.* 47, 2008, 4521–4524.
61. C. Laoire, S. Mukerjee, E.J. Plichta, M.A. Hendrickson and K.M. Abraham, Rechargeable lithium/TEGDME- $LiPF_6$/O_2 battery, *J. Electrochem. Soc.* 158, 2011, A302–A308.
62. J. Hassoun, F. Croce, M. Armand and B. Scrosati, Investigation of the O_2 electrochemistry in a polymer electrolyte solid-state cell, *Angew. Chem. Int. Ed.* 50, 2011, 2999–3002.
63. S.A. Freunberger, Y. Chen, Z. Peng, J.M. Griffin, L.J. Hardwick, F. Bardé, P. Novák and P.G. Bruce, Reactions in the rechargeable lithium–O_2 battery with alkyl carbonate electrolytes, *J. Am. Chem. Soc.* 133, 2011, 8040–8047.
64. W. Xu, J. Xiao, J. Zhang, D. Wang and J.-G. Zhang, Optimization of nonaqueous electrolytes for primary lithium/air batteries operated in ambient environment, *J. Electrochem. Soc.* 156, 2009, A773–A779.
65. T. Kuboki, T. Okuyama, T. Ohsaki and N. Takami, Lithium-air batteries using hydrophobic room temperature ionic liquid electrolyte, *J. Power Sources* 146, 2005, 766–769.
66. S.A. Freunberger, Y. Chen, N.E. Drewett, L.J. Hardwick, F. Bardé and P.G. Bruce, The lithium–oxygen battery with ether-based electrolytes, *Angew. Chem. Int. Ed.* 50, 2011, 8609–8613.
67. H. Wang and K. Xie, Investigation of oxygen reduction chemistry in ether and carbonate based electrolytes for Li–O_2 batteries, *Electrochim. Acta* 64, 2012, 29–34.
68. C.O. Laoire, S. Mukerjee, K.M. Abraham, E.J. Plichta and M.A. Hendrickson, Influence of nonaqueous solvents on the electrochemistry of oxygen in the rechargeable lithium–air battery, *J. Phys. Chem. C* 114, 2010, 9178–9186.

69. R. Black, S.H. Oh, J.-H. Lee, T. Yim, B. Adams and L.F. Nazar, Screening for superoxide reactivity in Li–O$_2$ batteries: Effect on Li$_2$O$_2$/LiOH crystallization, *J. Am. Chem. Soc.* 134, 2012, 2902–2905.
70. Y.-C. Lu, D.G. Kwabi, K.P.C. Yao, J.R. Harding, J. Zhou, L. Zuin and Y. Shao-Horn, The discharge rate capability of rechargeable Li–O$_2$ batteries, *Energy Environ. Sci.* 4, 2011, 2999–3007.
71. H.-G. Jung, J. Hassoun, J.-B. Park, Y.-K. Sun and B. Scrosati, An improved high-performance lithium–air battery, *Nat. Chem.* 4, 2012, 579–585.
72. H.-G. Jung, H.-S. Kim, J.-B. Park, I.-H. Oh, J. Hassoun, C.S. Yoon, B. Scrosati and Y.-K. Sun, A transmission electron microscopy study of the electrochemical process of lithium–oxygen cells, *Nano Lett.* 12, 2012, 4333–4335.
73. F.T. Wagner, B. Lakshmanan and M.F. Mathias, Electrochemistry and the future of the automobile, *J. Phys. Chem. Lett.* 1, 2010, 2204–2219.
74. Y. Shao, S. Park, J. Xiao, J.-G. Zhang, Y. Wang and J. Liu, Electrocatalysts for non-aqueous lithium–air batteries: Status, challenges, and perspective, *ACS Catalysis* 2, 2012, 844–857.
75. B.D. McCloskey, R. Scheffler, A. Speidel, D.S. Bethune, R.M. Shelby and A.C. Luntz, On the efficacy of electrocatalysis in nonaqueous Li–O$_2$ batteries, *J. Am. Chem. Soc.* 133, 2011, 18038–18041.
76. A. Débart, J. Bao, G. Armstrong and P.G. Bruce, An O$_2$ cathode for rechargeable lithium batteries: The effect of a catalyst, *J. Power Sources* 174, 2007, 1177–1182.
77. Y.-C. Lu, H.A. Gasteiger, E. Crumlin, R. McGuire and Y. Shao-Horn, Electrocatalytic activity studies of select metal surfaces and implications in Li-air batteries, *J. Electrochem. Soc.* 157, 2010, A1016–A1025.
78. Y.-C. Lu, H.A. Gasteiger, M.C. Parent, V. Chiloyan and Y. Shao-Horn, The influence of catalysts on discharge and charge voltages of rechargeable Li–oxygen batteries, *Electrochem. Solid-State Lett.* 13, 2010, A69–A72.
79. Y.-C. Lu, Z. Xu, H.A. Gasteiger, S. Chen, K. Hamad-Schifferli and Y. Shao-Horn, Platinum–gold nanoparticles: A highly active bifunctional electrocatalyst for rechargeable lithium–air batteries, *J. Am. Chem. Soc.* 132, 2010, 12170–12171.
80. V. Giordani, S.A. Freunberger, P.G. Bruce, J.-M. Tarascon and D. Larcher, H$_2$O$_2$ decomposition reaction as selecting tool for catalysts in Li–O$_2$ cells, *Electrochem. Solid-State Lett.* 13, 2010, A180–A183.
81. R. Black, J.-H. Lee, B. Adams, C.A. Mims and L.F. Nazar, The role of catalysts and peroxide oxidation in lithium–oxygen batteries, *Angew. Chem. Int. Ed.* 52, 2013, 392–396.
82. S.H. Oh, R. Black, E. Pomerantseva, J.-H. Lee and L.F. Nazar, Synthesis of a metallic mesoporous pyrochlore as a catalyst for lithium–O$_2$ batteries, *Nat. Chem.* 4, 2012, 1004–1010.
83. Z. Peng, S.A. Freunberger, Y. Chen and P.G. Bruce, A reversible and higher-rate Li–O$_2$ battery, *Science* 337, 2012, 563–566.
84. R. Kou, Y. Shao, D. Wang, M.H. Engelhard, J.H. Kwak, J. Wang, V.V. Viswanathan et al. Enhanced activity and stability of Pt catalysts on functionalized graphene sheets for electrocatalytic oxygen reduction, *Electrochem. Commun.* 11, 2009, 954–957.
85. E. Yoo, T. Okata, T. Akita, M. Kohyama, J. Nakamura and I. Honma, Enhanced electrocatalytic activity of Pt subnanoclusters on graphene nanosheet surface, *Nano Lett.* 9, 2009, 2255–2259.

86. Y. Shao, S. Zhang, C. Wang, Z. Nie, J. Liu, Y. Wang and Y. Lin, Highly durable graphene nanoplatelets supported Pt nanocatalysts for oxygen reduction, *J. Power Sources* 195, 2010, 4600–4605.
87. S.M. Choi, M.H. Seo, H.J. Kim and W.B. Kim, Synthesis of surface-functionalized graphene nanosheets with high Pt-loadings and their applications to methanol electrooxidation, *Carbon* 49, 2011, 904–909.
88. H.-G. Jung, Y.S. Jeong, J.-B. Park, Y.-K. Sun, B. Scrosati and Y.J. Lee, Ruthenium-based electrocatalysts supported on reduced graphene oxide for lithium-air batteries, *ACS Nano* 7, 2013, 3532–3539.
89. G. Girishkumar, B. McCloskey, A.C. Luntz, S. Swanson and W. Wilcke, Lithium–air battery: Promise and challenges, *J. Phys. Chem. Lett.* 1, 2010, 2193–2203.
90. J. Hassoun, H.-G. Jung, D.-J. Lee, J.-B. Park, K. Amine, Y.-K. Sun and B. Scrosati, A metal-free, lithium-ion oxygen battery: A step forward to safety in lithium-air batteries, *Nano Lett.* 12, 2012, 5775–5779.
91. J. Kumar, S.J. Rodrigues and B. Kumar, Interface-mediated electrochemical effects in lithium/polymer-ceramic cells, *J. Power Sources* 195, 2010, 327–334.
92. B. Kumar, J. Kumar, R. Leese, J.P. Fellner, S.J. Rodrigues and K.M. Abraham, A solid-state, rechargeable, long cycle life lithium–air battery, *J. Electrochem. Soc.* 157, 2010, A50–A54.
93. B. Kumar and J. Kumar, Cathodes for solid-state lithium–oxygen cells: Roles of Nasicon glass-ceramics, *J. Electrochem. Soc.* 157, 2010, A611–A616.

4

Hollow Mesoporous Carbon with Hierarchical Nanoarchitecture in Electrochemical Energy Storage and Conversion

Min-Sik Kim, Dae-Soo Yang, Min Young Song, Jung Ho Kim and Jong-Sung Yu

CONTENTS

4.1 Introduction 163
4.2 Structural Features of Hollow Mesoporous Carbon (HMC) 166
4.3 Electrode Material in Li Ion Battery (LIB) 168
4.4 Electrode Material in Electrical Double Layer Capacitor (EDLC) 176
4.5 Catalyst Support in Fuel Cell 180
4.6 Conclusions 183
Acknowledgements 185
References 185

4.1 Introduction

Depletion of the world's natural resources has been continuously destabilising and threatening the present fossil fuel-based energy economy. In addition, a continuous increase in greenhouse gases is a serious global threat. Therefore, the urgency for energy research requires the development of clean energy sources at a much higher level than that presently in force. Consequently, the exploitation of renewable energy resources is increasing worldwide, particularly in the area of wind and solar power energy plants (PEPs). Efficient use of these resources requires high-efficiency energy storage systems. Electrochemical systems, such as batteries and supercapacitors, which can efficiently store and deliver energy on demand in hybrid electric vehicles (HEVs), plug-in electric vehicles (PEVs) and portable consumer electronic equipments, are playing a crucial role in this field. The effectiveness of batteries in PEPs is directly related to their content in energy efficiency and lifetime. In recent decades, secondary Li ion batteries (LIBs) have evolved as a foremost energy source for all kinds of consumer electronic products as well as electric/hybrid vehicles due to their high electromotive force and high energy density [1,2]. However, at present, many potential electrode materials are experiencing slow Li ion diffusion and high resistance at the interface of electrode/electrolyte at high charge–discharge rates [3,4], and their rate capability, dominated mainly by the diffusion rate of Li ion and the electron transfer in electrode materials, must be improved considerably [5].

In order to improve the transport of Li ions in electrode, various nanostructured materials with high surface area, nanoscale size and/or nanoporous structure have been widely investigated [6–9]. In particular, nanostructured porous carbon materials have received great attention [10–18]. Nanostructured carbon materials have demonstrated good electrode properties in diversified energy conversion and/or storage systems such as low-temperature fuel cells [19–26], hydrogen storage systems [27,28], solar cells [29–31] and batteries as well [32–37]. Many nanostructured carbon materials such as ordered mesoporous carbon CMK-3, CNTs, graphenes and hierarchical porous carbons have demonstrated greatly enhanced Li storage capacity and/or improved rate capability compared with commercial graphite anode [32–37]. However, it is still highly desirable to explore new materials with further enhanced Li storage capacity, especially at high charge–discharge rates.

The electrochemical capacitor (EC) has evolved as a promising candidate for high energy storage due to its attractive advantages such as high power density (10^3–10^4 W kg^{-1}), long cycle life (>10^6 cycles), pulse power supply, low maintenance cost, simplicity and better safety compared to secondary batteries [38,39]. A promising application for EC is in electric vehicles, where high power density is needed during acceleration, and the energy can be recovered during braking. Microporous activated carbon materials are used widely, and capable to deliver high capacitance and power density due to their very high surface area and pore volume as well as low cost [40–42]. Their superb performance in EC has a close relation to the relative amount of both micropores (less than 2.0 nm) and mesopores (2.0 nm–50 nm) and their connectivity. Microporous materials having only micropores, possess smaller 'accessible' surface area for the formation of EDLC despite of the high surface area, and thus are often found to exhibit lower specific capacitance than expected. It is important to have more accessible microporous surface area rather than simply high microporous surface area. Therefore, a carbon structure with sufficient mesopore volume and well-ordered connectivity with micropores would be desirable, as the larger pore size in the framework can facilitate the transport of organic electrolyte to reach all available surface area.

Various groups have reported promising EC electrode materials using hierarchical nanostructured carbons with well-organised pore structures, large specific surface areas and, particularly, an interconnected pore network. Portet et al. has reported 146 F g^{-1} for zeolite-templated hollow-core mesoporous carbon synthesised by chemical vapor deposition (CVD) method [43]. Murali et al. has also reported hollow carbon capsules with specific capacitance of

95–122 F g^{-1} measured using different ionic liquids [44]. Similar hollow mesoporous carbon (HMC) capsules derived from thioether-derived bridged organosilica were also reported with specific capacitance of 121 F g^{-1} [45].

The proton exchange membrane fuel cell (PEMFC) has demonstrated great promise as a renewable and environment-friendly energy source. Pt and Pt-based alloys have been the dominant electrocatalytic elements in both anode and cathode catalyst layers. However, commercialisation of the PEMFC technology is mainly hindered by the high cost of Pt, susceptibility of Pt to CO poisoning and sluggish oxygen reduction reaction (ORR) in the cathode. Catalyst support technology has been proven to be an effective strategy to lower Pt usage and enhance CO tolerance of the supported catalyst. Carbon is among the most promising catalyst supports, with special features such as high inertness under harsh chemical and electrochemical conditions, high surface area and electrical conductivity, well-developed porosity, adequate water-handling capability and low cost. For example, carbon black Vulcan XC-72 (VC) has been frequently used as a catalyst support in low-temperature fuel cells. However, the VC contains a large quantity of primary micropores of less than 1 nm diameter, and many Pt nanoparticles trapped in the micropores were not involved in the electrochemical reactions on electrodes due to the absence of the triple-phase boundaries. Various novel carbon materials have been investigated as catalyst support, such as carbon nanotubes [46,47], graphitic carbon nanofibres [48], mesostructured carbon materials [49–55], macroporous carbons [56–58] and carbon microbeads [59]. These carbon-supported catalysts have shown enhanced catalytic activity toward ORR in PEMFC compared to those supported by VC.

Novel HMC with unique nanostructure such as hollow macroporous core/mesoporous shell has been developed through nanocasting technique [60,61]. Although in most cases HMC is categorised as mesoporous nanostructured carbons rather than macroporous, its hollow macroporous core plays a key role in fast mass transport [62], different to other mesoporous nanostructured carbons. Furthermore, the hollow macroporous cores that are connected and open to the outer mesoporous shell can serve as an electrolyte reservoir, facilitating mass transport and enabling HMC to perform better than other nanostructured carbons with meso-microporosity or macro-mesoporosity [28].

The first HMC with spherical macroporous core was developed by our research group using submicrometre-size spherical solid core–mesoporous shell (SCMS) silica as the template [60]. Discussed below are surface structural features such as the sizes of hollow core and of mesopores in the shell, the thickness of the mesoporous shell and other highly dependent parameters. Specific surface area and pore volume play a pivotal role in determining physiochemical, electrochemical and photochemical properties of the HMC materials. The HMC capsules are expected to demonstrate considerably improved performance in energy applications such as rechargeable LIBs, ECs and fuel cells, owing to their unique structural characteristics, particularly 3D interconnected nanostructure with hierarchical porosity, not only providing large specific surface area for high catalytic activity and storage capacity, but also granting highly developed hierarchical macro/mesoporosity for fast mass transport.

4.2 Structural Features of Hollow Mesoporous Carbon (HMC)

HMC is a very interesting porous carbon material with unique hierarchical macro/mesoporous spherical morphology. HMCs with various core sizes and/or shell thicknesses can be fabricated through the independent control of the core sizes and/or shell thicknesses of the SCMS silica templates [28,62,63], while the micro- and mesoporosity of the HMCs can be controlled to some extent by the source type and amount of carbon precursor incorporated into the SCMS silica template. HMC with different core shapes (nonspherical) have been synthesised through nanocasting techniques [61,64]. A key factor for the synthesis of HMCs with diverse shapes and sizes lies in the fabrication of SCMS silica replica templates.

Spherical carbon capsules with a hollow macroporous core/mesoporous shell nanostructure can be synthesised by a 'nanocasting' technique using the submicron SCMS silica spheres as templates [60,65]. A typical synthesis route for the HMC, as illustrated in the schematic diagram in Figure 4.1, includes impregnation of carbon precursor into the SCMS template followed by polymerisation and carbonisation of the precursor, and removal of the template to get template-free carbon replica. Solid silica spheres are used as starting material. The silica sphere can be produced in various sizes by controlling the amount of tetraethylorthosilicate (TEOS) added into the aqueous ammonia. Thus, the SCMS silica can be produced in various sizes and shell thicknesses, by using the solid silica spheres with various sizes and adjusting the molar ratio of n-octadecyl trimethoxysilane (C_{18}-TMS) to TEOS, respectively, according to an established procedures [60,65].

Mesosize- and microsize-thick silica walls can be formed in the shell of the SCMS silica from the interaction of C_{18}-TMS and TEOS, which are the primary sources of meso- and micropores of the replicated HMC. In general, the higher ratio of C_{18}-TMS to TEOS tends to increase the size of the mesopore and decrease the silica wall thickness in the shell of the SCMS silica, thus to some extent increasing the micropore volume in the corresponding HMC [28,62]. A carbon precursor can be incorporated into the mesoporous channels separated by silica walls in the shell of the SCMS silica, and after carbonisation of the polymerised carbon precursor and removal of SCMS silica, HMCs with macroporous hollow core in combination with meso/microporous shell can be produced (with various surface area, pore volume and meso- and microporosity) (see Figure 4.1).

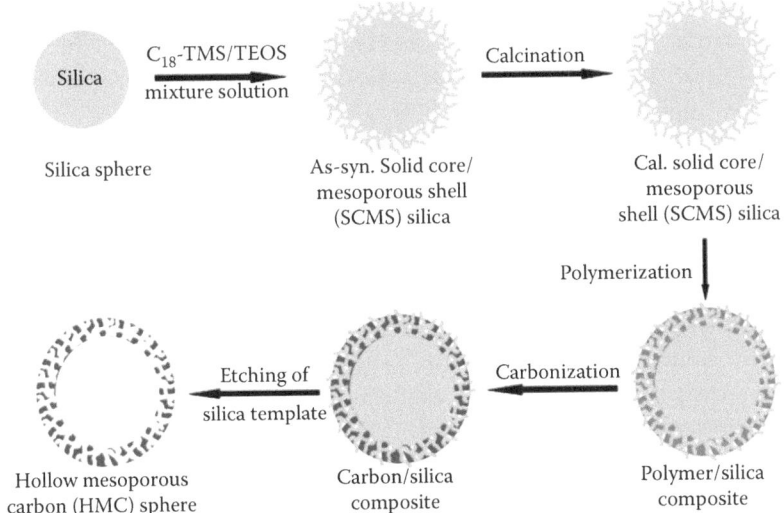

FIGURE 4.1
(See colour insert.) A typical scheme for the synthesis of HMC.

Figure 4.2 shows the SEM (a), ultra-high-resolution SEM (b) and TEM (c, d, e, f) images of HMC spherical capsules with different sizes. As revealed by the lower-magnification SEM image shown in Figure 4.2a, HMC capsules are fabricated as uniform individual discrete particles. Figure 4.2b reveals mesopores developed in the shell of an HCM sphere. A magnified UHR-SEM image in the inset of Figure 4.2b shows the distribution of uniform openings of 2~3 nm

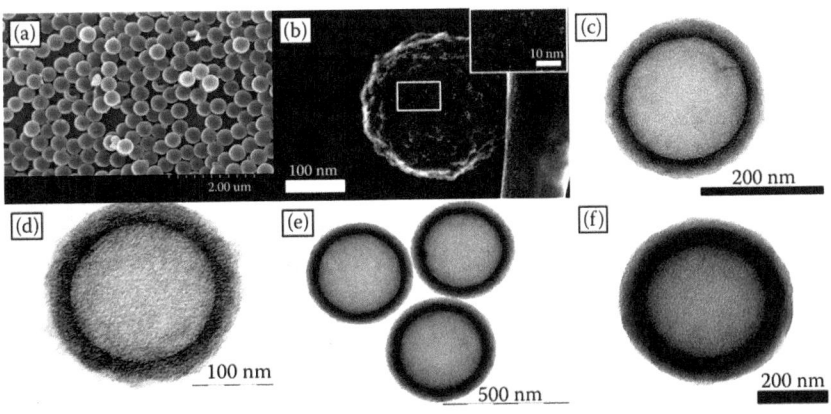

FIGURE 4.2
Representative (a) SEM, (b) UHR-SEM and TEM images for the HMC: (c) $HMC_{200/30}$, (d) $HMC_{180/40}$, (e) $HMC_{340/40}$ and (f) $HMC_{340/70}$.

mesopores in the shell. The TEM images in Figure 4.2c through f reveal the HMC spherical capsules with different hollow core sizes and shell thicknesses. Evidently, the core size and shell thickness of the HMC can be managed independently by controlling the core size and shell thickness of the SCMS template. This can be realised by two separate synthesis processes, consisting of the solid core formation by the Stöber method, and the subsequent formation of the mesoporous shell by the Kaiser approach. The solid core size is mainly controlled by the amount of added TEOS and the reaction time at the Stöber step, while the shell thickness is adjusted by the amount of TEOS and C_{18}-TMS at the Kaiser step. As evident from Figure 4.2c through f, HMCs with various core sizes and/or shell thicknesses can be fabricated. A slight shrinkage in core size and shell thickness of the HMC is always observed compared to those of the corresponding parent SCMS silica. $HMC_{180/40}$ ($HMC_{x/y}$: x and y stand for the core size and the shell thickness in nm, respectively) capsules were obtained from $SCMS_{190/42}$ spheres. Mesopores and micropores of the HMC are directly generated from the silica walls in the mesoporous shell of the SCMS silica. In addition, carbonisation of polymerised carbon precursor also intrinsically produces some micropores in the carbonised body.

4.3 Electrode Material in Li Ion Battery (LIB)

Figure 4.3 shows representative SEM and TEM images and typical N_2 adsorption–desorption isotherms along with derived pore size distribution (PSD) for the HMC and CMK-3, respectively. CMK-3 is one of the well-known ordered mesoporous carbons, prepared by nanocasting of SBA-15 mesoporous silica template with a hexagonal mesopore arrangement. Both HMC and CMK-3 are applied as anode materials for LIB in this work. Uniform discrete HMC particles with a particle size of 120 ± 10 nm are shown in the SEM image of Figure 4.3a. The TEM image (Figure 4.3b) reveals that the HMC has a hollow macroporous core of ca. 60 nm in diameter and a shell thickness of ca. 30 nm, which are slightly smaller than the core size (i.e. 65 nm) and the mesoporous shell thickness (i.e. 33 nm) of the SCMS silica. The slight shrinkage in the core size and shell thickness of the HMC compared with those of the parent SCMS silica was mainly caused by the heat treatment at high temperature.

Figure 4.3d and e shows representative SEM and TEM images of as-synthesised CMK-3 with rod-like morphology having a mesopore channel length of ca. 750–800 nm and a cross-sectional diameter of ca. 500 nm. The CMK-3 retains the inverse morphology of the corresponding SBA-15 silica template with similar rod shape, which influences the size and shape of the resulting CMK-3 [66]. The TEM image in Figure 4.3e reveals the alternating ordered arrangement of uniform carbon fibres and pores between the carbon fibres, which form the framework structure of rod-shaped CMK-3.

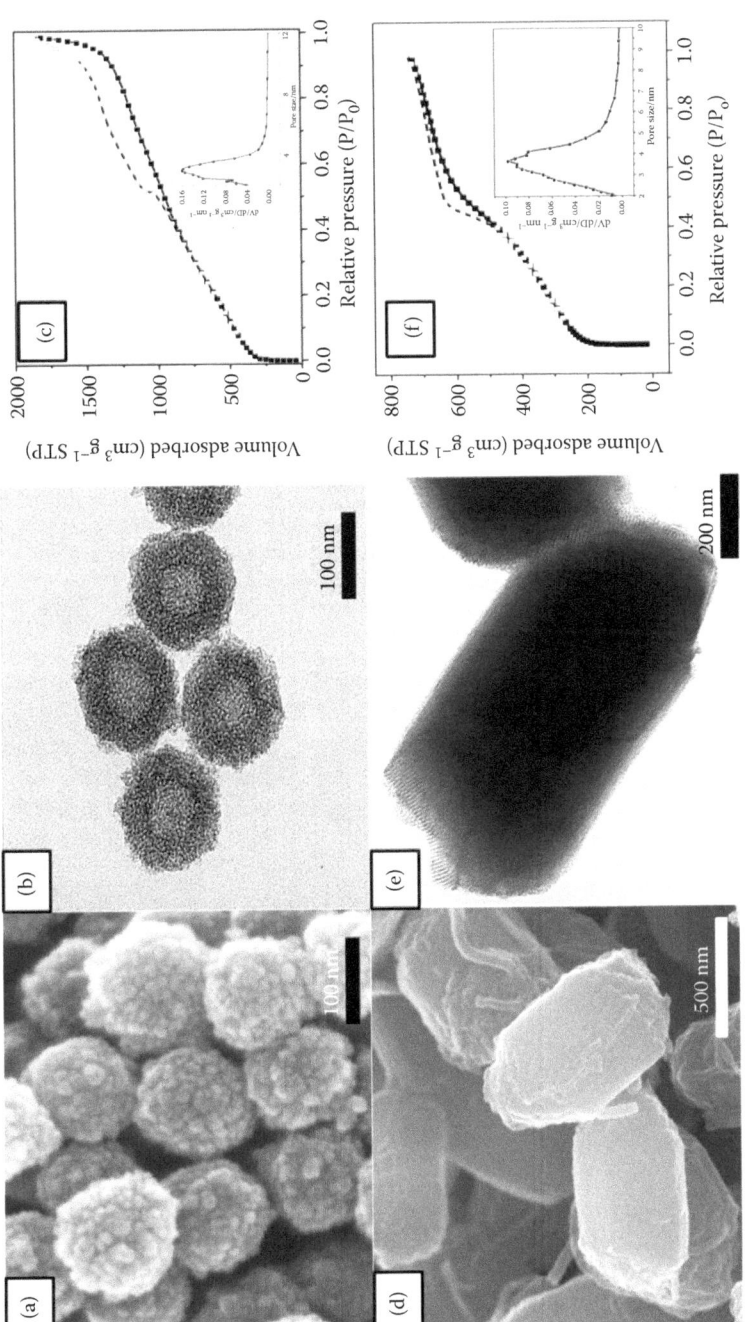

FIGURE 4.3
Typical SEM (a and d) and TEM (b and e) images and the N_2 adsorption–desorption isotherms (c and f) at 77 K along with PSD for $HMC_{60/30}$ (top) and CMK-3 (bottom), respectively.

The nitrogen adsorption–desorption isotherms shown in Figure 4.3c and f can be classified as a type IV isotherm with a type H2 hysteresis typical of mesoporous materials according to the International Union of Pure and Applied Chemistry (IUPAC) nomenclature. The narrow PSD demonstrates that both HMC and CMK-3 have uniform mesopores in the framework. This is in good agreement with those observed from the TEM and SEM images shown in Figure 4.3. The pore size was estimated to be ca. 3.5 nm from the PSD maximum, and a Brunauer–Emmett–Teller (BET) surface area of 2418 m^2 g^{-1} was determined for the HMC capsules (Figure 4.3c). In addition to the mesopores generated from the removal of the mesosize-thick silica walls in the silica/carbon composite, a fraction of micropores (ca. 22% of the total volume) formed during the carbonisation of the carbon precursor and partially from the microsize-thick silica walls in the silica/carbon composite were also found in the shell of the HMC as indicated by a big increase in adsorbed nitrogen at the low-pressure micropore region of the isotherms. Thus, it is clear that each HMC capsule possesses hierarchical nanostructure composed of macropores (i.e. hollow macroscopic core) in combination with mesopores and micropores in the mesoporous shell. In contrast, there is no such well-defined meso/macroporous hierarchical nanostructure for the CMK-3 as evident from the SEM and TEM images shown in Figure 4.3d and e, and the nitrogen isotherm shown in Figure 4.3f. BET measurement also reveals a microporous volume of 0.56 cm^3 g^{-1} for CMK-3, which is ca. 32 v/v% of the total pore volume (i.e. 1.77 cm^3 g^{-1}), a high BET surface area of 1228 m^2 g^{-1}, and uniform mesopores of ca. 3.9 nm in the framework. After all, the large surface area and mesopore volume, particularly the unique hierarchical macro/mesoporous nanostructure, are expected to provide HMC with enhanced Li storage capacity and improved rate capability. Table 4.1 summarises the structural parameters of CMK-3, graphite and HMC used as anode in this LIB application.

The electrochemical behaviour of LIBs based on the various carbon anode materials was studied in a CR2032 coin-type cell (Hohsen Corp., Japan). The complete cell preparation steps were performed in an argon glove box with the oxygen and the humidity level of 1 ppm or less, respectively, within the chamber. A pure Li metal foil (purity, 99.9% and 150 mm thick) was used as a reference electrode and counter electrode. A working anode electrode was made as follows. Carbon powder such as HMC, CMK-3 or graphite was

TABLE 4.1

Structural Properties of HMC and CMK-3

Sample	S_{BET} (m^2 g^{-1})	V_{total} (cm^3 g^{-1})	V_{meso} (cm^3 g^{-1})	PSD (nm)
CMK-3	1228	1.77	1.21	3.9
HMC$_{60/30}$	2418	2.12	1.65	3.5
Graphite	312	0.94	0.42	~

mixed with acetylene black (as a conductivity enhancer) and PVdF (poly-(vinylidene fluoride)) as a binder at a weight ratio of 8:1:1 in a solvent (i.e. NMP (N-methyl-2-pyrrolidone)). The slurry was uniformly pasted with 30 mm thickness on a Cu foil. The as-prepared working electrodes were dried at 120°C in a vacuum oven and pressed under a pressure around 4000 psi. For all the coin cells, 1.0 M LiPF$_6$ in ethylene carbonate (EC)–dimethyl carbonate (DMC) (1:1 in volume) was used as an electrolyte, and a typical polypropylene–polyethylene material (Celgard 2400) used as a separator.

The Li$^+$ insertion/extraction reactions of an HMC$_{60/30}$ electrode were studied by cyclic voltammetry (CV). Figure 4.4a depicts the CV plots of an HMC capsule electrode measured between 2.5 and 0.0 V at the scanning rate of 0.1 mV s^{-1}. In the first scanning cycle, there is a broad cathodic reduction wave starting from ca. 1.25 V with a peak located at around 0.75 V, which can be attributed to the formation of a solid-electrolyte interphase (SEI) passivation layer on the surface of the carbon electrode [66–68], due to the reaction of lithium with the electrolyte. The SEI layer becomes stable under subsequent lithium insertion and extraction, which is evidenced by the considerably reduced current response and the disappearance of these peaks from the second cycle.

In a recent review, Goodenough reported that electrons and ions flow from the anode to cathode (Li) upon discharge, while during charge, electrons and ions are forced by an applied electric field to flow from the cathode to the anode [69]. Figure 4.4b shows galvanostatic charging–discharging plots at 100 mA g^{-1} during the first 10 cycles. A very high first lithiation (charging) capacity of ca. 3159 mAh g^{-1} is measured. A voltage plateau starting from around 0.9 V (vs. Li/Li$^+$) was observed with a specific capacity of ca. 1500 mAh g^{-1} at the first lithiation. Evidently, the large initial irreversible capacity of ca. 1728 mAh g^{-1} observed in the first cycle is mainly attributable to the SEI formation resulting from ultra-large surface of the HMC. A reversible Li storage

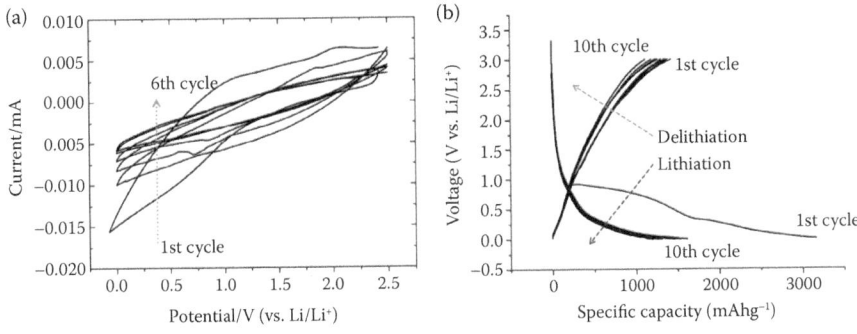

FIGURE 4.4
(a) CV plots during the first six cycles for HMC$_{60/30}$ capsules as anode in 1.0 M LiPF$_6$ (EC/DMC = 1:1 (v/v), scan rate: 0.1 mV s^{-1}) and (b) first 10 galvanostatic charge/discharge curves at 100 mA g^{-1} for HMC capsules as anode.

capacity of ca. 1431 mAh g^{-1} was observed for the HMC from the 2nd cycle to the 10th cycle, suggesting a good cycling performance for the HMC electrode.

Although the initial columbic efficiency is only 42% for the first cycle, it quickly increases to 86% during the 2nd cycle. The columbic efficiency reaches over 90% after 10 cycles and over 95% after 30 cycles. Initial low columbic efficiencies have been frequently reported in LIBs for porous carbonaceous materials with high specific surface areas, which are mainly attributed to their high specific surface areas and irreversible lithium insertion into special positions such as at the vicinity of residual H atoms, resulting in the decomposition of electrolytes and the formation of the SEI films at the electrode/electrolyte interface [36].

Figure 4.5 shows representative galvanostatic plots obtained in the second charging–discharging cycle for various carbon materials, that is, HMC, CMK-3 and the commercial graphite at a charge/discharge rate of 100 mA g^{-1}. The reversible Li storage capacities determined from the second discharge process are ca. 1387, 748 and 305 mAh g^{-1} for the HMC, CMK-3 and the commercially available graphite, respectively. Evidently, the Li storage capacities for the HMC corresponds to 3.7 times that of the theoretical capacity of graphite (i.e. 372 mAh g^{-1}) [70] and also much higher than that observed for the CMK-3. In general, nanostructured porous carbon materials have exhibited Li storage capacity in the range of 200–1100 mA g^{-1} [10–18]. Zhou et al. reported 1100 mAh g^{-1} at 100 mA g^{-1} for ordered mesoporous carbon (CMK-3) as a LIB anode material [10]. Stein et al. achieved a LIB capacity of 435 mAh g^{-1} at 0.1 charge rate (C rate) (1 C = 372 mA g^{-1}) using three-dimensionally ordered macroporous carbon [15]. Ordered multimodal porous carbon as an anode material revealed a capacity of ca. 900 mAh g^{-1} at 100 mA g^{-1} [35]. To the best

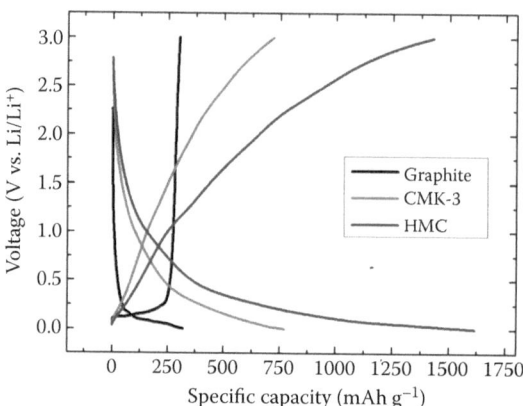

FIGURE 4.5
Galvanostic charge/discharge curves at 100 mAh g^{-1} for respective commercial graphite, CMK-3 and HMC capsules at the second cycle, in which discharge curves are labelled from left to right.

of our knowledge, this HMC demonstrates the largest Li ion storage capacity among nanostructured carbon materials studied so far as anode. Specific capacities of HMC and other previously reported carbon-based materials are summarised with corresponding current density in Table 4.2.

Figure 4.6a shows the galvanostatic cycling behaviour at 100 mA g^{-1} for various carbon materials. HMC was found to be very stable up to 80 cycles with a slight decrease in the reversible Li storage capacity. Compared with other carbon materials, HMC not only reveals a much higher initial reversible capacity but also much higher Li storage capacity at the 80th cycle (i.e. the HMC, CMK-3 and graphite electrode retain a specific capacity of ca. 1141, 513 and 245 mAh g^{-1} after 80 cycles). Although the initial columbic efficiency is as low as 41.3% due to the very large irreversible capacity loss from the formation of an SEI film [71], the columbic efficiency increases sharply to 86%

TABLE 4.2

Specific Discharge Capacity of Various Carbon Materials at First Cycle

Anode Material	Specific Capacity (mAh g^{-1})	Constant Current Density (mA g^{-1})	Reference
HMC$_{60/30}$	1431	100	[84]
OMPC[a]	903	100	[35]
CMK-3	1100	100	[10]
NGHC[b]	713	74	[36]
MWCNT	305	17	[72]
SWCNT	595	50	[73]
Graphite	372 (theoretical capacity)		[49]

[a] Ordered multimodal porous carbon.
[b] Nanographene-constructed hollow carbon.

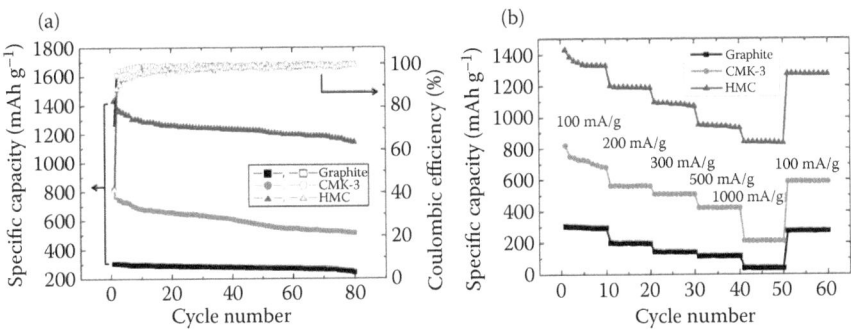

FIGURE 4.6
(a) Cycling performance and coulombic efficiency of commercial graphite, CMK-3 and HMC capsules at a specific current of 100 mA g^{-1}. (b) Rate performances of commercial graphite, CMK-3 and HMC capsules at different current densities from 100 to 1000 mA g^{-1} and then back to 100 mA g^{-1}.

during the 2nd cycle, and reaches over 95% after 30 cycles, which is in the same level as the other carbon materials.

Sustainable good rate capability and reversibility are the essential electrochemical features of LIBs to power the high-energy applications such as electric vehicle [72,73]. Figure 4.6b illustrates comparative rate performance of the commercial graphite, the CMK-3 and the HMC anode materials at different current densities. The battery based on the HMC anode demonstrates an excellent rate capability, still delivering a high fraction of its total capacity even at a rate as high as 1000 mA g^{-1}. Furthermore, after returning to the initial 100 mA g^{-1} regime, the electrode resumes to near the full original capacity, further confirming the exceptional reversible capacity of this electrode to maintain its integrity for prolonged cycles at high rates.

LIB performance depends mostly on three major parameters—Li ion storage capacity, cyclability and rate capability—which are strongly dependent on the pore structure of the active electrode materials. To further understand the relationship between the Li ion storage performance and the pore structure of the active electrode material, electrochemical impedance spectroscopy (EIS) measurements were conducted after the 10th charge–discharge cycle at a rate of 1000 mA g^{-1} for active electrode materials, that is, HMC and CMK-3. Representative Nyquist plots of these two materials are shown in Figure 4.7. It is evident that both electrodes reveal a depressed semicircle in the regime of high-medium frequency along with an inclined line at low frequency. The semicircle is mainly attributable to the summation of contact resistance, SEI resistance (R_{SEI}) and charge transfer resistance (R_{ct}), while the inclined line corresponds to the typical Warburg behaviour, which is related to the diffusion

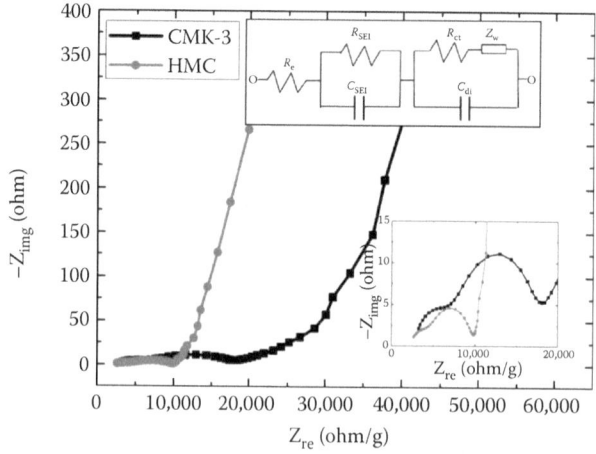

FIGURE 4.7
Typical Nyquist plots of HMC and CMK-3 electrodes; zoomed Nyquist plots in high frequency regime and equivalent circuit (inset).

TABLE 4.3

Cycling and Rate Performance, and Kinetic Parameters Derived from the Nyquist Plots for the Graphite, CMK-3 and HMC

Sample	1st Capacity (mAh g^{-1})	80th Capacity (mAh g^{-1})	Rate Capability (mAh g^{-1})		R_{SEI} (Ohm)	R_{ct} (Ohm)	Z_w (Ohm)
			1000 mA g^{-1}	100 mA g^{-1}			
Graphite	306.6	245.0	41.2	275.0	—	—	—
CMK-3	818.9	513.5	210.8	586.5	9.47	22.91	29.05
HMC	1431.1	1141.4	842.1	1277.4	8.49	3.71	12.9

of lithium within the porous carbon material [36,74]. Generally, the contact resistance is small, which can be neglected [75].

An equivalent circuit as shown in the inset of Figure 4.7 is proposed for the analysis of the impedance spectra. By fitting the impedance spectra to the proposed equivalent circuit using the code Zview, some significant kinetic parameters are derived and listed in Table 4.3, including R_{SEI}, R_{ct} and Warburg impedance (Z_W). Although HMC exhibits a R_{SEI} of 8.49, close to CMK-3, it reveals a much smaller R_{ct} of 3.71 than the latter (i.e. ca. 16% of the R_{ct} of CMK-3), implying a faster charge transfer reaction for lithium insertion/extraction and facile charge transfer at the HMC electrode/electrolyte interface. Furthermore, smaller Z_W observed for HMC suggests much faster mass transport for lithium within the pore network of HMC compared with CMK-3. The smaller R_{ct} and Z_W observed for HMC are mainly attributable to its unique nanostructure, which consists of a well-developed 3D interconnected macroporous core with an open mesoporous shell, not only facilitating fast mass transport due to the shortened Li ion diffusion distance but also reducing the volume change during the charge/discharge cycling, resulting from the buffer effect of the unique core/shell nanostructure. Yang et al. also reported an excellent cycling performance for a 3D porous material with macro-mesoporosities due to the small volume change and almost complete preservation of the 3D pore structure [36]. In addition, the large surface area of HMC and the abundant micropores provide large quantities of sites for Li ion adsorption, resulting in high Li storage capacity especially at relatively low charge/discharge rates, in which most of the micropores are accessible to Li species (ions, atoms) if the sizes of the micropores are larger than Li species. However, at a high rate, only a small portion of these micropores are still accessible to Li species and contribute to Li storage if there is no fast pathway constructed for fast mass transport by larger pores (i.e. mesopores and macropores, especially macropores). This is one of the main reasons why HMC greatly outperforms CMK-3 at a high rate because the latter lacks macropores. Owing to their large surface areas and the abundant micropores available in HMC and CMK-3, the capacitive contribution to the high Li capacity and high rate capability is also of our interest, and further research work will be conducted along this direction.

4.4 Electrode Material in Electrical Double Layer Capacitor (EDLC)

Electrochemical capacitor (EC), also referred to as EDLC, has also evolved as a promising candidate for high-energy storage due to their attractive advantages such as high power density (10^3–10^4 W kg^{-1}), long cycle life (>10^6 cycles), low maintenance cost, simplicity and better safety compared to secondary batteries [38,39]. The HMC capsules with a particle size of 510 ± 10 nm (Figure 4.8a) were used as electrode materials for the EC. The TEM image in Figure 4.8b reveals a hollow macroporous core of ca. 330 nm diameter and a shell with a thickness of ca. 90 nm for the synthesised HMC. Commercial NAS carbon (Norit A Supra) was used for comparison.

The nitrogen adsorption–desorption isotherms of the HMC capsules and NAS carbon are shown in Figure 4.8c. The isotherm of the HMC was classified as a type IV with H2-type hysteresis, typical of mesoporous materials and also reveals a narrow PSD, proving the fact shown by TEM images

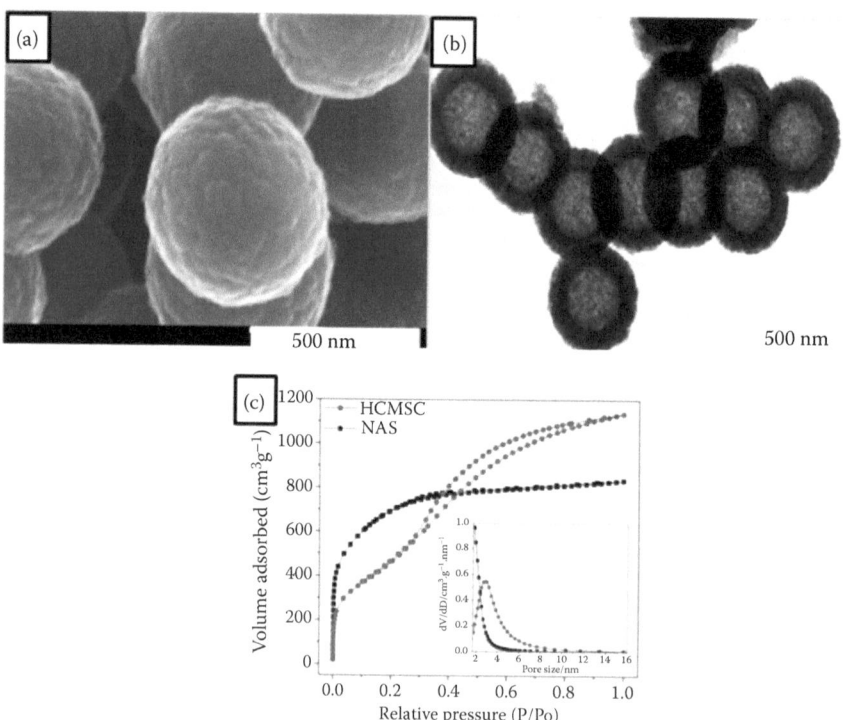

FIGURE 4.8
(a) FE-SEM image, (b) TEM image of HMC$_{330/90}$ and (c) N$_2$ adsorption–desorption isotherm of HMC and NAS. The inset in (c) shows PSD curves of the HMC and NAS determined on the basis of adsorption isotherms.

TABLE 4.4
Surface Structural Parameters Derived from N_2 Adsorption-Desorption Isotherms for HMC and NAS

Sample	BET Surface Area ($m^2\,g^{-1}$)	Total Pore Volume ($cm^3\,g^{-1}$)	Micropore Volume ($cm^3\,g^{-1}$)	Ratio of Micropore (%)	Average Pore Diameter (nm)
$HMC_{330/90}$	1667	1.77	0.53	29.9	3.46
NAS	2170	1.07	0.96	89.8	1.97

that the HMC has uniform mesopores in the shell. On the other hand, NAS reveals isotherm with a high amount of nitrogen sorption at a low relative pressure characteristic of microporous materials. The estimated pore size determined from the PSD maximum was found to be ca. 3.46 nm along with a BET surface area of 1667 $m^2\,g^{-1}$ for the HMC. The total pore volume (V_{Total}) of HMC was measured to be 1.77 $cm^3\,g^{-1}$, among which a significantly high mesopore volume (ca. 1.24 $cm^3\,g^{-1}$, i.e. 70% of V_{Total}) was found to be generated from the removal of the mesosize-thick silica walls in the silica/carbon composite. On the other hand, NAS possesses a much higher surface area of ca. 2170 $m^2\,g^{-1}$ and a total pore volume of 1.07 $cm^3\,g^{-1}$, ca. 90% of which is from micropores. Table 4.4 summarises the surface structural properties of the HMC and NAS carbon studied in this work.

All electrochemical characterisations of the HMC were performed using symmetrical two-electrode configuration with 1.0 M tetraethylammonium tetrafluoroborate (Et_4NBF_4) in acetonitrile (AN) as electrolyte. The two electrode test cell configuration with organic electrolyte is found to provide more reliable and practical data for evaluating a material's performance for ECs [76,77]. Figure 4.9a shows the typical galvanostatic charge–discharge

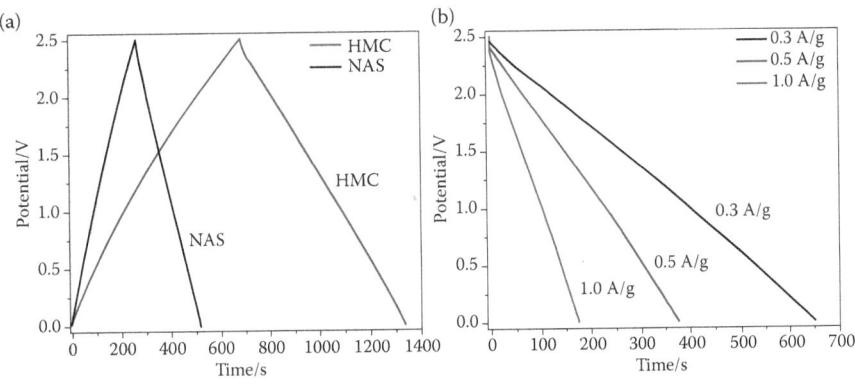

FIGURE 4.9
(a) Galvanostatic charge–discharge profiles for HMC and NAS carbon at current density of 0.3 A g^{-1} and (b) galvanostatic discharge curves for HMC at different current densities.

profile of the HMC and NAS carbon between 0 and 2.5 V potential window at 0.3 A g^{-1} current density. A high specific capacitance value of 162 F g^{-1} was measured for the HMC capsules from a linear fit of the discharge curve. It can be seen that the charge–discharge behaviour of the HMC shows almost linear variation of voltage with time as in ideal shown by electrical double layer capacitor. This high specific capacitance is attributed to the large active surface area of the HMC electrode. Most of the active surface area of the HMC electrode could be accessed by organic electrolyte ions due to the large amount of mesopore channels, which allows the facile movement of electrolyte for easy access to the smaller micropores, which are inaccessible otherwise. Additionally, a unique network of 3D-interconnected large interstitial empty space between spherical carbon structures and macropore open to mesopores in the shells of HMC capsules also facilitate the movement of big organic electrolyte ions within the mesopores, and thereby lead to shorter discharge time and consequently high capacitance value. However, in the case of commercial NAS carbon, although the total surface area is higher than HMC, significant quantities of micropores (approximately 90% of the total pore volume as shown in Figure 4.8c) restricts the electrolyte movement, and significant portions of the available surfaces might remain inaccessible for large organic electrolyte ions, leading to a lower specific capacitance value of 62 F g^{-1} at the same current density of 0.3 A g^{-1}. The linearity of charge–discharge profile for HMC was well maintained even with higher current density as shown in Figure 4.9b.

High rate capability is equally important for the application of EDLC in high-energy systems such as HEV [78–82]. The EDLC performances of the HMC capsules and NAS carbon at higher current density were recorded by continuous increase in current density (initially 0.3–0.5 and then to 1.0 A g^{-1}) after every five charge–discharge cycles, and the representative data are

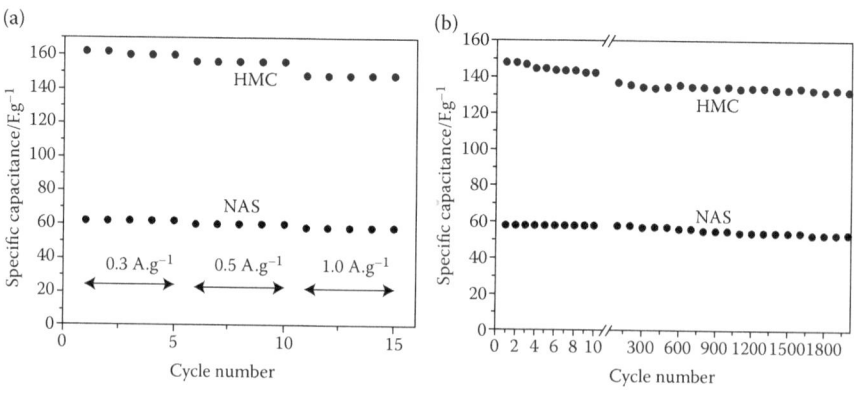

FIGURE 4.10
(a) Capacitive rate performance at different current densities. (b) Cyclic performance of HMC and NAS carbon at 1.0 A g^{-1} for 2000 cycles.

presented in Figure 4.10a. The specific capacitance of the HMC capsules are found to be stable as the cycle progresses, with only slight decrease in capacitance from 156 to 148 F g^{-1} when the current density is increased from 0.5 A to 1.0 A g^{-1}. The decrease in capacitance at higher current density may be ascribed to the disordered nature of mesopore structure of the shell, which may restrict faster movement of electrolyte ions to reach adsorption sites to some extent. However, the NAS electrode illustrates a relatively smaller decrease in specific capacity (60 F g^{-1} at 0.5 A g^{-1} and 58 F g^{-1} at 1.0 A g^{-1}). This small difference in rate performance shows that, in spite of the larger specific surface area available in the NAS carbon, the abundant ultramicropores below 0.7 nm are not accessible to organic electrolyte ions, and thus specific capacitance is rather less sensitive to current density. On the other hand, HMC has more adsorption sites available to accommodate organic electrolyte ions through the mesoporous entrances at low current density. Though the commercial NAS carbon has shown better stability at high rate than the HMC with increasing current density, the capacitance of HMC is still more than double that of NAS at a higher current density of 1.0 A g^{-1}. This enhanced performance clearly illustrates the importance of mesopore volume and pore characteristics in carbon nanostructure for better EDLC performance. Table 4.5 summarises the comparative capacitive performances of our synthesised HMC capsules and other reported porous carbon materials tested using organic electrolytes.

The capacitance retention of the HMC electrode was also studied by recording charge–discharge profiles for 2000 cycles at 1.0 A g^{-1}, and the corresponding data are presented in Figure 4.10b. Despite the initial decrease in capacitance for the first 100 cycles, probably due to the attainment of equilibrium electrode potentials, the HMC has shown very good cycle stability up to 2000 cycles and retains 88% of the initial capacitance. The hollow macroporous core, which is connected and open to the outer mesoporous

TABLE 4.5

Comparison of Specific Capacitances of HMC and Other Reported Porous Carbon Materials Tested in Organic Electrolyte as Electrochemical Capacitor

	Specific Capacitance (F g^{-1})	Current Density (A g^{-1})	Reference
HMC$_{330/90}$	162	0.3	[83]
HMC$_{330/90}$	148	1.0	[83]
Carbide-derived carbon	138	1.0	[78]
Mesoporous carbon spheres	97	0.5	[79]
Zeolite-templated porous carbon	146	0.25	[43]
Mesoporous carbon capsule	111	1.0	[44]
Hierarchical porous carbon	137	1.0	[80]
Templated mesoporous carbon	86	1.0	[81]
Mesoporous carbon from sucrose	116	0.5	[82]

shell, has assisted in minimising the diffusion distance during the charge/discharge cycling, ensuring good cycling performance. On the other hand, the mesoporous/microporous channels in the shell open to the macroporous core and interstitial void provide favourable sites for diffusion and adsorption of electrolyte ions.

4.5 Catalyst Support in Fuel Cell

HMC-supported Pt (20 wt%) catalysts were synthesised at room temperature through the impregnation method using $H_2PtCl_6 \cdot 6H_2O$ (Aldrich) as a metal precursors and $NaBH_4$ as a reducing agent. Details of synthesis and experimental can be found in an earlier work [84]. HMC capsules of ca. 360 nm diameter were prepared as shown in Figure 4.11a and used as support materials for the deposition of Pt nanoparticles. From the TEM image in Figure 4.11b, it can be observed that the core size and shell thickness of HMC were ca. 280 and 40 nm, respectively. As shown in Figure 4.11c, nitrogen adsorption–desorption isotherms of HMC revealed type IV isotherm with type H2 hysteresis along with narrow PSD. The pore size was estimated from the PSD maximum as 3.0 nm. The HMC capsules exhibited a high surface area of 917 m^2 g^{-1} and a total pore volume of 1.1 cm^3 g^{-1}, which are mainly attributable to the presence of the mesopores (2 nm < pore size < 50 nm) in the shell (mesopore volume: 0.80 cm^3 g^{-1}). A fair comparison among the different carbon supported catalyst was based on 20 wt% Pt loading: (i) Pt/HMC; (ii) Pt/VC and (iii) Pt/VC (E-TEK) (commercially available). Table 4.6 summarises the structural parameters for the HMC and the Vulcan carbon (VC) obtained from the N_2 adsorption–desorption isotherms conducted at 77 K.

Figure 4.11d shows the TEM image of as-prepared Pt (20 wt%)/HMC catalyst prepared using impregnation-$NaBH_4$ method. It is found that the Pt nanoparticles are homogenously dispersed as small, spherical and uniform dark spots over the entire surface of the HMC support. The TEM image also indicates that the structural integrity of the HMC was pretty well preserved even after catalyst loading, indicating the structural rigidity of the HMC as catalyst support. In contrast, Pt nanoparticles of homemade Pt (20 wt%)/VC (Figure 4.11e) show more aggregation and less uniform dispersion of the Pt nanoparticles. Higher uniformity and smaller particle size of Pt nanoparticles for the HMC-supported Pt catalyst stem from its higher surface area and highly developed mesoporosity. For comparison, the TEM image of the commercially available E-TEK Pt (20 wt%)/VC catalyst is shown in Figure 4.11f. Smaller Pt nanoparticles with higher uniformity were observed for the E-TEK catalyst than the homemade Pt (20 wt%)/VC catalyst, which is probably attributable to the differences in catalyst preparation methods or procedures.

Hollow Mesoporous Carbon with Hierarchical Nanoarchitecture

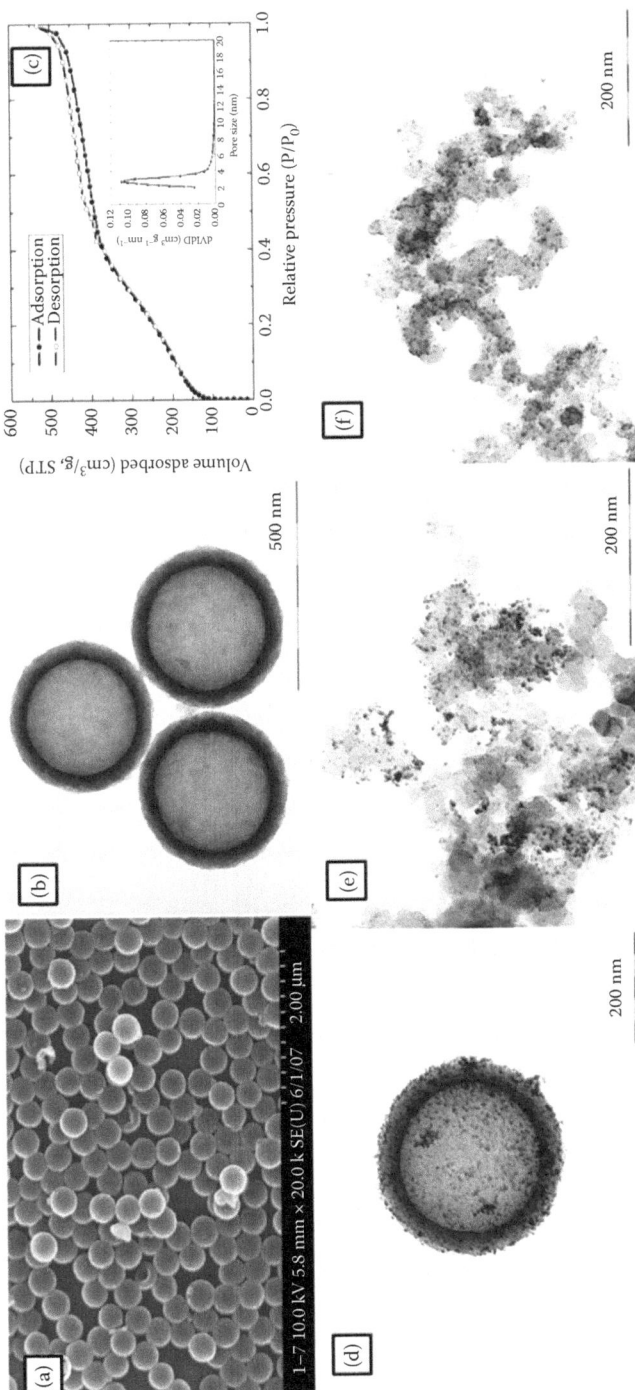

FIGURE 4.11
Representative (a) SEM, (b) TEM images and (c) N_2 adsorption–desorption isotherms obtained at 77 K and derived pore size distribution (PSD) for the $HMC_{280/40}$. TEM images for (d) as-prepared Pt (20 wt%) supported on HMC, (e) Pt (20 wt%) supported on commercial carbon black VC and (f) Pt (20 wt%)/VC (E-TEK).

TABLE 4.6
Structural Parameters for HMC and Commercial Carbon Black VC

Sample	BET Surface Area (m² g⁻¹)	V_{total} (cc g⁻¹)	V_{meso} (cc g⁻¹)	V_{micro} (cc g⁻¹)	Pore Size (nm)
$HMC_{280/40}$	917	1.10	0.80	0.30	3.0
Vulcan carbon (VC)	230	0.31	0.22	0.09	~

Single cells were constructed to evaluate catalytic activities towards ORR and fuel cell performance of various carbon-supported Pt (20 wt%) cathode catalysts [14,23–25]. E-TEK VC-supported PtRu (20 wt%) was used as an anode catalyst for all single-cell tests. Figure 4.12a shows the constant-current cell polarisation and power density plots for the PEMFCs using various carbon-supported Pt (20 wt%) catalysts at 60°C, with O_2-fed cathode mode. At low current density, H_2-fueled fuel cell polarisation is under electrochemical activation control, which is primarily caused by the sluggish kinetic, intrinsic to the ORR at the cathode surface. In this case, the HMC-supported Pt cathode catalyst shows lower polarisation voltage loss than the other VC-supported Pt catalysts, suggesting that the Pt/HMC cathode catalyst possesses the higher electrocatalytic activity towards ORR. The maximum power density at 60°C with O_2-fed cathode mode for the Pt/HMC is ca. 444 mW cm², corresponding to a 83% enhancement compared with the in-house Pt/VC (ca. 243 mW cm²) and of 59% compared with the Pt/VC (E-TEK) (ca. 280 mW cm²). The Pt/VC catalyst prepared under the same conditions as the Pt/HMC showed even lower power density than that of the commercial E-TEK Pt/VC. This can be mainly ascribed to the poorer Pt nanoparticle dispersion and wider Pt nanoparticle size distribution of the in-house Pt/VC catalyst than other catalysts. The Pt/HMC catalyst exhibited superior performance over the commercial E-TEK catalyst. This can be attributed solely to the effect of the HMC support.

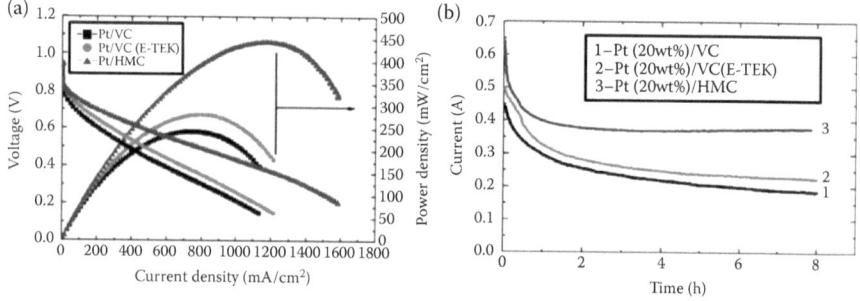

FIGURE 4.12
(a) Cell polarisation and power density plots and (b) chronoamperograms (0.75 V) obtained at 60°C with O_2-fed cathode mode for the H_2-fueled PEMFCs using various carbon-supported Pt (20 wt%) cathode catalysts.

Figure 4.12b shows the chronoamperograms at 0.75 V for 8 h at 60°C with O_2-fed cathode mode for various carbon-supported Pt (20 wt%) cathode catalysts. It was found that the HMC-supported Pt catalyst exhibited both higher initial and final current response within the test period, suggesting it was significantly more active with respect to ORR than those VC-supported. A steady state was reached after about 4 h for the Pt/HMC catalyst, while the current response continued to decay for those VC-supported, indicating the former is more stable under the same cathode working conditions. Faster ORR kinetics is observed on the Pt/HMC catalyst than on Pt/VC. After 8 h, the reaction current was ca. 377 mA for the Pt/HMC, corresponding to an improvement of 67% and 103% when compared with Pt/VC (E-TEK) (226 mA), and in-house Pt/VC (186 mA), respectively. The enhancement in the electrocatalytic activity and fuel cell performance is attributable to the unique structural properties of the HMC support. First, the higher surface area and larger pore volume of the HMC capsules favour a greater degree of catalyst dispersion. Second, the well-combined hierarchical nanoporous structures with the mesopores in the shell open to the outer interstitial space and to the inner hollow macroporous core provide efficient diffusion of the reactants and products. In contrast, mass transport may be less efficient in commercial carbon black VC owing to the presence of randomly distributed pores of various sizes.

4.6 Conclusions

HMC capsules with hierarchical macro/mesoporous nanoarchitecture are prepared and explored as material for electrochemical energy storage and conversion such as anode in LIB, an electrode in EC and a support material for fuel cell catalysts. HMC capsules with hollow macroporous core/mesoporous shell nanostructure were synthesised by a 'nanocasting' technique using SCMS silica spheres as template. Through size control of the SCMS silica template, it is easy to synthesise HMC materials with varied core size and shell thickness, which eventually result in different surface areas, pore volumes and meso- and microporosity for HMCs [28]. Unique hierarchical structural characteristics of HMC, such as large surface area and mesopore volume—particularly the multimodal porosity composed of well-developed 3D interconnected hollow macropore and mesopores embedded in the shells along with 3D large interconnected interstitial volume between HMC particles—make HMC ideal electrode materials for various electrochemical energy devices. With these distinctive structural characters, HMC as a support facilitates fast transport of reagents, including electrolytes, reactants and products in electrochemical application, resulting in enhanced electrochemical performance in terms of energy output, rate capability and durability when compared to commercial carbon, graphite and ordered mesoporous carbon.

HMC capsules were applied as the anode for the LIB. HMC capsules have demonstrated higher Li storage capacity, and outperformed the commercial graphite and other common porous CMK-3 carbon materials in terms of cycling and rate performance. The macroporous hollow core serves as an efficient Li storage and buffer reservoir, which not only shortens the diffusion path of Li ions but also reduces the volume change during charge–discharge, especially at high rates, thereby improving cycling performance and rate capability. In particular, HMC exhibits over 58% capacity retention at the high rate of 1000 mA g^{-1} and also retains 89% of the initial capacity when returning to the initial low current rate (100 mA g^{-1}). The HMC capsules have shown almost 80% retention of initial capacity after 80 cycles at 100 mA g^{-1}. As demonstrated by EIS spectra, HMC has low charge transfer resistance and Warburg impedance, thus faster mass transport of lithium within their porous HMC network [85].

HMC was also explored as an electrode material for an EC. The porous surface of HMC lead to facile charge movement and electrolyte diffusion and, therefore, most of the surfaces could be accessed by bigger organic electrolyte ions. These properties enabled the HMC materials to demonstrate a very high specific capacitance of 162 F g^{-1} at 0.3 A g^{-1} in a practical two-electrode symmetric system with organic electrolyte. HMC capsules have not only shown specific capacitance of 148 F g^{-1} at higher current density of 1.0 A g^{-1} with excellent 91% retention of initial capacitance but also outperforms the commercial activated carbon by a factor more than two, and demonstrates very good cyclic performance with almost 88% retention of the initial capacitance value up to 2000 charge–discharge cycles at 1.0 A g^{-1} [83].

High surface area, large pore volume, uniform particle size and well-developed three-dimensionally interconnected hierarchical porosity made HMC an ideal candidate for catalyst support in low-temperature H_2-fueled PEM fuel cells. As a result, the HMC-supported Pt (20 wt%) catalyst exhibited an excellent electrocatalytic oxygen reduction activity and enhanced current and power density than the commercial VC-supported Pt (20 wt%) catalysts.

The significant improvement in electrochemical performances of the HMC as an anode in the LIB, electrode material for the EC and cathode catalyst support for Pt in H_2-fueled PEMFC are attributed to unique structural properties of the HMC. In particular, the 3D-interconnected nanostructure with hierarchical porosity, providing not only large specific surface area and high mesopore volume for high specific capacitance and homogeneous dispersion of nano-sized Pt and Pt-based alloy nanoparticles but also highly developed hierarchical macro/mesoporosity for fast mass transport.

We believe that this systematic study will help to understand the correlation between surface parameters and LIB, EDLC and PEMFC performance, which is much needed to develop high-energy applications in accordance with present-day requirement. The development of new HMCs with various core sizes and shell thicknesses is in progress. In particular, the synthesis is directed to make hollow core size as small as possible to greatly increase the surface area. Relationship of mesopore size and mesopore shell thickness with electrochemical performance will be studied. Further enhancement in electrochemical energy generation and storage is expected from these HMC with better-optimised structural characters [86]. In addition, high-surface-area nanostructured HMC materials can be doped with heteroatoms such as N and/or P or made as composite with metal oxide (i.e. SnO_2) to improve the stability and performance of electrochemical devices, which is currently under investigation [87].

Acknowledgements

This work was supported by NRF Grant (NRF 2010-0029245) and Global Frontier R&D Program on Center for Multiscale Energy System (NRF 2011-0031571) funded by the Ministry of Education, Science and Technology through the National Research Foundation of Korea. We also thank the KBSIs at Jeonju, Chuncheon and Daejeon for SEM, TEM and XRD measurements.

References

1. B.-B. Owens, W.-H. Smyrl and J.-J. Xu, R&D on lithium batteries in the USA: High-energy electrode materials, *J. Power Sources* 81–82, 1999, 150–155.
2. K.-S. Kang, Y.-S. Meng, J. Breger, C.-P. Grey and G. Ceder, Electrodes with high power and high capacity for rechargeable lithium batteries, *Science* 311, 2006, 977–980.
3. J.-M. Tarascon and M. Armand, Issues and challenges facing rechargeable lithium batteries, *Nature* 414, 2001, 359–367.
4. J. Maier, Nanoionics: Ion transport and electrochemical storage in confined systems, *Nat. Mater.* 4, 2005, 805–815.
5. L. Taberna, S. Mitra, P. Poizot, P. Simon and J.M. Tarascon, High rate capabilities Fe_3O_4-based Cu nano-architectured electrodes for lithium-ion battery applications, *Nat. Mater.* 5, 2006, 567–573.
6. A.S. Arico, P. Bruce, B. Scrosati, J.-M. Tarascon and W. Van Schalkwijk, Nanostructured materials for advanced energy conversion and storage devices, *Nat. Mater.* 4, 2005, 366–377.
7. J.-S. Sakamoto and B. Dunn, Hierarchical battery electrodes based on inverted opal structures, *J. Mater. Chem.* 12, 2002, 2859–2861.
8. G. Derrien, J. Hassoun, S. Panero and B. Scrosati, Nanostructured Sn–C composite as an advanced anode material in high-performance lithium-ion batteries, *Adv. Mater.* 19, 2007, 2336–2340.

9. J. Li, Y. Zhao, N. Wang and L. Guan, A high performance carrier for SnO$_2$ nanoparticles used in lithium ion battery, *Chem. Commun.* 47, 2011, 5238–5240.
10. H.-S. Zhou, S.M. Zhu, M. Hibino, I. Honma and M. Ichihara, Lithium storage in ordered mesoporous carbon (CMK-3) with high reversible specific energy capacity and good cycling performance, *Adv. Mater.* 15, 2003, 2107–2111.
11. M. Yoshio, H. Wang and K. Fukuda, Spherical carbon-coated natural graphite as a lithium-ion battery-anode material, *Angew. Chem. Int. Ed.* 42, 2003, 4203–4206.
12. B. Fang, M.-S. Kim, S.-H. Hwang and J.-S. Yu, Colloid-imprinted carbon with tailored nanostructure as an unique anode electrocatalyst support for formic acid oxidation, *Carbon* 46, 2008, 876–883.
13. K.T. Lee, Y.S. Jung and S.M. Oh, Synthesis of tin-encapsulated spherical hollow carbon for anode material in lithium secondary batteries, *J. Am. Chem. Soc.* 125, 2003, 5652–5653.
14. B. Fang, M.-S. Kim, J.-H. Kim, M.-Y. Song, Y.-J. Wang, H.-J. Wang, D.P. Wilkinson and J.-S. Yu, High Pt loading on functionalized multiwall carbon nanotubes as a highly efficient cathode electrocatalyst for proton exchange membrane fuel cells, *J. Mater. Chem.* 21, 2011, 8066–8073.
15. Z. Wang, F. Li, N.S. Ergang and A. Stein, Effects of hierarchical architecture on electronic and mechanical properties of nanocast monolithic porous carbons and carbon–carbon nanocomposites, *Chem. Mater.* 18, 2006, 5543–5553.
16. K.T. Lee, J.C. Lytle, N.S. Ergang, S.M. Oh and A. Stein, Synthesis and rate performance of monolithic macroporous carbon electrodes for lithium-ion secondary batteries, *Adv. Funct. Mater.* 14, 2005, 547–556.
17. Y. Zhou, Y. Kim, C. Jo, J. Lee and S. Yoon, A novel mesoporous carbon–silica–titania nanocomposite as a high performance anode material in lithium ion batteries, *Chem. Commun.* 47, 2011, 4944–4946.
18. B. Fang, M.-W. Kim, S.-Q. Fan, J.-H. Kim, D.P. Wilkinson, J.-J. Ko and J.-S. Yu, Facile synthesis of open mesoporous carbon nanofibers with tailored nanostructure as a highly efficient counter electrode in CdSe quantum-dot-sensitized solar cells, *J. Mater. Chem.* 21, 2011, 8742–8748.
19. J.H. Kim and J.-S. Yu, Erythrocyte-like hollow carbon capsules and their application in proton exchange membrane fuel cell, *Phys. Chem. Chem. Phys.* 12, 2010, 15301–15308.
20. G.M. Olyveira, J.H. Kim, M.V.A. Martins, R.M. Iost, K.N. Chaudhari, J.-S. Yu and F.N. Crespilho, Flexible carbon cloth electrode modified by hollow core-mesoporous shell carbon as a novel efficient bio-anode for biofuel cell, *J. Nanosci. Nanotechnol.* 12, 2012, 356–360.
21. J.H. Kim, B. Fang, M.Y. Song and J.-S. Yu, Topological transformation of thioether-bridged organosilicas into nanostructured functional materials, *Chem. Mater.* 24, 2012, 2256–2264.
22. K.-Y. Chan, J. Ding, J.-W. Ren, S.-A. Cheng and K.-Y. Tsang, Supported mixed metal nanoparticles as electrocatalysts in low temperature fuel cells, *J. Mater. Chem.* 14, 2004, 505–516.
23. B. Fang, M.-S. Kim and J.-S. Yu, Hollow core/mesoporous shell carbon as a highly efficient catalyst support in direct formic acid fuel cell, *Appl. Catal. B* 84, 2008, 100–105.
24. B. Fang, J.-H. Kim, M.-S. Kim and J.-S. Yu, Ordered hierarchical nanostructured carbon as a highly efficient cathode catalyst support in proton exchange membrane fuel cell, *Chem. Mater.* 21, 2009, 789–796.

25. J.-H. Kim, B.-Z. Fang, M.-S. Kim, S.-B. Yoon, T.-S. Bae, D.R. Ranade and J.-S. Yu, Facile synthesis of bimodal porous silica and multimodal porous carbon as an anode catalyst support in proton exchange membrane fuel cell, *Electrochim. Acta* 55, 2010, 7628.
26. J.-H. Kim, B. Fang, M. Kim and J.-S. Yu, Hollow spherical carbon with mesoporous shell as a superb anode catalyst support in proton exchange membrane fuel cell, *Catal. Today* 146, 2009, 25–30.
27. B. Fang, H.-S. Zhou and I. Honma, Ordered porous carbon with tailored pore size for electrochemical hydrogen storage application, *J. Phys. Chem. B* 110, 2006, 4875–4880.
28. B. Fang, J.-H. Kim, M.-S. Kim and J.-S. Yu, Controllable synthesis of hierarchical nanostructured hollow core/mesopore shell carbon for electrochemical hydrogen storage, *Langmuir* 24, 2008, 12068–12072.
29. G.S. Paul, J.H. Kim, M.-S. Kim, K. Do, J. Ko and J.-S. Yu, Different hierarchical nanostructured carbons as counter electrodes for CdS quantum dot solar cells, *ACS Appl. Mater. Interfaces* 4, 2012, 375 – 381.
30. B. Fang, S.-Q. Fan, J.-H. Kim, M.-S. Kim, M.-W. Kim, N.K. Chaudhari, J.-J. Ko and J.-S. Yu, Incorporating hierarchical nanostructured carbon counter electrode into metal-free organic dye-sensitized solar cell, *Langmuir* 26, 2010, 11238–11243.
31. S.-Q. Fan, B. Fang, J.-H. Kim, J.-J. Kim, J.-S. Yu and J.-J. Ko, Hierarchical nanostructured spherical carbon with hollow core/mesoporous shell as a highly efficient counter electrode in CdSe quantum-dot-sensitized solar cells, *Appl. Phys. Lett.* 96, 2010, 063501–063503.
32. E.M. Sorensen, S.J. Barry, H.K. Jung, J.M. Rondinelli, J.T. Vaughey and K.R. Poeppelmeier, Three-dimensionally ordered macroporous $Li_4Ti_5O_{12}$: Effect of wall structure on electrochemical properties, *Chem. Mater.* 18, 2006, 482–489.
33. N.S. Ergang, J.C. Lytle, K.T. Lee, S.M. Oh, W.H. Smyrl and A. Stein, Photonic crystal structures as a basis for a three-dimensionally interpenetrating electrochemical-cell system, *Adv. Mater.* 18, 2006, 1750–1753.
34. Z. Wen and J. Li, Hierarchically structured carbon nanocomposites as electrode materials for electrochemical energy storage, conversion and biosensor systems, *J. Mater. Chem.* 19, 2009, 8707–8713.
35. B. Fang, M.-S. Kim, J.-H. Kim, S.-M. Lim and J.-S. Yu, Ordered multimodal porous carbon with hierarchical nanostructure for high Li storage capacity and good cycling performance, *J. Mater. Chem.* 20, 2010, 10253–10259.
36. S. Yang, X. Feng, L. Zhi, Q. Cao, J. Maier and K. Müllen, Nanographene-constructed hollow carbon spheres and their favorable electroactivity with respect to lithium storage, *Adv. Mater.* 22, 2010, 838–842.
37. E. Frackowiak and F. Beguin, Carbon materials for the electrochemical storage of energy in capacitors, *Carbon* 39, 2001, 937–950.
38. B.E. Conway, *Electrochemical Supercapacitors: Scientific Fundamentals and Technological Applications*, Plenum Publishers, New York, 1999.
39. A. Balducci, R. Dugas, P. Taberna, P. Simon, D. Plee, M. Mastragostino and S. Passerini, High temperature carbon–carbon supercapacitor using ionic liquid as electrolyte, *J. Power Sources* 165, 2007, 922–927.
40. G.A. Snook, P. Kao and A.S. Best, Conducting-polymer-based supercapacitor devices and electrodes, *J. Power Sources* 196, 2011, 1–12.

41. S.I. Lee, S. Mitani, C.W. Park, S.H. Yoon, Y. Korai and I. Mochida, Electric double-layer capacitance of microporous carbon nano spheres prepared through precipitation of aromatic resin pitch, *J. Power Sources* 139, 2005, 379–383.
42. H. Tamai, M. Kouzu, M. Morita and H. Yasuda, Highly mesoporous carbon electrodes for electric double-layer capacitors, *Electrochem. Solid-State Lett.* 6, 2003, A214–A217.
43. C. Portet, Z. Yang, Y. Korenblit, Y. Gogotsi, R. Mokaya and G. Yushinc, Electrical double-layer capacitance of zeolite-templated carbon in organic electrolyte, *J. Electrochem. Soc.* 156, 2009, A1–A6.
44. S. Murali, D.R. Dreyer, P. Valle-Vigón, M.D. Stoller, Y. Zhu, C. Morales, A.B. Fuertes, C.W. Bielawski and R.S. Ruoff, Mesoporous carbon capsules as electrode materials in electrochemical double layer capacitors, *Phys. Chem. Chem. Phys.* 13, 2011, 2652–2655.
45. B. Fang, J.H. Kim, M.-S. Kim, A. Bonakdarpour, A. Lam, D.P. Wilkinson and J.-S. Yu, Fabrication of hollow core carbon spheres with hierarchical nanoarchitecture for ultrahigh electrical charge storage, *J. Mater. Chem.* 22, 2012, 19031–19038.
46. N. Giordano, E. Passalacqua, L. Pino, A.S. Arico, V. Antonucci, M. Vivaldi and K. Kinoshita, Analysis of platinum particle size and oxygen reduction in phosphoric acid, *Electrochim. Acta* 36, 1991, 1979–1984.
47. Y. Takasu, T. Kawaguchi, W. Sugimoto and Y. Murakami, Effects of the surface area of carbon support on the characteristics of highly-dispersed PtRu particles as catalysts for methanol oxidation, *Electrochim. Acta* 48, 2003, 3861–3868.
48. C.A. Bessel, K. Laubernds, N.M. Rodriguez and R.T.K. Baker, Graphite nano fibers as an electrode for fuel cell applications, *J. Phys. Chem. B* 105, 2001, 1115–1118.
49. L. Zhang, J. Zhang, D.P. Wilkinson and H. Wang, Progress in preparation of non-noble electrocatalysts for PEM fuel cell reactions, *J. Power Sources* 156, 2006, 171–182.
50. P.A. Simonov, V.A. Likholobov, A. Wieckowski, E.R. Savinova and C.G. Vayenas, Eds.; *Catalysis and Electrocatalysis at Nanoparticle Surfaces*, Marcel Dekker, New York, 2003, 409.
51. G.S. Chai, S.B. Yoon, J.-S. Yu, J.-H. Choi and Y.-E. Sung, Ordered porous carbons with tunable pore sizes as catalyst supports in direct methanol fuel cell, *J. Phys. Chem. B* 108, 2004, 7074–7079.
52. J.-S. Yu, S. Kang, S.B. Yoon and G. Chai, Fabrication of ordered uniform porous carbon networks and their application to a catalyst supporter, *J. Am. Chem. Soc.* 124, 2002, 9382–9383.
53. G.S. Chai, S.B. Yoon and J.-S. Yu, Highly efficient anode electrode materials for direct methanol fuel cell prepared with ordered and disordered arrays of carbon nanofibers, *Carbon* 43, 2005, 3028–3031.
54. Y. Shao, G. Yin, Y. Gao and P. Shi, Durability study of Pt/C and Pt/CNTs catalysts under simulated PEM fuel cell conditions, *J. Electrochem. Soc.* 153, 2006, A1093–A1097.
55. M. Watanabe, H. Sei and P. Stonehart, The influence of platinum crystallite size on the electroreduction of oxygen, *J. Electroanal. Chem.* 261, 1989, 375.
56. V. Raghuveer and A. Manthiram, Mesoporous carbons with controlled porosity as an electrocatalytic support for methanol oxidation, *J. Electrochem. Soc.* 152, 2005, A1504–1510.

57. V. Rao, P.A. Simonov, E.R. Savinova, G.V. Plaksin, S. Cherepanova, G. Kryukova and U. Stimming, The influence of carbon support porosity on the activity of PtRu/Sibunit anode catalysts for methanol oxidation, *J. Power Sources* 145, 2005, 178–187.
58. K.W. Park, Y.E. Sung, S. Han, Y. Yun and T. Hyeon, Origin of the enhanced catalytic activity of carbon nanocoil-supported PtRu alloy electrocatalysts, *J. Phys. Chem. B* 108, 2004, 939–944.
59. Z.Q. Tian, S.P. Jiang, Y.M. Liang and P.K. Shen, Synthesis and characterization of platinum catalysts on multiwalled carbon nanotubes by intermittent microwave irradiation for fuel cell applications, *J. Phys. Chem. B* 110, 2006, 5343–5350.
60. S.-B. Yoon, K. Sohn, J.-Y. Kim, C.-H. Shin, J.-S. Yu and T. Hyeon, Fabrication of carbon capsules with hollow macroporous core/mesoporous shell structures, *Adv. Mater.* 14, 2002, 19–21.
61. J.-S. Yu, S.-B. Yoon, Y.-J. Lee and K.-B. Yoon, Fabrication of bimodal porous silicate with silicalite-1 core/mesoporous shell structures and synthesis of non spherical carbon and silica nano cases with hollow core/mesoporous shell structures, *J. Phys. Chem. B* 109, 2005, 7040–7045.
62. J.-H. Kim, B.-Z. Fang, S.-B. Yoon and J.-S. Yu, Hollow core/mesoporous shell carbon capsule as an unique cathode catalyst support in direct methanol fuel cell, *Appl. Catal., B* 88, 2009, 368–375.
63. M.-S. Kim, S.-B. Yoon, K.-N. Sohn, J.-Y. Kim, C.-H. Shin, T.-H. Hyeon and J.-S. Yu, Synthesis and characterization of spherical carbon and polymer capsules with hollow macroporous core and mesoporous shell structures, *Microporous Mesoporous Mater.* 63, 2003, 1–9.
64. Q.-M. Ji, S.-B. Yoon, J.P. Hill, A. Vinu, J.-S. Yu and K. Ariga, Layer-by-layer films of dual-pore carbon capsules with designable selectivity of gas adsorption, *J. Am. Chem. Soc.* 131, 2009, 4220–4221.
65. S.-B. Yoon, J.-H. Kim, F. Kooli, C.W. Lee and J.-S. Yu, Synthetic control of ordered and disordered arrays of carbon nanofibers from SBA-15 silica templates, *Chem. Commun.* 2003, 14, 1740–1741.
66. S. Kang, Y.-B. Chae and J.-S. Yu, HCl as a key parameter in size-tunable synthesis of SBA-15 silica with rodlike morphology, *J. Nanosci. Nanotechnol.* 9, 2009, 527–532.
67. M. Endo, C. Kim, T. Karaki, N. Nishimura, M.J. Matthews, S.D.M. Brown and M.S. Dresselhaus, Anode performance of a Li ion battery based on graphitized and B-doped milled mesophase pitch-based carbon fibers, *Carbon* 37, 1999, 561–568.
68. H. Xia, S. Tang and L. Lu, Properties of amorphous Si thin film anodes prepared by pulsed laser deposition, *Mater. Res. Bull.* 42, 2007, 1301–1309.
69. J.B. Goodenough, Evolution of strategies for modern rechargeable batteries, *Acc. Chem. Res.* 46, 2013, 1053–1061.
70. J. Shim and K.A. Striebel, The dependence of natural graphite anode performance on electrode density, *J. Power Source* 130, 2004, 247–253.
71. W.A. Van Schalkwijk and B. Scrosati, Eds. *Advanced in Lithium-Ion Batteries*, Kluwer Academic/Plenum Publishers, New York, 7, 2002.
72. E. Frackowiak, S. Gautier, H. Gaucher, S. Bonnamy and F. Beguin, Electrochemical storage of lithium in multiwalled carbon nanotubes, *Carbon* 37, 1999, 61–69.
73. B. Gao, C. Bower, J.D. Lorentzen, L. Fleming, A. Kleinhammes, X.P. Tang, L.E. McNeil, Y. Wu and O. Zhou, Enhanced saturation lithium composition in ball-milled single-walled carbon nanotubes, *Chem. Phys. Lett.* 327, 2000, 69–75.

74. J. Jamnik and J. Maier, Treatment of the impedance of mixed conductors equivalent circuit model and explicit approximate solutions, *J. Electrochem. Soc.* 146, 1999, 4183–4188.
75. C.S. Wang, A.J. Appleby and F.E. Little, Electrochemical impedance study of initial lithium ion intercalation into graphite powders, *Electrochim. Acta* 46, 2001, 1793–1813.
76. M.D. Stoller and R.S. Ruoff, Best practice methods for determining an electrode material's performance for ultracapacitors, *Energy Environ. Sci.* 3, 2010, 1294–1301.
77. V. Khomenko, E. Frackowiak and F. Beguin, Determination of the specific capacitance of conducting polymer/nanotubes composite electrodes using different cell configurations, *Electrochim. Acta* 50, 2005, 2499–2506.
78. J. Chmiola, G. Yushin, Y. Gogotsi, C. Portet, P. Simon and P.L. Taberna, Anomalous increase in carbon capacitance at pore sizes less than 1 nanometer, *Science* 313, 2006, 1760–1763.
79. Q. Li, R. Jiang, Y. Dou, Z. Wu, T. Huang, D. Feng, J. Yang, A. Yu and D. Zhao, Synthesis of mesoporous carbon spheres with a hierarchical pore structure for the electrochemical double-layer capacitor, *Carbon* 49, 2011, 1248–1257.
80. Z. Zheng and Q. Gao, Decomposition of NH_3BH_3 at sub-ambient pressures: A combined thermogravimetry–differential thermal analysis–mass spectrometry study, *J. Power Sources* 196, 2011, 1615–1618.
81. A.B. Fuertes, G. Lota, T.A. Centeno and E. Frackowiak, Templated mesoporous carbons for supercapacitor application, *Electrochim. Acta* 50, 2005, 2799–2805.
82. C.V. Guterl, S. Saadallah, K. Jurewicz, E. Frackowick, M. Reda, J. Parmentier, J. Patarin and F. Beguin, Supercapacitor electrodes from new ordered porous carbon materials obtained by a templating procedure, *Mater. Sci. Eng. B* 108, 2004, 148–155.
83. D. Bhattacharjya, M.-S. Kim, T.-S. Bae and J.-S. Yu, High performance supercapacitor prepared from hollow mesoporous carbon capsules with hierarchical nanoarchitecture, *J. Power Sources* 244, 2013, 799–805.
84. B. Fang, J.H. Kim, C. Lee and J.-S. Yu, Hollow macroporous core/mesoporous shell carbon with a tailored structure as a cathode electrocatalyst support for proton exchange membrane fuel cells, *J. Phys. Chem. C* 112, 2008, 639–645.
85. M.-S. Kim, B. Fang, J.H. Kim, D.-S. Yang, Y.K. Kim, T.-S. Bae and J.-S. Yu, Ultrahigh Li storage capacity demonstrated by hollow carbon capsules with hierarchical nanoarchitecture, *J. Mater. Chem.* 21, 2011, 19362–19367.
86. B. Fang, M.-S. Kim, J.H. Kim and J.-S. Yu, Hierarchical nanostructured carbons with meso-macroporosity: Design, characterization and applications, *Acc. Chem. Res.* 46, 2013, 1397–1406.
87. D.-S. Yang, D. Bhattacharjya, S. Inamdar, J. Park and J.-S. Yu, Phosphorus-doped ordered mesoporous carbons with different channel lengths as efficient metal-free electrocatalysts for oxygen reduction reaction in alkaline media, *J. Am. Chem. Soc.* 134, 2012, 16127–16130.

5
Layer-Structured Cathode Materials for Energy Storage

Bohang Song, Man On Lai and Li Lu

CONTENTS

5.1 Introduction 191
5.2 Conventional Layer-Structured Materials 193
    5.2.1 $LiCoO_2$ 193
    5.2.2 $LiNiO_2$ and $LiMnO_2$ 195
    5.2.3 $LiNi_{1/2}Mn_{1/2}O_2$ and $LiNi_{1/3}Co_{1/3}Mn_{1/3}O_2$ 196
5.3 Integrated Layer-Structured Materials 200
5.4 Li Diffusion in Layered Structures 204
5.5 Methodology of Synthesis and Characterisations 205
5.6 Layered Cathode for All-Solid-State Thin-Film Lithium Ion Battery 206
5.7 Nanoscale Mapping of Li^+ Diffusion in All-Solid-State Thin-Film Lithium Ion Battery 209
5.8 Conclusions 214
References 215

5.1 Introduction

Fossil fuels, hydropower, nuclear fission and other sources of energy are deployed to satisfy the needs of modern technologies and benefits of everyone's daily life. However, irrespective of whatever means are used to produce energy, energy storage always becomes a significant issue and presents a great challenge to scientists and engineers. Therefore, in recent decades, energy-storage devices that are based on electrochemical principles, such as batteries and supercapacitors, have been drawing intense interests from fundamental research to practical industrial applications. Among advanced energy-storage conceptions, lithium-ion batteries (LiBs) are remarkable systems due to their high-energy density, high-power density and no gaseous exhaust with considerable reliability [1]. Consequently, such power storage has been playing an important role in various fields of applications such as portable electronics and telecommunication equipment, and is even considered as a serious contender for the next-generation hybrid electric vehicles and electric vehicles [2]. To meet the needs of new application devices, low-cost, safe, environmental-friendly, good gravimetric energy density, volume energy density, capacity and rate capability are the targets for improvements [3].

Since the realisation of $LiCoO_2$ as cathode materials of commercial LiBs in 1990, researchers have been trying to find alternative cathode materials for the secondary LiBs, mainly in consideration of certain drawbacks out of traditional $LiCoO_2$, such as toxicity, unsafe at high temperature, high cost and low specific capacity. Starting from the framework design, novel materials possessing spinel (LiM_2O_4) and olivine ($LiMPO_4$) structures (M = transition metal) [4] were proposed to exploit their advantages. Spinel crystals such as $LiMn_2O_4$, have high operating voltage of around 4 V, low cost benefiting from manganese and better safety over a wider temperature range, while olivine crystals such as $LiFePO_4$ exhibit acceptable cost with remarkable cycling performance despite their limited volume density. On the basis of the assumption of one-electron reaction, the structural essence of spinel ($LiMn_2O_4$) and olivine ($LiFePO_4$) shows low theoretical capacities of only 148 and 170 mAh/g, respectively. In comparison, layer-structured cathode materials possessing the formula of $LiMO_2$ (M = transition metal) such as $LiCoO_2$ have theoretical capacity of about 274 mAh/g. The charge/discharge curves and corresponding specific capacities of different types of cathodes, and the $Li_4Ti_5O_{12}$ anode are summarised in Figure 5.1 [5]. This feature of high specific capacity therefore

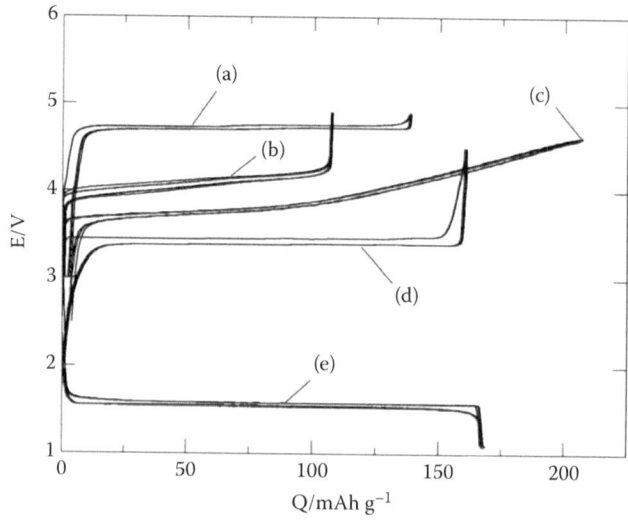

FIGURE 5.1
Charge and discharge curves of (a) $Li(Ni_{0.5}Mn_{1.5})O_4$, (b) $LiAl_xMn_{2-x}O_4$, (c) $LiCo_{1/3}Ni_{1/3}Mn_{1/3}O_2$, (d) $LiFePO_4$ cathodes and (e) $Li_4Ti_5O_{12}$ anode. (Reprinted from T. Ohzuku and R.J. Brodd, *J. Power Sources* 174, 2007, 449–456. With permission.)

makes the family of layered cathode materials (curve c) more competitive than other types of cathodes (curves a, b and d). To take advantage of the high capacity of layered materials, structural stability should be maintained even for deep deintercalation of lithium ions. For example, only 140 mAh/g reversible capacity could be obtained in $LiCoO_2$ by applying 4.2 V (vs. Li^0/Li^+) as the charging cutoff voltage [6]; otherwise, the framework of such structure will rapidly collapse leading to drastic loss of capacity upon cycling.

5.2 Conventional Layer-Structured Materials

5.2.1 $LiCoO_2$

$LiCoO_2$ has served as the commercial cathode material with graphite as the anode material for LiBs since 1991, and has dominated more than 90% of worldwide market [5]. Generally, $LiCoO_2$ exhibits several advantages such as high operating voltage (3.9 V vs. Li^0/Li^+), high specific reversible capacity (140 mAh/g), and long cycle life (more than 500 cycles).

The layered $LiCoO_2$ has rhombohedral-type Bravais lattice which belongs to the R3̄m pace group (α-$NaFeO_2$). This O_3 layered crystal structure consists of close-packed oxygen layers stacked in the sequence of ABCABC… with cationic cobalt and lithium sheets alternatively residing at octahedral sites between the oxygen layers [6–8], as can be seen from Figure 5.2a and b. With

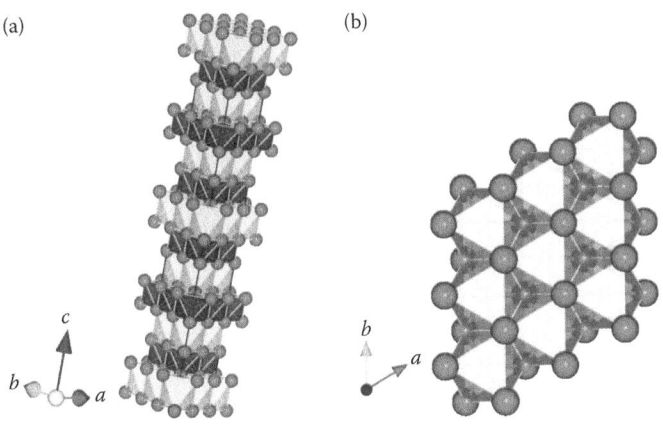

FIGURE 5.2
(See colour insert.) Models of superstructural $LiCoO_2$. (a) The red spheres indicate oxygen ions, the green octahedrons indicate the positions of lithium ions and the blue octahedrons indicate the positions of cobalt ions, (b) view along the c-axis of (a) the trigonal arrangement of oxygen ions is similar with hexagonal close packed (HCP) structure where lithium and cobalt ions locate at the triangular centres alternatively.

the extraction of lithium ions when charging, vacancies leave behind at octahedral sites within the lithium layers. Since the delithiation process promotes in the layered structure Li_xCoO_2, where x deceases from 1 to 0, a certain amount of phase transformation may occur for lower free energy. The final phase that exists at the end of the delithiation process is the CoO_2 in O_1 form [6,7].

However, $LiCoO_2$ suffers from a significant drawback. If the charging cutoff voltage is set to be higher than 4.2 V which means more than 50% of lithium needs to be extracted from the local structure, the layered framework will be unstable, leading to a drastic phase transformation coupled with a lattice distortion from hexagonal to monoclinic symmetry. As can be seen from Figure 5.3 [6], the variations in a and c lattice parameters indicate an appearance of the monoclinic phase when $x = 0.46$ and 0.22 in case of delithiated Li_xCoO_2 ($0 \leq x \leq 1$). This irreversible phase transition results in rapid capacity loss upon cycling. On the basis of detailed TEM (transmission electron microscopy)

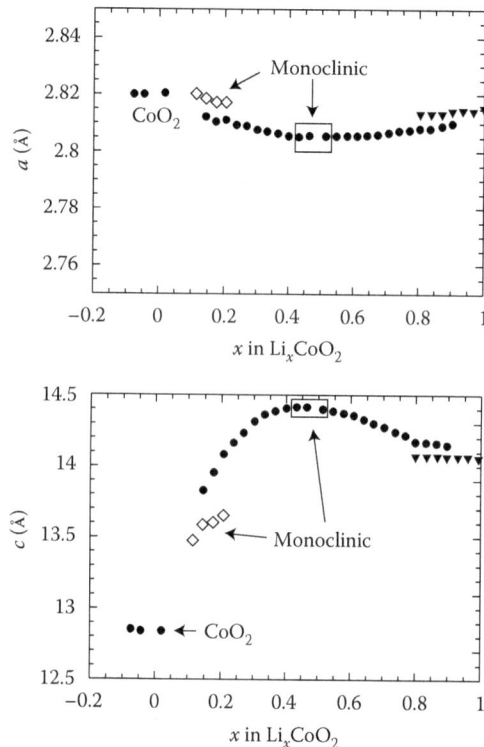

FIGURE 5.3
Variation of a and c lattice parameters characterised by *in situ* XRD during electrochemical lithium extraction of $LiCoO_2$. Triangles and circles represent hexagonal phases, and diamonds are monoclinic phases. (Reprinted with permission from G.G. Amatucci, J.M. Tarascon and L.C. Klein, CoO_2, the end member of the Li_xCoO_2 solid solution, *J. Electrochem. Soc.* 143, 1996, 1114–1123. Copyright 1996, The Electrochemical Society.)

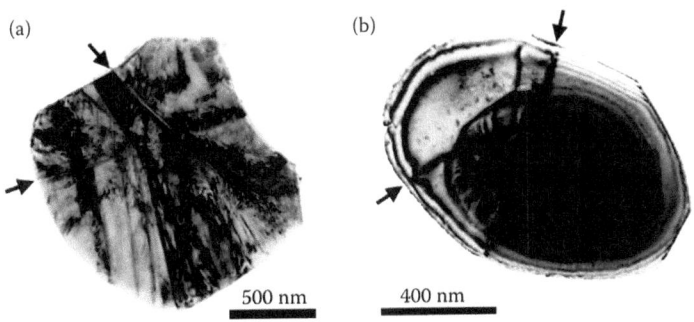

FIGURE 5.4
(a) Severely damaged LiCoO$_2$ particle from a cycled cathode showed microfracture (arrows) and strong diffraction contrast from internal strain and extended defects. (b) Another fractured LiCoO$_2$ particle also contained an unstrained region apparently relieved of stress by the fracture event. (Reprinted with permission from H.F. Wang et al. TEM study of electrochemical cycling-induced damage and disorder in LiCoO$_2$ cathodes for rechargeable lithium batteries, *J. Electrochem. Soc.* 146, 1999, 473–480. Copyright 1999, The Electrochemical Society.)

investigation, an internal strain-induced fracture could be easily observed in an extensive-cycled LiCoO$_2$ particle, as shown in Figure 5.4 [9]. Such cumulative damage upon cycling may also be responsible for property degradation.

As a result of the structural stability, the practical capacity obtained from layered Li$_x$CoO$_2$ is only 140 mAh/g, which limits its application in future high-energy, high-power demands. In addition, LiCoO$_2$ suffers from toxic, high cost and safety issue. Therefore, alternative layered structural materials were developed.

5.2.2 LiNiO$_2$ and LiMnO$_2$

Lithium nickel oxide, which is isostructural to LiCoO$_2$, has also been widely investigated. LiNiO$_2$ is not commercially utilised due to its serious cation mixing between lithium and nickel because of similar ion radius ($r_{Ni^{2+}} = 0.69$ Å, $r_{Li^+} = 0.76$ Å). This cation mixing significantly reduces the diffusion coefficient of lithium, and hence the power capability. Additionally, low lithium contents in the structure appear to be unstable because of the high oxygen partial pressure at equilibrium state [4]. These features make pure LiNiO$_2$ unsuitable for direct use as a cathode material, even if it has the advantages of low cost and environmental compatibility.

LiMnO$_2$ is another isostructural host with LiCoO$_2$ which is recognised as a promising cathode material due to its low-cost manganese, environmental-friendly and relatively stable structure with high initial capacity such as 190 mAh/g (2–4.25 V vs. Li^0/Li$^+$). However, the removal of 50% of the lithium leads to a phase transition from a layered to a spinel structure in local LiMnO$_2$, involving 25% of manganese ions transferring from transition metal layers into neighbouring alkali metal layers and the remaining

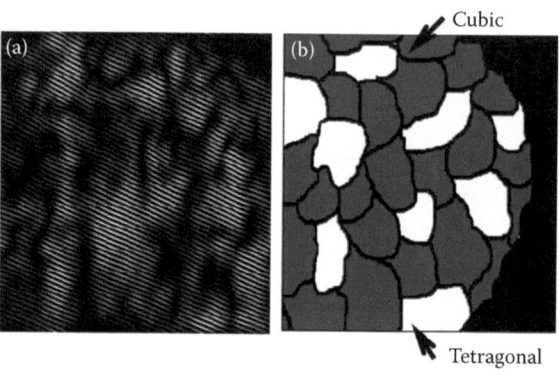

FIGURE 5.5
(a) TEM image of nanostructured $LiMn_2O_4$ spinel obtained on cycling layered $LiMnO_2$. Fourier-filtered image highlights the nano-domain structure of average dimensions of 50–70 Å. (b) A schematic representation of the nano-domain structure of $LiMn_2O_4$ spinel derived from layered $LiMnO_2$, showing cubic and tetragonal nano-domains. (From P.G. Bruce, B. Scrosati and J.M. Tarascon, Nanomaterials for rechargeable lithium batteries, *Angew. Chem. Int. Ed.* 2008, 47, 2930–2946. Copyright Wiley-VCH Verlag GmbH & Co. KGaA. Reproduced with permission.)

octahedral sites, while lithium is displaced into adjacent tetrahedral sites [10]. Such phase transformation incorporated with the Jahn–Teller distortion effect reduces the structural stability of layered $LiMnO_2$ [11,12] and, hence, limits its cycling performance. However, the spinel $Li_xMn_2O_4$ ($0 \leq x \leq 2$) nano-domain structure, *in situ* formed in the non-stoichiometric-layered $LiMnO_2$ is able to spontaneously switch between a cubic spinel and a tetragonal spinel, as illustrated in Figure 5.5 [13]. Consequently, it leads to a dramatic improvement in capacity retention from 61% after 10 cycles (no nano-domain structure) to 97% after 50 cycles (with nano-domain structure).

5.2.3 $LiNi_{1/2}Mn_{1/2}O_2$ and $LiNi_{1/3}Co_{1/3}Mn_{1/3}O_2$

$LiNi_{1/2}Mn_{1/2}O_2$ was first reported in 2001 [14]. With high reversible capacity close to 200 mAh/g (2.5–4.5 V vs. Li/Li$^+$), and limited capacity loss in initial cycles (as can be seen in Figure 5.6 [15]), $LiNi_{1/2}Mn_{1/2}O_2$ attracts interests for development [16–18]. $LiNi_{1/2}Mn_{1/2}O_2$ possesses the classic α-$NaFeO_2$ ($R\bar{3}m$) symmetrical layered structure, in which the valence states of transition metals can be accommodated as Ni^{2+} and Mn^{4+}, as confirmed by x-ray adsorption spectroscopy [16] and *ab initio* calculation [18]. During charge/discharge cycling, only Ni^{2+} ions participate in redox reaction to provide electrons, experiencing reversible change of valence states between 2+ and 4+ in the composition range of $0 < x < 1$ $Li_xNi_{1/2}Mn_{1/2}O_2$. Although the previous works on $LiNiO_2$ indicate that serious cation mixing between Li^+ and Ni^{2+} could result in structural instability when cycling [19], the presence of additional tetravalent manganese indeed stabilizes this structure, and no irreversible phase transformation took

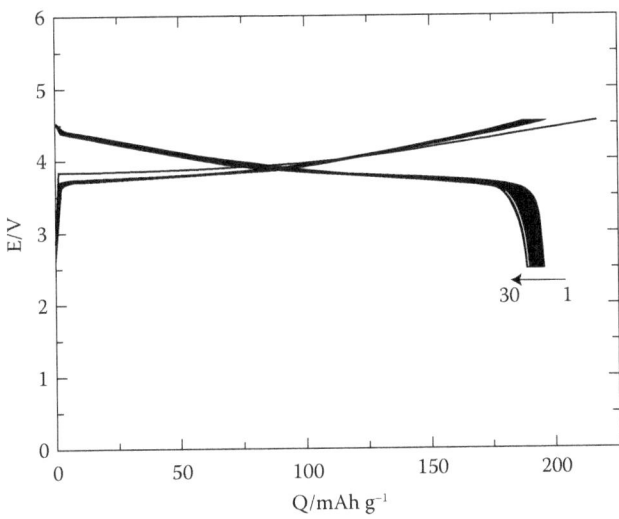

FIGURE 5.6
Charge and discharge curves of a Li/LiNi$_{1/2}$Mn$_{1/2}$O$_2$ cell operated in voltages of 2.5–4.5 V at a rate of 0.17 mA cm^{-2} for 30 cycles. (Reprinted from *J. Power Sources*, 119, Y. Makimura and T. Ohzuku, Lithium insertion material of LiNi$_{1/2}$Mn$_{1/2}$O$_2$ for advanced lithium-ion batteries, 156–160, Copyright 2003, with permission from Elsevier.)

place during the lithiation/delithiation process [14,16]. Since manganese is always tetravalent during delithiation/lithiation, Li$_x$Ni$_{1/2}$Mn$_{1/2}$O$_2$ is able to suppress either spinel-like phase transitions or Jahn–Teller distortion effects, both of which could be ascribed to Mn$^{3.5+}$ [20]. With the effect of similar ion radius between lithium and nickel, 8–10% cation disorder can always occur between 3a and 3b sites, leading to poor rate performance. In this regard, some strategies were proposed to reduce the ratio of cation disorder, such as ion-exchange method. Kang et al. [21] succeeded in decreasing the amount of cation mixing to 4.3%. This ion-exchange method remarkably improved the rate capability of LiNi$_{1/2}$Mn$_{1/2}$O$_2$, attaining 183 mAh/g at a 6 C rate (1 C = 280 mA/g), as well as better thermal stability compared to LiCoO$_2$ and LiNiO$_2$ [15]. In addition, Schougaard et al. [22] managed to reduce nickel content in the lithium layer to 5.6% by adjusting the composition of LiNi$_{0.5+x}$Mn$_{0.5-x}$O$_2$, hence, controlling the charge state of nickel. Nevertheless, such LiNi$_{1/2}$Mn$_{1/2}$O$_2$ structures with less cation disorder remain difficult to synthesise, in particular, for industrial applications.

LiNi$_{1/3}$Co$_{1/3}$Mn$_{1/3}$O$_2$ was also reported in 2001 [23], and has shown particularly promising electrochemical properties and structural stability. This material can be recognised as a complex solid solution in the ratio of LiCoO$_2$:LiNiO$_2$:LiMnO$_2$ = 1:1:1, or in the ratio of LiCoO$_2$:LiNi$_{1/2}$Mn$_{1/2}$O$_2$ = 1:2. In fact, all the transition metals—nickel, cobalt and manganese—can be accommodated in the transition metal layers homogeneously without phase separation [24]. In other words, nickel and manganese ions substitute partial cobalt ions, that is, occupying the original cobalt octahedral sites sequentially in transition metal layers, forming $(Ni_{1/3}Co_{1/3}Mn_{1/3})$ O_2-type superlattice based on triangular lattice of sites as illustrated in Figure 5.7 [25]. From another point of view, $(Ni_{1/2}Mn_{1/2})$ which are partially substituted by Co have a beneficial effect on the local ordering arrangement, since it is likely that the trivalent cobalt will disturb the charge ordering, and electronic interaction between manganese and nickel ions [26]. Such ordering arrangements significantly reduced the cation mixing from 8–10% in $LiNi_{1/2}Mn_{1/2}O_2$ to 1–6% in $LiNi_{1/3}Co_{1/3}Mn_{1/3}O_2$, which further improved its electrochemical performance. $LiNi_{1/3}Co_{1/3}Mn_{1/3}O_2$ has the same α-$NaFeO_2$ layered structure as $LiNi_{1/2}Mn_{1/2}O_2$, while the valence states of nickel, cobalt and manganese and cobalt are 2+, 3+ and 4+, respectively [25,27]. During charge/discharge cycling, the redox reactions, including Ni^{2+}/Ni^{4+} with $Co^{3+}/Co^{3.6+}$ in the voltage range between 2.5 and 4.5 V lead to a reversible capacity close to 200 mAh/g without obvious phase transformation in the Li-vacancy-arranged structure [28].

Owing to the complex distribution among these three cations in transition metal layers, the homogeneity at the atomic level in terms of chemical composition of cobalt, nickel and manganese is the key in the preparation of the electrochemical-preferred $LiNi_{1/3}Co_{1/3}Mn_{1/3}O_2$, where the traditional mechanical mixing techniques may not be suitable for synthesis [29].

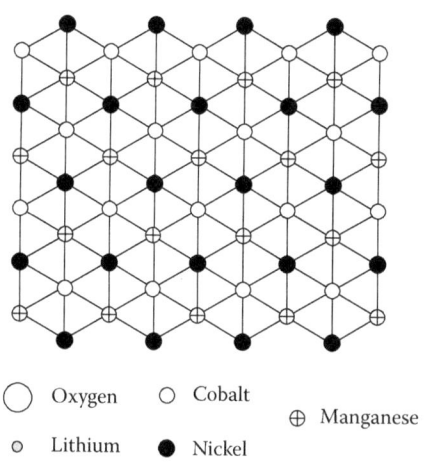

FIGURE 5.7
A schematic illustration on the crystal of having a superstructural $(Ni_{1/3}Co_{1/3}Mn_{1/3})O_2$ layer based on the triangular basal net of sites in the α-$NaFeO_2$-type structure. (Reprinted with permission from Y. Koyama et al. Solid-state chemistry, and electrochemistry of $LiCo_{1/3}Ni_{1/3}Mn_{1/3}O_2$ for advanced lithium-ion batteries—I. First-principles calculation on the crystal and electronic structures, *J. Electrochem. Soc.* 151, 2004, A1545–A1551. Copyright 2004, The Electrochemical Society.)

Therefore, the so-called co-precipitation method was introduced by many research groups to obtain greater homogeneity and better electrochemical properties, as can be seen from Figure 5.8 [29–31]. In addition, the modified sol–gel method to prepare macroporous structure also leads to a remarkable increase in cycle life, as shown in Figure 5.9 [32]. Based on this consideration on homogeneity, a lot of other synthesis methods were applied for preparing this kind of material [33–35].

The side reaction between solid materials and the electrolyte is unavoidable. The SEI (solid electrolyte interface) always plays a significant role in determining the electrochemical properties, depending on both polarisation effect and hydrofluoric (HF) acid corrosion [36]. In consideration of the capacity loss upon cycling based on SEI effect, coating nonreactive carbon or metal oxides on $LiNi_{1/3}Co_{1/3}Mn_{1/3}O_2$ particles by microwave plasma chemical vapour deposition [37], chemical solution [38], and slurry spray-drying method [39]; or AlF_3 coating by chemical solution [40], ZrO_2 coating by chemical solution [41,42], $Al(OH)_3$ coating by chemical solution [43] and Al_2O_3 coating by sol–gel method [44,45] were developed. Furthermore, the core–shell particle with high-capacity structure as the core and electrochemically stable structure as the shell [46], as well as ultrathin direct atomic layer coating strategies [47] were used to improve cycling performance tremendously, as shown in Figure 5.10 [46].

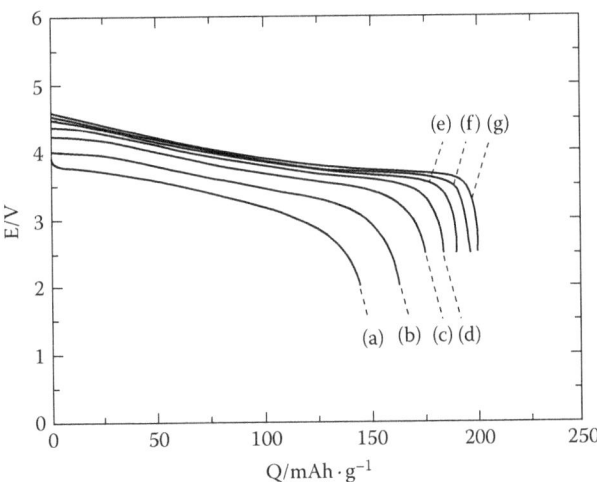

FIGURE 5.8
Rate capability tests on a $Li/LiCo_{1/3}Ni_{1/3}Mn_{1/3}O_2$ cell operated at 30°C. The cell was charged at 0.6 mA cm^{-2} and then held at 4.6 V for 4 h, followed by discharge at different current densities: (a) 19.2 mA cm^{-2} (2400 mA g^{-1} based on $LiCo_{1/3}Ni_{1/3}Mn_{1/3}O_2$ sample weight), (b) 12.4 (1600), (c) 6.4 (800), (d) 3.2 (400), (e) 1.6 (200), (f) 0.8 (100) and (g) 0.4 (50). (Reprinted with permission from N. Yabuuchi et al. Solid-state chemistry and electrochemistry of $LiCo_{1/3}Ni_{1/3}Mn_{1/3}O_2$ for advanced lithium-ion batteries, *J. Electrochem. Soc.* 152, 2005, A1434–A1440. Copyright 2005, The Electrochemical Society.)

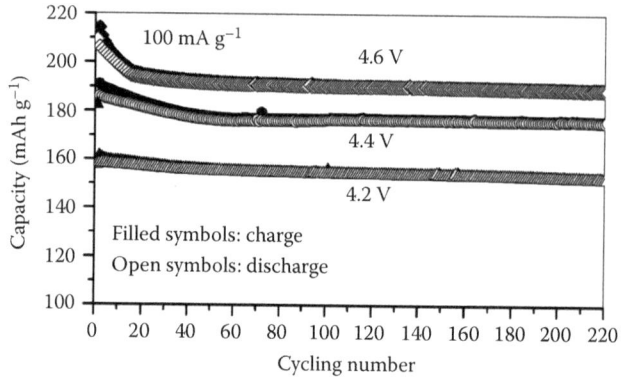

FIGURE 5.9
Performance of macroporous Li(Ni$_{1/3}$Co$_{1/3}$Mn$_{1/3}$)O$_2$ synthesized from sol-gel method at the current density of 100 mAg^{-1} cycled to different upper cutoff potentials of 4.2, 4.4 and 4.6 V, respectively. (From K.M. Shaju and P.G. Bruce, Macroporous Li(Ni$_{1/3}$Co$_{1/3}$Mn$_{1/3}$)O$_2$: A high-power and high-energy cathode for rechargeable lithium batteries, *Adv. Mater.* 2006, 18, 2330–2334. Copyright Wiley-VCH Verlag GmbH & Co. KGaA. Reproduced with permission.)

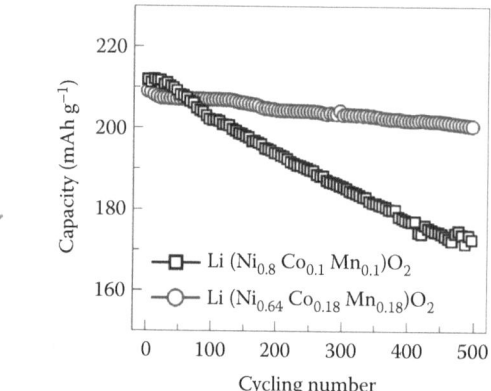

FIGURE 5.10
Cycling performance at 1 C rate (75 mA, corresponds to 190 mAg^{-1}) of laminated-type LiBs with Li(Ni$_{0.8}$Co$_{0.1}$Mn$_{0.1}$)O$_2$ and core–shell Li(Ni$_{0.64}$Co$_{0.18}$Mn$_{0.18}$)O$_2$ (upper cutoff voltage of 4.2 V). (Reprinted from Y.K. Sun et al., *Nat. Mater.* 8, 2009, 320–324. With permission.)

5.3 Integrated Layer-Structured Materials

Lithium-rich layered structural materials are of more interests mainly due to their high reversible capacity of more than 250 mAh/g at ambient temperature and even higher such as 300 mAh/g at 55°C [48] at low current density. The design strategy is based on an assumption of integrated composites, or solid solution in the form of layered–layered integrated $x\text{Li}_2\text{M}'\text{O}_3 \cdot (1-x)$ LiMO_2 or layered–spinel-integrated $x\text{Li}_2\text{M}'\text{O}_3 \cdot (1-x)\text{LiM}_2\text{O}_4$, where M' can be manganese, chromium and titanium; M indicates the commonly used transition metals, and x is the molar ratio between the two compounds [26]. The manganese-involved $x\text{Li}_2\text{MnO}_3 \cdot (1-x)\text{LiMO}_2$ structures show more advantages than other systems since the component Li_2MnO_3 itself is electrochemically active as a cathode material [49], while others such as Li_2ZrO_3 and Li_2TiO_3 are electrochemically inactive as electrode materials. Usually, researchers use three different forms of chemical formula to describe different aspects of the same material [50], as tabulated in Table 5.1. Taking $x\text{Li}_2\text{MnO}_3 \cdot (1-x)$ LiMO_2 as an example, Li_2MnO_3 can be written as $\text{Li}(\text{Li}_{1/3}\text{Mn}_{2/3})\text{O}_2$ by dividing the original formula by 1.5; such kind of layered–layered integrated materials may be comprehended as a doping strategy, thus leading to a formula such as $\text{Li}(\text{Li}_{1/3-x}\text{Mn}_{2/3-y}\text{M}_{x+y})\text{O}_2$, where M can be transition metals such as cobalt and nickel. Among these integrated structural materials, cobalt-free $\text{Li}(\text{Li}_{1/3-2x/3}\text{Ni}_x\text{Mn}_{2/3-x/3})\text{O}_2$ is the most widely studied group due to its high reversible capacity of more than 200 mAh/g [51].

Compatible integration between monoclinic Li_2MnO_3 (C2/m symmetry), and rhombohedral LiMnO_2 (R-3m symmetry) phase can be ascribed to the close interlayer spacing of $(001)_{\text{monoclinic}}$ and $(003)_{\text{rhombohedral}}$ which is around 0.47 nm [26]. Such integration has also been observed in bright-field scanning transmission electron microscopy (STEM) images of $\text{Li}_{1.2}\text{Ni}_{0.2}\text{Mn}_{0.6}\text{O}_2$ (LNMO) nanoparticle [52], which shows plate-like shape with clear surface facets (Figure 5.11a). The particles are composed of two components of trigonal LiMO_2 phase with R-3m symmetry (Figure 5.11b), and monoclinic Li_2MnO_3 phase with C2/m symmetry (Figure 5.11c). Since Ni^{2+} and Li^+ have similar radius, the cations of Ni^{2+} and Li^+ occupy at $3a$ and $3b$ sites in LiMO_2-layered phase, whereas the Mn ions reside only at $3a$ site (Figure 5.11b). On the other hand, Ni^{2+} cations substitute large amounts of Li^+ cations and small portions of Mn^{4+} cations (Figure 5.11d). These crystallographic models have been confirmed by the high-angle annular dark-field Z-contrast image as Figure 5.11e where the highest contrast of transition metal layers is revealed. Figure 5.11f clearly shows the oxygen columns and the transition metal layers on the two sides. A cation-ordered region in Li_2MnO_3-like phase is also observed as [100] zone projection in Figure 5.11g. In this well-organised structure, two bright Mn columns are separated by 0.14 nm as the centre of two adjacent Mn/Ni dumbbells is spaced by 0.42 nm [53,54] (Figure 5.11h).

TABLE 5.1
Different Expressions of $\text{Li}(\text{Li}_{1/3-2x/3}\text{Ni}_x\text{Mn}_{2/3-x/3})\text{O}_2$ When $x = 0.2$

	Integrated Composite	Chemical Composition	Structural Order
Chemical formula	$0.5\text{Li}_2\text{MnO}_3 \cdot 0.5\text{LiNi}_{0.5}\text{Mn}_{0.5}\text{O}_2$	$\text{Li}_{1.5}\text{Ni}_{0.25}\text{Mn}_{0.75}\text{O}_{2.5}$	$\text{Li}(\text{Li}_{0.2}\text{Ni}_{0.2}\text{Mn}_{0.6})\text{O}_2$

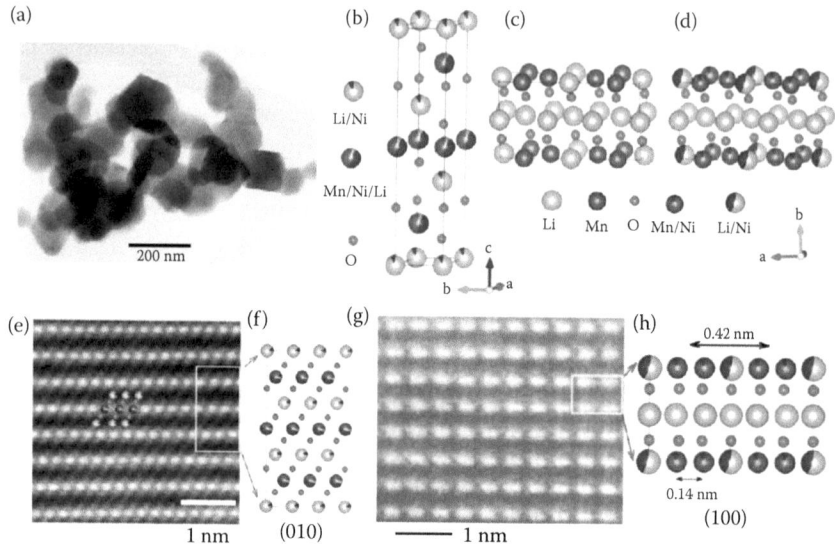

FIGURE 5.11
(See colour insert.) (a) Overview of bright-field STEM image of the as-obtained LMNO nanoparticles. (b) Crystal model for the LiMO$_2$ (M refers to Mn and Ni) R-3m parent phase based on a Li(Mn$_{0.5}$Ni$_{0.5}$)O$_2$ structure with lattice parameters: a = b = 2.887 Å, c = 14.29 Å, α = β = 90° and γ = 120°. (c) Crystal model for Li$_2$MO$_3$ C2/m parent phase based on Li$_2$MnO$_3$ with lattice parameters: a = 4.926 Å, b = 8.527 Å, c = 5.028 Å, α = γ = 90° and β = 109.22°. (d) Partially cation-ordered Li$_2$MO$_3$ C2/m phase based on Ni-containing Li$_2$MnO$_3$. (e) Z-contrast image of one sample region corresponding to (f) [010] zone projection of the LiMO$_2$ R-3m model. (g) Crystal region corresponding to (h) [100] zone projection of the Ni-doped Li$_2$MnO$_3$ C2/m phase. (Reprinted with permission from M. Gu et al. Conflicting roles of nickel in controlling cathode performance in lithium ion batteries, *Nano Letters*, 12, 2012, 5186–5191. Copyright 2013, American Chemical Society.)

When charging these Li-rich layered cathodes, an anomalous charge plateau is present when the charge potential is around 4.4 V versus Li$^+$/Li. This is because the Fermi level is shifted into the oxygen valence band at this delithiated state, leading to a removal of electrons from O^{2-} species accompanied with the extraction of Li$^+$ [55]. Overall, this electrochemical activation process of the Li$_2$MnO$_3$ component involves extraction of two moles of Li with one mole of O (Li$_2$O) from the local structure, resulting in layered MnO$_2$ as the host structure to be relithiated. However, the oxygen release is an irreversible process, which leads to oxygen vacancies in the framework. Therefore, only 1 mol Li is able to return to the MnO$_2$ framework which is subsequently reversible. In consideration of integrated xLi$_2$MnO$_3 \cdot (1 - x)$LiMO$_2$, when the charge potential is lower than 4.4 V versus Li$^+$/Li, the redox couples involving M works when accompany with lithium extraction from local $(1 - x)$ LiMO$_2$ regions. This first charge/discharge process of xLi$_2$MnO$_3 \cdot (1 - x)$ LiMO$_2$ is illustrated in Figure 5.12 [26] and the corresponding chemical reaction can be written as the following in Equations 5.1 through 5.3:

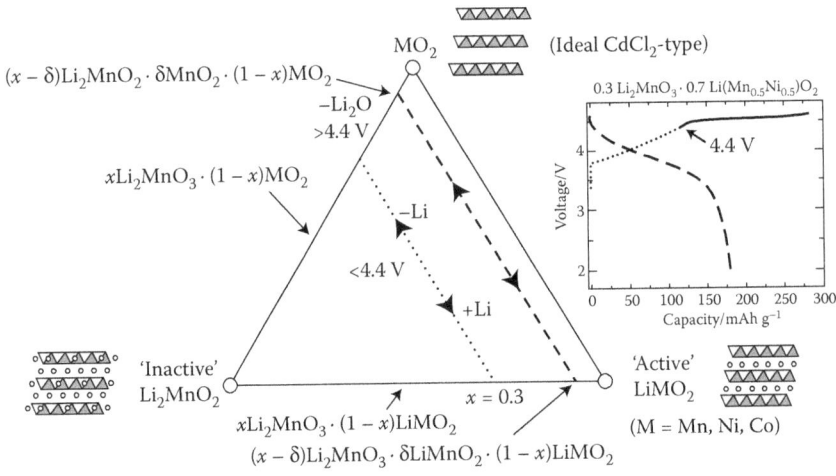

FIGURE 5.12
Compositional phase diagram showing the electrochemical reaction pathways for an $x\text{Li}_2\text{MnO}_3 \cdot (1 - x)\text{LiMO}_2$ material. The processes of Equations 5.1 through 5.3 correspond to the correlated chemical reactions as follows.

$$x\text{Li}_2\text{MnO}_3 \cdot (1 - x)\text{LiMO}_2 \rightarrow x\text{Li}_2\text{MnO}_3 \cdot (1 - x)\text{MO}_2 + (1 - x)\text{Li} \quad (5.1)$$

$$x\text{Li}_2\text{MnO}_3 \cdot (1 - x)\text{MO}_2 \rightarrow x\text{MnO}_2 \cdot (1 - x)\text{MO}_2 + x\text{Li}_2\text{O} \quad (5.2)$$

$$x\text{MnO}_2 \cdot (1 - x)\text{MO}_2 + \text{Li} \rightarrow x\text{LiMnO}_2 + (1 - x)\text{LiMO}_2 \quad (5.3)$$

Li_2MnO_3-stabilised integrated materials exhibit reasonable stability and no obvious structural change even with a charge cutoff voltage as high as 4.8 V versus Li^+/Li, while it is inevitable in the case of LiCoO_2. The higher reversible capacity of 250 mAh/g means nearly one mole of lithium extraction for every transition metal ion. The reason for the high stability of the structure is the contribution derived from the Li_2MnO_3 component, or alternatively speaking, excess Li in the transition metal layers termed $\text{Li}(\text{Li}_{1/3-2x/3}\text{Ni}_x\text{Mn}_{2/3-x/3})\text{O}_2$.

As shown in Figure 5.13 [26], during the process of delithiation, the depletion of octahedral lithium ions decreases the extent of framework stability. However, a compensation of reduced bonding energy could be provided by newly formed chemical bonds between neighboured oxygen and tetrahedral lithium ions. These lithium ions are transported from the transition metal layers mixed by both lithium and manganese ions in Li_2MnO_3 (or written as $\text{Li}(\text{Li}_{1/3}\text{Mn}_{2/3})\text{O}_2$) component [26]. As a result, this additional chemical bond is able to sustain the local structure from collapse. This mechanism hypothesis has been experimentally observed by magic-angle spinning nuclear magnetic resonance (MAS NMR) [56], and has also been confirmed by theoretical calculation [57].

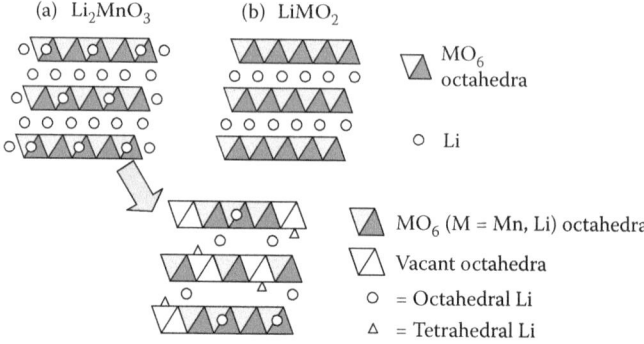

FIGURE 5.13
The schematic illustration of region Li_2MnO_3 (a) and region $LiMO_2$ (b) from layered-like to spinel-like configuration transition during delithiation process in $xLi_2MnO_3 \cdot (1-x)LiMO_2$.

Nevertheless, several problems still need to be solved before practical application of these integrated compounds such as layered–layered $xLi_2MnO_3 \cdot (1-x)LiMO_2$. The oxygen release during the first cycle [55] leads to serious safety problem. Structural evolution upon insertion/reinsertion of Li ions decreases the discharge plateau over long cycles. Regarding the low capacity even at low current density (500 mA/g), conductive surface coating [58–60] and hydrothermal-induced nano-shaped particles [50,61,62] have been applied to improve the low rate capability of $xLi_2MnO_3 \cdot (1-x)LiMO_2$ materials. The cause of low rate capability remains to be solved.

5.4 Li Diffusion in Layered Structures

The rate performance of cathode materials is highly related to the speed of lithium diffusion in the layered structure. Based on the *ab initio* calculation from generalized gradient approximation density functional theory (GGA DFT), the tetrahedral site hop (TSH) mechanism proposed by Van der Ven et al. [63] is widely accepted to explain the dominant factors that affect lithium diffusion in Li_xMO_2 systems. In this proposed mechanism, lithium migration from the original site to an adjacent octahedral vacancy site has to pass through an intermediate tetrahedral site, and the activation barrier for this migration is determined by the nearby transition metal cation, as illustrated in Figure 5.14b [63]. Another interesting discovery revealed by first-principle calculation is that the activation energy of Li diffusion is also strongly dependent on the distance between oxygen layers, that is, $d(LiO_2)$. Therefore, Li motion is so sensitive to the spacing between oxygen layers that a very small change (~0.02 Å) results in a 20- to 30-meV increase in the activation barrier

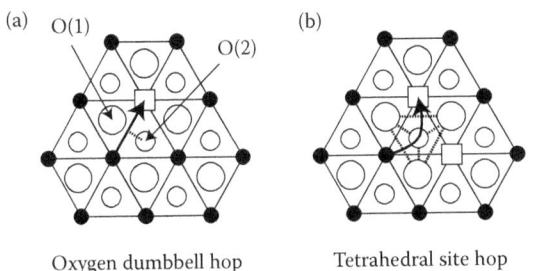

FIGURE 5.14
The two lithium migration paths as seen looking down on the Li_xCoO_2 crystal structure along the c-axis. The triangular grid corresponds to the lithium sites, the filled circles are lithium ions, the large empty circles are oxygen ions in the plane directly above the lithium plane and the small empty circles are oxygen ions directly below the lithium plane. □ refer to lithium vacancies. (a) Oxygen dumbbell hop (ODH). (b) TSH. (Reprinted with permission from A. Van der Ven and G. Ceder, Lithium diffusion in layered Li_xCoO_2, *Electrochem. Solid State Lett.* 3, 2000, 301–304. Copyright 2000, The Electrochemical Society.)

[21]. As a result, reducing the Li/Ni disorder of $LiNi_{1/2}Mn_{1/2}O_2$ is significantly beneficial in improving the rate performance.

5.5 Methodology of Synthesis and Characterisations

The conventional $LiCoO_2$ is simply synthesised by heating the mixture of lithium salt with cobalt salt at a high temperature such as 900°C [8]. However, the unique structural features of $LiNiO_2$ and $LiMnO_2$ make the synthesis procedures more complicated than the traditional $LiCoO_2$. For example, Ohzuku et al. [64] used the mixture of $NiCO_3$ and $LiNiO_3$ as starting materials, followed by preheating them at 600°C for 15 h under an oxygen flow. Then the pressed pellets were sintered at 750°C for 24 h under an oxygen flow to obtain the pure phase of $LiNiO_2$. The layered and stoichiometric $LiMnO_2$ was first synthesised by ion exchange from isostructural $NaMnO_2$ in Bruce's group [12], as the previous attempts involving the traditional use of the aqueous solution for synthesis failed. To obtain the pure phase of $LiNi_{1/2}Mn_{1/2}O_2$ with less cation mixing between Li and Ni ions, the ion-exchange method was first applied by Kang et al. [21]. The $LiNi_{1/2}Mn_{1/2}O_2$ was alternatively prepared with $LiOH \cdot H_2O$ and $Ni_{1/2}Mn_{1/2}(OH)_2$ as raw materials, followed by heating the mixture pellets at 1000°C in air for 5 h in Ohzuku's group [14]. This preparation route involving the multiple transitional metals hydroxide was also applied in the synthesis of $LiNi_{1/3}Co_{1/3}Mn_{1/3}O_2$ [23]. These precursors of hydroxide are the key to obtaining the pure phase with homogeneity of all transition metals at the atomic level. Furthermore, the derivative multiple transition metals in the layered structure such as $LiNi_xCo_yMn_{1-x-y}O_2$ are also based on this precursor method, which could be called co-precipitation method. In terms of integrated layered cathodes with the $xLi_2M'O_3 \cdot (1-x)$ $LiMO_2$ formula (both M' and M indicate transition metals), co-precipitation methods [26,65], sol–gel methods [66], combustion methods [67], as well as ion-exchange methods [68] have been successfully applied to obtain the well-crystallised products with controllable particle size. Owing to the similar ordering of atoms in the layered structure (i.e. close-packed oxygen atoms consisting of a framework with ABCABC... stacking sequence, with the transition metal ions and lithium ions alternatively residing at octahedral sites in between the oxygen layers), this group of materials have similar x-ray diffraction (XRD) patterns as shown in Figure 5.15 [64]. However, the distortion feature from ideal rhombohedral to monoclinic symmetry arising from the presence of the active Mn^{3+} in layered $LiMnO_2$ leads to a different diffraction pattern, as shown in Figure 5.16 [69].

In terms of characterisation on solid-state chemistry in the layered group of cathodes, the x-ray photoelectron spectroscopy (XPS) technique is always performed to investigate the chemical valence states of different transition metals. The initial valence states of transition metals in as-prepared cathodes are highly associated with the redox pairs and electrochemical stability. In Table 5.2 [70–77], the previous works on characterising such surface properties of the commonly used layered cathode materials using XPS technique are summarised.

5.6 Layered Cathode for All-Solid-State Thin-Film Lithium Ion Battery

With rapid development in micro/nano-electromechanical system (MEMS/NEMS) technologies, energy consumptions of these electronic devices are extremely low [78]. In addition, the requirement of compactness in size and dimension would mean conventional LiBs are unable to meet the needs, leading to the development of all-solid-state thin-film lithium-ion microbatteries [79–83]. Figure 5.17 shows the schematic diagram of the microbattery setup. Basically, it is the same as the traditional bulk battery. The main difference is the replacement of the liquid electrolyte with solid electrolyte, which allows thin-film microbatteries to possess high-energy density (250 $Whkg^{-1}$), excellent cycling performance (>10,000 cycles), small self-discharge and intrinsic safety. Since a solid electrolyte is used, very often, Li metal is used as the anode; thus, a high voltage (1.5 V) can be achieved [84]. However, ionic conductivity and stability of the solid electrolytes are still questionable as most of the solid electrolytes have significantly low ionic conductivity with only 10^{-6} S cm^{-1} order [82,85]. If Li metal is used, an interfacial layer between the Li metal and electrolyte must be inserted since most of the electrolytes will be reduced when the potential is below about 0.5 V.

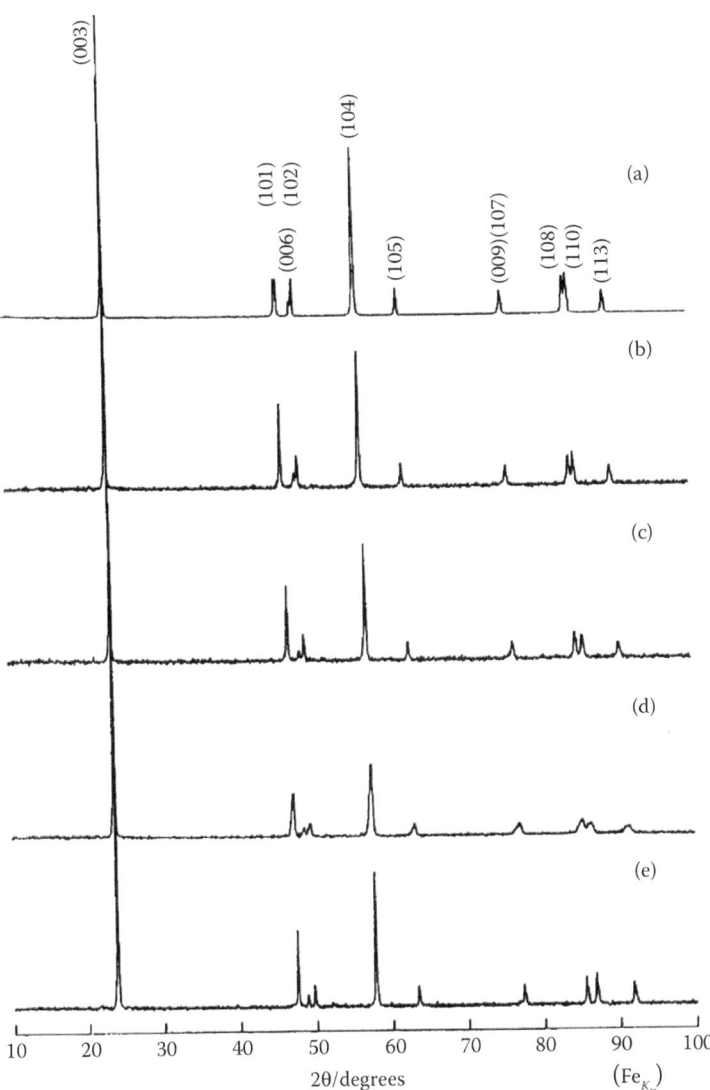

FIGURE 5.15
XRD patterns of $LiNi_{1-x}Co_xO_2$: (a) $LiNiO_2$ prepared by the nitrate method, (b) $LiNi_{3/4}Co_{1/4}O_2$ prepared by the precursor method, (c) $LiNi_{1/2}Co_{1/2}O_2$, (d) $LiNi_{1/4}Co_{3/4}O_2$ and (e) $LiCoO_2$. The samples (c) through (e) were prepared by the carbonate method. (Reprinted from *Electrochim. Acta*, 38, T. Ohzuku et al., Comparative study of $LiCoO_2$, $LiNi_{1/2}Co_{1/2}O_2$ and $LiNiO_2$ for 4-Volt secondary lithium cells, 1159–1167, Copyright 1993, with permission from Elsevier.)

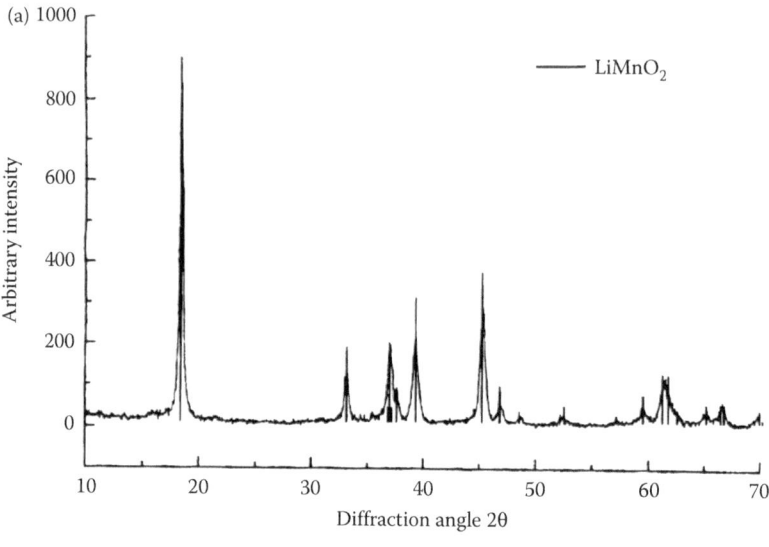

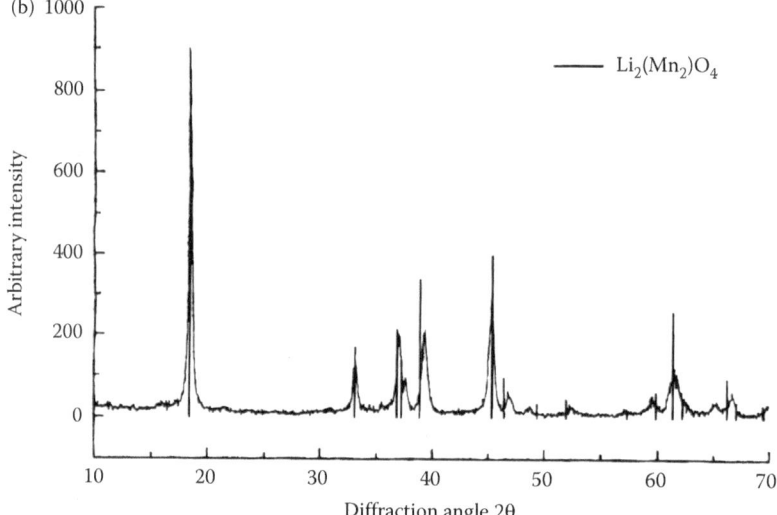

FIGURE 5.16
The powdered XRD pattern of the parent LiMnO$_2$ compound with simulated x-ray patterns of (a) layered LiMnO$_2$ (C2/m) and (b) lithiated–spinel Li$_2$(Mn$_2$)O$_4$ (F4$_1$/*ddm*). (Reprinted with permission from Y. Shao-Horn et al. Thackeray, Structural characterization of layered LiMnO$_2$ electrodes by electron diffraction and lattice imaging, *J. Electrochem. Soc.* 146, 1999, 2404–2412. Copyright 1999. The Electrochemical Society.)

TABLE 5.2

Comparison of Binding Energies of Co, Ni and Mn in Different Types of Layered Cathode Materials

	Binding Energy(eV)	Compound	Reference
Co^{3+}/Co^{4+} ($2p_{3/2}$)	779.6/–	$LiCoO_2$	Dupin et al. [70]
	779.6/–	$LiCoO_2$	Moses et al. [71]
	779.5/–	$LiNi_{1/3}Mn_{1/3}Co_{1/3}O_2$	Shaju et al. [31]
	779.9/–	$Li_{1.2}Mn_{0.54}Ni_{0.13}Co_{0.13}O_2$	Wu et al. [72]
Ni^{2+}/Ni^{3+} ($2p_{3/2}$)	–/854.7	$LiNiO_2$	Moses et al. [71]
	–/855.4	$LiNiO_2$	Zhou et al. [73]
	853.9/856.0	$LiNi_{1/2}Mn_{1/2}O_2$	Shaju et al. [74]
	854.0/855.5	$LiNi_{1/3}Mn_{1/3}Co_{1/3}O_2$	Shaju et al. [31]
	854.3/855.5	$Li_{1.2}Mn_{0.54}Ni_{0.13}Co_{0.13}O_2$	Song et al. [75]
Mn^{3+}/Mn^{4+} ($2p_{3/2}$)	641.5/642.4	$LiMnO_2$	Regan et al. [76]
	641.6/–	$LiMnO_2$	Su et al. [77]
	641.1/642.0	$LiNi_{1/2}Mn_{1/2}O_2$	Shaju et al. [74]
	641.0/642.2	$LiNi_{1/3}Mn_{1/3}Co_{1/3}O_2$	Shaju et al. [31]
	–/642.0	$Li_{1.2}Mn_{0.54}Ni_{0.13}Co_{0.13}O_2$	Wu et al. [72]

Source: Adapted from J.C. Dupin et al. *Thin Solid Films* 384, 2001, 23–32; A.W. Moses et al. *Appl. Surf. Sci.* 253, 2007, 4782–4791; Y. Wu, A.V. Murugan and A. Manthiram, *J. Electrochem. Soc.* 155, 2008, A635–A641; Y.K. Zhou et al., *Mater. Sci. Eng. A: Struct. Mater. Prop. Microstruct. Process.* 335, 2002, 260–267; K.M. Shaju, G.V.S. Rao and B.V.R. Chowdari, *Electrochim. Acta* 48, 2003; B.H. Song et al., *Phys. Chem. Chem. Phys.* 14, 2012, 12875–12883; E. Regan et al., *Surf. Interface Anal.* 27, 1999, 1064–1068; Z. Su et al., *J. Power Sources* 189, 2009, 411–415.

Figure 5.18 shows a typical all-solid-state thin-film LiB consisting of layer-structured $LiNi_{1/3}Co_{1/3}Mn_{1/3}O_2$ as the positive electrode, and TiO_2 as the negative electrode, separated by LiPON forming a TiO_2/LiPON/$LiNi_{1/3}Co_{1/3}Mn_{1/3}O_2$ full cell (~1.5 μm) [86]. Because of relatively low ionic conductivity caused by the solid electrolyte, high polarisation is very often observed. As shown in Figure 5.19a, the first cycle charge/discharge curves of the TiO_2/LiPON/$LiNi_{1/3}Co_{1/3}Mn_{1/3}O_2$ full cell are cycled between 0.0 and 3.0 V at room temperature. The cycle performance of the full cell reveals comparable or even better stability than the bulk cells. After 100 cycles, about 92% of its initial discharge capacity (33.35 μAhcm^{-2}μm^{-1}) can be maintained (Figure 5.19b).

5.7 Nanoscale Mapping of Li$^+$ Diffusion in All-Solid-State Thin-Film Lithium Ion Battery

Furthermore, a newly developed scanning probe microscopy (SPM) technique, 'Band-excitation Electrochemical Strain Microscopy' (BE-ESM) was

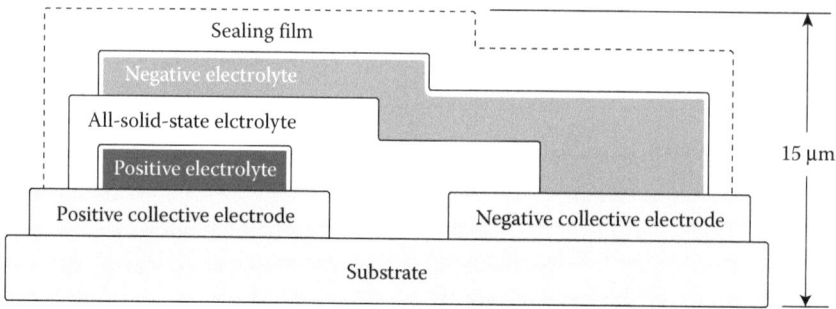

FIGURE 5.17
Schematic drawing of the cross-sectional layout of all-solid-state thin-film LiBs.

FIGURE 5.18
A cross-sectional image of all-solid-state Li-ion battery. (Reprinted with permission from J. Zhu et al. Nanoscale Mapping of Lithium-Ion Diffusion in a Cathode within an All-Solid-State Lithium-Ion Battery by Advanced Scanning Probe Microscopy Techniques, *ACS Nano*, 7, 2013, 1666–1675. Copyright 2013, American Chemical Society.)

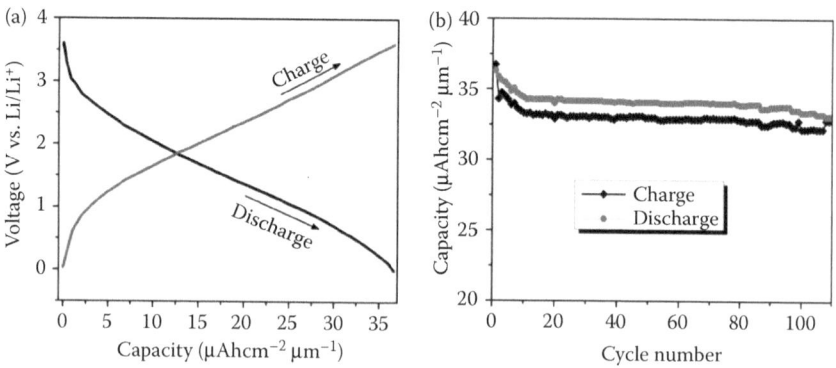

FIGURE 5.19
(a) The first cycle galvanostatic charge/discharge curves and (b) cycling performance of TiO_2/LiPON/$LiNi_{1/3}Co_{1/3}Mn_{1/3}O_2$ full cell over 100 cycles. (Reprinted with permission from S. Kondo, All solid-state lithium secondary battery with highly ion conductive glassy electrolyte, in: M. Wakihara, O. Yamamoto, (eds.), *Lithium Ion Batteries: Fundamentals and Performance*, Wiley-VCH, 1998, pp. 199–217. Copyright 2013, American Chemical Society.)

applied to obtain nanoscale mapping of Li+ diffusion and redistribution induced by the tip bias. First, BE-ESM mapping was conducted on the as-deposited $LiNi_{1/3}Co_{1/3}Mn_{1/3}O_2$ thin-film cathode. The structure of this cathode material has been confirmed to be a hexagonal α-$NaFeO_2$-type structure with the preferred crystallographic orientation of (003) and (104) by XRD. Figure 5.20a shows a high-resolution topographical image of the polycrystalline $LiNi_{1/3}Co_{1/3}Mn_{1/3}O_2$ thin-film cathode as synthesised, which consists of columnar grains with the average grain size of ~100 nm. The surface roughness (rms) is around ~12 nm, due to the presence of large cavities and grain boundaries. Panels (b) through (d) of Figure 5.20 show the BE-ESM mapping of the resonance amplitude, resonance quality factor (Q-factor) and resonance frequency, respectively. Features less than 10 nm can be clearly observed from the images, providing the high-resolution imaging capability of the BE-ESM technique. Figure 5.20b shows that the amplitude on resonance frequency is not homogeneous across the sample surface and changes abruptly at boundary-like features, indicating clear variations of ionic mobility and electrochemical activity across grain boundaries (also shown in Figure 5.20e).

Additionally, it is found that some 'nanospots' concentrate at certain grain boundaries and show strong amplitude signal (deep purple colour), indicating enhanced Li-ion diffusion and accumulation in these regions. Lithium ions are more concentrated at grain boundary regions. It is noted that the

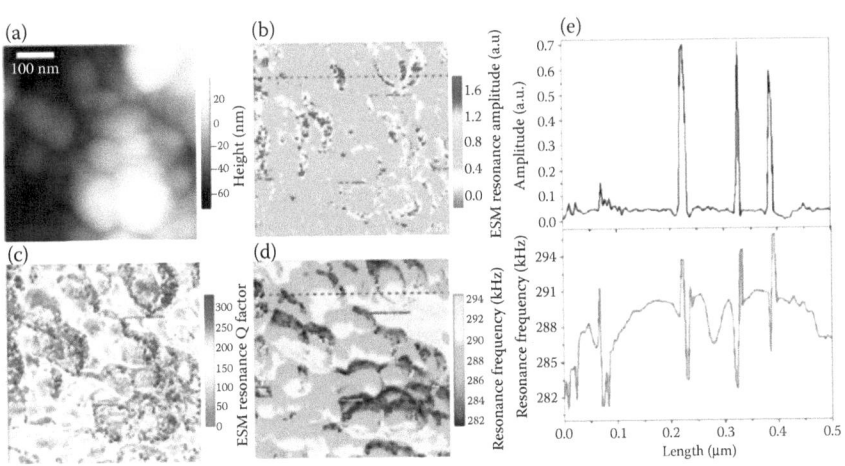

FIGURE 5.20
(See colour insert.) (a) Surface topography, BE-ESM maps of (b) resonance amplitude, (c) resonance Q-factor, (d) contact resonance frequency and (e) line section profiles in amplitude and resonance frequency, corresponding to the lines in (b) and (d) of the $LiNi_{1/3}Co_{1/3}Mn_{1/3}O_2$ thin-film cathode. (Reprinted with permission from J. Zhu et al. Nanoscale mapping of lithium-ion diffusion in a cathode within an all-solid-state lithium-ion battery by advanced scanning probe microscopy techniques, *ACS Nano* 7, 2013, 1666–1675. Copyright 2013, American Chemical Society.)

amplitude of oscillatory surface displacement, which is defined as the electrochemical strain, is directly related to the changes of Li-ion concentration induced by the high-frequency tip bias. Therefore, the electrochemical strain mapping can provide important information on the variations of bias-induced Li-ion diffusion and redistribution, also establishing the direct relationship between the preferred Li-ion diffusion paths and microstructure of the $LiNi_{1/3}Co_{1/3}Mn_{1/3}O_2$ thin-film cathode at the nanoscale.

Another feature of BE-ESM is the capability to map the surface mechanical properties at the nanoscale, which is similar to that obtained from atomic force acoustic microscopy. More specifically, Q-factor (peak width in the BE-ESM signal) is a measurement of the dissipative energy due to the tip–sample interactions, providing fundamental information on mechanical dissipative properties, while the resonance frequency is a measurement of the conservative tip–sample interactions, providing information on local elastic properties of the sample surface. As shown in Figure 5.20c, the resonance Q-factor mapping shows very strong variations (~300) between the grains and the grain boundaries, corresponding well to the surface topography differences. It is observed that Q-factor decreases at grain boundaries (red colour) and increases at areas within the individual grains (blue colour); this indicates higher energy dissipation at the topographical depressions and lower energy dissipation at the protrusions.

Similarly, the resonance frequency mapping (Figure 5.20d) also largely corresponds to the surface topographical features. It is observed that the resonance frequency increases at topographical depressions (yellow colour) and decreases within grains, but at the area adjacent to the grain boundaries (blue colour). The variation of resonance frequency due to the grain structure and surface roughness is on the order of ~15 kHz compared with ~290 kHz drive frequency. Therefore, the topographical depressions (e.g. grain boundaries) can increase the effective tip–sample contact area, resulting in higher contact stiffness (large resonance frequency) and larger energy dissipation (small Q-factor). In addition, Figure 5.20e shows line section profiles in both amplitude and resonance frequency images. It is obvious that the contact resonance frequency slightly changes within individual grains, but abruptly increases across the grain boundary regions, corresponding to high amplitude of surface displacement; hence, the high Li-ion concentration and enhanced ionic diffusion at these regions. In summary, the variations of electrochemical strain induced by ionic diffusion have been evidently observed on $LiNi_{1/3}Co_{1/3}Mn_{1/3}O_2$ cathode surfaces, indicating inhomogeneous distribution of Li-ion diffusion and intercalation paths. For the thin-film cathode, most of the high Li-ion concentration areas are localised at grain-boundary-like features and surface defects, respectively.

The structure of the electrode material gradually changes during charge/discharge cycling, resulting in the capacity fading and electrochemical degradation. Therefore, to investigate the effects of galvanostatic cycling on the variations of electrochemical strain as well as the Li-ion diffusion dynamics, BE-ESM is further conducted on $LiNi_{1/3}Co_{1/3}Mn_{1/3}O_2$ thin-film cathode at different cycling stages. Panels (a) through (c) of Figure 5.21 show the BE-ESM mapping of $LiNi_{1/3}Co_{1/3}Mn_{1/3}O_2$ thin-film cathodes after 10, 50 and 100 charge/discharge cycles, respectively. In Figure 5.21, the columns from left to right are resonance amplitude images, resonance Q-factor images and line section profiles from the selected lines in the amplitude image. Compared with the BE-ESM mapping on the as-deposited $LiNi_{1/3}Co_{1/3}Mn_{1/3}O_2$ thin-film cathode (Figure 5.20a), the variations of electrochemical strain during charge/discharge cycling are clearly observed. First, grain-boundary-related nanospots in Figure 5.21b, which are associated with highly localised Li-ion diffusion and redistribution, have disappeared after the cycling process. After 100 cycles, besides some deep cavities, even the BE-ESM-enhanced signal at the grain boundaries cannot be clearly observed (Figure 5.21c). Second, the as-deposited sample has higher BE-ESM response (in absolute value) compared to that of the cycled samples, as shown in the amplitude scale bar. Third, through the comparison of amplitude line section profiles, the variations of amplitude between topographical depressions and protrusions have become smaller with increasing number of cycles, corresponding to the lower contrast in the amplitude images. All these gradual changes are shown to be functions of cycling number, and it can be explained by the degradation in electrochemical activity upon Li-ion intercalation/deintercalation processes in the $LiNi_{1/3}Co_{1/3}Mn_{1/3}O_2$ thin-film cathode, especially at boundary regions.

Through the comparison of the as-deposited (Figure 5.21a) and aged $LiNi_{1/3}Co_{1/3}Mn_{1/3}O_2$ thin-film cathodes (after 100 cycles) (Figure 5.21c), it is obvious that the decrease of Li-ion concentration is localised at the grain boundary regions, which illustrates the weak enhancement of the Li-ion diffusion and accumulation at these regions after cycling, corresponding to reduced electrochemical activity after cycling process. In addition, the decrease of contrast at grain boundaries in BE-ESM amplitude is more evident after the first 10 cycles (Figure 5.20e), meaning intense electrochemical degradation during the initial charge/discharge cycles.

Furthermore, combining resonance amplitude and Q-factor images, the correlation between electrochemical strain and energy dissipation can be locally established. As shown in Figures 5.20 and 5.21, the scanned region in the as-deposited $LiNi_{1/3}Co_{1/3}Mn_{1/3}O_2$ thin-film cathode has a higher Q-factor and also a relatively larger amplitude on average (indicated in the scale bar). It can be understood that a higher Q-factor corresponds to smaller energy dissipation, indicating smaller energy barriers for Li-ion intercalation/deintercalation for the as-deposited sample. Since the energy barriers increase with the increase of cycling number, Li-ion intercalation/deintercalation becomes more difficult, resulting in lower Li-ion diffusivity and electrochemical activity, and thus smaller electrochemical strain. In general, the variations of BE-ESM amplitude during the charge/discharge cycling process can be explained such factors as (i) differences in Li-ion diffusion coefficient, (ii) differences in sample mechanical properties and (iii) differences in sample electrical properties, such as surface conductivity.

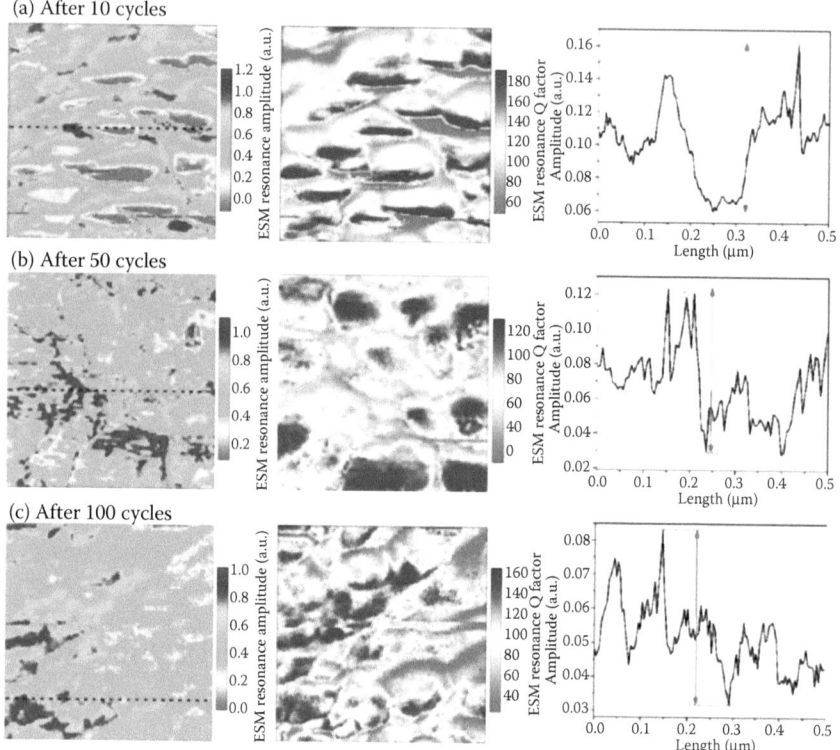

FIGURE 5.21
(See colour insert.) BE-ESM maps of resonance amplitude and Q-factor of the $LiNi_{1/3}Co_{1/3}Mn_{1/3}O_2$ thin-film cathode at different cycling stages: (a) 10 cycles, (b) 50 cycles and (c) 100 cycles, as well as the line section profiles of resonance amplitude. The left column is the amplitude map, the middle column is the Q-factor map and the right column is the line section profile of the amplitude maps, corresponding to the lines in the left column. (Reprinted with permission from J. Zhu et al. Nanoscale mapping of lithium-ion diffusion in a cathode within an all-solid-state lithium-ion battery by advanced scanning probe microscopy techniques, *ACS Nano* 7, 2013, 1666–1675. Copyright 2013, American Chemical Society.)

5.8 Conclusions

In summary, a layered group of cathode materials is still attractive to researchers and battery industries mainly due to their promising capacities, being higher than other groups. Table 5.3 [87–92] shows the theoretical capacities of various cathode materials based on an assumption that all lithium ions can be delithiated from the structures in response to their respective redox couples, compared to the reversible capacities at low rates obtained from experiments along with corresponding specific surface areas tested by Brunauer–Emmett–Teller (BET) method. Despite the original attraction

TABLE 5.3

Theoretical Capacities and Reversible Capacities (mAh/g) for Different Cathode Materials

Cathode Materials	Theoretical Capacities (mAh/g)	Reversible Capacities (mAh/g)	Specific Surface Area (m^2/g)	Reference
$LiCoO_2$	274	140	2.0	[87]
$LiMn_2O_4$	148	125	18	[88]
$LiFePO_4$	170	168	51	[89]
$LiNi_{1/2}Mn_{1/2}O_2$	280	180	8.9	[90]
$LiNi_{1/3}Co_{1/3}Mn_{1/3}O_2$	278	175	3.9	[91]
$Li_{1.2}Ni_{0.2}Mn_{0.6}O_2$	315	274	4.6	[92]
$Li_{1.2}Ni_{0.13}Co_{0.13}Mn_{0.54}O_2$	314	275	7.5	[66]

Source: Adapted from M.V. Reddy, G.V.S. Rao and B.V.R. Chowdari, *Electrochim. Acta* 50, 2005, 3375–3382; I. Taniguchi et al., *Solid State Ionics* 146, 2002, 239–247; M. Konarova and I. Taniguchi, *J. Power Sources* 195, 2010, 3661–3667; X.Y. Wang et al., *Electrochim. Acta* 56, 2011, 4065–4069; Z.D. Huang et al., *J. Mater. Chem.* 21, 2011, 10777–10784; D.K. Lee et al., *J. Power Sources* 162, 2006, 1346–1350.

from higher theoretical capacity of layer-structured materials, these materials are still limited by (i) the structural instability at deep delithiation state which affects cycling performance and leads to safety concerns, (ii) poor rate capability due to phase segregation and (iii) SEI layer which dramatically reduces its electronic and ionic conductivity. Doping and coating strategies in a layered group are always considered as effective ways to alleviate these problems; for instance, increase reversible capacity at higher cutoff voltages, improve thermal stability, reduce side reactions with electrolytes and, hence, improve the cycle life. In addition, layered materials have been proven to be effective as the cathode component in an all-solid-state thin-film Li-ion battery, which exhibits promising electrochemical properties.

References

1. J.M. Tarascon and M. Armand, Issues and challenges facing rechargeable lithium batteries, *Nature* 414, 2001, 359–367.
2. M. Armand and J.M. Tarascon, Building better batteries, *Nature* 451, 2008, 652–657.
3. J.B. Goodenough and Y. Kim, Challenges for rechargeable Li batteries, *Chem. Mater.* 22, 2010, 587–603.
4. M.S. Whittingham, Lithium batteries and cathode materials, *Chem. Rev.* 104, 2004, 4271–4301.
5. T. Ohzuku and R.J. Brodd, An overview of positive-electrode materials for advanced lithium-ion batteries, *J. Power Sources* 174, 2007, 449–456.

6. G.G. Amatucci, J.M. Tarascon and L.C. Klein, CoO_2, the end member of the Li_xCoO_2 solid solution, *J. Electrochem. Soc.* 143, 1996, 1114–1123.
7. A.V.D. Ven, M.K. Aydinol, G. Ceder, G. Kresse and J. Hafner, First-principles investigation of phase stability in Li_xCoO_2, *Phys. Rev.* B58, 1998, 2975–2987.
8. K. Mizushima, P.C. Jones, P.J. Wiseman and J.B. Goodenough, Li_xCoO_2 (0<x<1) A new cathode material for batteries of high energy density, *Mater. Res. Bull.* 15, 1980, 783–789.
9. H.F. Wang, Y.I. Jang, B.Y. Huang, D.R. Sadoway and Y.T. Chiang, TEM study of electrochemical cycling-induced damage and disorder in $LiCoO_2$ cathodes for rechargeable lithium batteries, *J. Electrochem. Soc.* 146, 1999, 473–480.
10. P.G. Bruce, A.R. Armstrong and R.L. Gitzendanner, New intercalation compounds for lithium batteries: Layered $LiMnO_2$, *J. Mater. Chem.* 9, 1999, 193–198.
11. A.D. Robertson, A.R. Armstrong and P.G. Bruce, Layered $Li_xMn_{1-y}Co_yO_2$ intercalation electrodes—Influence of ion exchange on capacity and structure upon cycling, *Chem. Mater.* 13, 2001, 2380–2386.
12. A.R. Armstrong and P.G. Bruce, Synthesis of layered $LiMnO_2$ as an electrode for rechargeable lithium batteries, *Nature* 381, 1996, 499–500.
13. P.G. Bruce, B. Scrosati and J.M. Tarascon, Nanomaterials for rechargeable lithium batteries, *Angew. Chem. Int. Ed.* 47, 2008, 2930–2946.
14. T. Ohzuku and Y. Makimura, Layered lithium insertion material of $LiNi_{1/2}Mn_{1/2}O_2$: A possible alternative to $LiCoO_2$ for advanced lithium-ion batteries, *Chem. Lett.* 30, 2001, 744–745.
15. Y. Makimura and T. Ohzuku, Lithium insertion material of $LiNi_{1/2}Mn_{1/2}O_2$ for advanced lithium-ion batteries, *J. Power Sources* 119, 2003, 156–160.
16. W.S. Yoon, Y. Paik, X.Q. Yang, M. Balasubramanian, J. McBreen and C.P. Grey, Investigation of the local structure of the $LiNi_{0.5}Mn_{0.5}O_2$ cathode material during electrochemical cycling by x-ray absorption and NMR spectroscopy, *Electrochem. Solid State Lett.* 5, 2002, A263–A266.
17. W.S. Yoon, C.P. Grey, M. Balasubramanian, X.Q. Yang and J. McBreen, In situ x-ray absorption spectroscopic study on $LiNi_{0.5}Mn_{0.5}O_2$ cathode material during electrochemical cycling, *Chem. Mater.* 15, 2003, 3161–3169.
18. J. Reed and G. Ceder, Charge, potential, and phase stability of layered $LiNi_{0.5}Mn_{0.5}O_2$, *Electrochem. Solid State Lett.* 5, 2002, A145–A148.
19. C. Delmas, J.P. Peres, A. Rougier, A. Demourgues, F. Weill, A. Chadwick, M. Broussely, F. Perton, P. Biensan and P. Willmann, On the behavior of the Li_xNiO_2 system: An electrochemical and structural overview, *J. Power Sources* 68, 1997, 120–125.
20. Y.I. Jang, B.Y. Huang, H.F. Wang, D.R. Sadoway and Y.M. Chiang, Electrochemical cycling-induced spinel formation in high-charge-capacity orthorhombic $LiMnO_2$, *J. Electrochem. Soc.* 146, 1999, 3217–3223.
21. K.S. Kang, Y.S. Meng, J. Breger, C.P. Grey and G. Ceder, Electrodes with high power and high capacity for rechargeable lithium batteries, *Science* 311, 2006, 977–980.
22. S.B. Schougaard, J. Breger, M. Jiang, C.P. Grey and J.B. Goodenough, $LiNi_{0.5+\delta}Mn_{0.5-\delta}O_2$—A high-rate, high-capacity cathode for lithium rechargeable batteries, *Adv. Mater.* 18, 2006, 905–909.
23. T. Ohzuku and Y. Makimura, Layered lithium insertion material of $LiCo_{1/3}Ni_{1/3}Mn_{1/3}O_2$ for lithium-ion batteries, *Chem. Lett.* 30, 2001, 642–643.

24. Y. Koyama, Y. Makimura, I. Tanaka, H. Adachi and T. Ohzuku, Systematic research on insertion materials based on superlattice models in a phase triangle of $LiCoO_2$–$LiNiO_2$–$LiMnO_2$. I. First-principles calculation on electronic and crystal structures, phase stability and new $LiNi_{1/2}Mn_{1/2}O_2$ material, *J. Electrochem. Soc.* 151, 2004, A1499–A1506.
25. Y. Koyama, N. Yabuuchi, I. Tanaka, H. Adachi and T. Ohzuku, Solid-state chemistry, and electrochemistry of $LiCo_{1/3}Ni_{1/3}Mn_{1/3}O_2$ for advanced lithium-ion batteries—I. First-principles calculation on the crystal and electronic structures, *J. Electrochem. Soc.* 151, 2004, A1545–A1551.
26. M.M. Thackeray, S.H. Kang, C.S. Johnson, J.T. Vaughey, R. Benedek and S.A. Hackney, Li_2MnO_3-stabilized $LiMO_2$ (M = Mn, Ni, Co) electrodes for lithium-ion batteries, *J. Mater. Chem.* 17, 2007, 3112–3125.
27. B.J. Hwang, Y.W. Tsai, D. Carlier and G. Ceder, A combined computational/experimental study on $LiNi_{1/3}Co_{1/3}Mn_{1/3}O_2$, *Chem. Mater.* 15, 2003, 3676–3682.
28. N. Yabuuchi, Y. Makimura and T. Ohzuku, Solid-state chemistry and electrochemistry of $LiCo_{1/3}Ni_{1/3}Mn_{1/3}O_2$ for advanced lithium-ion batteries III. Rechargeable capacity and cycleability, *J. Electrochem. Soc.* 154, 2007, A314–A321.
29. N. Yabuuchi, Y. Koyama, N. Nakayama and T. Ohzuku, Solid-state chemistry and electrochemistry of $LiCo_{1/3}Ni_{1/3}Mn_{1/3}O_2$ for advanced lithium-ion batteries, *J. Electrochem. Soc.* 152, 2005, A1434–A1440.
30. S.H. Park, H.S. Shin, S.T. Myung, C.S. Yoon, K. Amine and Y.J. Sun, Synthesis of nanostructured $LiNi_{1/3}Co_{1/3}Mn_{1/3}O_2$ via a modified carbonate process, *Chem. Mater.* 17, 2005, 6–8.
31. K.M. Shaju, G.V.S. Rao and B.V.R. Chowdari, Performance of layered $LiNi_{1/3}Co_{1/3}Mn_{1/3}O_2$ as cathode for Li-ion batteries, *Electrochim. Acta* 48, 2002, 145–151.
32. K.M. Shaju and P.G. Bruce, Macroporous $LiNi_{1/3}Co_{1/3}Mn_{1/3}O_2$: A high-power and high-energy cathode for rechargeable lithium batteries, *Adv. Mater.* 18, 2006, 2330–2334.
33. S.T. Myung, M.H. Lee, S. Komaba, N. Kumagai and Y.K. Sun, Hydrothermal synthesis of layered $LiNi_{1/3}Co_{1/3}Mn_{1/3}O_2$ as positive electrode material for lithium secondary battery, *Electrochim. Acta* 50, 2005, 4800–4806.
34. B. Lin, Z.Y. Wen, Z.H. Gu and H.H. Huang, Morphology and electrochemical performance of $LiNi_{1/3}Co_{1/3}Mn_{1/3}O_2$ cathode material by a slurry spray drying method, *J. Power Sources* 175, 2008, 564–569.
35. S.H. Park, C.S. Yoon, S.G. Kang, H.S. Kim, S.I. Moon and Y.K. Sun, Synthesis and structural characterization of layered $LiNi_{1/3}Co_{1/3}Mn_{1/3}O_2$ cathode materials by ultrasonic spray pyrolysis method, *Electrochim. Acta* 49, 2004, 557–563.
36. S.T. Myung, K. Amine and Y.K. Sun, Surface modification of cathode materials from nano- to microscale for rechargeable lithium-ion batteries, *J. Mater. Chem.* 20, 2010, 7074–7095.
37. M.L. Marcinek, J.W. Wilcox, M.M. Doeff and R.M. Kostecki, Microwave plasma chemical vapor deposition of carbon coatings on $LiNi_{1/3}Co_{1/3}Mn_{1/3}O_2$ for Li-ion battery composite cathodes, *J. Electrochem. Soc.* 156, 2009, A48–A51.
38. H.S. Kim, M. Kong, K. Kim, I.J. Kim and H.B. Gu, Effect of carbon coating on $LiNi_{1/3}Mn_{1/3}Co_{1/3}O_2$ cathode material for lithium secondary batteries, *J. Power Sources* 171, 2007, 917–921.

39. B. Lin, Z.Y. Wen, J.D. Han and X.W. Wu, Electrochemical properties of carbon-coated $LiNi_{1/3}Co_{1/3}Mn_{1/3}O_2$ cathode material for lithium-ion batteries, *Solid State Ionics* 179, 2008, 1750–1753.
40. Y.K. Sun, S.W. Cho, S.W. Lee, C.S. Yoon and K. Amine, AlF_3-coating to improve high voltage cycling performance of $LiNi_{1/3}Co_{1/3}Mn_{1/3}O_2$ cathode materials for lithium secondary batteries, *J. Electrochem. Soc.* 154, 2007, A168–A172.
41. Y.Y. Huang, J.T. Chen, J.F. Ni, H.H. Zhou and X.X. Zhang, A modified ZrO_2-coating process to improve electrochemical performance of $LiNi_{1/3}Co_{1/3}Mn_{1/3}O_2$, *J. Power Sources* 188, 2009, 538–545.
42. S.K. Hu, G.H. Cheng, M.Y. Cheng, B.J. Hwang and R. Santhanam, Cycle life improvement of ZrO_2-coated spherical $LiNi_{1/3}Co_{1/3}Mn_{1/3}O_2$ cathode material for lithium ion batteries, *J. Power Sources* 188, 2009, 564–569.
43. S.B. Jang, S.H. Kang, K. Amine, Y.C. Bae and Y.K. Sun, Synthesis and improved electrochemical performance of $Al(OH)_3$-coated $LiNi_{1/3}Mn_{1/3}Co_{1/3}O_2$ cathode materials at elevated temperature, *Electrochim. Acta* 50, 2005, 4168–4173.
44. S.T. Myung, K. Izumi, S. Komaba, Y.K. Sun, H. Yashiro and N. Kumagai, Role of alumina coating on Li–Ni–Co–Mn–O particles as positive electrode material for lithium-ion batteries, *Chem. Mater.* 17, 2005, 3695–3704.
45. Y. Kim, H.S. Kim and S.W. Martin, Synthesis and electrochemical characteristics of Al_2O_3-coated $LiNi_{1/3}Co_{1/3}Mn_{1/3}O_2$ cathode materials for lithium ion batteries, *Electrochim. Acta* 52, 2006, 1316–1322.
46. Y.K. Sun, S.T. Myung, B.C. Park, J. Prakash, I. Belharouak and K. Amine, High-energy cathode material for long-life and safe lithium batteries, *Nat. Mater.* 8, 2009, 320–324.
47. Y.S. Jung, A.S. Cavanagh, L.A. Riley, S.H. Kang, A.C. Dillon, M.D. Groner, S.M. George and S.H. Lee, Ultrathin direct atomic layer deposition on composite electrodes for highly durable and safe Li-ion batteries, *Adv. Mater.* 22, 2010, 2172–2176.
48. C.S. Johnson, N.C. Li, C. Lefief and M.M. Thackeray, Anomalous capacity and cycling stability of $xLi_2MnO_3 \cdot (1-x)LiMO_2$ electrodes (M = Mn, Ni, Co) in lithium batteries at 50 degrees C, *Electrochem. Commun.* 9, 2007, 787–795.
49. A.D. Robertson and P.G. Bruce, Mechanism of electrochemical activity in Li_2MnO_3, *Chem. Mater.* 15, 2003, 1984–1992.
50. H.X. Deng, I. Belharouak, R.E. Cook, H.M. Wu, Y.K. Sun and K. Amine, Nanostructured lithium nickel manganese oxides for lithium-ion batteries, *J. Electrochem. Soc.* 157, 2010, A447–A452.
51. Z.H. Lu, D.D. MacNeil and J.R. Dahn, Layered cathode materials $Li(Ni_xLi_{1/3-2x/3}Mn_{2/3-x/3})O_2$ for lithium-ion batteries, *Electrochem. Solid State Lett.* 4, 2001, A191–A194.
52. M. Gu, I. Belharouak, A. Genc, Z.G. Wang, D.P. Wang, K. Amine, F. Gao. et al. Conflicting roles of nickel in controlling cathode performance in lithium ion batteries, *Nano. Letters*, 12, 2012, 5186–5191.
53. K.A. Jarvis, Z.Q. Deng, L.F. Allard, A. Manthiram and P.J. Ferreira, Atomic structure of a lithium-rich layered oxide material for lithium-ion batteries: Evidence of a solid solution, *Chem. Mater.* 23, 2011, 3614–3621.
54. J. Bareno, C.H. Lei, J.G. Wen, S.H. Kang, I. Petrov and D.P. Abraham, Local structure of layered oxide electrode materials for lithium-ion batteries, *Adv. Mater.* 22, 2010, 1122–1127.
55. A.R. Armstrong, M. Holzapfel, P. Novak, C.S. Johnson, S.H. Kang, M.M. Thackeray and P.G. Bruce, Demonstrating oxygen loss and associated structural

reorganization in the lithium battery cathode Li(Ni$_{0.2}$Li$_{0.2}$Mn$_{0.6}$)O$_2$, *J. Am. Chem. Soc.* 128, 2006, 8694–8698.
56. C.P. Grey, W.S. Yoon, J. Reed and G. Ceder, Electrochemical activity of Li in the transition-metal sites of O$_3$Li[Li$_{(1-2x)/3}$Mn$_{(2-x)/3}$Ni$_x$]O$_2$, *Electrochem. Solid State Lett.* 7, 2004, A290–A293.
57. K. Kang and G. Ceder, Factors that affect Li mobility in layered lithium transition metal oxides, *Phys. Rev. B* 74, 2006, 094105.
58. Y. Wu, A.V. Murugan and A. Manthiram, Surface modification of high capacity layered Li(Li$_{0.2}$Mn$_{0.54}$Ni$_{0.13}$Co$_{0.13}$)O$_2$ cathodes by AlPO$_4$, *J. Electrochem. Soc.* 155, 2008, A635–A641.
59. S.H. Kang and M.M. Thackeray, Enhancing the rate capability of high capacity xLi$_2$MnO$_3$ · (1 − x)LiMO$_2$ (M = Mn, Ni, Co) electrodes by Li–Ni–PO$_4$ treatment, *Electrochem. Commun.* 11, 2009, 748–751.
60. J. Liu, B. Reeja-Jayan and A. Manthiram, Conductive surface modification with aluminum of high capacity layered Li(Li$_{0.2}$Mn$_{0.54}$Ni$_{0.13}$Co$_{0.13}$)O$_2$ cathodes, *J. Phys. Chem. C* 114, 2010, 9528–9533.
61. J. Cho, Y. Kim and M.G. Kim, Synthesis and characterization of Li(Ni$_{0.41}$Li$_{0.08}$Mn$_{0.51}$)O$_2$ nanoplates for Li battery cathode material, *J. Phys. Chem. C* 111, 2007, 3192–3196.
62. M.G. Kim, M. Jo, Y.S. Hong and J. Cho, Template-free synthesis of Li(Ni$_{0.25}$Li$_{0.15}$Mn$_{0.6}$)O$_2$ nanowires for high performance lithium battery cathode, *Chem. Commun.* 2009, 218–220.
63. A. Van der Ven and G. Ceder, Lithium diffusion in layered Li$_x$CoO$_2$, *Electrochem. Solid State Lett.* 3, 2000, 301–304.
64. T. Ohzuku, A. Ueda, M. Nagayama, Y. Iwakoshi and H. Komori, Comparative study of LiCoO$_2$, LiNi$_{1/2}$Co$_{1/2}$O$_2$ and LiNiO$_2$ for 4-Volt secondary lithium cells, *Electrochim. Acta* 38, 1993, 1159–1167.
65. B.H. Song, M.O. Lai and L. Lu, Influence of Ru substitution on Li-rich 0.55Li$_2$MnO$_3$ · 0.45LiNi$_{1/3}$Co$_{1/3}$Mn$_{1/3}$O$_2$ cathode for Li-ion batteries, *Electrochim. Acta* 80, 2012, 187–195.
66. J.M. Zheng, X.B. Wu and Y. Yang, A comparison of preparation method on the electrochemical performance of cathode material Li(Li$_{0.2}$Mn$_{0.54}$Ni$_{0.13}$Co$_{0.13}$)O$_2$ for lithium ion battery, *Electrochim. Acta* 56, 2011, 3071–3078.
67. G.Y. Kim, S.B. Yi, Y.J. Park and H.G. Kim, Electrochemical behaviors of Li(Li$_{(1-x)/3}$Mn$_{(2-x)/3}$Ni$_{x/3}$Co$_{x/3}$O$_2$ cathode series (0<x<1 synthesized by sucrose combustion process for high capacity lithium ion batteries, *Mater. Res. Bull.* 43 (2008) 3543-3552.
68. D. Kim, S.H. Kang, M. Balasubramanian and C.S. Johnson, High-energy and high-power Li-rich nickel manganese oxide electrode materials, *Electrochem. Commun.* 12, 2010, 1618–1621.
69. Y. Shao-Horn, S.A. Hackney, A.R. Armstrong, P.G. Bruce, R. Gitzendanner, C.S. Johnson and M.M. Thackeray, Structural characterization of layered LiMnO$_2$ electrodes by electron diffraction and lattice imaging, *J. Electrochem. Soc.* 146, 1999, 2404–2412.
70. J.C. Dupin, D. Gonbeau, H. Benqlilou-Moudden, P. Vinatier and A. Levasseur, XPS analysis of new lithium cobalt oxide thin-films before and after lithium deintercalation, *Thin Solid Films* 384, 2001, 23–32.
71. A.W. Moses, H.G.G. Flores, J.G. Kim and M.A. Langell, Surface properties of LiCoO$_2$, LiNiO$_2$ and LiNi$_{(1-x)}$Co$_x$O$_2$, *Appl. Surf. Sci.* 253, 2007, 4782–4791.

72. Y. Wu, A.V. Murugan and A. Manthiram, Surface modification of high capacity layered Li(Li$_{0.2}$Mn$_{0.54}$Ni$_{0.13}$Co$_{0.13}$)O$_2$ cathodes by AlPO$_4$, *J. Electrochem. Soc.* 155, 2008, A635–A641.
73. Y.K. Zhou, J. Huang, C.M. Shen and H.L. Li, Synthesis of highly ordered LiNiO$_2$ nanowire arrays in AAO templates and their structural properties, *Mater. Sci. Eng. A: Struct. Mater. Prop. Microstruct. Process.* 335, 2002, 260–267.
74. K.M. Shaju, G.V.S. Rao and B.V.R. Chowdari, X-ray photoelectron spectroscopy and electrochemical behaviour of 4 V cathode, Li(Ni$_{1/2}$Mn$_{1/2}$)O$_2$, *Electrochim. Acta* 48, 2003, 1505–1514.
75. B.H. Song, Z.W. Liu, M.O. Lai and L. Lu, Structural evolution and the capacity fade mechanism upon long-term cycling in Li-rich cathode material, *Phys. Chem. Chem. Phys.* 14, 2012, 12875–12883.
76. E. Regan, T. Groutso, J.B. Metson, R. Steiner, B. Ammundsen, D. Hassell and P. Pickering, Surface and bulk composition of lithium manganese oxides, *Surf. Interface Anal.* 27, 1999, 1064–1068.
77. Z. Su, Z.W. Lu, X.P. Gao, P.W. Shen, X.J. Liu and J.Q. Wang, Preparation and electrochemical properties of indium- and sulfur-doped LiMnO$_2$ with orthorhombic structure as cathode materials, *J. Power Sources* 189, 2009, 411–415.
78. N.J. Dudney and B.J. Neudecker, Solid state thin-film lithium battery systems, *Curr. Opin. Solid State Mater.* 4, 1999, 479–482.
79. J.B. Bates, N.J. Dudney, B. Neudecker, A. Ueda and C.D. Evans, Thin-film lithium and lithium-ion batteries, *Solid State Ionics* 135, 2000, 33–45.
80. J.B. Bates, G.R. Gruzalski, N.J. Dudney, C.F. Luck and X.H. Yu, Rechargeable thin-film lithium microbatteries, *Solid State Technol.* 36, 1993, 59.
81. J.B. Bates, D. Lubben and N.J. Dudney, Thin film LiMn$_2$O$_4$ batteries, *IEEE Aero. El. Syst. Mag.* 10, 1995, 30–32.
82. J.F.M. Oudenhoven, L. Baggetto and P.H.L. Notten, All-solid-state lithium-ion microbatteries: A review of various three-dimensional concepts, *Adv. Energy Mater.* 1, 2011, 10–33.
83. J.L. Souquet and M. Duclot, Thin film lithium batteries, *Solid State Ionics* 148, 2002, 375–379.
84. A. Patil, V. Patil, D. Wook Shin, J.-W. Choi, D.-S. Paik and S.-J. Yoon, Issue and challenges facing rechargeable thin film lithium batteries, *Mater. Res. Bull.* 43, 2008, 1913–1942.
85. S. Kondo, All solid-state lithium secondary battery with highly ion conductive glassy electrolyte, in: M. Wakihara, O. Yamamoto, (eds.), *Lithium Ion Batteries: Fundamentals and Performance*, Wiley-VCH, 1998, pp. 199–217.
86. J. Zhu, L. Lu and K.Y. Zeng, Nanoscale mapping of lithium-ion diffusion in a cathode within an all-solid-state lithium-ion battery by advanced scanning probe microscopy techniques, *ACS Nano* 7, 2013, 1666–1675.
87. M.V. Reddy, G.V.S. Rao and B.V.R. Chowdari, Cathodic behaviour of NiO-coated Li(Ni$_{1/2}$Mn$_{1/2}$)O$_2$, *Electrochim. Acta* 50, 2005, 3375–3382.
88. I. Taniguchi, C.K. Lim, D. Song and M. Wakihara, Particle morphology and electrochemical performances of spinel LiMn$_2$O$_4$ powders synthesized using ultrasonic spray pyrolysis method, *Solid State Ionics* 146, 2002, 239–247.
89. M. Konarova and I. Taniguchi, Synthesis of carbon-coated LiFePO$_4$ nanoparticles with high rate performance in lithium secondary batteries, *J. Power Sources* 195, 2010, 3661–3667.

90. X.Y. Wang, H. Hao, J.L. Liu, T. Huang and A.S. Yu, A novel method for preparation of macroporous lithium nickel manganese oxygen as cathode material for lithium ion batteries, *Electrochim. Acta* 56, 2011, 4065–4069.
91. Z.D. Huang, X.M. Li, S.W. Oh, B.A. Zhang, P.C. Ma and J.K. Kim, Microscopically porous, interconnected single crystal $LiNi_{1/3}Co_{1/3}Mn_{1/3}O_2$ cathode material for lithium ion batteries, *J. Mater. Chem.* 21, 2011, 10777–10784.
92. D.K. Lee, S.H. Park, K. Amine, H.J. Bang, J. Parakash and Y.K. Sun, High capacity $Li(Li_{0.2}Ni_{0.2}Mn_{0.6})O_2$ cathode materials via a carbonate co-precipitation method, *J. Power Sources* 162, 2006, 1346–1350.

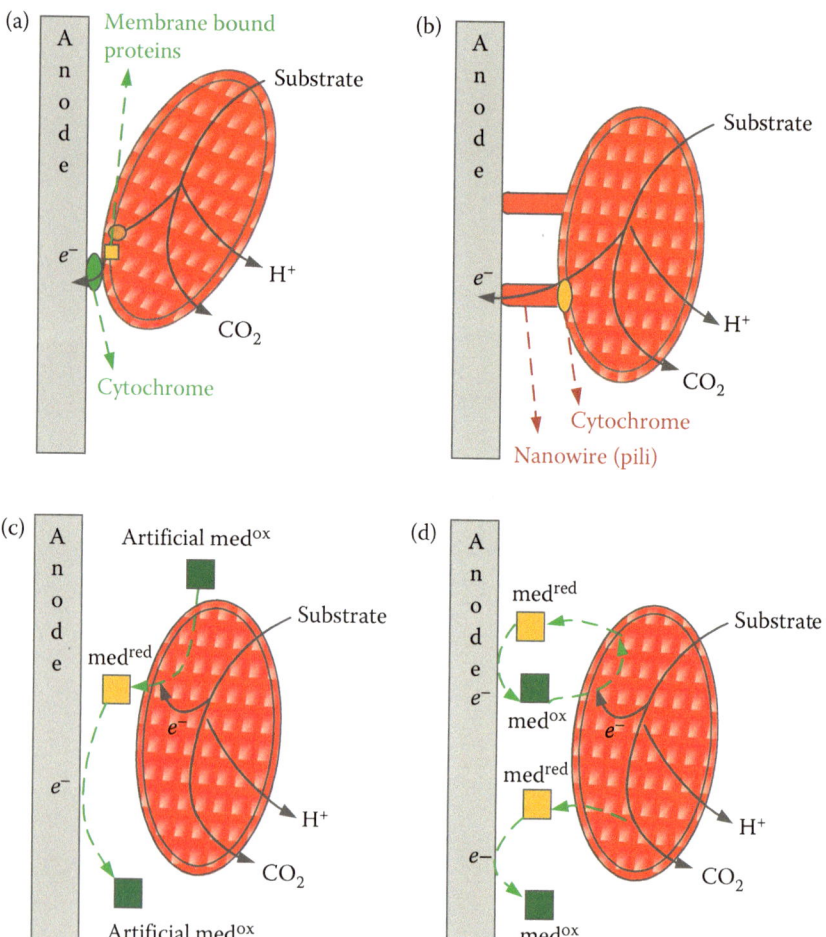

FIGURE 2.6
See text for caption.

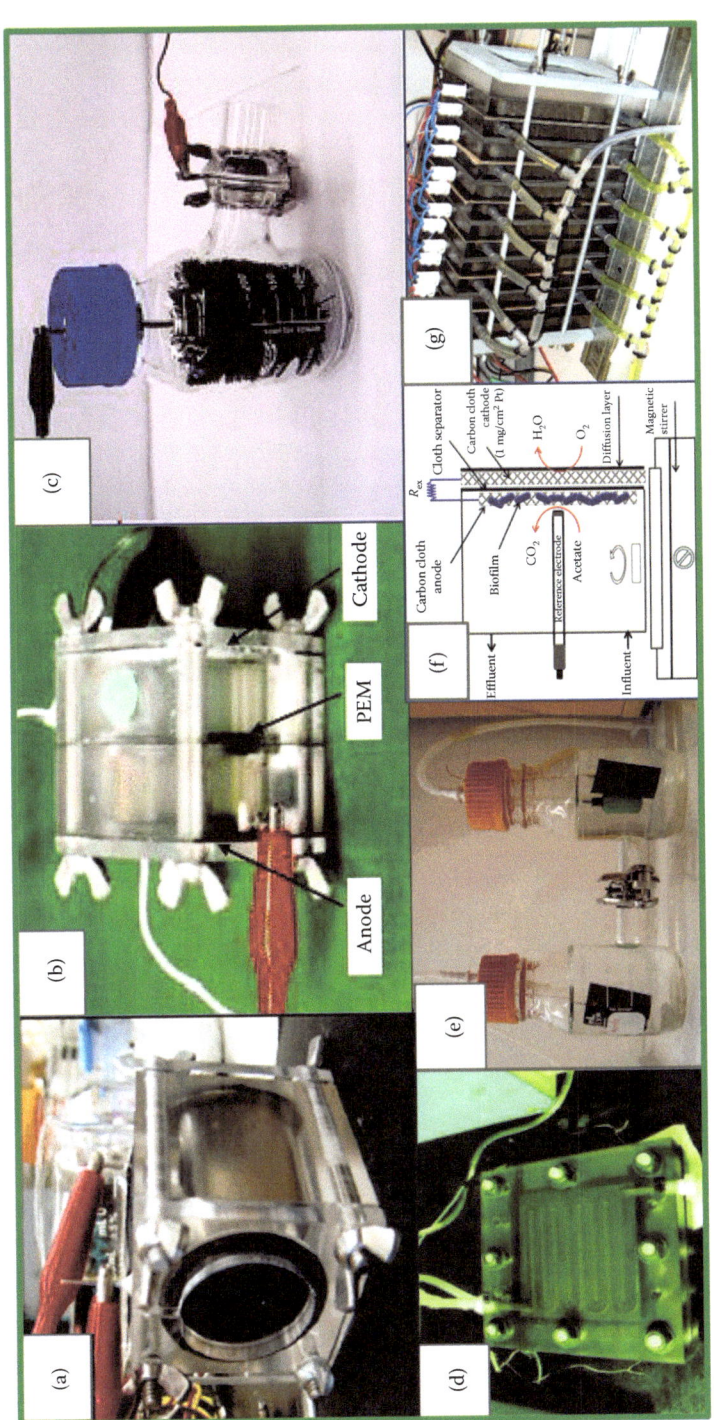

FIGURE 2.9
See text for caption.

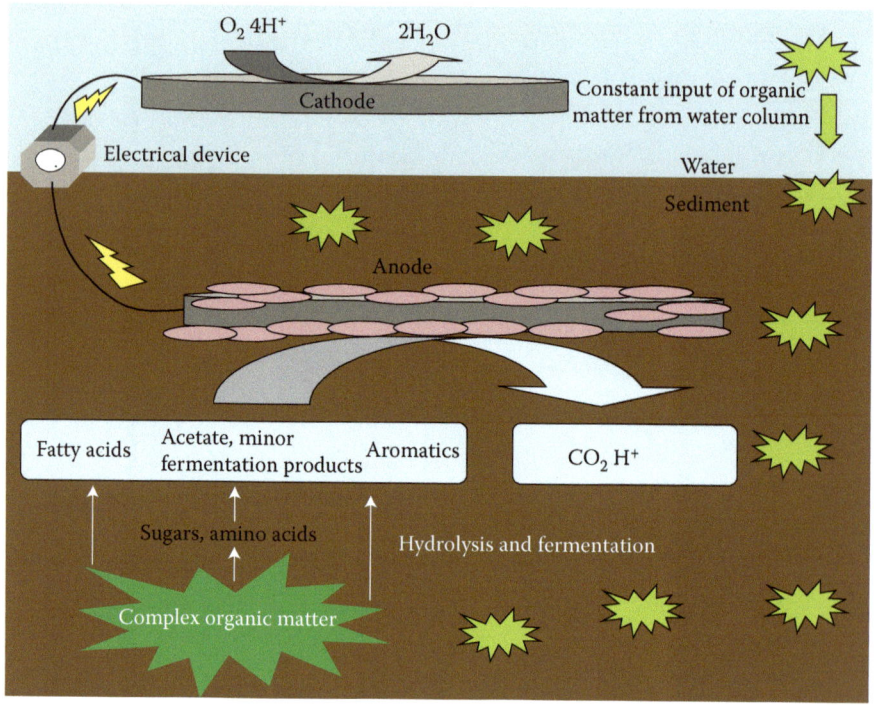

FIGURE 2.10
See text for caption.

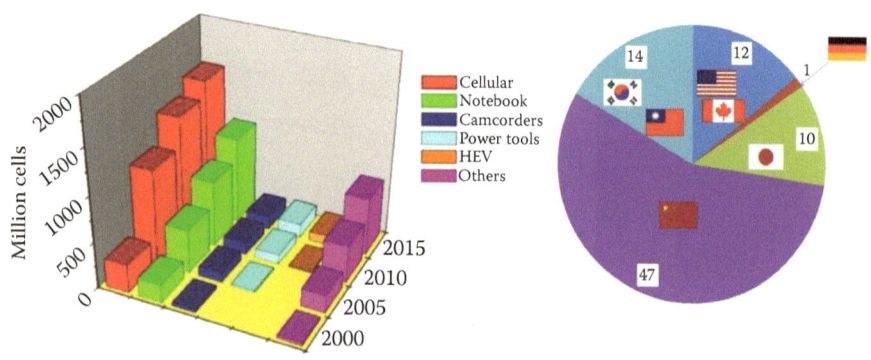

FIGURE 3.5
See text for caption.

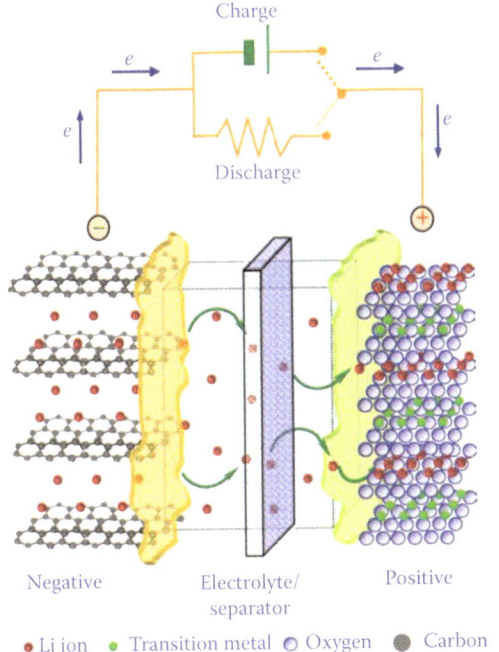

FIGURE 3.11
See text for caption.

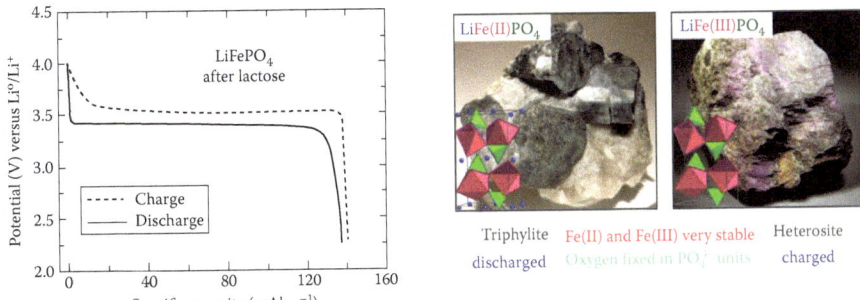

FIGURE 3.23
See text for caption.

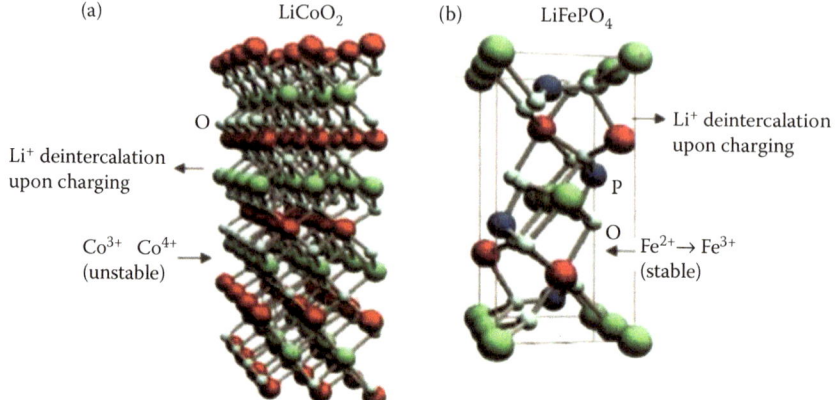

FIGURE 3.24
See text for caption.

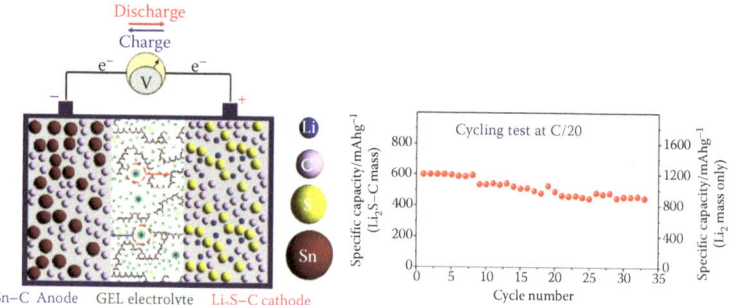

FIGURE 3.30
See text for caption.

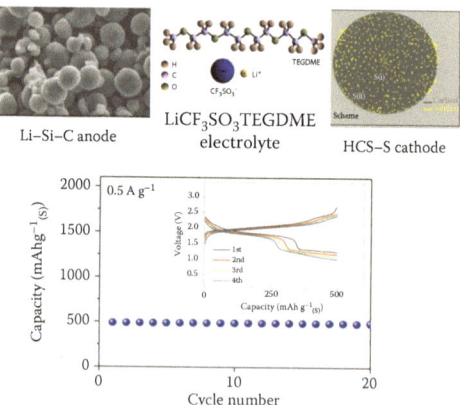

FIGURE 3.31
See text for caption.

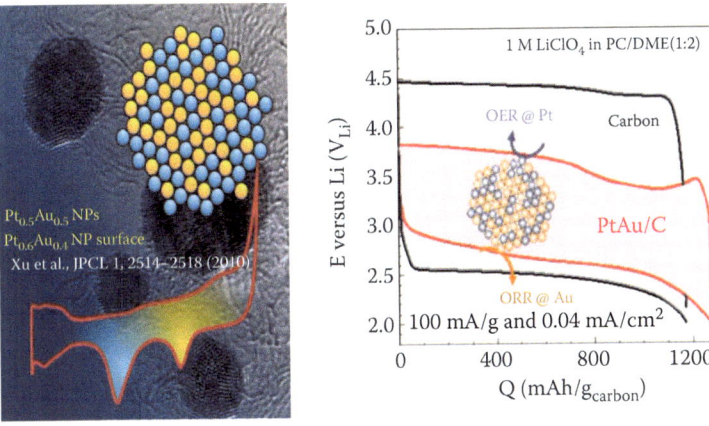

FIGURE 3.36
See text for caption.

TABLE 3.1

Summary of Characteristics and Properties of Anode Materials Designed for Advanced Lithium-Ion Batteries

Material	Theoretical capacity (mAhg^{-1})	Advantages	Issues	Structures
$Li_4Ti_5O_{12}$	170	Negligible volume expansion, low cost, stable electrochemical operation and high thermal stability in the charged–discharged state, already in large-scale production	Low specific capacity, high voltage	
Li_xSn_y	990	High capacity, low cost	Large volume changes. Nanostructures required	
Li_xSi_y	2000	Very high capacity, low cost	Large volume changes. Nanostructures required production announced for 2012 (Panasonic)	
$Sn_xCo_yC_z$	500	High capacity, commercially tested	Cycle life to be demonstrated	
C, MCMB	370	Field tested operation	Operation outside electrolyte stability	

TABLE 3.2

Summary of Characteristics and Properties for Electrolyte Materials Designed for Lithium-Ion Batteries

Material	Advantages	Issues	
Liquid organic solutions	High conductivity over a wide temperature range. Liquid state.	Narrow electrochemical stability window. Safety.	
Gel-type membranes	High conductivity in a wide temperature range	Solid–liquid hybrid configuration; liquid retention	
PEO-LiX, membrane	Solvent-free, fully solid configuration; low cost	Temperature-dependent conductivity; low lithium ion transfer number and limited oxidation stability	
Ionic liquids	Non-flammable. High thermal stability. High conductivity	Scarce compatibility with low-voltage anodes	

TABLE 3.3

Characteristics and Properties for Cathode Materials Designed for Lithium-Ion Batteries as Alternatives to the Common $LiCoO_2$

Material	Theoretical capacity ($mAhg^{-1}$)	Advantages	Issues	Structure
$LiNi_{0.5}Mn_{1.5}O_4$	145	High voltage, good specific capacity	Operation within the limits of the electrolyte stability domain	
$LiNi_{1/3}Co_{1/3}Mn_{1/3}O_2$	290	High capacity, high rate	Structure retention upon cycling	
$LiFePO_4$	170	Basic low cost, intrinsic safety and environmental compatibility	Low electronic conductivity, low tap density and needs C coating	
$LiCoO_2$	140	Field-tested operation	High cost	

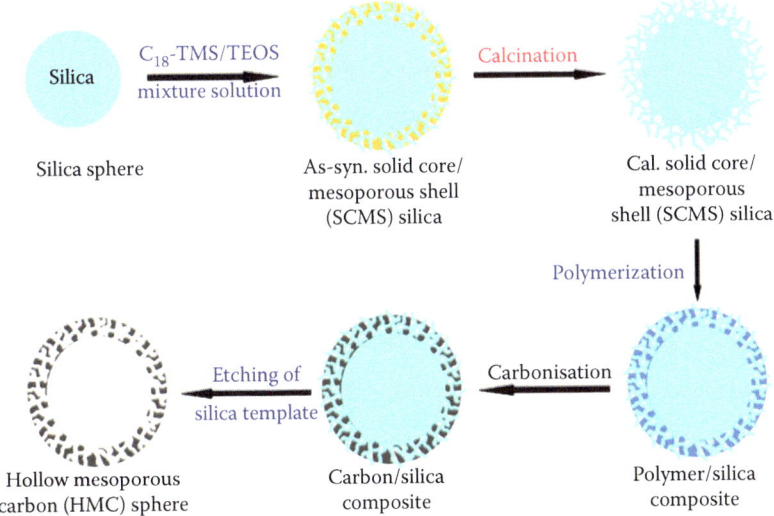

FIGURE 4.1
See text for caption.

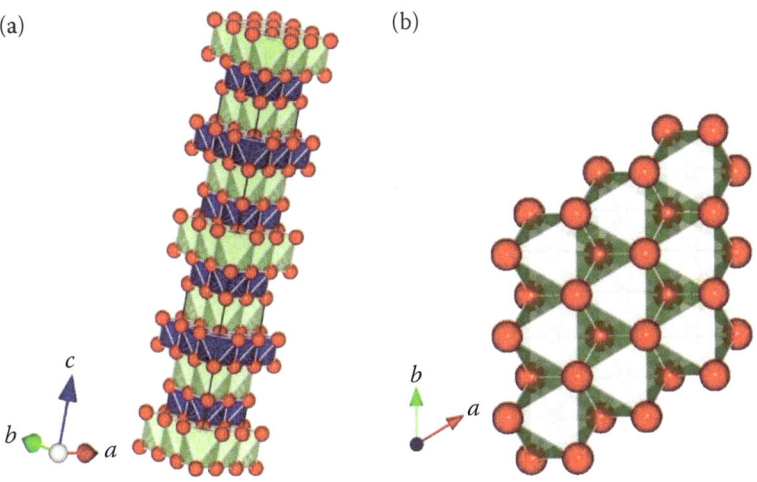

FIGURE 5.2
See text for caption.

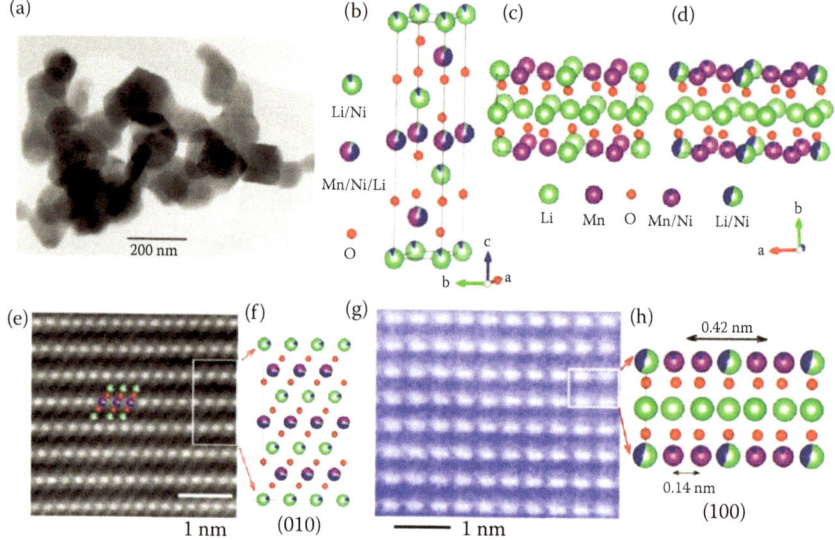

FIGURE 5.11
See text for caption.

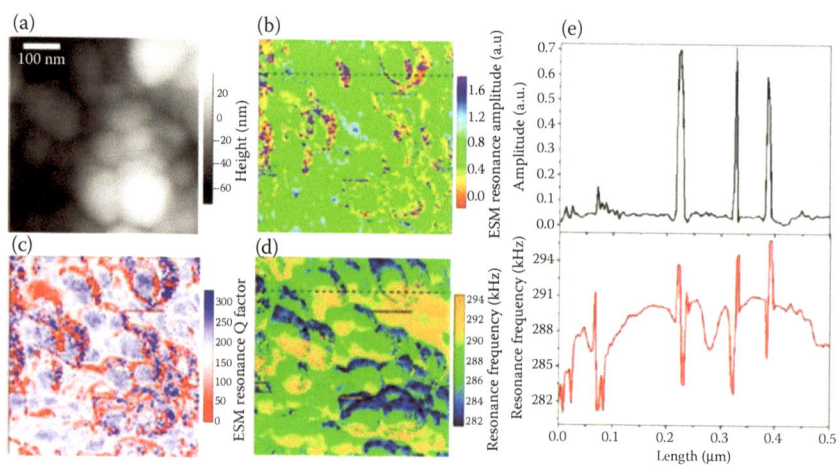

FIGURE 5.20
See text for caption.

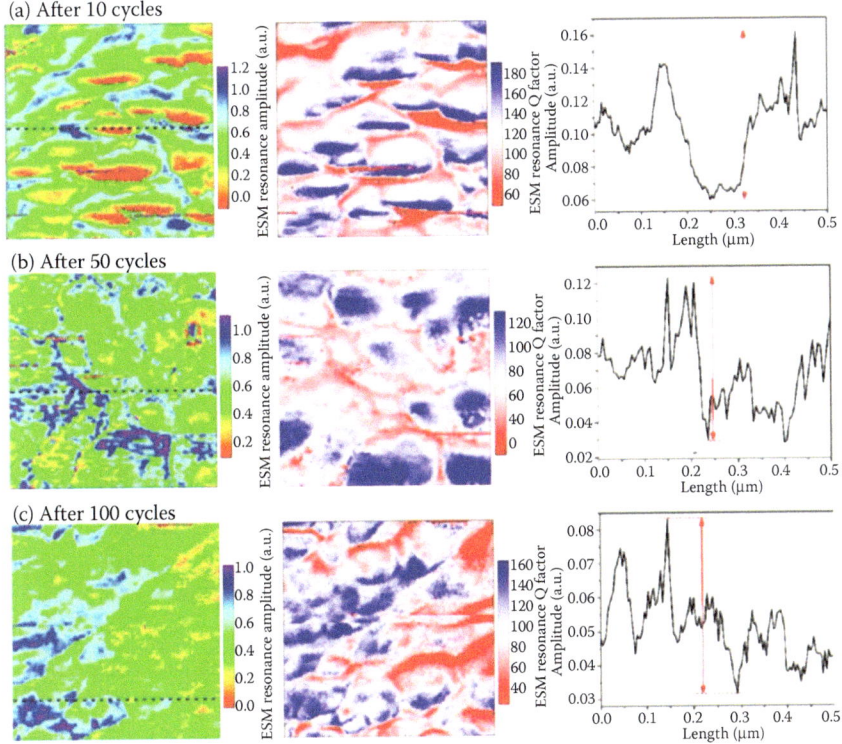

FIGURE 5.21
See text for caption.

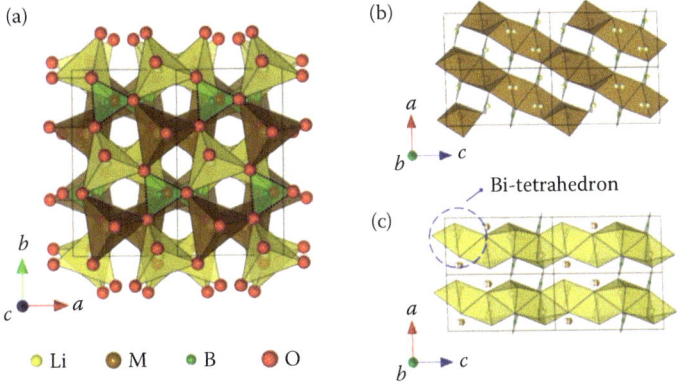

FIGURE 6.1
See text for caption.

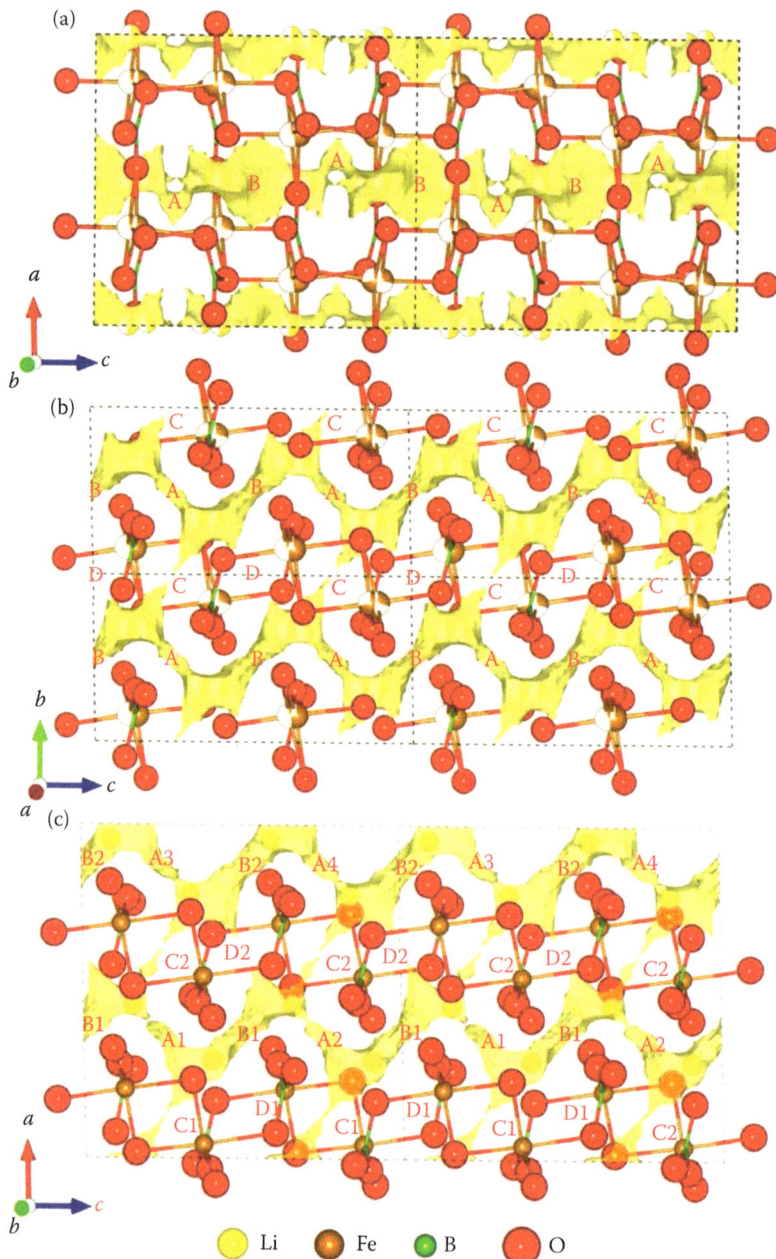

FIGURE 6.6
See text for caption.

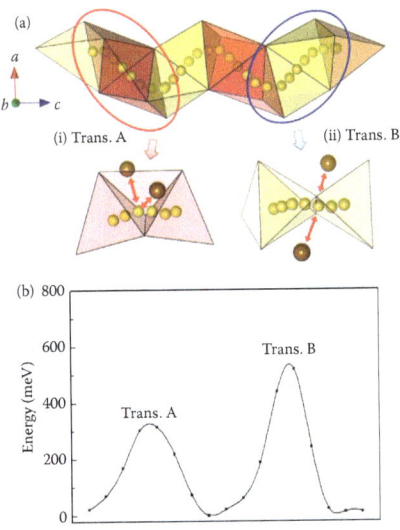

FIGURE 6.7
See text for caption.

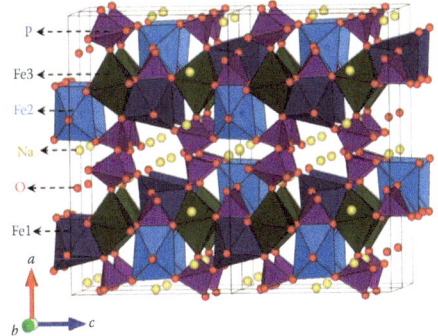

FIGURE 6.9
See text for caption.

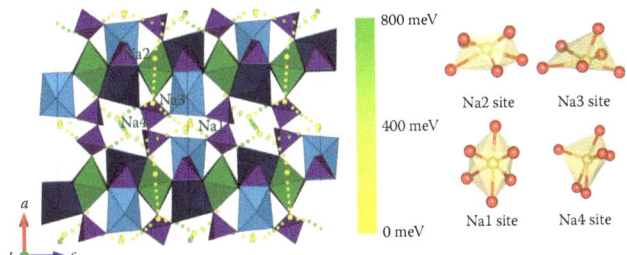

FIGURE 6.11
See text for caption.

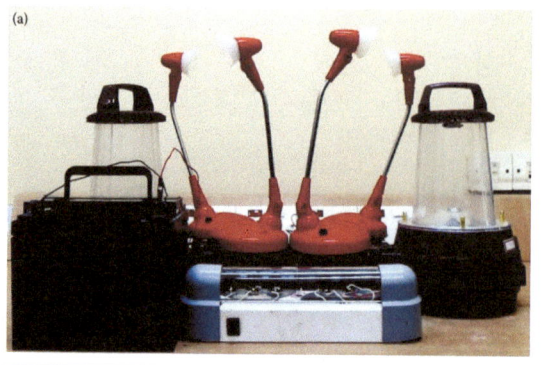
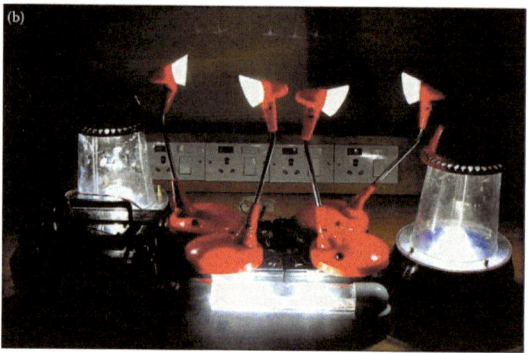

FIGURE 8.13
See text for caption.

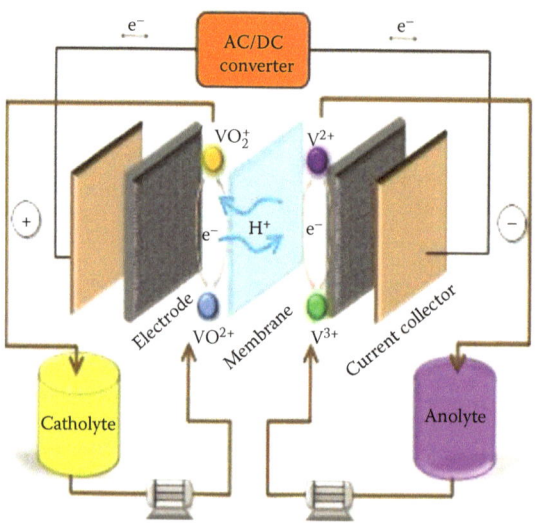

FIGURE 9.2
See text for caption.

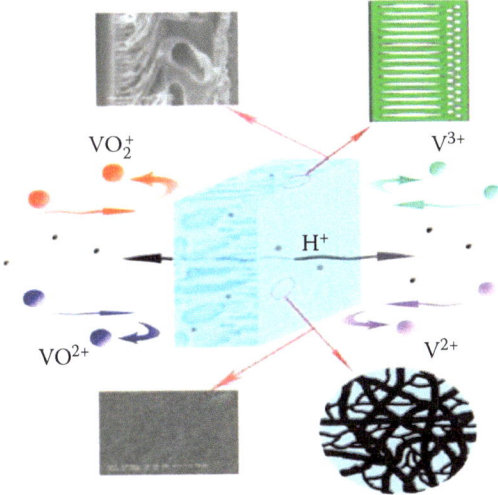

FIGURE 9.13
See text for caption.

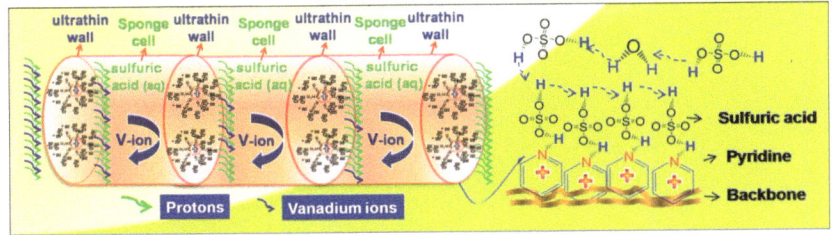

FIGURE 9.16
See text for caption.

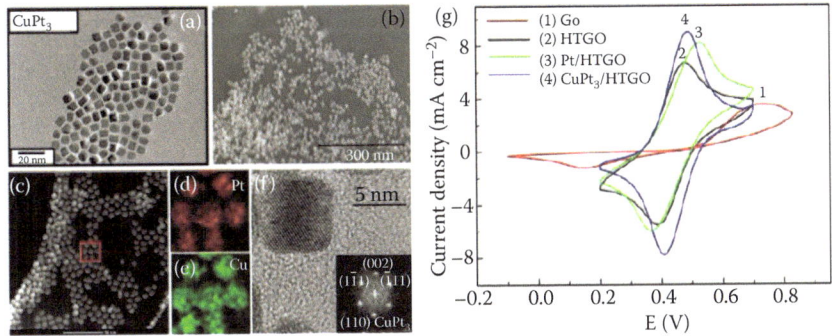

FIGURE 9.22
See text for caption.

FIGURE 9.30
See text for caption.

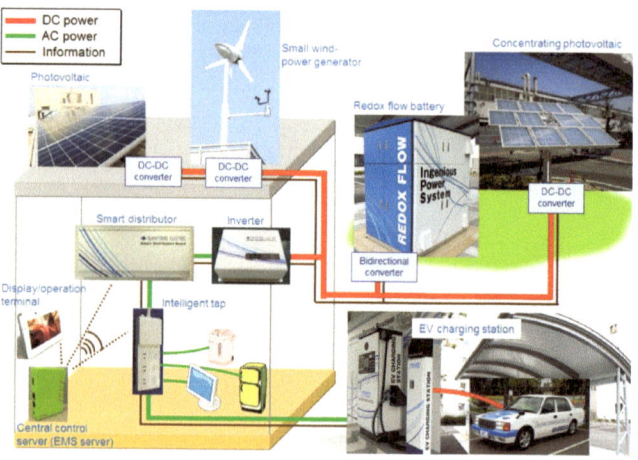

FIGURE 9.35
See text for caption.

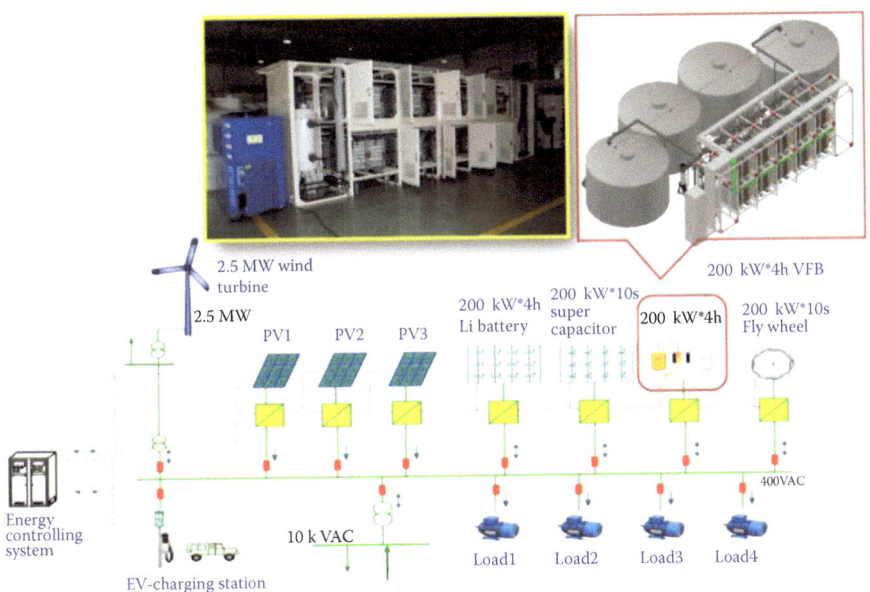

FIGURE 9.37
See text for caption.

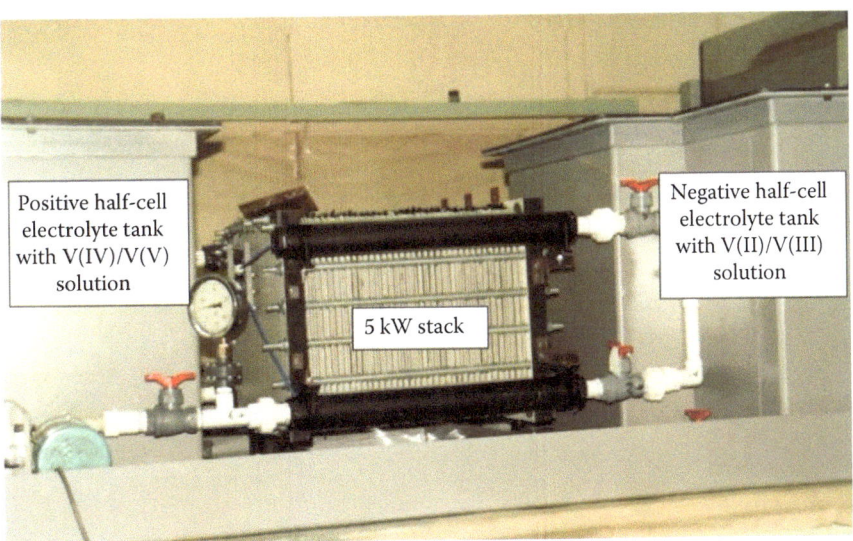

FIGURE 10.1
See text for caption.

6

First-Principles Approach for Cathode Design and Characterisation

Dong-Hwa Seo, Inchul Park and Kisuk Kang

CONTENTS

6.1 Introduction 223
6.2 Methodology 224
    6.2.1 First-Principles Calculations 224
    6.2.2 Application of First-Principles Calculations to Cathode Design 228
        6.2.2.1 Lithium/Sodium Intercalation Voltage 228
        6.2.2.2 Li/Na Ion Diffusion 228
6.3 Case Study on Polyanionic Cathode Materials with First Principles: Lithium Metal Borates 229
    6.3.1 Crystal Structure and Volume Change with Delithiation 230
    6.3.2 Electronic Structure and Redox Reaction with Delithiation 232
    6.3.3 Lithium Intercalation Voltages of Lithium Metal Borates 236
    6.3.4 Lithium Ion Diffusion 238
6.4 Application of First-Principle Studies to Search for New Cathode Materials 244
6.5 Conclusions 248
References 249

6.1 Introduction

The exhaustion of oil resources and other environmental problems have provoked tremendous interest in energy-storage systems for renewable energy [1]. The Li rechargeable battery is one of the most advanced energy-storage systems, due to its high-energy density. The $LiCoO_2$–graphite-based system has served as the major power system of portable electric devices, such as mobile phones and laptop computers. With the high cost and the unreliable safety and stability of the current systems, these systems are also inadequate for electric vehicle and large-scale power-storage systems. In this regard, intensive research efforts have been focussed on developing alternative electrode materials with high stability, energy and power for these applications

[2–7]. Recently, the first-principles calculation technique has been used as a new tool for exploring novel materials and understanding electrode behaviours, and is proven to be effective, not only providing a fundamental understanding of electrode properties but also leading to better design for new electrode materials. It is now possible to predict voltage, Li diffusivity, electronic conductivity, phase stability and other related properties of unknown electrode materials with reasonable accuracy; these key properties offer insight for screening material groups in the discovery of new electrode materials.

In this chapter, we will review the first-principles approach to the characterisation and design of cathode materials for rechargeable batteries, with a particular focus on polyanionic materials. Polyanionic materials such as phosphates [2,8–14], pyrophosphates [15–17], silicates [18–21], fluorophosphates [4,22], fluorosulphates [5,7,23,24] and borates [25–33] have been proposed as promising Li-rechargeable battery cathodes for large applications. The strong covalent bonds of a polyanion make a structure stable during charge/discharge [9,18]. Moreover, the electrochemical properties of polyanionic cathode materials can be tuned by combining the appropriate polyanions and transition metals (TMs). Li intercalation potentials, Li storage capacity, thermal stability and other factors are critically dependent on the polyanion species [34,35]. Thus, there are ongoing research efforts to find the next-generation cathodes among polyanionic materials by combining various TMs and polyanions [36].

This chapter begins with a short introduction of the first-principles computational methods and continues with the application of first principles to electrode characterisation. Next, we illustrate how first-principles calculations can be used to characterise various intrinsic material properties with a case study of $LiMBO_3$ (M = Mn, Fe and Co). Based on the understanding of material properties at the atomic scale, we predict the most important cathode properties, such as cyclability, energy density and power capability. Finally, we will demonstrate with an example on how our first-principles approach can be extended to the design of novel cathode materials for new energy-storage chemistry and Na-rechargeable batteries that have high stability and power capability. We believe that the first-principles approach offers new opportunities for the discovery of novel electrode materials.

6.2 Methodology

6.2.1 First-Principles Calculations

First-principles (or *ab initio* in Latin, which means 'from the beginning') calculations are distinct from other computational methods such as empirical molecular dynamics simulations. All other computational methods require prior determination of parameters from experiments or quantum mechanical calculations. However, first-principles calculations do not require any parameters because they start directly at the level of established laws of physics. This method only requires the numbers of nuclear charges and electrons as input, without any empirical models or fitting parameters.

First-principles calculations obtain wavefunctions of materials, which contain all the information of a given system, by solving the time-dependent Schrödinger equation. However, the time-independent Schrödinger equation is sufficient for obtaining the key properties of electrode materials for energy-storage systems, such as voltage and conductivity. The time-independent Schrödinger equation is as follows:

$$H\Psi = E\Psi \tag{6.1}$$

In Equation 6.1, H is the Hamiltonian operator, Ψ is the wavefunction and E is the total energy of the system. The Hamiltonian operator H for a system consisting of N electrons and M nuclei can be expressed as follows:

$$H = -\sum_{i=1}^{N}\frac{1}{2}\nabla_i^2 - \sum_{A=1}^{M}\frac{1}{2M_A}\nabla_A^2 - \sum_{i=1}^{N}\sum_{A=1}^{M}\frac{Z_A}{r_{iA}} + \sum_{i=1}^{N}\sum_{j>i}^{N}\frac{1}{r_{ij}} + \sum_{A=1}^{M}\sum_{B>A}^{M}\frac{Z_A Z_B}{R_{AB}} \tag{6.2}$$

where A, B and i, j denote the nuclei and electrons, respectively, Z_A and Z_B are the atomic numbers of the nuclei, and R_{AB} (or r_{ij}) is the distance between A and B (or i and j). The first and second terms are the kinetic operators of the electrons and nuclei, respectively. The third, fourth and fifth terms describe the Coulombic interactions between nuclei and electrons, multiple electrons, and multiple nuclei, respectively.

For many body systems, solving the Schrödinger equation is intractable because of a formidable number of terms in the equation except for a simple system such as hydrogen atom. Therefore, it is necessary to introduce some approximations to obtain the wavefunctions for the system. The first approximation is the Born–Oppenheimer approximation. Within this approximation, the Hamiltonian operator of the Schrödinger equation (Equation 6.2) can be split into two parts, the nuclear and electronic Hamiltonian operators. The nuclear Hamiltonian is then treated as a constant because the mass of nuclei is much larger than that of an electron; thus, the motions of nuclei are regarded negligible compared to those of electrons. Therefore, the kinetic energies of the nuclei, and the Coulombic interaction term between the nuclei can be treated as a constant. The wavefunctions of electrons can be calculated by solving the Schrödinger equation with the following electronic Hamiltonian operator:

$$H_{\text{electron}} = -\sum_{i}\frac{1}{2}\nabla_i^2 - \sum_{i}\sum_{A}\frac{Z_A}{r_{iA}} + \sum_{i}\sum_{j>i}\frac{1}{r_{ij}} \tag{6.3}$$

The total energy of the system can be calculated by the electronic and nuclear Hamiltonian operators

$$E_{\text{total}} = E_{\text{electron}} + E_{\text{nuc}} \qquad (6.4)$$

where E_{nuc} can be obtained using the kinetic operation of nuclei and the interaction between them, as follows:

$$E_{nuc} = -\sum_{A=1}^{M} \frac{1}{2M_A} \nabla_A^2 - \sum_{A}\sum_{B>A} \frac{Z_A Z_B}{R_{AB}} \qquad (6.5)$$

To describe the electrode material properly, the wavefunction of the Schrödinger equation should contain the spin rate. Thus, two spin functions* α(ω) and β(ω) were introduced into the wavefunction. Since they satisfy the Pauli exclusion principle, Slater determinants[†] were used to satisfy the opposite spin requirements for the wavefunction.

Even with the simplification on the nuclear parts, the solution of an N-electron Hamiltonian remains too complicated due to the correlating interactions between electrons. In this respect, the Hartree–Fock method was introduced, which assumes an approximated interaction between electrons; each electron interacts with the potential field formed by the other electrons. Within the Hartree–Fock method, the system can be described by one-electron Schrödinger equations that can be solved with the variational method. As a result, the complicated N-body problem can be replaced by simple one-body problems with the Hartree–Fock method.

A different approach to solving time-independent Schrödinger equations is density functional theory (DFT) [37], which can be less complicated due to its use of the three-coordinate electron density, compared to the 4N coordinate electron wavefunction in the Hartree–Fock method. The essence of DFT is that 'all properties of all states are formally determined by the ground state density', and 'the ground state energy of the system is a functional to the ground state density', as stated in the Hohenberg–Kohn theorem [38,39]. The ground state energy can be expressed as follows:

$$E[\rho] = T[\rho] + V_{ee}[\rho] + \int \rho(\vec{r}) V_{Ne}(\vec{r}) d\vec{r} \qquad (6.6)$$

where ρ is the electron density, and each successive term represents kinetic energy, interaction between electrons and Coulombic potential between

---

* Since wave functions must be anti-symmetric in the interchange of any two electrons, one-electron functions are composed of a spatial orbital, ψ(r) and spin functions, α(s) or β(s). Ψ(r, s) = ψ(r)α(s) or ψ(r)β(s). Spin functions are orthonormal, that is, <α|α> = <β|β> = 1 and <α|β> = < β|α> = 0.
† Slater determinant describes the wavefunction of a multi-fermionic system that satisfies anti-symmetry requirements and Pauli principles, which state that Ψ(2, 1) = –Ψ(1, 2).

electrons and nuclei. The first and second terms are called universal functionals, $F[\rho]$. If the correct universal functionals are known, the ground state energy of the system can be calculated with the variational method. However, the correct universal functional is unknown and an approximated form of $F[\rho]$ was introduced by Kohn and Sham [40] as follows:

$$F[\rho] = T_s[\rho] + J[\rho] + E_{xc}[\rho] \tag{6.7}$$

where $T_s[\rho]$ is the kinetic energy of non-interacting electrons and $J[\rho]$ is a classical Coulombic repulsion known as the Hartree term. The last term $E_{xc}[\rho]$ is the exchange–correlation energy of electrons, which includes the difference in kinetic energies between non-interacting electrons and real electrons as well as $J[\rho]$, arising from the correlations between electrons.

The exact $E_{xc}[\rho]$ is still unknown, but some approximations have been suggested. Kohn and Sham [40] suggested the local density approximation (LDA), which is the most common approximation to exchange–correlation energy and can be written as follows:

$$E_{xc}(r) = \int \rho(\vec{r}) \varepsilon_{xc}(\rho) d\vec{r} \tag{6.8}$$

where ε_{xc} is the exchange–correlation energy per electron in a uniform electron gas. Nevertheless, most battery materials that typically contain TMs in oxide frameworks have non-uniform electron density. Thus, the generalised gradient approximation (GGA) is usually chosen because it can treat the non-local electron density system by taking into account the gradient of the electron density [41]. However, GGA and LDA cannot fully describe the structural and electronic properties of TM oxides, because of abnormally strong electron correlations within the d state of the TM. Therefore, we added the Hubbard-type U parameter to GGA (GGA + U), which accurately calculates structural and electronic properties of TM oxides [42–44].

In most of the cases in this chapter, first-principles calculations were performed on the basis of spin-polarised DFT using a GGA within the Perdew–Burke–Ernzerhof (PBE) functional [41]. A plane-wave basis set and the projector-augmented wave (PAW) method, as implemented in the Vienna ab initio simulation package (VASP), were used [45]. PAW potentials have been widely applied and show good predictive capabilities for battery materials [46–54]. The incomplete cancellation of the self-interaction of GGA or LDA is often reported to result in large errors, especially for systems featuring strong localisation of the metal d orbitals, such as TM polyoxianion materials [49,50,55]. Thus, the GGA + U approach was employed in the rotationally invariant scheme presented by Lichtenstein and coworkers [44] to calculate structural and electronic properties of $LiMBO_3$ and $Na_4Fe_3(PO_4)_2(P_2O_7)$ [32,56].

6.2.2 Application of First-Principles Calculations to Cathode Design

6.2.2.1 Lithium/Sodium Intercalation Voltage

The equilibrium intercalation voltage is determined by the chemical potential difference of guest ions (Li or Na in most cases) in the cathode and anode. The chemical potential of a guest ion is the partial derivative of the free energy of the material with respect to the concentration of the guest ion. The open-circuit voltage at intercalation level x is obtained as follows:

$$V(x) = -\frac{\mu_A^{cathode}(x) - \mu_A^{anode}}{ze} \qquad (6.9)$$

where A represents the guest ion, $\mu_A^{cathode}$ is the chemical potential of the guest ion in the cathode, μ_A^{anode} is the chemical potential of the guest ion in the anode, z is the charge on the guest ion in the electrolyte and e is the electron charge. If the anode is a pure metal of the guest ion, μ_A^{anode} is constant; so, the voltage only depends on changes in the chemical potential on $\mu_A^{cathode}$, the average voltage between compositions x_1 and x_2 is easily obtained by integrating Equation 6.9 over that range, as follows:

$$V = -\frac{\Delta G}{(x_2 - x_1)ze} \qquad (6.10)$$

where ΔG is the Gibbs free energy for the reaction between compositions x_1 and x_2 in the electrode [19,46,48,49,57–60]. The reaction free energy can be written as $\Delta G = \Delta E + P\Delta V - T\Delta S$. The changes of internal energy ΔE can be obtained from first-principles calculations of each material at 0 K. The term $P\Delta V$ is generally neglected for a solid-state reaction (~10^{-5} eV) where changes in volume are usually small, whereas ΔE is on the order of 10^0 eV/fu. The term $T\Delta S$ can also be neglected as it is only on the order of thermal energy, which is approximately 10^{-2} eV at room temperature. Therefore, ΔG can be replaced by ΔE in Equation 6.10 with only a small error [59–61].

Sometimes, not only the average voltage but also the voltage profile of electrodes is of great interest. To obtain the precise voltage profile, predictions of the configurations of guest ions in the intermediate structure are needed. By comparing the first-principles energies of all the available configurations with the same composition, we can find the most stable configuration. The stability of a configuration can be evaluated by comparing its energy to the linear combination of the energies of other compositions using the convex hull construction [62].

6.2.2.2 Li/Na Ion Diffusion

Li/Na ion diffusion in electrode materials can be estimated by first-principles calculation. In transition state theory, the ion diffusivity (D) can be estimated as

$$D = a^2 v^* \exp\left(\frac{-E_{act}}{k_B T}\right) \qquad (6.11)$$

where a is the hopping distance, v^* is the attempt frequency, E_{act} is the activation barrier for hopping, k_B is the Boltzmann constant and T is the temperature [48,51,63,64]. In practical calculations, v^* is generally assumed to be ~10^{12} Hz, which is in the range of phonon frequencies, and a corresponds to lattice parameters [32,63]. The nudged elastic band (NEB) method is introduced to determine the activation barrier for Li/Na hopping between stable Li/Na sites along plausible diffusion pathways [65,66]. For these calculations, a Li/Na ion is allowed to diffuse in the supercells of materials, where the initial Li/Na hopping trajectory is obtained by linear interpolation between the initial and final states of the ion hopping [67]. All lattice parameters are fixed at the fully lithiated/delithiated or sodiated/desodiated state, but all the internal degrees of freedom are relaxed during NEB calculations. Alkali ions and electron/hole pairs are strongly bound in polyanionic materials within GGA+U approximation [13]. To exclude the interaction between Li/Na hopping and charge transfer, the GGA approximation for NEB calculations is often used instead of the GGA+U approximation [68].

The plausible Li diffusion pathways of electrode materials can be more easily estimated by the bond-valence (BV) method with less accuracy [69,70]. The BV sum of the structure can be calculated by summing up the BV of Li at each grid point to all oxygen ions within certain radius. The energy of Li at each point increases with the BV mismatch from the ideal Li value (+1); so, the Li ion may be found in the areas with low BV mismatch in the structure; the Li ion migrates through the low BV mismatch pathway [33,69,71]. The three-dimensional BV mismatch map can be visualised. Additionally, the feasibility of Li diffusion through a diffusion pathway can be predicted by the threshold values permitting connections between Li sites in the BV mismatch map. Li species in the diffusion pathway with lower threshold values might diffuse faster [33,71]. Since the distances between Li$^+$ and O^{2-} are only considered in the BV sum, the electrostatic repulsions by adjacent cations and the lattice relaxation induced by diffusive Li ions cannot be considered; thus, accurate activation barriers are not easily obtained with the BV method [33]. Therefore, NEB calculations are used to determine activation barriers of all the plausible diffusion pathways derived from the BV method.

6.3 Case Study on Polyanionic Cathode Materials with First Principles: Lithium Metal Borates

Lithium metal borates (LiMBO$_3$) are attractive cathode materials for Li-rechargeable batteries, as they have approximately 30% higher theoretical capacities (~220 mAh g^{-1}) than olivine LiMPO$_4$ (~170 mAh g^{-1}). However, despite their high theoretical capacities, the first reported experiments of LiMBO$_3$ (M = Mn, Fe and Co) showed very small capacities at even very low current densities (~9 mAh g^{-1} at C/250 for LiFeBO$_3$) at room temperature [27]. Abouimrane et al. [28] applied carbon nanopainting to LiFeBO$_3$ to improve electronic conductivity and decreased particle size, which resulted in large capacity 158 mAh g^{-1} at 80°C. Yamada et al. [30] also demonstrated that a near-theoretical capacity of LiFeBO$_3$ can be achieved under moderate current density at room temperature by nano-sizing and avoiding surface poisoning.

In this section, first-principles findings on the intrinsic structural, electronic and electrochemical properties of lithium metal borates are presented in a comparative study of three isotopic LiMBO$_3$ (M = Mn, Fe and Co) as a case study from Ref. 32. Furthermore, Li diffusion paths, Li diffusivity, and anti-site defects are discussed as the possible causes of the slow kinetics of lithium metal borates.

6.3.1 Crystal Structure and Volume Change with Delithiation

LiFeBO$_3$, LiMnBO$_3$ and LiCoPO$_3$ have an isotopic crystal structure with space group C2/c, as shown in Figure 6.1 [25–27]. Although LiMnBO$_3$ has a high-temperature hexagonal polymorph phase, the monoclinic polymorph of LiMnBO$_3$ is energetically more stable than the hexagonal phase; therefore, monoclinic LiMnBO$_3$ was adopted as the ground state structure [31]. The oxygen atoms are arranged in distorted hexagonal close-packed planes perpendicular to the c-axis in LiMBO$_3$. TM ions are slightly off-centre to the trigonal bipyramidal sites; their sites are split into M1 and M2 sites, as shown in Figure 6.1a, with occupancies of 0.5/0.5 for Mn, 0.72/0.28 for Fe and 0.56/0.44 for Co in LiMBO$_3$ [25–27]. Adjacent M1 and M2 sites cannot be simultaneously occupied by TM ions because they are very close to each other (~0.3 Å). Although Li ions seem to be found in a trigonal bipyramidal arrangement of oxygen atoms, Li coordination is actually tetrahedral because the ions are not centered in trigonal bipyramidal coordination, as shown in Figure 6.1b [25–27,33]. Owing to asymmetric coordination of the Li sites, they are also split into Li1 and Li2 sites with occupancies of approximately 0.5/0.5 [25–27]. Li ions will also alternately occupy only one of the adjacent Li1 and Li2 sites due to the short distance between them (~ 0.6 Å). MO$_5$ hexahedra are connected by sharing two opposite edges with neighbouring MO$_5$ [−101], as shown in Figure 6.1a. LiO$_5$ hexahedra also form chains by edge-sharing [001], as shown in Figure 6.1b. MO$_5$ and LiO$_5$ chains are interconnected by trigonal planar BO$_3$, as shown in Figure 6.1.

A more recent study of LiFeBO$_3$ by Janssen et al. [33] refined it in the four-dimensional superspace group C2/c($\alpha 0\gamma$)00, with $\alpha = 1/2$ and $\gamma = 0$. The Fe and Li sites are not split in their modulated superstructure, and Li ions are at long-range-ordered tetrahedral sites. This modulation of Li sites can slightly

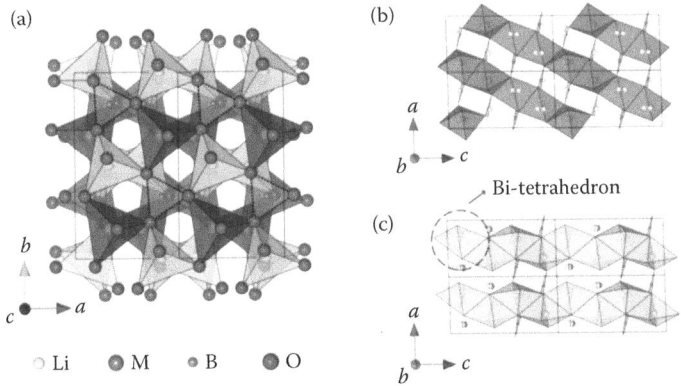

FIGURE 6.1
(See colour insert.) (a) Crystal structure of monoclinic LiMBO$_3$ (M = Mn, Fe and Co). (b) MO$_5$ and (c) LiO$_5$ chains are interconnected by trigonal planar BO$_3$. BO$_3$ is highlighted with bold dashed line. (Reprinted with permission from D.-H. Seo et al., First-principles study on lithium metal borate cathodes for lithium rechargeable batteries, *Phys. Rev. B* 83, 2011, 205127. Copyright 2011, American Physical Society.)

alter the computational results of lattice parameters, operating voltage and Li migration pathways, which will also be discussed later.

Experimentally reported crystal structure of LiMBO$_3$ (M = Mn, Fe and Co) with C2/c monoclinic supercells ($2a \times b \times c$) was used as a starting point of calculations [25–27]. After fully relaxation with first-principles calculations, the lattice parameters of LiMBO$_3$ were in agreement with the reported experimental values within 2%, as shown in Table 6.1 [25–27]. The modulation of LiO$_4$ along the c-direction in LiFeBO$_3$ barely affects the lattice parameters. Although the experimental lattice parameters of fully delithiated states of lithium metal borates are yet to be reported, they could be predicted by first-principles calculation [32]. The lattice parameters of MBO$_3$ are also summarised in Table 6.1. With delithiation, a lattice parameters increase, whereas b and c lattice parameters decrease for all LiMBO$_3$. The changes in the lattice parameters are negligible, except for c of LiMnBO$_3$, which was correlated with the change in the electronic configuration of the Mn ion at the trigonal bipyramidal site, as shown in Section 6.3.2. Nevertheless, the volume changes with delithiation derived from first-principles calculations are 2.0% for LiMnBO$_3$, 1.4% for LiFeBO$_3$ and 1.6% for LiCoBO$_3$ (Table 6.1), which are remarkably small changes compared to those of other commercialised cathodes, such as LiFePO$_4$ (6.5%), LiMnPO$_4$ (6.1%) and LiMn$_2$O$_4$ (6.4%) [30,57,72]. Such a small volume change indicates minimal distortion between lithiated and delithiated phases during cycling and contributes to the good cycle life of LiMBO$_3$. Indeed, the experimental report by Yamada et al. [30] showed excellent cyclability in Li$_x$FeBO$_3$ with only 2% volume change ($0.15 \leq x \leq 1$).

TABLE 6.1
Calculated and Experimental Lattice Parameters of Li_xMBO_3

	a (Å)	b (Å)	c (Å)	β (°)	ΔV (%)
$LiMnBO_3$	5.225	9.013	10.446	91.82	2.0
$MnBO_3$	5.249	8.971	10.135	91.11	
exp-$LiMnBO_3$ (Ref. 25)	5.19	8.95	10.37	91.80	
$LiFeBO_3$	5.230	9.024	10.236	91.36	1.4
Modulated $LiFeBO_3$	5.229	8.997	10.254	91.60	
$FeBO_3$	5.241	8.957	10.150	90.92	
exp-$LiFeBO_3$ (Ref. 27)	5.16	8.92	10.19	91.36	2.0 (Ref. 30)
$LiCoBO_3$	5.164	8.898	10.189	91.41	1.6
$CoBO_3$	5.199	8.852	10.005	90.24	
exp-$LiCoBO_3$ (Ref. 26)	5.13	8.84	10.10	91.36	

Source: Adapted with permission from D.-H. Seo et al., First-principles study on lithium metal borate cathodes for lithium rechargeable batteries, *Phys. Rev. B* 83, 2011, 205127. Copyright 2011, American Physical Society.

6.3.2 Electronic Structure and Redox Reaction with Delithiation

The electronic structure is crucial for understanding the redox reaction mechanism in electrode materials. The oxidation state of ions in Li_xMBO_3 (M = Mn, Fe and Co) at $x = 1$ and $x = 0$ could provide insights to the redox reaction mechanism during charge/discharge. The oxidation states of ions can be determined by integrating the electron spin in a sphere around the target ion [12,73]. Figure 6.2 shows the net spin moment integrated as a function of the distance from the M (Mn, Fe or Co), B and O ion cores. The calculated net spin moments of B^{3-} and O^{2-} ions in Li_xMBO_3 are very small, and the changes are negligible with delithiation, as they have no unpaired spins. As shown in Figure 6.2a, the integrated spin density of Mn ions steeply increases due to the unpaired spin of the d-states of the Mn ion, but then reaches a plateau value where the contribution from oxygen ions begins [12]. The plateau value of Mn ions in $LiMnBO_3$ is about +4.77, which agrees well with the number of unpaired electron spins of Mn^{2+}(+5) at a high-spin state. A slight underestimation of the moment of the Mn ion compared to the expected value is typically observed in computations on olivine and layered structures due to the transfer of the moments of oxygen ions [12,73]. The plateau value of Mn ions in $MnBO_3$ is about +3.98, which is also close to the unpaired electron spin counts of Mn^{3+}(+4) at a high-spin state. This result indicates that Mn ions in Li_xMnBO_3 are oxidised from 2+ to 3+ with delithiation. Similarly, the change in oxidation states of Fe and Co in Li_xMBO_3 (M = Fe and Co) with delithiation can be

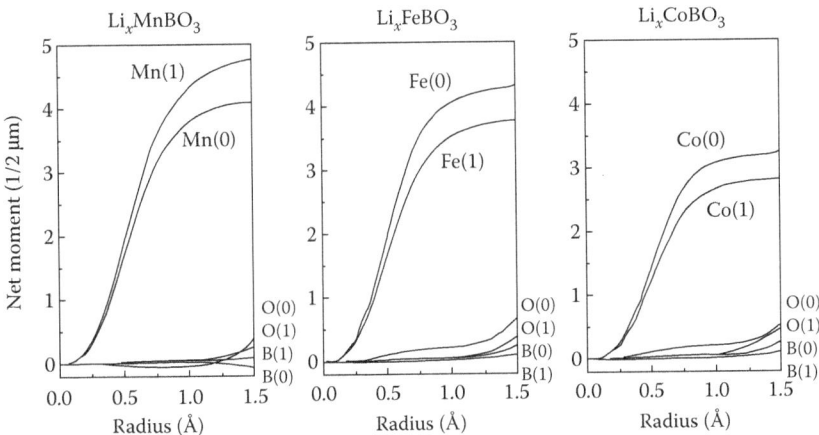

FIGURE 6.2
Integrated spin density as a function of distance from the M, B and O ion cores in Li_xMBO_3 (M = Mn, Fe and Co) at $x = 1$ and $x = 0$. The numbers (1 or 0) in parentheses after each atom refer to x.

estimated from the integrated spin density, as shown in Figure 6.2b and c. The net spin moments are about +3.79 and +2.85 for Fe and Co in $LiMBO_3$ (M = Fe and Co), respectively, which correspond to the unpaired electron spin counts of Fe^{2+} (+4) and Co^{2+} (+3) at high-spin states. With delithiation, net moments are observed to decrease to +4.32 and +3.23 for Fe and Co in MBO_3 (M = Fe and Co), respectively. These values are also close to the unpaired electron spin counts of Fe^{3+} (+5) and Co^{3+} (+4) at high-spin states, indicating that the oxidation states of Fe and Co in Li_xMBO_3 shift from 2+ to 3+ upon delithiation with relatively large redox contributions from oxygen.

Electron configurations of Li_xMBO_3 (M = Mn, Fe and Co) at $x = 1$ and $x = 0$ are also estimated from the spin states of TM ions. The 3d bands of TM ions in trigonal bipyramidal sites are split into the e'' (d_{xz}, d_{yz}), e' (d_{xy}, $d_{x^2-y^2}$), and a' (d_{z^2}) bands by crystal field theory [74], as illustrated in Figure 6.3a. Considering the band splitting and calculated net moments, the five up-spin electrons of the Mn ion in $LiMnBO_3$ occupy the e'', e' and a' bands at the high-spin state, as shown in Figure 6.3b. When a Li ion is removed from $LiMnBO_3$, the electron in the highest occupied state band, a', will be extracted simultaneously. For $LiFeBO_3$, five up-spin electrons occupy the e'', e' and a' bands, while one down-spin electron occupies the e'' band at the high-spin state, as shown in Figure 6.3b. The down-spin electron will be extracted from the e'' band of $LiFeBO_3$ during delithiation. Similarly, two down-spin electrons occupy the e'' band for $LiCoBO_3$, and one of these two electrons will be extracted upon delithiation, as shown in Figure 6.3b.

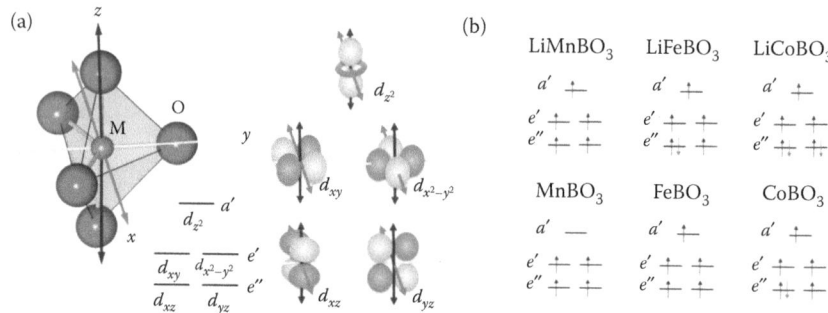

FIGURE 6.3
(a) MO$_5$ trigonal bipyramid and schematic energy levels of the 3d bands of the TM ion in trigonal bipyramidal coordination and (b) schematic energy levels and occupied electrons of Li$_x$MBO$_3$ (x = 1, 0, M = Mn, Fe and Co). The 3d bands are split into the e'' (d_{xz}, d_{yz}), e' (d_{xy}, $d_{x^2-y^2}$) and a' (d_{z^2}) bands by the electrostatic field generated by five oxygen ions. (Reprinted with permission from D.-H. Seo et al., First-principles study on lithium metal borate cathodes for lithium rechargeable batteries, *Phys. Rev. B* 83, 2011, 205127. Copyright 2011, American Physical Society.)

The changes in the electronic structures are correlated with the structural evolution of Li$_x$MBO$_3$ during delithiation. As the electron in the d_{z^2} orbital in the a' band is removed from LiMnBO$_3$ with delithiation, the electrostatic repulsion with the oxygen 2p orbital along the z-axis in Figure 6.3a will be greatly relieved. Therefore, Mn–O bonds along the z-axis are shortened approximately from 2.270 to 1.908 Å; these reductions are much larger than those of other Mn–O bonds (around 2.116–2.011 Å) along the x- or y-axes. Thus, the c lattice parameter, which is parallel to the z-axis of Li$_x$MnBO$_3$, significantly decreases with delithiation. On the contrary, the electron is extracted from the orbital of the e'' band (d_{xz}, d_{yz}) of LiFeBO$_3$ and LiCoBO$_3$ during delithiation, which does not overlap with oxygen ions, as shown in Figure 6.3. The Fe–O bonds along the z-axis contract from 2.244 to 2.071 Å, and Co–O bonds contract from 2.256 to 2.085 Å, which are much smaller contractions than those of Li$_x$MnBO$_3$. Therefore, there is no significant change in c lattice parameters of LiFeBO$_3$ and LiCoBO$_3$ during delithiation.

Understanding the electronic structure of Li$_x$MBO$_3$ (M = Mn, Fe and Co) is vital for designing electrode materials because electronic conductivity is an important aspect of high-power batteries. The band gaps of Li$_x$MBO$_3$ are determined from the total density of states (DOS) of Li$_x$MBO$_3$ (M = Mn, Fe and Co) at x = 1 and x = 0, as shown in Figure 6.4. We adopted U$_P$ (U values in phosphates) as reference U values for the calculation (Mn: 4.5 eV, Fe: 4.3 eV and Co: 5.7 eV). The calculated band gaps of Li$_x$MBO$_3$ increase with U in the GGA+U scheme; for example, the band gap of LiFeBO$_3$ is 3.190 eV with U$_P$, 2.970 eV with U$_P$–0.3 eV and 2.682 eV with U$_P$–0.6 eV [32]. Therefore, the quantitative determination of the band gaps of Li$_x$MBO$_3$ should be done carefully [19,55]. Nevertheless, the band gaps of borates (Li$_x$MBO$_3$) are

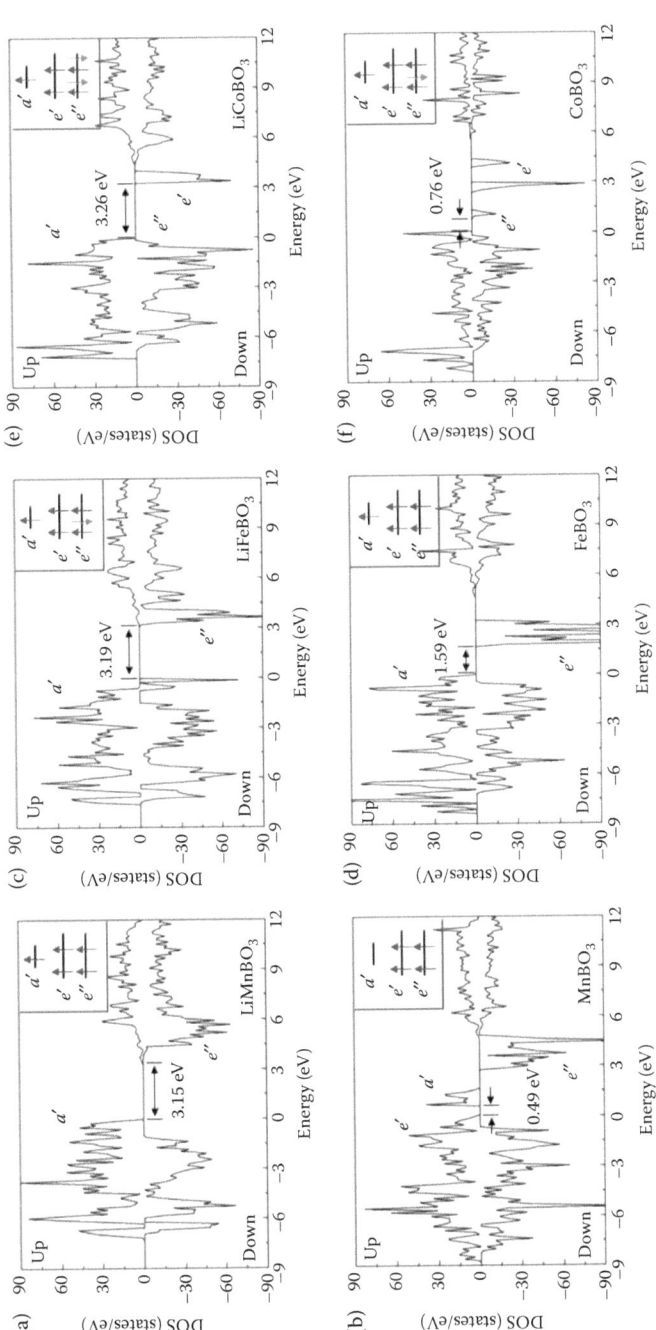

FIGURE 6.4
Density of state of Li_xMBO_3 ($x = 1, 0, M = Mn, Fe, Co$). The highest occupied bands, lowest unoccupied bands and band gaps are presented. The inset shows estimated electronic configurations. (a) $LiMnBO_3$, (b) $MnBO_3$, (c) $LiFeBO_3$, (d) $FeBO_3$, (e) $LiCoBO_3$, (f) $CoBO_3$. (Reprinted with permission from D.-H. Seo et al., First-principles study on lithium metal borate cathodes for lithium rechargeable batteries, *Phys. Rev. B* 83, 2011, 205127. Copyright 2011, American Physical Society.)

generally smaller than those of phosphates (Li$_x$MPO$_4$) calculated with the same scheme and U value [50]. The band gap is 3.7 eV for LiFePO$_4$ and 1.9 eV for delithiated FePO$_4$, which is approximately 0.5 eV higher than those of LiFeBO$_3$. However, the band gaps of Li$_x$MBO$_3$ are still large enough to prevent thermal excitation of an electron/hole pair over the band gap at room temperature; thus, the carrier concentration of Li$_x$MBO$_3$ will be determined by its Li ion content. Therefore, band gaps are not expected to significantly contribute to the electronic conductivity of borates. Instead, the small polaron localised around a TM ion is likely to determine the electronic conductivity of borates. This is similar to other insulating intercalation materials, such as olivine phosphates [50].

6.3.3 Lithium Intercalation Voltages of Lithium Metal Borates

A cathode material with high operating voltage is generally desirable for high-energy density. The average voltages of the intercalation materials can be determined by the free energies of the reactants and products, as explained in Section 6.2.2.1. The overall intercalation reaction of Li$_x$MBO$_3$ can be written as

$$\text{LiMBO}_3 \leftrightarrow \text{MBO}_3 + \text{Li}^+ + e^- \qquad (6.12)$$

For this reaction, the average voltages can be determined simply by using the following equation:

$$<V> = \frac{-[E(\text{LiMBO}_3) - E(\text{MBO}_3) - E(\text{Li}^+)]}{F} \qquad (6.13)$$

where E is the DFT energy of the fully relaxed ground state structure of the reactants or products, and F is the Faraday constant. As GGA+U results greatly depend on the U parameter [57], the average voltages of Li$_x$MBO$_3$ (M = Mn, Fe and Co) change with various U values from 0 to 7 eV, as shown in Figure 6.5. The average voltages of Li$_x$MBO$_3$ tend to increase with U value. The calculated average voltages of Li$_x$FeBO$_3$ are in agreement with experiments (3.0 V) near $U_P \sim U_P$-0.6 eV, which was our choice of U range. However, average voltages of Li$_x$FeBO$_3$ calculated with GGA (U = 0) are quite underestimated compared to experimental results, as shown in Table 6.2. Janssen et al. demonstrated that the modulation of LiO$_4$ along the c-direction stabilises LiFeBO$_3$ by 12 meV/fu, and the modulation disappears after delithiation, reflecting that the effect of structural modulation on the average voltage of LiFeBO$_3$ is negligible. When compared with different TM ions in LiMBO$_3$, a systematic shift of average voltages is observed. The average voltages were higher by about 0.6 and 1.1 V for LiMnBO$_3$ and LiCoBO$_3$, respectively, than that of LiFeBO$_3$, as shown in Figure 6.5. The redox potentials

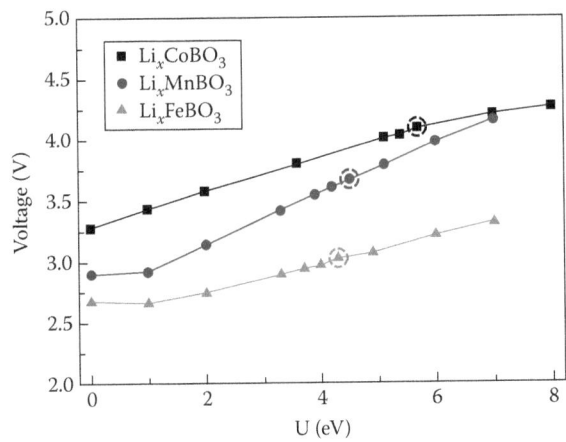

FIGURE 6.5
Calculated average voltage of Li$_x$MBO$_3$ (M = Mn, Fe, Co) versus Li as a function of various U values. Dashed circles indicate the average voltages calculated with U$_P$ values. (Adapted with permission from D.-H. Seo et al., First-principles study on lithium metal borate cathodes for lithium rechargeable batteries, *Phys. Rev. B* 83, 2011, 205127. Copyright 2011, American Physical Society.)

of Mn^{2+}/Mn^{3+} and Co^{2+}/Co^{3+} are generally higher than that of Fe^{2+}/Fe^{3+} in polyanionic cathode materials; hence, systematic shifts of average voltages in LiMBO$_3$ for different TM ions are also consistent with other polyanionic cathode materials [19]. The calculated average voltages of borates (LiMBO$_3$) are approximately 0.4 V lower than those of phosphates (LiMPO$_4$) using the same computational methods [49,57]. The different redox potentials of M^{2+}/M^{3+} with polyanions can be explained by the inductive effect. The less electronegative boron atom (2.04 in Pauling units) attracts fewer electrons from the oxygen atom through M–O–X bonds (M = Mn, Fe, Co, X = B, P) than does the phosphorous atom (2.19 in Pauling units), which leads to a strong M–O covalent nature. The strong covalency of the M–O bond generally lowers the redox potentials of M^{2+}/M^{3+} [8,75,76]; thus, the average voltages of LiMBO$_3$ are lower than those of LiMPO$_4$. The relatively low operating voltages of

TABLE 6.2
Calculated Average Voltage of Li$_x$MBO$_3$ (M = Mn, Fe and Co) with Different U Values and Experimental Average Voltage of LiFeBO$_3$

	LiFeBO$_3$ (V)	LiMnBO$_3$ (V)	LiCoBO$_3$ (V)
U$_p$	3.02	3.67	4.09
U$_p$-0.3 eV	2.97	3.61	4.04
U$_p$-0.6 eV	2.94	3.55	4.01
GGA (U = 0 eV)	2.67	2.90	3.28
exp.	~3.0 (Ref. 30)		

TABLE 6.3

Average Voltage, Specific Capacity and Specific Energy Density of $LiMBO_3$ and $LiMPO_4$ (M = Mn, Fe and Co)

	$LiFeBO_3$	$LiMnBO_3$	$LiCoBO_3$
Voltage (V)	3.02	3.67	4.09
Capacity (mAh g^{-1})	220	222	215
Energy density (Wh kg^{-1})	664	815	879
	$LiFePO_4$	$LiMnPO_4$	$LiCoPO_4$
Voltage (V)	3.5 (Ref. 8)	4.1 (Ref. 75)	4.8 (Ref. 75)
Capacity (mAh g^{-1})	170	171	167
Energy density (Wh kg^{-1})	595	701	802

$LiFeBO_3$ are disadvantageous for practical purposes in terms of the energy density. However, the theoretical capacities of $LiMBO_3$ are higher by approximately 50 mAh g^{-1} than those of the corresponding $LiMPO_4$, which in turn compensates for the lower voltage, allowing the theoretical energy densities of lithium metal borates (660–880 Wh kg^{-1}) to be 10% higher than those of olivine phosphates (595–802 Wh kg^{-1}) [8,77,78], as shown in Table 6.3. Additionally, $LiMnBO_3$ and $LiCoBO_3$ exhibit attractive operating voltages as cathodes for the conventional Li cell.

6.3.4 Lithium Ion Diffusion

Fast Li ion diffusion within the crystal structure is essential for an electrode material in a high-power Li-rechargeable battery. Lithium ion mobility in Li_xMBO_3 (M = Mn, Fe and Co) is estimated by calculating the activation barrier for Li ion hopping. The plausible diffusion pathways in Li_xMBO_3 can be visualised by the BV model, as described in Section 6.2.2.2. Figure 6.6a and b shows the BV mismatch maps of the $LiFeBO_3$ crystal structure reported by Legagneur et al. [27]. The most probable Li diffusion pathway is within the LiO_4 chains along the c-direction in $LiFeBO_3$, which are depicted as yellow isosurfaces of the low BV mismatch for Li in Figure 6.6. There are two different transition states (A and B) for Li migration along the c-direction, which come from two symmetrically different types of edge sharing between LiO_5 bipyramids in a [001] chain in Figure 6.7a. These types repeatedly appear in an ABAB sequence along these zigzag-like diffusion pathways, as shown in Figures 6.6a and b, and 6.7a. The space of the low BV mismatch pathway for transition state A is much wider than that for transition state B at isovalue ±0.1 valence unit (v.u.), as shown in Figure 6.6a and b. Additionally, the threshold value of transition state B (0.011 v.u.) is slightly lower than that of transition state A (0.084 v.u.), indicating that Li hopping is easier through transition state B than through transition state A. The diffusion pathway

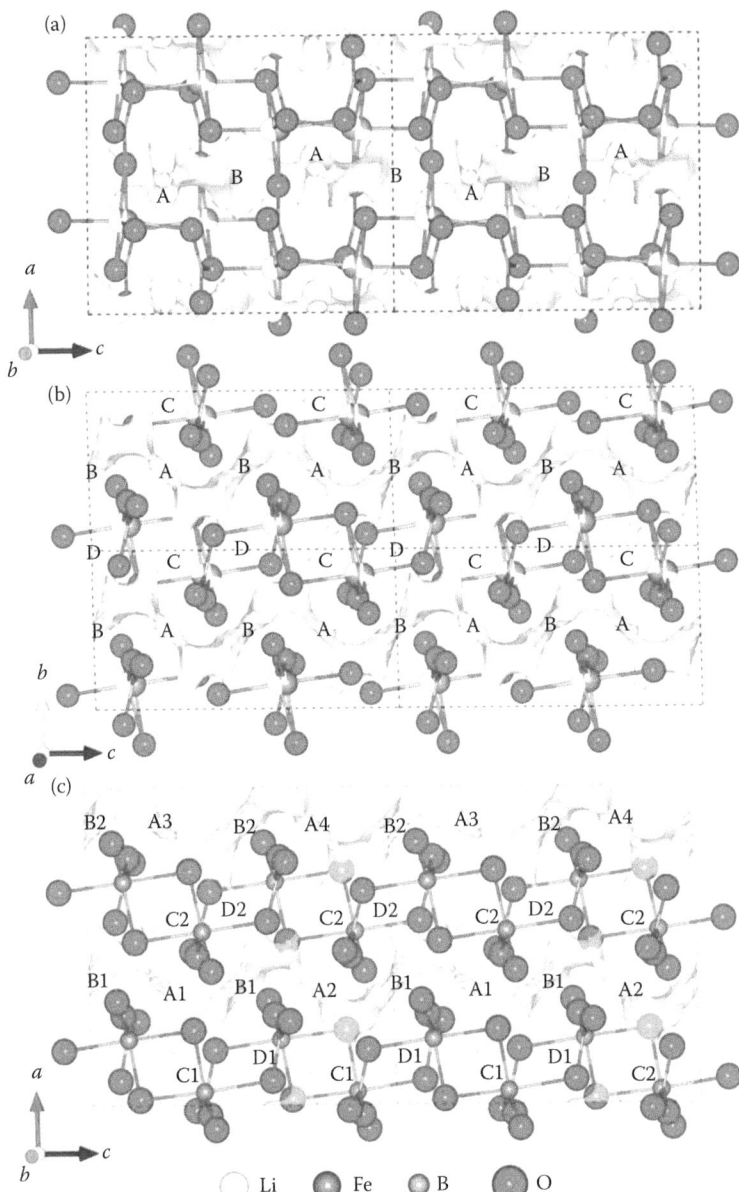

FIGURE 6.6
(**See colour insert.**) The BV mismatch map for Li in (a) and (b) LiFeBO$_3$ crystal structure reported by Legagneur et al. (isovalue ±0.10 valence unit) and (c) modulated LiFeBO$_3$ crystal structure reported by Janssen et al. (isovalue ±0.12 valence unit). Yellow isosurfaces of the BV mismatch for Li represent the possible diffusion pathways. There are four plausible transition states (A, B, C and D) for Li migration in LiFeBO$_3$. Thick dashed black lines denote the unit cells of LiFeBO$_3$. The BV mismatch map was calculated by the 3DBVSMAPPER programme and drawing.

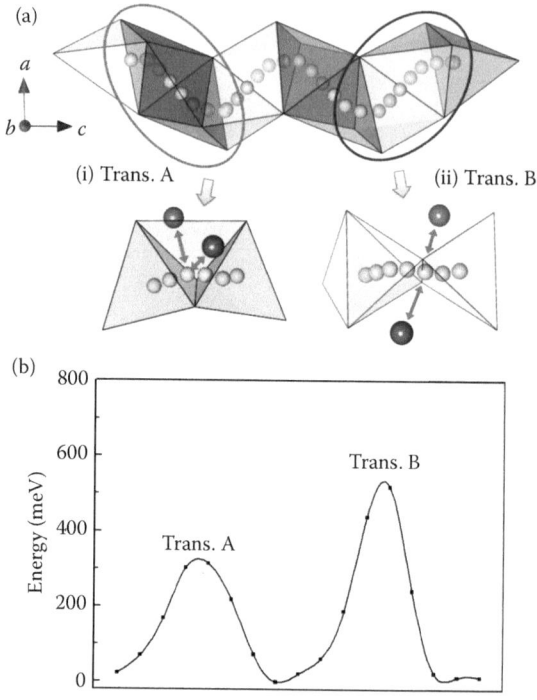

FIGURE 6.7
(See colour insert.) Trajectories and activation barriers of the Li hopping in the diffusion path along the c-direction of LiFeBO$_3$. (a) (i) The transition state of A type (red) and (ii) the transition state of B type (yellow), and (b) their activation energies for Li diffusion along the c-direction of LiFeBO$_3$. Yellow circles indicate Li ions and gold circles indicate TM ions and the points of tetrahedrons are oxygen ions. Red arrows indicate the electrostatic repulsion between Fe ions and Li ions. (Reprinted with permission from D.-H. Seo et al., First-principles study on lithium metal borate cathodes for lithium rechargeable batteries, *Phys. Rev. B* 83, 2011, 205127. Copyright 2011, American Physical Society.)

through the third transition state C along the (011) direction is also possible in LiFeBO$_3$. However, an expected high activation barrier for transition state C from the very narrow space of the low BV mismatch pathway and high threshold value (over 1 v.u.) would prevent appreciable diffusion through the path.

To estimate activation barriers for Li migration along plausible diffusion pathways, NEB calculations were performed. Figure 6.7b shows Li trajectories along the c-direction in Li$_x$FeBO$_3$ ($x \approx 1$) and their activation barriers. The activation barrier is 315 meV for transition state A and 519 meV for transition state B in Li$_x$FeBO$_3$. When the Li ion migrates through transition state A, Li hops from its original tetrahedral site to the other through the face-shared tetrahedral site, as shown in the red tetrahedron in Figure 6.7a(i). The Li ions pass through the intermediate tetrahedral site slightly off-centre due to electrostatic repulsion by neighbouring Fe ions at transition state A. In the case of transition state B, the BV model of LiFeBO$_3$ suggested that Li ions migrate through two bisymmetric LiO$_4$ tetrahedra, as shown in Figure 6.6a.

However, the NEB calculation shows that Li migration through the edge-shared tetrahedral sites is energetically more favourable than Li migration through one of the bisymmetric LiO$_4$ tetrahedra, as shown by the yellow tetrahedron in Figure 6.7a(ii). This result is attributed to electrostatic repulsion from neighbouring Fe ions at the intermediate tetrahedral site, which was neglected in the BV model. The distance between the centre of the intermediate tetrahedral site and the nearest Fe ion is only approximately 2.06 Å (~2.46 Å for transition state A). To minimise the repulsion from Fe ions, a Li ion passes across the edge of LiO$_4$, while pushing out two oxygen ions of the edge (from 3.28 to 3.65 Å at the transition state), as shown in Figure 6.7a(ii). This motion results in the notably higher activation barrier of transition state B than that of transition state A.

The discrepancy in results between NEB and BV calculations is attributed to correlations between Li and adjacent ions during migration that are neglected in the BV calculation [33]. The electrostatic repulsions between Li and adjacent Fe and B cations and relaxations of the sublattice around the Li ion can be considered in NEB calculations. Therefore, first-principles based on the NEB method are more accurate to calculate the activation barrier for ion hopping in the crystal structure than the BVS method, while the BVS method is an easy and fast computation method to predict possible diffusion pathways.

As expected from the BV mismatch pathway, diffusion through transition state C showed a significantly higher activation barrier of approximately 1580 meV. The transition state C includes a distorted octahedral site that face shares with two FeO$_5$; thus, it is energetically unstable due to the strong electrostatic repulsion from nearby cations (Fe and B), as shown in Figure 6.8. We also calculated the activation barrier for transition state D, which is not connected by isosurfaces of low BV mismatch, as shown in Figure 6.6b. The activation barrier for transition state D (1347 meV) is still too high to overcome at room temperature, implying Li ions are not likely to crossover between LiO$_4$ chains. Hence, LiFeBO$_3$ is expected to be a one-dimensional (1-D) lithium diffuser similar to olivine electrode materials.

The effect of Li modulation on Li diffusion in Li$_x$FeBO$_3$ is discussed in the following paragraph. Janssen et al. claimed that the diffusion pathway in LiO$_4$ chains along the c-direction splits into two different pathways with Li modulation. The activation barriers for Li diffusion will be higher in the modulated structure of LiFeBO$_3$ than in its unmodulated analogue based on its BV sum analyses [33]. Figure 6.6c shows BV mismatch maps of modulated LiFeBO$_3$, in which transition states A and B are split into A1, A2, A3 and A4, and B1 and B2, respectively. The threshold values of the A1, A2, A3, A4, B1 and B2 transition states in modulated LiFeBO$_3$ are 0.122, 0.116, 0.126, 0.113, 0.054 and 0.036 v.u., respectively, which are slightly higher than those of the A (0.084 v.u.) and B (0.011 v.u.) transition states in unmodulated LiFeBO$_3$,

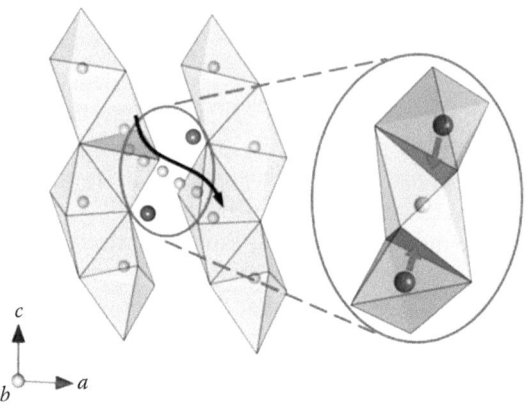

FIGURE 6.8
The trajectory of Li hopping across LiO_5 channels. The C transition state of Li migration is highlighted in the red circle. Yellow circles indicate Li ions and gold circles indicate Fe ions. Red arrows indicate the electrostatic repulsion between Fe ions and Li ions. (Reprinted with permission from D.-H. Seo et al., First-principles study on lithium metal borate cathodes for lithium rechargeable batteries, *Phys. Rev. B* 83, 2011, 205127. Copyright 2011, American Physical Society.)

as shown in Table 6.4. The activation barriers for Li migration through these transition states were calculated in modulated $LiFeBO_3$ with the NEB method. The activation barriers were 361, 321, 352 and 318 meV for the A1, A2, A3 and A4 transition states, respectively, which are slightly higher than 315 meV, the transition state A in unmodulated $LiFeBO_3$. The activation barriers are 429 meV for transition state B1 and 413 meV for transition state B2, which are lower than that of transition state B (519 meV) in unmodulated $LiFeBO_3$. The shift of O_3 positions in the Li-modulated structure [33] appears

TABLE 6.4

Activation Barriers and Threshold Values of Various Transition States

Transition State	E_a (meV)	Threshold Value (v.u.)	Transition State	E_a (meV)	Threshold Value (v.u.)
A	315	0.084	B	519	0.011
A1	361	0.122	B1	429	0.054
A2	352	0.126	B2	426	0.035
A3	321	0.116			
A4	318	0.113			
C	1580	>1	D	1347	>1
C1	1798	0.118	D1	1318	>1
C2	1720	0.009	D2	1241	>1

Note: A, B, C and D transition states are for the unmodulated $LiFeBO_3$ and A1, A2, A3, A4, B1, B2, C1, C2, D1 and D2 transition states are for the modulated $LiFeBO_3$, as depicted in Figure 6.6.

to increase the activation barriers of the A and B transition states. Activation barriers of type B are higher than those of type A; so, Li hopping through the B transition state will be a rate-limiting step for Li diffusion along the [001] direction in Li$_x$FeBO$_3$. Therefore, if the effect of Li modulation on Li diffusion is considered, the Li diffusion along a Li channel would be faster than is expected in unmodulated LiFeBO$_3$. The transition states C and D are also split into C1, C2, D1 and D2, and the activation barriers change from 1580 to 1798 and 1720 meV for transition states C1 and C2, and from 1347 to 1319 and 1241 meV for transition states D1 and D2, which are still too high to be overcome at room temperature and hopping across Li channels remains difficult. Li modulation does not open a new diffusion pathway but slightly alters activation barriers from those of the unmodulated model.

The Li diffusions in LiMnBO$_3$ and LiCoBO$_3$ could be investigated in a similar manner. Activation barriers are approximately 355 meV for type A and approximately 491 meV for type B in Li$_x$MnBO$_3$; they are approximately 366 meV for type A and approximately 479 meV for type B in Li$_x$CoBO$_3$. The activation barriers do not vary much among different TM borates for each type of diffusion in Li$_x$MBO$_3$ (M = Mn, Fe and Co). Activation barriers of type B in Li$_x$MBO$_3$ are approximately 150 meV higher than those of type A; so, hopping through type B will generally be the rate-limiting step for Li diffusion along the [001] direction in Li$_x$MBO$_3$ (M = Mn, Fe and Co).

Activation barriers of Li diffusion in Li$_x$MBO$_3$ (M = Mn, Fe and Co) are higher than those of conventional cathode materials such as spinel, olivine and layered cathode materials calculated within a similar scheme. The reported Li activation barriers are ~300 meV in olivine cathodes [63], ~250 meV in LiCoO$_2$ [48,51] and ~350 meV in the LiNi$_{0.5}$Mn$_{1.5}$O$_4$ spinel structure [79]. These values are approximately 100 meV lower than the observed value in LiMBO$_3$. Using Equation 6.10, the diffusivity of Li$_x$MBO$_3$ can be approximated to be on the order of 10^{-12} (cm^2 s^{-1}) at room temperature. Assuming the same prefactor of diffusivity, and the average activation barriers of Li$_x$MBO$_3$ are 500 meV, Li diffusivity in borates at room temperature is expected to be smaller by three orders of magnitude than those of conventional cathodes such as olivine phosphates and spinel oxides. However, the diffusivity is high enough to allow Li to diffuse over micrometer distances in an hour at room temperature.

Nevertheless, channel blocking of 1-D Li diffusion pathways by the antisite defect, which is the exchange of sites between Li and TM ions, can easily cause poor kinetics in LiMBO$_3$. When the 1-D Li channel is blocked by TM ions, Li must take a detour through other pathways with high activation barriers [63]. Malik et al. [80] demonstrated that even a 1% anti-site defect in bulk LiFePO$_4$ can significantly increase the number of blocked Li sites and impede fast Li diffusion along 1-D Li channels. First-principles calculations showed that the anti-site energies are 747 meV for LiMnBO$_3$ and 549 meV for LiFeBO$_3$, which are comparable to those of LiMnPO$_4$ (758–802 meV) [81] and LiFePO$_4$ (515–550 meV) [80]. The relatively low anti-site energies of LiMnBO$_3$

and LiFeBO$_3$ are attributed to similar coordination and size between Li and TM ions. Both Li and TM ions have trigonal bipyramidal coordination in LiMBO$_3$, as mentioned above, and the ionic radii of Li$^+$, Mn^{2+} and Fe^{2+} are 0.59, 0.66 and 0.63 Å, respectively, in four coordinations [82]. Anti-site defects are experimentally observed to be approximately 2% in LiMnPO$_4$ [81] and up to 7% in LiFePO$_4$ [83,84]; so, a similar number of anti-site defects are expected for LiMnBO$_3$ and LiFeBO$_3$ depending on the synthetic conditions, inducing a possible kinetics problem. Therefore, lithium metal borate compounds should be prepared carefully to prevent the formation of anti-site defects at the synthesis level. Nano-sizing is also very effective for improving the kinetics of borates because the short channels of nanoparticles contain fewer defects to block the Li diffusion path than the channels in large particles [80].

In this section, it has been shown that first-principles studies can provide detailed information on the structural, electronic and electrochemical properties of electrode materials, from the case study of LiMBO$_3$ (M = Mn, Fe and Co) [32]. On the basis of understanding their intrinsic properties, LiMBO$_3$ was predicted to have reasonably high-energy densities and exhibit small volume changes during battery cycling, which imply enhanced sustainability as electrodes in long-term battery cycling. The investigation of the electronic structure identified that a small polaron is likely to be the main conductor of Li$_x$MBO$_3$, similar to olivine materials. Furthermore, Li diffusion studies with bond-valence sums (BVS) and the NEB method showed zigzag 1-D Li diffusion pathways in Li$_x$MBO$_3$ with reasonably low activation barriers for Li motion. The relatively low anti-site energy for Li-M site exchange, however, indicated that the metal ions in the Li site can block 1-D Li diffusion paths.

6.4 Application of First-Principle Studies to Search for New Cathode Materials

In this section, we further demonstrate how the predictive capabilities of first-principles calculations can lead to the discovery of new cathode materials from the case study of new iron-containing cathode for Na-rechargeable batteries [56]. The search for alternatives to replace the current layered oxides has generated considerable research activities. Materials that include the naturally abundant iron redox couple and a strong covalent bond that prevents oxygen evolution are desirable options. A group of composition minerals Na$_4$M$_3$(PO$_4$)$_2$(P$_2$O$_7$) (M = Mn, Co, Ni and Mg) [85,86] was considered as a new possible candidate. However, the iron analogue of this phase was not reported before. While experimental research on unknown materials can be difficult, first-principles calculations could relatively easily predict the key properties such as voltage, volume change with charge/discharge, and diffusivity of alkali ions for a hypothetical material. Therefore, the possibility of the iron phase of this group of minerals as a cathode material for rechargeable batteries was examined with first-principles calculations. For the calculations, the experimentally reported crystal structure of $Na_4Co_3(PO_4)_2(P_2O_7)$, space group $Pn21a$, was employed by substituting cobalt with iron [85]. After full relaxation with first-principles calculations, the lattice parameters of $Na_4Fe_3(PO_4)_2(P_2O_7)$ were found to be $a = 18.328$ Å, $b = 6.591$ Å and $c = 10.719$ Å [56].

Figure 6.9 shows a schematic representation of the $Na_4Fe_3(PO_4)_2(P_2O_7)$ crystal structure. This structure has three distinguishable FeO_6 octahedra that compose layers parallel to the bc plane. A diphosphate group connects these layers along the a-axis and provides large tunnels between the layers. Inside the tunnels, two symmetrically distinguishable Na sites are present. There are two additional symmetrically distinguishable Na sites formed by the seven coordinated NaO_7 polyhedra.

As mentioned in Section 6.2.2.1, the intercalation voltage of an alkali ion can be calculated from the energy difference between initial and final states of a material. In most cases, the prediction of average voltage can be accomplished by simply calculating the energy difference between a fully occupied phase such as $LiMBO_3$ and an empty phase, such as MBO_3. However, in the case of $Na_{4-x}Fe_3(PO_4)_2(P_2O_7)$, the calculation is somewhat complicated, since the compound has four different sodium sites per formula unit, but

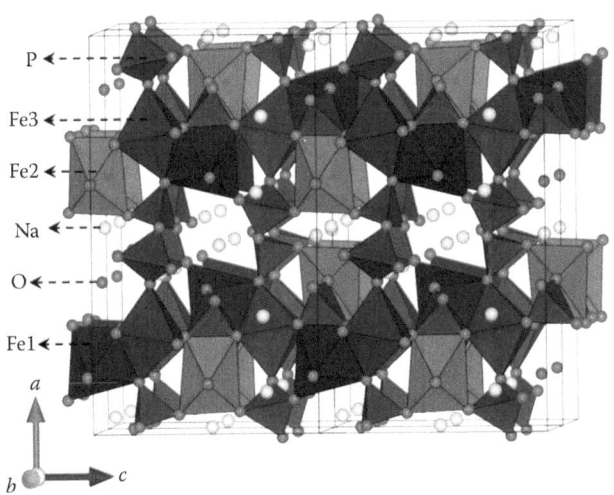

FIGURE 6.9
(See colour insert.) Schematic representation of the $Na_4Fe_3(PO_4)_2(P_2O_7)$ crystal structure. (Reprinted with permission from H. Kim et al., New iron-based mixed-polyanion cathodes for lithium and sodium rechargeable batteries: Combined first principles calculations and experimental study, *J. Am. Chem. Soc.* 134, 2012, 10369–10372. Copyright 2012, American Chemical Society.)

only three sodium atoms can exit the structure during charging because the material employs only three Fe^{2+}/Fe^{3+} redox couples. Therefore, the most stable configuration of intermediate states such as $x = 1$, 2 or 3 should be identified, to determine the precise voltage profile and phase reaction. To obtain the most stable configurations of $Na_xFe_3(PO_4)_2(P_2O_7)$ ($x = 1 \sim 4$) for compositions $x = 1$, 2 and 3, and ground state structures with corresponding Na/vacancy orderings, numerous Na/vacancy configurations need to be considered.

The inset in Figure 6.10 shows the formation energies of various Na/vacancy configurations as a function of sodium content. The square markers represent the energies of possible Na/vacancy configurations. The dashed line, which connects the most stable configurations, is called a convex hull [62,87]. The convex hull depicts the lowest energy trajectory as a function of sodium content. The thermodynamic instability of any phase could be qualified by evaluating the energy above the hull; its magnitude indicates the degree of instability of the compound [62,87]. The intercalation voltage can also be obtained from the calculated energies of the most stable configurations. The average voltage was calculated to be 3.32 V, in agreement with experimental data (solid line), with three average voltage steps, 2.97, 3.26 and 3.83 V (see the dashed line in Figure 6.10). The lattice parameters at $x = 1$, i.e. $NaFe_3(PO_4)_2(P_2O_7)$, were determined to be $a = 17.734$ Å, $b = 6.415$ Å and

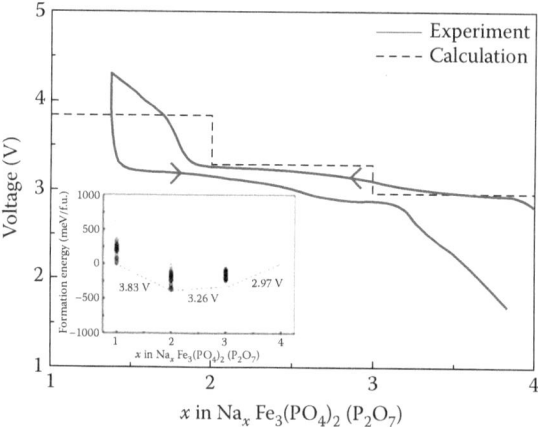

FIGURE 6.10
The first principles calculated average voltage at each region and galvanostatic charge/discharge profile of $Na_4Fe_3(PO_4)_2(P_2O_7)$ under a C/40 rate (inset). Calculated convex hull for $Na_xFe_3(PO_4)_2(P_2O_7)$. Each black square represents on calculated Na^+ ordering. Voltages were calculated from the stable configurations from the convex hull. (Reprinted with permission from H. Kim et al., New iron-based mixed-polyanion cathodes for lithium and sodium rechargeable batteries: Combined first principles calculations and experimental study, *J. Am. Chem. Soc.* 134, 2012, 10369–10372. Copyright 2012, American Chemical Society.)

First-Principles Approach for Cathode Design and Characterisation 247

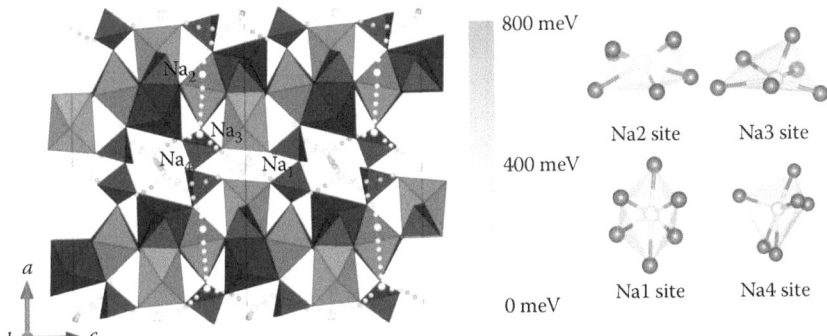

FIGURE 6.11
(See colour insert.) The 3D sodium diffusion paths in the $Na_4Fe_3(PO_4)_2(P_2O_7)$ structure and its activation barriers. (Reprinted with permission from H. Kim et al., New iron-based mixed-polyanion cathodes for lithium and sodium rechargeable batteries: Combined first principles calculations and experimental study, *J. Am. Chem. Soc.* 134, 2012, 10369–10372. Copyright 2012, American Chemical Society.)

$c = 10.668$ Å. From the lattice parameter for $x = 1$ and 4, the volume change is <6%, which is considerably lower than that of other rechargeable battery cathode materials.

The capability of Na intercalation was further confirmed from the activation barriers for sodium migration along the channels from several NEB calculations (see Section 6.2.2.2). Figure 6.11 shows sodium trajectories in $Na_xFe_3(PO_4)_2(P_2O_7)$ ($x \approx 4$) and its activation barriers by colour gradient from yellow to green. As shown in Figure 6.11 and Table 6.5, all sodium sites are connected with reasonably low activation barriers (<800 meV), while a large

TABLE 6.5
Sodium Vacancy Activation Barriers in the $Na_4Fe_3(PO_4)_2(P_2O_7)$ Structure

From (Site)	To (Site)	Activation Energy (meV)
Na1	Na1	256
Na1	Na4	599
Na2	Na4	481
Na3	Na1	540
Na2	Na3	344
Na3	Na4	685
Na4	Na4	544

Source: Reprinted with permission from H. Kim et al., New iron-based mixed-polyanion cathodes for lithium and sodium rechargeable batteries: Combined first principles calculations and experimental study, *J. Am. Chem. Soc.* 134, 2012, 10369–10372. Copyright 2012, American Chemical Society.

TABLE 6.6

Activation Barriers for Li Diffusion in the Cathode Materials

	Activation Energy (meV)	Reference
$LiFePO_4$	~270	[63]
$LiCoO_2$	~250	[48,51]
$LiNi_{0.5}Mn_{1.5}O_4$	~350	[79]
$LiFeBO_3$	519	Section 6.3.4, Figure 6.7

tunnel along the b-axis (connecting Na1 sites) shows an exceptionally low activation barrier (<300 meV). These activation barriers are comparable to those of lithium ions in conventional cathode materials (Table 6.6). These low activation barriers imply that sodium can easily diffuse through all directions.

In light of the promising prediction from first-principles study [56], $Na_4Fe_3(PO_4)_2(P_2O_7)$ was experimentally synthesised and shown that reversible Na intercalation was possible at predicted voltages.

In this section, it was shown that first-principles studies could provide the basic electrochemical properties of cathode material candidates. With simple calculations, intrinsic electrochemical properties of new electrode materials such as average voltage, lattice and volume change during charging, and activation barriers for cation diffusion, can be predicted without experiments. This approach, prediction from first-principles and subsequent experimental development, can be helpful in searching for new cathode materials and as guidelines for further experimental research.

6.5 Conclusions

In this chapter, we presented the first-principles approach to the characterisation and design of polyanionic cathode materials. The structural evolution, electronic properties, voltages and ion diffusivities of polyanion materials could be calculated at the atomic/electronic level. On the basis of an in-depth understanding of the intrinsic properties, electrode properties such as cyclability, energy density and power capability could be predicted. Electrode properties of new cathode materials with hypothetical crystal structures could also be predicted using first-principles calculations.

The structures and electrochemical behaviours of new hypothetical polyanionic materials can be complex; thus, development and optimisation of new electrode materials with experiments alone can be difficult. To explore the less-understood materials, first-principles-based computations can be a powerful tool for predicting the key properties prior to synthesis. Calculated voltages and activation energies for Li diffusion are a good starting point for finding promising electrodes for Li-rechargeable batteries. Moreover, first-principles-based multi-scale computation methods in combination with empirical methods, such as the cluster expansion method and Monte Carlo simulations, allow us to understand the thermodynamic properties and phase diagrams of electrode materials. This strategy can be applied to the cathodes and anodes of various types of batteries, such as sodium- and magnesium-rechargeable batteries. We believe that first-principles calculations present opportunities for discovering novel electrode materials for energy storage.

References

1. M. Armand and J.M. Tarascon, Building better batteries, *Nature* 451, 2008, 652–657.
2. C. Delacourt, P. Poizot, J.-M. Tarascon and C. Masquelier, The existence of a temperature-driven solid solution in Li_xFePO_4 for $0 \leq x \leq 1$, *Nat. Mater.* 4, 2005, 254–260.
3. K. Kang, Y.S. Meng, J. Breger, C.P. Grey and G. Ceder, Electrodes with high power and high capacity for rechargeable lithium batteries, *Science* 311, 2006, 977–980.
4. B.L. Ellis, W.R.M. Makahnouk, Y. Makimura, K. Toghill and L.F. Nazar, A multifunctional 3.5 V iron-based phosphate cathode for rechargeable batteries, *Nat. Mater.* 6, 2007, 749–753.
5. P. Barpanda, M. Ati, B.C. Melot, G. Rousse, J.N. Chotard, M.L. Doublet, M.T. Sougrati, S.A. Corr, J.C. Jumas and J.M. Tarascon, A 3.90 V iron-based fluorosulphate material for lithium-ion batteries crystallizing in the triplite structure, *Nat. Mater.* 10, 2011, 772–776.
6. Y.-U. Park, D.-H. Seo, B. Kim, K.-P. Hong, H. Kim, S. Lee, R.A. Shakoor, K. Miyasaka, J.-M. Tarascon and K. Kang, Tailoring a fluorophosphate as a novel 4 V cathode for lithium-ion batteries, *Sci. Rep.* 2, 704, 2012.
7. N. Recham, J.N. Chotard, L. Dupont, C. Delacourt, W. Walker, M. Armand and J.M. Tarascon, A 3.6 V lithium-based fluorosulphate insertion positive electrode for lithium-ion batteries, *Nat. Mater.* 9, 2010, 68–74.
8. A.K. Padhi, K.S. Nanjundaswamy and J.B. Goodenough, Phospho-olivines as positive-electrode materials for rechargeable lithium batteries, *J. Electrochem. Soc.* 144, 1997, 1188–1194.
9. A. Yamada, S.C. Chung and K. Hinokuma, Optimized $LiFePO_4$ for lithium battery cathodes, *J. Electrochem. Soc.* 148, 2001, A224–A226.
10. A. Yamada, Y. Kudo and K.Y. Liu, Phase diagram of $Li_1-x(Mn_yFe_{1-y})PO_4$ ($0 \leq x$, $y \leq 1$), *J. Electrochem. Soc.* 148, 2001, A1153–A1158.
11. S.Y. Chung, J.T. Bloking and Y.M. Chiang, Electronically conductive phospho-olivines as lithium storage electrodes, *Nat. Mater.* 1, 2002, 123–128.
12. H. Gwon, D.-H. Seo, S. Kim, J. Kim and K. Kang, Combined first-principle calculations and experimental study on multi-component olivine cathode for lithium rechargeable batteries, *Adv. Funct. Mater.* 19, 2009, 3285–3292.

13. D.-H. Seo, H. Gwon, S. Kim, J. Kim and K. Kang, Multicomponent olivine cathode for lithium rechargeable batteries: A first-principles study, *Chem. Mater.* 22, 2010, 518–523.
14. J. Kim, D.-H. Seo, S.-W. Kim, Y. Park and K. Kang, Mn based olivine electrode material with high power and energy, *Chem. Commun.* 46, 2010, 1305–1307.
15. H. Kim, S. Lee, Y.-U. Park, H. Kim, J. Kim, S. Jeon and K. Kang, Neutron and x-ray diffraction study of pyrophosphate-based $Li_{2-x}MP_2O_7$ (M = Fe, Co) for lithium rechargeable battery electrodes, *Chem. Mater.* 23, 2011, 3930–3937.
16. J.M. Clark, S.-I. Nishimura, A. Yamada and M.S. Islam, High-voltage pyrophosphate cathode: Insights into local structure and lithium-diffusion pathways, *Angew. Chem.* 124, 2012, 13326–13330.
17. R.A. Shakoor, H. Kim, W. Cho, S.Y. Lim, H. Song, J.W. Lee, J.K. Kang, Y.-T. Kim, Y. Jung and J.W. Choi, Site-specific transition metal occupation in multicomponent pyrophosphate for improved electrochemical and thermal properties in lithium battery cathodes: A combined experimental and theoretical study, *J. Am. Chem. Soc.* 134, 2012, 11740–11748.
18. A. Nytén, A. Abouimrane, M. Armand, T. Gustafsson and J.O. Thomas, Electrochemical performance of Li_2FeSiO_4 as a new Li-battery cathode material, *Electrochem. Commun.* 7, 2005, 156–160.
19. M.E. Arroyo-de Dompablo, M. Armand, J.M. Tarascon and U. Amador, On-demand design of polyoxianionic cathode materials based on electronegativity correlations: An exploration of the Li_2MSiO_4 system (M = Fe, Mn, Co, Ni), *Electrochem. Commun.* 8, 2006, 1292–1298.
20. N. Kuganathan and M.S. Islam, Li_2MnSiO_4 lithium battery material: Atomic-scale study of defects, lithium mobility, and trivalent dopants, *Chem. Mater.* 21, 2009, 5196–5202.
21. D.-H. Seo, H. Kim, I. Park, J. Hong and K. Kang, Polymorphism and phase transformations of $Li_{2-x}FeSiO_4$ ($0 \leq x \leq 2$) from first principles, *Phys. Rev. B* 84, 220106(R), 2011.
22. S.-W. Kim, D.-H. Seo, H. Kim, K.-Y. Park and K. Kang, A comparative study on Na_2MnPO_4F and Li_2MnPO_4F for rechargeable battery cathodes, *Phys. Chem. Chem. Phys.* 14, 2012, 3299–3303.
23. R. Tripathi, T.N. Ramesh, B.L. Ellis and L.F. Nazar, Scalable synthesis of tavorite $LiFeSO_4F$ and $NaFeSO_4F$ cathode materials, *Angew. Chem.* 49, 2010, 8738–8742.
24. T. Mueller, G. Hautier, A. Jain and G. Ceder, Evaluation of tavorite-structured cathode materials for lithium-ion batteries using high-throughput computing, *Chem. Mater.* 23, 2011, 3854–3862.
25. O.S. Bondareva, M.A. Simonov, Y.K. Egorov-Tismenko and N.V. Belov, The crystal structures of $LiZnBO_3$ and $LiMnBO_3$, *Sov. Phys. Crystallogr.* 23, 1978, 269–271.
26. Y. Piffard, K.K. Rangan, Y. An, D. Guyomard and M. Tournoux, Cobalt lithium orthoborate, $LiCoBO_3$, *Acta Crystallograph. Sect. C* 54, 1998, 1561–1563.
27. V. Legagneur, Y. An, A. Mosbah, R. Portal, A. Le Gal La Salle, A. Verbaere, D. Guyomard and Y. Piffard, $LiMBO_3$ (M = Mn, Fe, Co): Synthesis, crystal structure and lithium deinsertion/insertion properties, *Solid State Ionics* 139, 2001, 37–46.
28. A. Abouimrane, M. Armandand and N. Ravet, Carbon nano-painting: Application to non-phosphate oxyanions, eg. borates, in: K. Zaghib, C.M. Julien and J. Prakash, (eds.), *New Trends in Intercalation Compounds for Energy Storage Conversion*, Vol. PV 2003-20, The Electrochemical Society, Pennington, NJ, 2003, p. 15.

29. Y.Z. Dong, Y.M. Zhao, P. Fu, H. Zhou and X.M. Hou, Phase relations of Li_2O–FeO–B_2O_3 ternary system and electrochemical properties of $LiFeBO_3$ compound, *J. All. Compd.* 461, 2008, 585–590.
30. A. Yamada, N. Iwane, Y. Harada, S.-I. Nishimura, Y. Koyama and I. Tanaka, Lithium iron borates as high-capacity battery electrodes, *Adv. Mater.* 22, 2010, 3583–3587.
31. J.C. Kim, C.J. Moore, B. Kang, G. Hautier, A. Jain and G. Ceder, Synthesis and electrochemical properties of monoclinic $LiMnBO_3$ as a Li intercalation material, *J. Electrochem. Soc.* 158, 2011, A309–A315.
32. D.-H. Seo, Y.-U. Park, S.-W. Kim, I. Park, R. Shakoor and K. Kang, First-principles study on lithium metal borate cathodes for lithium rechargeable batteries, *Phys. Rev. B* 83, 2011, 205127.
33. Y. Janssen, D.S. Middlemiss, S.-H. Bo, C.P. Grey and P.G. Khalifah, Structural modulation in the high capacity battery cathode material $LiFeBO_3$, *J. Am. Chem. Soc.* 134, 2012, 12516–12527.
34. G. Ceder, G. Hautier, A. Jain and S.P. Ong, Recharging lithium battery research with first-principles methods, *MRS Bull.* 36, 2011, 185–191.
35. G. Ceder, Opportunities and challenges for first-principles materials design and applications to Li battery materials, *MRS Bull.* 35, 2010, 693–701.
36. S.-W. Kim, D.-H. Seo, X. Ma, G. Ceder and K. Kang, Electrode materials for rechargeable sodium-ion batteries: Potential alternatives for current lithium-ion batteries, *Adv. Energy Mater.* 2, 2012, 710–721.
37. R.O. Jones and O. Gunnarsson, The density functional formalism, its applications and prospects, *Rev. Mod. Phys.* 61, 1989, 689–746.
38. P. Hohenberg and W. Kohn, Inhomogeneous electron gas, *Phys. Rev.* 136, 1964, B864–B871.
39. W. Koch and M.C. Holthausen, *A Chemist's Guide to Density Functional Theory*, Vol. 2, Wiley-VCH, Weinheim, 2001.
40. W. Kohn and L.J. Sham, Self-consistent equations including exchange and correlation effects, *Phys. Rev.* 140, 1965, A1133–A1138.
41. J.P. Perdew, K. Burke and M. Ernzerhof, Generalized gradient approximation made simple, *Phys. Rev. Lett.* 77, 1996, 3865–3868.
42. A. Rohrbach, J. Hafner and G. Kresse, Electronic correlation effects in transition-metal sulfides, *J. Phys.: Condens. Matter* 15, 2003, 976.
43. V.I. Anisimov, J. Zaanen and O.K. Andersen, Band theory and Mott insulators: Hubbard U instead of Stoner I, *Phys. Rev. B* 44, 1991, 943.
44. V.I. Anisimov, F. Aryasetiawan and A. Lichtenstein, First-principles calculations of the electronic structure and spectra of strongly correlated systems: The LDA+U method, *J. Phys.: Condens. Matter* 9, 1997, 767.
45. G. Kresse and J. Furthmuller, Efficiency of ab-initio total energy calculations for metals and semiconductors using a plane-wave basis set, *Comput. Mater. Sci.* 6, 1996, 15–50.
46. G. Ceder, Y.M. Chiang, D.R. Sadoway, M.K. Aydinol, Y.I. Jang and B. Huang, Identification of cathode materials for lithium batteries guided by first-principles calculations, *Nature* 392, 1998, 694–696.
47. A. Van der Ven, M.K. Aydinol, G. Ceder, G. Kresse and J. Hafner, First-principles investigation of phase stability in Li_xCoO_2, *Phys. Rev. B* 58, 1998, 2975.
48. A. Van der Ven, G. Ceder, M. Asta and P.D. Tepesch, First-principles theory of ionic diffusion with nondilute carriers, *Phys. Rev. B* 64, 2001, 184307.

49. F. Zhou, M. Cococcioni, K. Kang and G. Ceder, The Li intercalation potential of LiMPO$_4$ and LiMSiO$_4$ olivines with M = Fe, Mn, Co, Ni, *Electrochem. Commun.* 6, 2004, 1144–1148.
50. F. Zhou, K. Kang, T. Maxisch, G. Ceder and D. Morgan, The electronic structure and band gap of LiFePO$_4$ and LiMnPO$_4$, *Solid State Commun.* 132, 2004, 181–186.
51. K. Kang and G. Ceder, Factors that affect Li mobility in layered lithium transition metal oxides, *Phys. Rev. B* 74, 2006, 094105.
52. D.-H. Seo, H. Kim, H. Kim, W.A. Goddard, III and K. Kang, The predicted crystal structure of Li$_4$C$_6$O$_6$, an organic cathode material for Li-ion batteries, from first-principles multi-level computational methods, *Energy Environ. Sci.* 4, 2011, 4938–4941.
53. R.A. Shakoor, D.-H. Seo, H. Kim, Y.-U. Park, J. Kim, S.-W. Kim, H. Gwon, S. Lee and K. Kang, A combined first principles and experimental study on Na$_3$V$_2$(PO$_4$)$_2$F$_3$ for rechargeable Na batteries, *J. Mater. Chem.* 22, 2012, 20535–20541.
54. H. Kim, D.-H. Seo, H. Kim, I. Park, J. Hong, K.-Y. Park and K. Kang, Multicomponent effects on the crystal structures and electrochemical properties of spinel-structured M$_3$O$_4$ (M = Fe, Mn, Co) anodes in lithium rechargeable batteries, *Chem. Mater.* 24, 2012, 720–725.
55. J.L.F. Da Silva, M.V. Ganduglia-Pirovano, J. Sauer, V. Bayer and G. Kresse, Hybrid functionals applied to rare-earth oxides: The example of ceria, *Phys. Rev. B* 75, 2007, 045121.
56. H. Kim, I. Park, D.-H. Seo, S. Lee, S.-W. Kim, W.J. Kwon, Y.-U. Park, C.S. Kim, S. Jeon and K. Kang, New iron-based mixed-polyanion cathodes for lithium and sodium rechargeable batteries: Combined first principles calculations and experimental study, *J. Am. Chem. Soc.* 134, 2012, 10369–10372.
57. F. Zhou, M. Cococcioni, C.A. Marianetti, D. Morgan and G. Ceder, First-principles prediction of redox potentials in transition-metal compounds with LDA + U, *Phys. Rev. B* 70, 2004, 235121.
58. A. Van der Ven and G. Ceder, Electrochemical properties of spinel Li$_x$CoO$_2$: A first-principles investigation, *Phys. Rev. B* 59, 1999, 742–746.
59. M.K. Aydinol, A.F. Kohan and G. Ceder, Ab initio calculation of the intercalation voltage of lithium-transition-metal oxide electrodes for rechargeable batteries, *J. Power Sources* 68, 1997, 664–668.
60. Y.S. Meng and M.E.A.-D. Dompablo, First principles computational materials design for energy storage materials in lithium ion batteries, *Energy Environ. Sci.* 2, 2009, 589–606.
61. M.K. Aydinol, A.F. Kohan, G. Ceder, K. Cho and J. Joannopoulos, Ab initio study of lithium intercalation in metal oxides and metal dichalcogenides, *Phys. Rev. B* 56, 1997, 1354.
62. A. Jain, G. Hautier, C. Moore, B. Kang, J. Lee, H. Chen, N. Twu and G. Ceder, A computational investigation of Li$_9$M$_3$(P$_2$O$_7$)$_3$(PO$_4$)$_2$ (M = V, Mo) as cathodes for Li ion batteries, *J. Electrochem. Soc.* 159, 2012, A622–A633.
63. D. Morgan, A. Van der Ven and G. Ceder, Li conductivity in Li$_x$MPO$_4$ (M = Mn, Fe, Co, Ni) olivine materials, *Electrochem. Solid State Lett.* 7, 2004, A30–A32.
64. A. Van der Ven, J. Bhattacharya and A.A. Belak, Understanding Li diffusion in Li-intercalation compounds, *Acc. Chem. Res.* 46(5), 2013, 1216–1225.
65. H. Jonsson, G. Mills, K.W. Jacobsen and B.J. Berne, in: D. Chandler, B.J. Berne, G. Ciccotti and D.F. Coker, (eds.), *Classical and Quantum Dynamics in Condensed Phase Simulations*, World Scientific, Singapore, 1998, pp. 385.

66. G. Henkelman, B.P. Uberuaga and H. Jonsson, A climbing image nudged elastic band method for finding saddle points and minimum energy paths, *J. Chem. Phys.* 113, 2000, 9901–9904.
67. S. Curtarolo, D. Morgan and G. Ceder, Accuracy of ab initio methods in predicting the crystal structures of metals: A review of 80 binary alloys, *CALPHAD* 29, 2005, 163–211.
68. S.P. Ong, V.L. Chevrier, G. Hautier, A. Jain, C. Moore, S. Kim, X. Ma and G. Ceder, Voltage, stability and diffusion barrier differences between sodium-ion and lithium-ion intercalation materials, *Energy Environ. Sci.* 4, 2011, 3680–3688.
69. S. Adams, Modelling ion conduction pathways by bond valence pseudopotential maps, *Solid State Ionics* 136–137, 2000, 1351–1361.
70. I.D. Brown, Recent developments in the methods and applications of the bond valence model, *Chem. Rev.* 109, 2009, 6858–6916.
71. M. Sale and M. Avdeev, 3DBVSMAPPER: A program for automatically generating bond-valence sum landscapes, *J. Appl. Crystallograph.* 45, 2012, 1054–1056.
72. H. Berg and J.O. Thomas, Neutron diffraction study of electrochemically delithiated $LiMn_2O_4$ spinel, *Solid State Ionics* 126, 1999, 227–234.
73. J. Reed, G. Ceder and A. Van Der Ven, Layered-to-spinel phase transition in Li_xMnO_2, *Electrochem. Solid State Lett.* 4(6), 2001, A78–A81.
74. A.E. Smith, H. Mizoguchi, K. Delaney, N.A. Spaldin, A.W. Sleight and M.A. Subramanian, Mn^{3+} in trigonal bipyramidal coordination: A new blue chromophore, *J. Am. Chem. Soc.* 131, 2009, 17084–17086.
75. A.K. Padhi, K.S. Nanjundaswamy, C. Masquelier, S. Okada and J.B. Goodenough, Effect of structure on the Fe^{3+}/Fe^{2+} redox couple in iron phosphates, *J. Electrochem. Soc.* 144, 1997, 1609–1613.
76. A.K. Padhi, K.S. Nanjundaswamy, C. Masquelier and J.B. Goodenough, Mapping of transition metal redox energies in phosphates with NASICON structure by lithium intercalation, *J. Electrochem. Soc.* 144, 1997, 2581–2586.
77. G. Li, H. Azuma and M. Tohda, $LiMnPO_4$ as the cathode for lithium batteries, *Electrochem. Solid State Lett.* 5, 2002, A135–A137.
78. K. Amine, H. Yasuda and M. Yamachi, Olivine $LiCoPO_4$ as 4.8 V electrode material for lithium batteries, *Electrochem. Solid State Lett.* 3, 2000, 178–176.
79. X. Ma, B. Kang and G. Ceder, High rate micron-sized ordered $LiNi_{0.5}Mn_{1.5}O_4$, *J. Electrochem. Soc.* 157, 2010, A925–A931.
80. R. Malik, D. Burch, M. Bazant and G. Ceder, Particle size dependence of the ionic diffusivity, *Nano Lett.* 10, 2010, 4123–4127.
81. S.-Y. Chung, S.-Y. Choi, S. Lee and Y. Ikuhara, Distinct configurations of antisite defects in ordered metal phosphates: Comparison between $LiMnPO_4$ and $LiFePO_4$, *Phys. Rev. Lett.* 108, 2012, 195501.
82. R.D. Shannon and C.T. Prewitt, Effective ionic radii in oxides and fluorides, *Acta Crystallogr.* B25, 1969, 925.
83. S. Yang, Y. Song, P.Y. Zavalij and M. Stanley Whittingham, Reactivity, stability and electrochemical behavior of lithium iron phosphates, *Electrochem. Commun.* 4, 2002, 239–244.
84. S.-Y. Chung, S.-Y. Choi, T. Yamamoto and Y. Ikuhara, Atomic-scale visualization of antisite defects in $LiFePO_4$, *Phys. Rev. Lett.* 100, 2008, 125502.
85. F. Sanz, C. Parada, J.M. Rojo and C. Ruíz-Valero, Synthesis, structural characterization, magnetic properties, and ionic conductivity of $Na_4M^{II}_3(PO_4)_2(P_2O_7)$ (M^{II} = Mn, Co, Ni), *Chem. Mater.* 13, 2001, 1334–1340.

86. R. Essehli, B. El Bali, S. Benmokhtar, H. Fuess, I. Svoboda and S. Obbade, Synthesis, crystal structure and infrared spectroscopy of a new non-centrosymmetric mixed-anion phosphate $Na_4Mg_3(PO_4)_2(P_2O_7)$, *J. All. Compd.* 493, 2010, 654–660.
87. G. Hautier, A. Jain, S.P. Ong, B. Kang, C. Moore, R. Doe and G. Ceder, Phosphates as lithium-ion battery cathodes: An evaluation based on high-throughput ab initio calculations, *Chem. Mater.* 23, 2011, 3495–3508.

7

Advanced Batteries and Improvements in Electrode Materials

Ashok Kumar Shukla, Vedam Ganesh Kumar
and Musuwathi Krishnamoorthy Ravikumar

CONTENTS

7.1 Overview for Common Batteries 256
7.2 Lead–Acid Batteries 257
7.3 Nickel-Based Batteries 261
    7.3.1 Nickel–Cadmium Batteries 262
    7.3.2 Nickel–Iron Batteries 264
    7.3.3 Nickel–Metal Hydride Batteries 266
7.4 Lithium–Ion Batteries 270
    7.4.1 Cathode Materials 271
    7.4.2 Anode Materials 286
        7.4.2.1 Lithium Metal 286
        7.4.2.2 Carbonaceous Materials 288
        7.4.2.3 Other Carbonaceous Materials 309
    7.4.3 Other Anode Materials 313
7.5 Electrolytes for Lithium Batteries 313
7.6 Outlook 315
References 316

The improvement in battery science and technology is a critical link in the transition from fossil fuels to alternative energy sources. In particular, the emerging field of nanotechnology promises not only higher-energy-density batteries but also rechargeable batteries with a longer lifetime. This chapter is an effort to elaborate on technological developments that are currently being undertaken for improving cathodes, anodes and electrolytes for rechargeable batteries by keeping the common characteristic of all these components as their nanoscale structure. After a brief summary of advances in lead–acid and nickel-based batteries, this chapter concludes with a brief summary on advances in lithium batteries, which are a major attraction for R&D at present.

7.1 Overview for Common Batteries

A battery is a device that converts the chemical energy stored in its active materials directly into electrical energy by means of an electrochemical oxidation–reduction reaction. A battery consists of two electrodes, an anode and a cathode separated by an electrolyte through which electrically charged particles but not the electrons can move. During the electrochemical operation of a battery, two chemical reactions take place at the same time. The reaction taking place at the anode is an oxidation reaction of the type: $R \rightarrow O + ne$, where R is the reduced species and O is the oxidised species, and n is the number of electrons. Since the electrons are generated at the anode, it is also referred as the negative electrode or negative plate. The chemical reaction taking place at the cathode is a reduction reaction of the type: $O' + ne \rightarrow R'$, where O' and R' are the oxidised and reduced species, respectively. Since the reaction results in the depletion of electrons, the cathode is also called the positive electrode or positive plate. When the battery is connected to an external circuit, the excess electrons from the anode flow through the circuit to the cathode. As the electrons move through the circuit, they lose energy. This energy may be used to create heat or light as in an electrical heater or light bulb, or to do work as in a motor. The flow of electrons results in a current and by convention the direction of the current flow is opposite to the direction of the electron flow. The energy released per unit charge while the current flows through the circuit is called voltage. The product of the current and the voltage is the power delivered to the circuit. When a battery delivers electric current to an external load, the active materials in the battery electrodes are converted into other materials at lower energy states and the battery is eventually discharged. During recharge, a storage battery behaves like an electrolytic cell where the active electrode materials are retrieved. A battery that can be recharged after its discharge is referred to as rechargeable (storage) battery or an accumulator [1].

In 1836, John Daniell unveiled a two-fluid battery to provide a constant and reliable source of electricity over a long period of time. In 1859, Plantè discovered the lead–acid battery that revolutionised the world. In 1866, Leclanché patented the dry cell that ignited the commercial interest in batteries. Today, batteries are at the centre stage of various technologies with applications ranging from cars to aerospace. At present, the world market for batteries is about US$100 billion [2].

There are about 100 odd elements known to us in the periodic table. Accordingly, nearly 5000 pair-wise combinations of single electrode reactions involving stable reactants and products can be theoretically envisaged leading to a like number of electrochemical cells. Ironically, however, more than a century of effort has resulted in only a few systems of practical importance and efforts are being expended to enhance their specific energy in conjunction with optimum power-density, minimum internal-resistance, maximum

Advanced Batteries and Improvements in Electrode Materials 257

charge-retention, mechanical strength and long cycle life. Accordingly, R&D on innovative battery chemistries continues.

In this chapter, we will highlight the improvements made possible in some of the battery systems and, in particular, lead–acid, nickel-based batteries and lithium-ion batteries (LIBs) through nanostructural design of electrode materials.

7.2 Lead–Acid Batteries

Historically, subsequent to exploratory studies on lead–acid systems by Daniel, Grove and Sindesten, practical lead–acid batteries began with the research and inventions of Raymond Gaston Planté in France as early as 1859 and, even today, lead–acid battery remains the most successful battery system ever developed. There are three types of lead–acid batteries in common use: (1) batteries with flooded or excess electrolyte, (2) low-maintenance lead–acid batteries with a large excess of electrolyte and (3) batteries with immobilised electrolyte and a pressure-sensitive valve, usually referred to as valve-regulated lead–acid (VRLA) or sealed lead–acid (SLA) batteries [3].

Electrochemical reactions taking place at the positive and negative plates of a lead–acid battery are as follows:

At the positive plate:

$$PbO_2 + H_2SO_4 + 2H^+ + 2e^- \underset{\text{charge}}{\overset{\text{discharge}}{\rightleftharpoons}} PbSO_4 + 2H_2O \ (E^\circ = 1.69V \text{ vs. SHE}) \tag{7.1}$$

At the negative plate:

$$Pb + H_2SO_4 \underset{\text{charge}}{\overset{\text{discharge}}{\rightleftharpoons}} PbSO_4 + 2H^+ + 2e^- \ (E^\circ = -0.36V \text{ vs. SHE}) \tag{7.2}$$

The net cell reaction is given by

$$PbO_2 + Pb + 2H_2SO_4 \underset{\text{charge}}{\overset{\text{discharge}}{\rightleftharpoons}} 2PbSO_4 + 2H_2O \ (E^\circ_{cell} = 2.05V) \tag{7.3}$$

Thermodynamic stability of the electrolyte requires (i) its lowest unoccupied state to have higher energy than the highest occupied state of the reductant, or (ii) its highest occupied state to have lower energy than the lowest unoccupied state of the oxidant (Figure 7.1). If either of these two conditions is violated, electrons may be transferred to or from the electrolyte to reduce or oxidise it. Therefore, thermodynamic stability of the electrolyte restricts the cell voltage (E_{cell}) to be lower than the thermodynamic window (E_g) of the electrolyte [4].

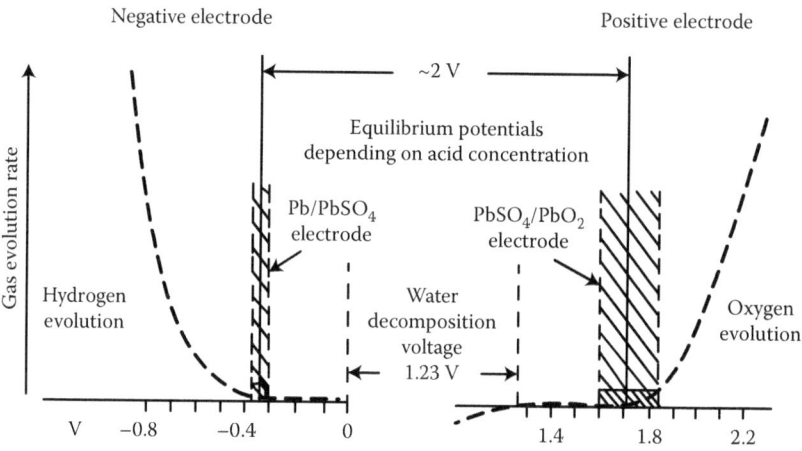

FIGURE 7.1
Operation of Pb–acid battery electrode reactions in conjunction with the stability of water present in the electrolyte. (Adapted from D. Pavlov, *Essentials of Lead-Acid Batteries*, India: SAEST, 2006.)

Both PbO_2 and Pb are thermodynamically unstable in sulphuric acid. The equilibrium potential of the $PbO_2/PbSO_4$ couple is more anodic to the O_2/H_2O couple by 0.46 V, and the equilibrium potential of $Pb/PbSO_4$ is more cathodic to the H_2/H^+ couple by 0.36 V as shown in Figure 7.1. Hence, even under open-circuit conditions, O_2 evolution at the positive plate and H_2 evolution at the negative plate can occur according to the reactions:

$$PbO_2 + H_2SO_4 \rightarrow PbSO_4 + H_2O + \tfrac{1}{2} O_2 \uparrow \qquad (7.4)$$

and

$$Pb + H_2SO_4 \rightarrow PbSO_4 + H_2 \uparrow \qquad (7.5)$$

Since the rates for reactions (7.4) and (7.5) increase with acid concentration, the lead–acid cell should not work in principle. In practice, however, the high O_2 and H_2 over-potentials on PbO_2 and Pb enable the respective electrodes to be charged before O_2 and H_2 evolve at a substantial rate. In the lead–acid cell, it is the poor kinetics at the plate–electrolyte interface that allows a cell voltage higher than the thermodynamic window (E_g) of 1.23 V between the lowest unoccupied molecular orbital (LUMO) of the H_2/H^+ redox energy and the highest occupied molecular orbital (HOMO) of the O_2/H_2O redox energy (Figure 7.1).

The positive plate in a lead–acid cell accepts a charge less efficiently than the negative plate. Therefore, O_2 and H_2 are evolved non-stoichiometrically during the recharge of a lead–acid battery with O_2 evolution occurring prior to H_2 evolution. The oxygen evolved at the positive plate is constantly reduced at the negative plate as follows:

$$O_2 + 2Pb + 2H_2SO_4 \rightarrow 2PbSO_4 + 2H_2O \tag{7.6}$$

This feature is the basis in the design of SLA batteries.

Factors seminal for realising a high-performance battery are (a) maximum specific energy, (b) maximum specific power, (c) maximum battery life (not only in environmental durability but also most importantly in cycle life, i.e. number of possible charges and discharges) and (d) low cost. The demand for cost-effective batteries has traditionally driven lead–acid battery manufacturers to use low-cost lead materials, which has limited the first three aforesaid performance characteristics. Given the long-standing use of lead metal, improvements in the battery current collector—grid, design and advances in lead alloys—appear to be reaching a plateau. Accordingly, the lead–acid battery industry has been working in obscure corners of the periodic table to find alloying elements that would help stabilise current collection. Improved manufacturing techniques have allowed more mechanically delicate and thinner lead-based alloys to be employed in high-speed production. The traditional lead–acid industry appears to be approaching the limits in plate thickness, alloy corrosion rates and active material pellet structures. A paradigm shift to new materials and/or new processes is desired to take lead–acid battery technology beyond evolutionary engineering.

Both 'corrosion' (on the positive plate) and 'sulphation' (on the negative plate) are the two key failure modes of today's lead–acid batteries, especially in varying temperature environments to which batteries are often subjected [5]. Failures due to corrosion accelerate as operational temperatures rise, and when the battery is left uncharged. To mitigate the effects of corrosion, most battery companies have focused their research on developing corrosion-resistant lead alloys and grid manufacturing processes that reduce the mechanical stresses in the as-manufactured grids. All battery manufacturers provide battery service life based on lead alloy and grid wire cross-sectional area, which is dictated by grid thickness and the corresponding plate thickness. Thicker grids provide longer life, but at the expense of power density, cost, weight and volume. Failures due to sulphation occur when a lead–acid battery is left on open-circuit stand or kept in a partially/fully discharged state; the lead sulphate formed in the discharge reaction re-crystallises to form larger, low-surface-area and non-conductive lead sulphate crystals that are often referred to as a 'hard lead sulphate', which blocks the conductive path needed for recharging. These crystals, especially those located away from the plate grid, often fail to convert back into the charged lead and lead dioxide active materials. Even a well-maintained battery will lose some of its capacity over time due to the continued growth of large lead sulphate crystals that are not fully recharged. These sulphate crystals with a density of 6.287 g/cm^3 happen to be larger in volume than the original paste, and would mechanically deform the plate by splitting the material apart [6,7].

Traditionally, electrodes for lead–acid battery are fabricated by the pasting method. Traditional pasted plates depend upon the linear diffusion of electrolytes into the plate pores from the separator or an inter-plate reservoir for delivery of electrical energy as a part of the discharge process. Discharge begins at the plate surfaces and progresses into the interior as long as electrolyte diffusion rates can support the load current. In low-rate applications, with present lead–acid technology, it is possible to achieve utilisation levels, that is, the ratio of the amount of the active material actually discharged to the total present in the plates, of about 50–60% [8,9]. The distances traversed by the electrolyte during a full discharge in starting–lighting–ignition (SLI) or thin-plate VRLA products are of the order of 1–3 mm and in thick-plate deep-cycle batteries, it can be as much as 4–6 mm. In such thick plates, the path length for the electrolyte to reach the interior of the electrodes is longer. During the deep discharge, the reaction products get deposited into the pores and, hence, block the electrolyte in accessing the active material. Accordingly, lead–acid batteries for high-rate applications employ thin plates with small plate spacing. But this design approach adds cost and weight due to relatively higher grid materials. Accordingly, there is a definite need to go beyond the traditional lead–acid construction by improving the traditional pasted plate (or tubular plate) lead–acid batteries.

The desirable characteristics for the advanced VRLA batteries are high energy density (volumetric and gravimetric), high power density (volumetric and gravimetric), good mechanical, chemical and structural stability, flat discharge curve to ensure continuous power through discharge process, high-rate-discharge capability, fast-recharge capability, recovery to full capacity after off-season storage, excellent cold-temperature capacity utilisation, high-temperature resilience, recovery to full capacity after discharge, long cycle life, low self-discharge, wide temperature range, low cost, high safety and minimal environmental impact. It is also imperative to mitigate positive-grid corrosion, negative-plate sulphation, electrolyte stratification and a thermal run-away for the longevity of these batteries.

Morales et al. have adopted a novel approach towards nanostructural design of active material for lead–acid battery. The positive plate material, PbO_2, has been prepared by the hydrolysis of the Pb precursor, $Pb(OAc)_4$. Figure 7.2 shows the SEM images of micro- and nanostructured PbO_2. The x-ray diffraction (XRD) patterns for the synthesised PbO_2 and commercial PbO_2 confirm the presence of $\beta\text{-}PbO_2$ in tetragonal crystal systems with an average crystallite size of 7 and 44 nm. The nanostructured PbO_2 showed a discharge capacity of ~160 mAh/g that corresponds to 75% utilisation of the active material as against 120 mAh/g that corresponds to 50% utilisation of the active material exhibited by micro-structured materials. Applying

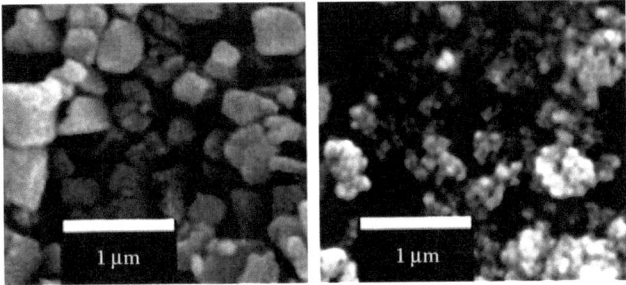

FIGURE 7.2
SEM micrographs for micro-structured and nanostructured PbO_2, the cathode active material of lead–acid cell. (Reprinted with permission from *J. Power Sources*, 158, J. Morales et al., Synthesis and characterization of lead dioxide activematerial for lead-acid batteries, 831–836, Copyright 2006, Elsevier.)

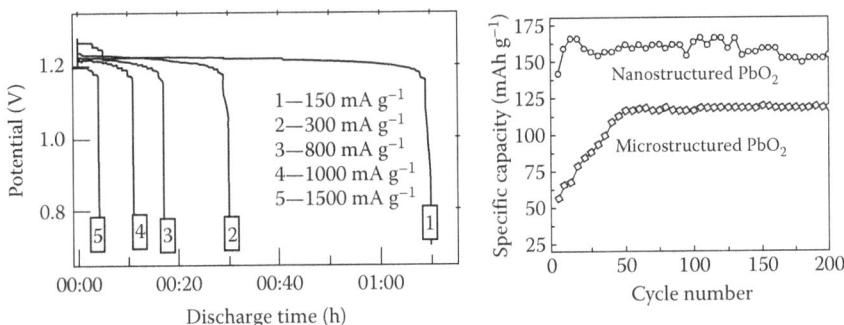

FIGURE 7.3
Typical galvanostatic discharge profiles of nanostructured PbO_2 and cycle life data for micro-structured and nanostructured PbO_2, the cathode active material of lead–acid cell. (Reprinted with permission from J. Morales et al., Nanostructured lead dioxide thin electrode, *Electrochem. Solid State Lett.* 7, 2004, A75–A77. Copyright 2004, The Electrochemical Society Inc.)

the nanostructured PbO_2 as cathode material for lead–acid batteries would improve battery performance by 25%, as the nanoparticles could hold a significant amount of structural water [10,11] (Figures 7.2 and 7.3).

7.3 Nickel-Based Batteries

Prominent nickel-based batteries are nickel–cadmium, nickel–iron and nickel–metal hydride batteries.

7.3.1 Nickel–Cadmium Batteries

Nickel–cadmium (Ni–Cd) batteries are developed in sealed, maintenance-free and vented cells with capacities ranging between 10 mAh and 100 Ah. Ni–Cd batteries used for cranking (capacity ~1000 Ah) are capable of delivering 8000 A of peak current. Ni–Cd batteries are characterised by their long life, continuous overcharge capability, relatively high charge and discharge rates, almost constant discharge voltage and the ability to operate at low temperatures [12].

The overall cell reaction is expressed as

$$2NiOOH + Cd + 2H_2O \underset{charge}{\overset{discharge}{\rightleftarrows}} 2Ni(OH)_2 + Cd(OH)_2 \quad (V_{cell} = 1.30 \text{ V}) \tag{7.7}$$

The operating principle of a sealed Ni–Cd cell is depicted in Figure 7.4. The active material for nickel positive electrodes is known for its phase transformation between various forms of oxidised and reduced states as expressed in Figure 7.5 [13].

Under deep-discharge conditions, due to inevitable differences in storage capacities of series-connected cells in a battery, gaseous hydrogen could both be evolved and consumed at the positive electrode, albeit at a low rate. Accordingly, repeated occurrence of over-discharge may cause internal pressure build-up and lead to cell explosion. To ensure proper operation of sealed Ni–Cd cells under a variety of operating conditions, the cells are designed

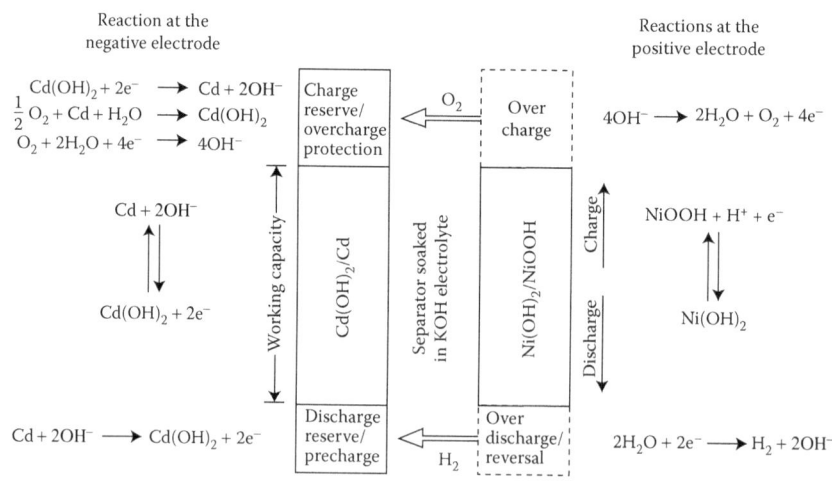

FIGURE 7.4
A schematic diagram of the electrochemical reactions during the charge, discharge, overcharge and over-discharge process of a Ni–Cd cell.

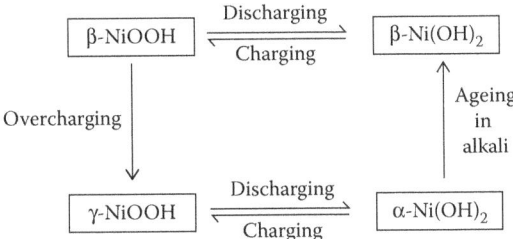

FIGURE 7.5
'Bode diagram' showing the transformations between oxidised and reduced phases of Ni(OH)$_2$/NiOOH electrode. (Reprinted with permission from *Electrochim. Acta*, 11, H. Bode, K. Dehmelt and J. Witte, Zur kenntnis der nickelhydroxidelektrode—I. Über das nickel (II)-hydroxidhydrat, 1079–1087, Copyright 1966, Elsevier.)

to be positive-limited such that only oxygen evolves, and diffuses to the cadmium electrode and combines with active cadmium to form cadmium hydroxide according to the reaction:

$$Cd + \tfrac{1}{2}O_2 + H_2O \rightarrow Cd(OH)_2 \qquad (7.8)$$

Recently, nanostructured Cd(OH)$_2$ has been synthesised. Gao et al. [14] reported the synthesis of Cd(OH)$_2$ nanowires by adding NH$_3 \cdot$H$_2$O to a solution of Cd(NO$_3$)$_2$. Such nanostructured Cd(OH)$_2$ are reported to show no memory effect besides offering a high capacity. As shown in Figure 7.6a, the shape of the discharge curve of Ni–Cd cell changes appreciably upon cycling at high C rates. The discharge profile at the beginning is flat. But with cycling, it appears to have a distinct slope owing to the possible formation of different phases of nickel oxyhydroxide during extensive cycling or continuous overcharge. Nickel–cadmium cells suffer from memory effect that leads to an apparent reduction in discharge voltage and capacity to a pre-determined cut-off voltage and results from highly repetitive shallow charge–discharge cycling with very little overcharge. The memory effect is reflected as a step in the discharge curve of a cell. The memory effect is completely reversible through a maintenance cycle consisting of a thorough discharge followed by a full and complete charge–overcharge [1]. Voltage depression is a phenomenon principally associated with sintered electrodes. During extended periods of overcharge, changes in the morphology and composition of the sintered nickel oxide electrodes can create a resistance effect that typically depresses the discharge voltage by up to about 100 mV per cell. Such behaviour is more pronounced at high operating temperatures.

Sealed Ni–Cd cells are inherently long lived. The performance of Ni–Cd cells is strongly dependent on the rate of discharge and temperature as shown in Figure 7.6b. The longest cell life is achieved between 5°C and 15°C. At low-temperature operations, cell capacity degradation occurs very slowly [15].

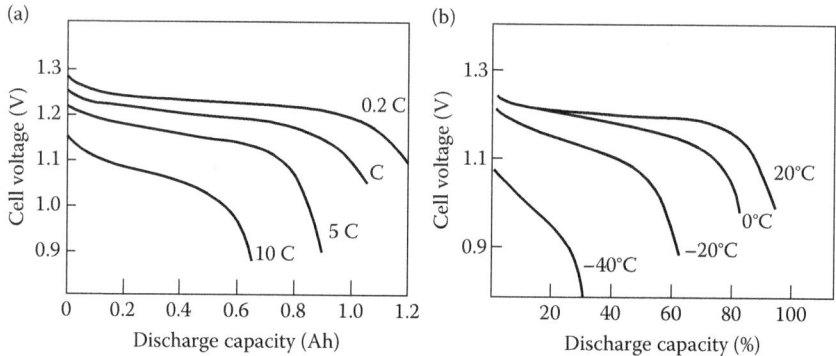

FIGURE 7.6
(a) Typical discharge profiles of a Ni–Cd cell at various rates at 20°C. (b) Typical discharge profiles of a Ni–Cd cell at ambient and low temperatures. (Reprinted with permission from *J. Power Sources*, 105, D. Ohms et al., New developments on high power alkaline batteries for industrial applications, 127–133, Copyright 2002, Elsevier.)

7.3.2 Nickel–Iron Batteries

The charge–discharge reactions of the Ni–Fe battery are

$$2NiOOH + Fe + 2H_2O \underset{\text{charge}}{\overset{\text{discharge}}{\rightleftarrows}} 2Ni(OH)_2 + Fe(OH)_2 \ (E^0 = 1.37 \text{ V}) \quad (7.9)$$

Under deep discharge, a Ni–Fe cell with a negative-limited configuration will undergo a further discharge reaction at a voltage, which is lower than that of the first step represented by reaction (7.10):

$$NiOOH + Fe(OH)_2 \underset{\text{charge}}{\overset{\text{discharge}}{\rightleftarrows}} Ni(OH)_2 + FeOOH \ (E^0 = 1.05 \text{ V}) \quad (7.10)$$

The cell reactions are highly reversible in the alkaline electrolyte, particularly if the discharge is limited to the first step. Typical two-step, charge–discharge curves for a Ni–Fe cell are shown in Figure 7.7 [16]. Discharge curves for commercial cells at different rates at 25°C and at varying temperatures are given in Figure 7.8.

Below about 1 C discharge rate, the cycle life of Ni–Fe batteries is found to be ~3000 cycles under normal conditions of use in industrial traction vehicles and railway carriage services that involve deep discharge between cycles, moderate vibrations, and shocks during regular duty schedules. Under similar conditions of usage, a calendar life of about 20 years has been realised for the Ni–Fe batteries. On the contrary, there is a decrease in performance at operational temperatures of about 45°C, nearly 1500 charge–discharge cycles with about 8 years of calendar life, primarily owing to the corrosion of

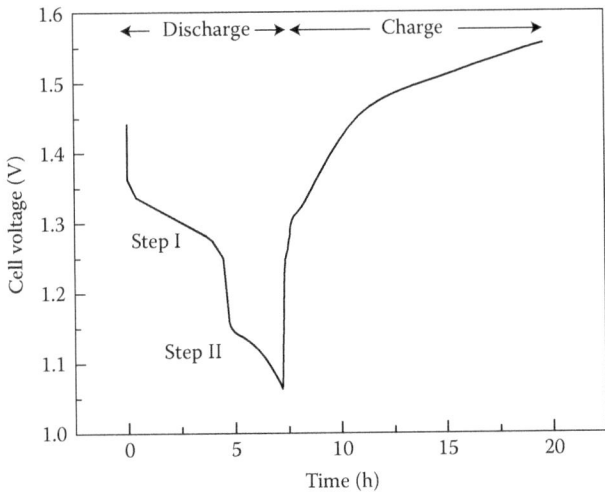

FIGURE 7.7
Typical two-step discharge and charge profile of Ni–Fe cell. (Reprinted with permission from *Encyclopedia of Electrochemical Power Sources*, Vol. 4. A.K. Shukla and B. Hariprakash, Nickel-iron, Amsterdam: Elsevier, pp. 522–527, Copyright 2009, Elsevier.)

alkaline iron electrodes at a faster rate than under ambient conditions. The wet shelf life in the discharged state exceeds 2 years. The battery provides the normal cycle life even after a continuous period of wet storage in its discharged state. The wet shelf-life can be extended to 10 years or more with the application of a reconditioning cycle every 6 months or so. The only routine

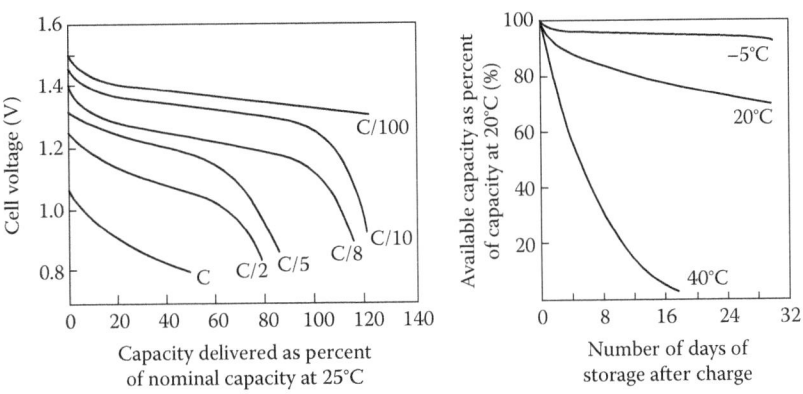

FIGURE 7.8
Typical galvanostatic discharge profiles of Ni–Fe cell at various discharging rates and temperatures. (Reprinted with permission from *Encyclopedia of Electrochemical Power Sources*, Vol. 4, A.K. Shukla and B. Hariprakash, Nickel-Iron, Amsterdam: Elsevier, pp. 522–527, Copyright 2009, Elsevier.)

maintenance operation is the addition of water to make up for the losses during overcharge [17].

Unlike Ni–Cd batteries, Ni–Fe batteries do not suffer from memory effect. The key problem in the development of Ni–Fe batteries is poisoning of the iron electrode. The iron electrode undergoes self-discharge as a result of its corrosion. It has been demonstrated that a substantial improvement in the overall performance of Ni–Fe cells is possible by electrocatalysis of the iron electrode reaction. Although complete suppression of hydrogen evolution appears to be difficult, efforts have been made to realise sealed Ni–Fe batteries by using efficient hydrogen–oxygen recombinant catalysts in the headspace of each cell.

The performance characteristics of laboratory-scale, 1.2 V/1 Ah, starved-electrolyte, sealed Ni–Fe cells, each with a recombinant catalyst, were found to be similar to those of vented cells of similar size [16]. About 60% of the nominal C/5 capacity is delivered at the C rate and about 120% at the C/10 rate. The low-temperature performance is very poor, for example, only about 10% of the nominal capacity at 1 C, due to the limited solubility of the reaction intermediates together with the increased resistance and viscosity of the electrolyte in conjunction with the retarded reaction kinetics at the electrodes. The self-discharge rate is about 1.5% of its nominal capacity per day at 25°C. The maximum catalyst-bed (catalyst on a porous mat) temperature reaches about 65°C at C-rate charging at 25°C [16]. This is due to the exothermic hydrogen–oxygen recombination reaction. The maximum pressure build-up inside the cells owing to the evolution of hydrogen and oxygen gases is found to be 110 kN/m^2. Laboratory-scale, sealed Ni–Fe batteries of size 6 V/1 Ah have also been developed. Sealed technology appears to be most appropriate for float-charge applications such as uninterruptible power supply (UPS) application in offshore platforms and petrochemical industries. Problems envisaged in developing commercial-grade, sealed Ni–Fe batteries include engineering difficulties in controlling the catalyst-bed temperature below 60°C and excessive internal heating of the cells during charge–discharge operations [18].

7.3.3 Nickel–Metal Hydride Batteries

At present, most commercial nickel–metal hydride (Ni–MH) batteries either use AB_5-type $MmNi_{3.2}Co_{1.0}Mn_{0.6}Al_{0.11}Mo_{0.09}$ (Mm:misch metal: 25wt% La, 50wt% Ce, 7wt% Pr, 18wt% Nd) or use AB_2-type $Ti_{0.51}Zr_{0.49}V_{0.70}Ni_{1.18}Cr_{0.12}$ alloys; AB_5- and AB_2-type alloys have different crystal structures wherein A and B sites are occupied by mixtures of elements. The commonly used misch metal alloys are capable of delivering a specific capacity of only 300 Ah/kg. Although these alloys utilise Ovshinsky's concept of disorder, they cannot be compared with commercial Ovonic transition metal alloys that approach a specific capacity of 400 Ah/kg and provide 80 Wh/kg of specific energy at cell level. This jump is only considered as a first threshold

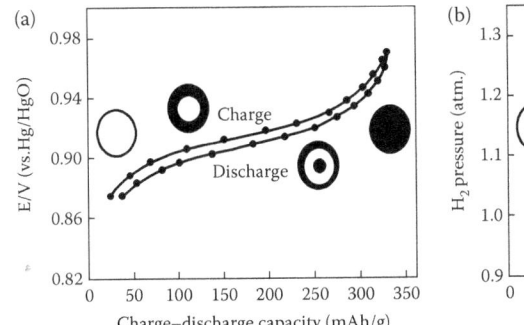 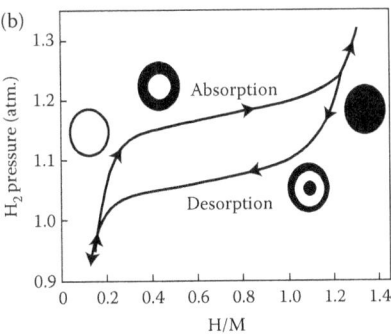

FIGURE 7.9
(a) Change of EMF upon hydrogen absorption–desorption expressed with the hydrogen content translated to charge–discharge capacity. (b) A typical pressure–composition isotherm of hydrogen storage alloy. Growth of MH phase from M and vice versa are depicted. (Reprinted with permission from *Encyclopedia of Electrochemical Power Sources*, Vol. 4, B. Hariprakash, A.K. Shukla and S. Venugoplan, Nickel-metal hydride: Overview. Amsterdam: Elsevier, pp. 494–501, Copyright 2009, Elsevier.)

to the projected specific capacity value of 700–1000 Ah/kg in future phases [19] (Figure 7.9).

Magnesium-based alloys have been projected as attractive materials for MH electrodes with specific capacity values as high as 1000 Ah/kg. However, these alloys are prone to corrosion in alkaline environment. The major problems associated with these alloys are the sluggish hydriding kinetics at room temperature and oxidation of the material under ambient environmental conditions (Figure 7.10).

$$\text{NiOOH} + \text{MH} \underset{\text{charge}}{\overset{\text{discharge}}{\rightleftarrows}} \text{Ni(OH)}_2 + \text{M} \quad (E_{cell} = 1.32 \text{ V vs. SHE}) \quad (7.11)$$

Nickel–metal hydride cells are generally available in 'AA', 'sub-C' and 'C' sizes. Prismatic cells up to 250 Ah are manufactured for EV application by Ovonic Battery Company (USA) and Gold Peak Industries (Hong Kong) [20,21]. The capacity that can be obtained from a Ni–MH cell is about two times of an equivalently sized Ni–Cd cell. Nickel–metal hydride cells can operate in the temperature range between −20°C and 45°C. But above 45°C, the charge efficiency falls steeply. The charge efficiency of Ni–MH cell is better than that for Ni–Cd cells for charge rates between C/10 and C/20 at 20°C. The internal pressure of the cell is generally observed to be below 50 psi for C/10 charge at 20°C. The typical profiles of internal pressure during charge at various rates are shown in Figure 7.11a; the cell pressure increases both at higher charge rates and at higher temperatures. Cylindrical cells are provided with a safety vent operating at around 400 psi. Owing to the

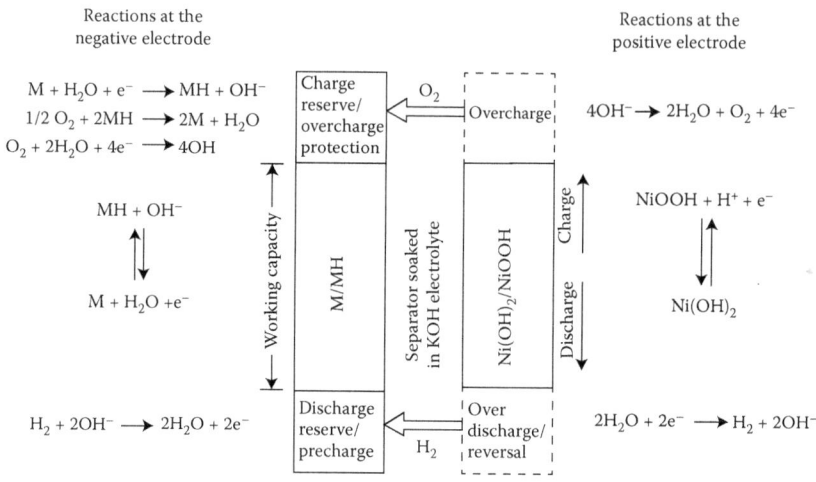

FIGURE 7.10
Electrochemical reactions occurring during the charge, discharge, over-charge and over-discharge process of a Ni–MH cell.

endothermic nature of the discharge process, the heat evolved during discharge is relatively less than Ni–Cd cells at discharge rates below 1 C. Joule heating masks the cooling because of endothermic desorption of hydrogen during discharge. Nickel–metal hydride cells can be discharged even at 5 C rate. The dependence of the discharge rate and the discharge capacity is depicted in Figure 7.11b.

The absorption of hydrogen during the cell charge is an exothermic process. But the rise in temperature becomes noticeable only from the point where

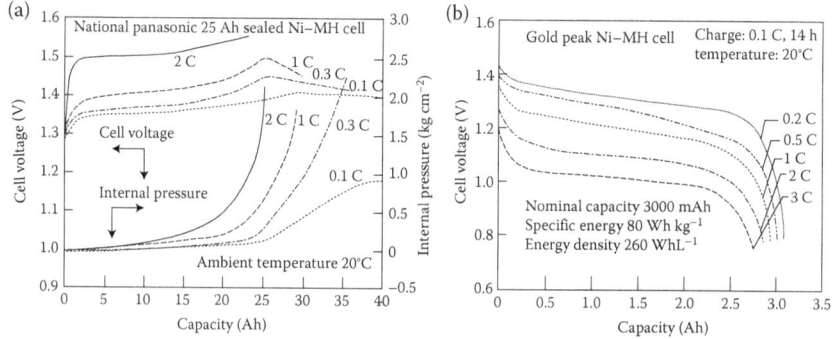

FIGURE 7.11
(a) Evolution of cell voltage and internal pressure with increasing charge inputs at various charging rates for a commercial 25 Ah Ni–MH cell, and (b) typical discharge profiles of a commercial 3 Ah Ni–MH cell at various rates at 20°C. (Reprinted with permission from *Encyclopedia of Electrochemical Power Sources*, Vol. 4, B. Hariprakash, A.K. Shukla and S. Venugoplan, Nickel-Metal hydride: Overview. Amsterdam: Elsevier, pp. 494–501. Copyright 2009, Elsevier.)

oxygen recombination begins to occur. During hydriding, the electronic conductivity of metal hydride decreases. If the overcharge is controlled, then the heat generation in the cells is mainly due to Joule heating. Cells can be charged at 1 C rate. The inflection in charge voltage is considered to be the most appropriate method for preventing overcharge and it improves the cycle life of the cell. A cycle life >22,000 cycles at 45°C with 50 A pulse charge–discharge to an SOC of 40–60% has been demonstrated [22].

A very common problem with hydrogen storage materials is severe volume expansion during the charge–discharge process, leading to cracking and pulverisation of the alloy and making it amenable to oxidation. In addition, dissolution of the alloy in the electrolyte contributes to capacity decay. Present materials science strategies are concentrated in combining different phases and micro-structures to overcome shortcomings associated with the bulk metal hydrides.

To increase rate capability, materials with high surface areas are being produced by powder metallurgy, mechanical alloying and chemical/electrochemical etching. Materials with high surface areas invariably suffer from lower cycle life owing to the increased oxidation of their surfaces. Metallurgical processes such as encapsulation (electroless plating of copper and nickel), doping with palladium and cerium, rapid solidification, and macro-alloying have shown only limited success in reducing the rate of degradation. Pure LaNi$_5$ electrodes in contact with potassium hydroxide undergo brittle fracture during hydriding, resulting in a rapid decay of capacity with cycling. Cobalt addition with aluminium or silicon has been found to significantly improve the cycling behaviour of Ni–MH cells.

Laboratories around the world have exhausted nearly all the elements in the periodic table to synthesise various AB$_2$ and AB$_5$ alloys to improve the cycle life and capacity of metal hydride cells. The major problem with metal hydride electrodes is that the function of the alloying elements either pristine or in combination with other alloying elements cannot be predicted with certainty. Conventional AB$_5$ and AB$_2$ alloys based on LaNi$_5$ and (Ti, Zr)Ni$_2$ have relatively low coulombic capacity values between 300 and 450 Ah/kg. Present research has focused on alloys such as TiZrNi$_2$ and Mg$_2$Ni as low-cost, lightweight and safer electrode alternatives, thereby increasing the energy density of the Ni–MH battery.

An amorphous structure appears to achieve high discharge capacities. These results indicate that the kinetics of hydriding/de-hydriding reactions of magnesium-based alloy electrodes can be greatly improved by ball milling and chemical coating. With a combination of modifications in the alloy composition and new methods of electrode preparation, discharge capacities between 630 and 780 Ah/kg have been achieved at a discharge current density of 50 A/kg for the magnesium-based Mg$_{1.9}$Al$_{0.1}$Ni$_{0.8}$Co$_{0.1}$Mn$_{0.1}$ alloy electrodes [23]. Efforts are therefore being directed towards maintaining particle-to-particle electrical contact by the use of polymer binders, compaction of porous nickel foam, surface plating, and doping and compaction with conductive powders. The specific energy of Ni–MH batteries can vary from 40 to 110 Wh/kg depending on the application. For example, where device runtime is paramount, Ni–MH batteries would not need high-power capability. On the other hand, for extremely high power charge and discharge requirements, factors that affect the specific energy of Ni–MH batteries are (a) extra current collection, (b) high N/P ratios (proportion of excess negative electrode capacity to positive electrode capacity) and (c) specific cell design and construction.

The specific energy of Ni–MH batteries is kept between 90 and 110 Wh/kg for portable power applications, 65 and 80 Wh/kg for EV batteries and 45 and 60 Wh/kg for hybrid electric vehicle (HEV) and other high-power applications. Although gravimetric energy density is imperative for advanced battery technologies, in many cases, volumetric energy density (Wh/L) happens to be more important. It has been demonstrated energy density values as high as 420 Wh/L can be achieved. Cost reduction is at the centre stage of Ni–MH battery development. In recent years, high-volume battery production has pushed the cost of Ni–MH batteries below that of Ni–Cd batteries [19].

7.4 Lithium–Ion Batteries

Rechargeable LIB technology was first commercialised by Sony Inc. in 1991. Since then, LIBs have become the energy source for most of mobile power applications such as handphones and laptops. Specific energy densities >150 Wh/kg have been commercially achieved with LIBs. R&D efforts are being expended to increase the capacity and operational performance of LIBs for cost-effective, inherently safe and thus commercially viable electric vehicles. The technological aspects include production, process development and integration, as these high-performance battery packs produced would need to address various energy specifications (Table 7.1). Other attractive attributes

TABLE 7.1

Technical requirements of Battery Packs for Cell Phones, Notebook PCs and Electric Bikes

	Cell Phones	NBPC	EV Motorbikes
Capacity	3–4 Wh	40–60 Wh	500–3000 Wh
Weight/volume	20–50 g (5–10 cc)	0.3–0.5 kg (200 cc)	10–15 kg
Control circuits	Simple	Smart circuits	Smart circuits
Charger	Not integrated	Not integrated	Integrated
Packaging	Simple	Rugged	Highly reliable

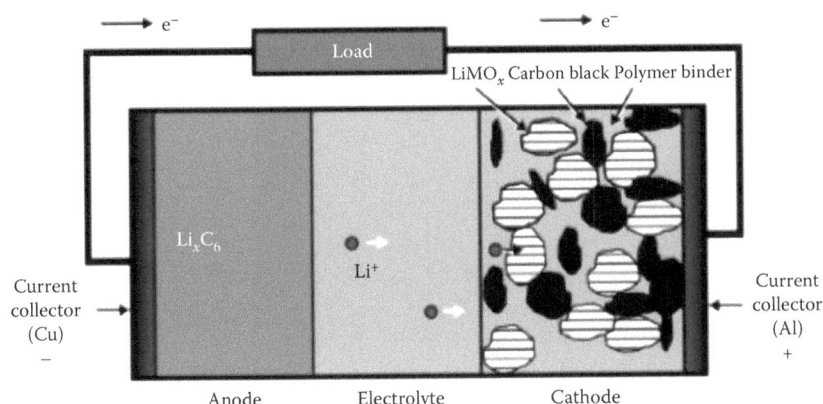

FIGURE 7.12
Schematic representation of the lithium-ion cell.

of LIBs are high power, high energy efficiency and low self-discharge with good cycle life [24].

In contrast to other battery systems, ionic current in the electrolyte is fully contributed by Li⁺ ions surrounded by non-aqueous solvent molecules. In addition, the anode and cathode active materials of the LIBs operate by the solid-state insertion/exertion of Li ions, different to the two-phase reactions found in lead–acid battery. During discharge of the lithium–ion cell, the following reactions occur at its anode and cathode as depicted in Figure 7.12. The process is reversed when the cell is charging.

Anode:

$$C_6Li_x \rightarrow C_6 + xLi^+ + xe^- \quad (7.12)$$

Cathode:

$$xLi^+ + xe^- + Li_{1-x}CoO_2 \rightarrow LiCoO_2 \quad (7.13)$$

Performance of an LIB is strongly affected by the amount of lithiation/de-lithiation of Li ions, during the charge and discharge of the battery, determined by the structure of the cathode and anode materials.

7.4.1 Cathode Materials

Nanostructured materials have become increasingly important in recent years owing to their unique mechanical, electrical and optical properties, endowed by confining the dimensions of such materials with the bulk and surface properties contributing to the overall behaviour. New nanostructured materials hold the key to fundamental advances in energy conversion and storage.

A large variety of nanomaterials have been synthesised and evaluated as cathode materials in non-aqueous electrolytes. These include chalcogenides, layered oxides, spinel oxides and olivines. Only a few of these compounds were found to be promising. Presently, $LiCoO_2$, Ni-substituted $LiCoO_2$ and $LiMn_2O_4$ are used in commercial Li-ion cells. Ni-substituted $LiCoO_2$, $LiFePO_4$, Al-substituted $LiNiO_2$, Co- and Ni-substituted $LiMnO_2$ and Ni- and Fe-substituted $LiMnO_2$ could find applications as cathodes in futuristic Li-ion cells, if they could meet long-term cyclability of >1000 cycles and storage characteristics. Table 7.2 summarises a number of cathode materials on its reversibility, gravimetric and volumetric capacity. Mostly, $LiCoO_2$ has been employed as a cathode material in commercial LIBs. However, $LiCoO_2$ has disadvantages in terms of cost, toxicity and low electrochemical capacity. By contrast, $LiNiO_2$ is iso-structural with $LiCoO_2$, less toxic and more economical. However, the synthesis of stoichiometric $LiNiO_2$ without any cation mixing is difficult. Furthermore, the cycling behaviour of $LiNiO_2$ is poor at high voltages and thermally unstable in its charged state due to its decomposition at elevated temperatures. Although $LiMn_2O_4$ has the benefits of low cost, high voltage, high thermal stability and non-toxicity, it exhibits capacity fade due to Jahn–Teller (JT) distortion, dissolution of Mn^{3+} from spinel structure and oxidation of Mn^{4+} in the electrolyte at high operating voltages.

In recent years, $LiFePO_4$ has emerged as a promising cathode material, particularly due to its lower cost and negligibly small toxicity. The inherently poor electronic conductivity of $LiFePO_4$ has opened up new avenues to investigate systematically, the possibility of enhancing the ionic/electronic

TABLE 7.2

LIB Cathode Materials with Emphasis on the Amount of Reversible Lithium Intercalation/De-Intercalation and Other Related Properties

Cathode Material	Molecular Weight	Density (g/cc)	Reversibility Range (Δx)	Gravimetric Capacity (mAh/g)	Volumetric Capacity (mAh/cc)
$Li_{1-x}TiS_2$	112.0	3.27	1.0	239	682
$Li_{1-x}MoS_2$	160.1	5.06	0.8	134	678
$Li_{1-x}V_2O_5$	181.9	3.36	1.0	147	495
$Li_{1-x}V_6O_{13}$	513.6	3.91	3.6	188	734
$Li_{1-x}MnO_2$	86.9	5.03	0.5	154	775
$Li_{1-x}NbSe_3$	329.8	8.70	3.0	244	2121
Li_xCoO_2	97.9	5.16	0.5	137	706
Li_xNiO_2	97.6	4.78	0.7	192	919
$Li_xMn_2O_4$	180.8	4.28	1.0	148	634
$LiFePO_4$	157.8	3.6	1.0	170	226

conductivity, electrochemical property and the rate capability of LiFePO$_4$ and related materials through innovative approaches.

The feasibility of insertion-compound electrodes for lithium batteries was demonstrated as early as 1976 by Stanley Whittingham using LiTiS$_2$ as cathode material with a capacity of 239 mAh/g. Ironically, however, the operating voltage of the Li$_x$TiS$_2$/Li cell was only ~2 V, a value achievable even with the lead–acid cell. Goodenough et al. were the first to design a LiCoO$_2$ cathode material that could exhibit ~4.0 V vs. Li/Li$^+$. Both LiCoO$_2$ and LiTiS$_2$ have the same layered crystalline structure [25,26], and LiCoO$_2$ is a layered material akin to α-NaFeO$_2$ (Figure 7.13).

Layered materials possess a two-dimensional structure that provides the path for intercalation/de-intercalation of lithium ions along the interlayer planes. First-row transition-metal oxides and chalcogenides used to be considered as seminal hosts for reversible intercalation/de-intercalation of lithium ions. But it is also desirable that the two-dimensional material possess a high Li insertion/de-insertion range (Δx). Usually achieving Δx value of 1/transition metal ion is considered as the maximum limit since it involves the change of the oxidation state of all metal ions by 1-unit.

The earlier host materials attempted for reversible lithium intercalation/de-intercalation were MoS$_2$, NiPS$_3$, V$_2$O$_5$, MoO$_3$, FeOCl and V$_6$O$_{13}$. Subsequently, LiVO$_2$, LiCrO$_2$, LiCoO$_2$, LiNiO$_2$ with layered oxide framework were used as Li intercalation/de-intercalation host materials. Weppner and Huggins [27] studied the Li-ion mobility and diffusion in the oxide phases between Li$_{0.2}$CoO$_2$ and Li$_{0.8}$CoO$_2$. The Li-ion diffusion coefficient was found to be ~5 × 10^{-9} cm^2/s at room temperature in these phases. Mizushima et al. [26] successfully synthesised layered oxides with high Li$^+$-ion mobility and voltage window achieving load-current densities of ~1 mA/cm^2.

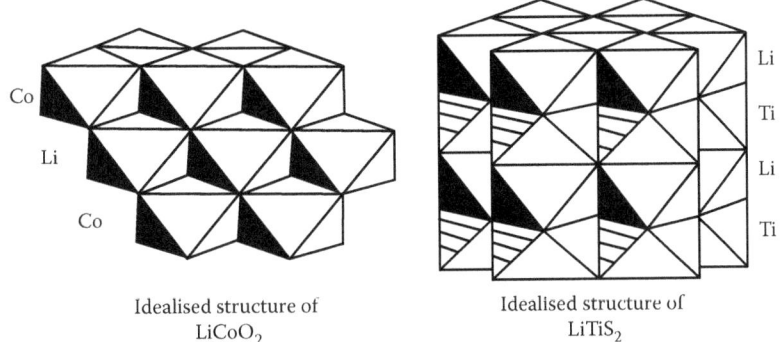

Idealised structure of LiCoO$_2$

Idealised structure of LiTiS$_2$

FIGURE 7.13
Comparison of crystal structures for LiCoO$_2$ and LiTiS$_2$. (Reprinted with permission from *Mater. Res. Bull.*, 15, K. Mizushima et al., Li$_x$CoO$_2$ (0 < x ~ 1): A new cathode material for batteries of high energy density, 783–789. Copyright 1980, Elsevier.)

LiCoO$_2$ has been synthesised by the spray-drying method. In this method, an aqueous acetate precursor solution consisting of Li$_2$CO$_3$ Co(OAc)$_2$ in dilute acetic acid is sprayed using a spray drier. LiCoO$_2$ obtained [28] is fairly crystalline in a preferred orientation of '003' layers, which indicate the particles grow in the direction perpendicular to transition metal and lithium layers. This preferred growth of the crystallites is evidenced by the increased ratio of 003/010 (hkl) reflections in XRD (Figure 7.14b). Moreover, these preferentially oriented crystallites give rise to higher capacity due to the reduction of diffusion path for the intercalation/de-intercalation perpendicular to the c-layer.

Jo et al. synthesised LiCoO$_2$ hydrothermally at 200°C and subsequently sintered at different temperatures (500°C, 700°C and 900°C), to obtain well-crystallised LiCoO$_2$ powders [29] (Figures 7.15 and 7.16). It is observed that the

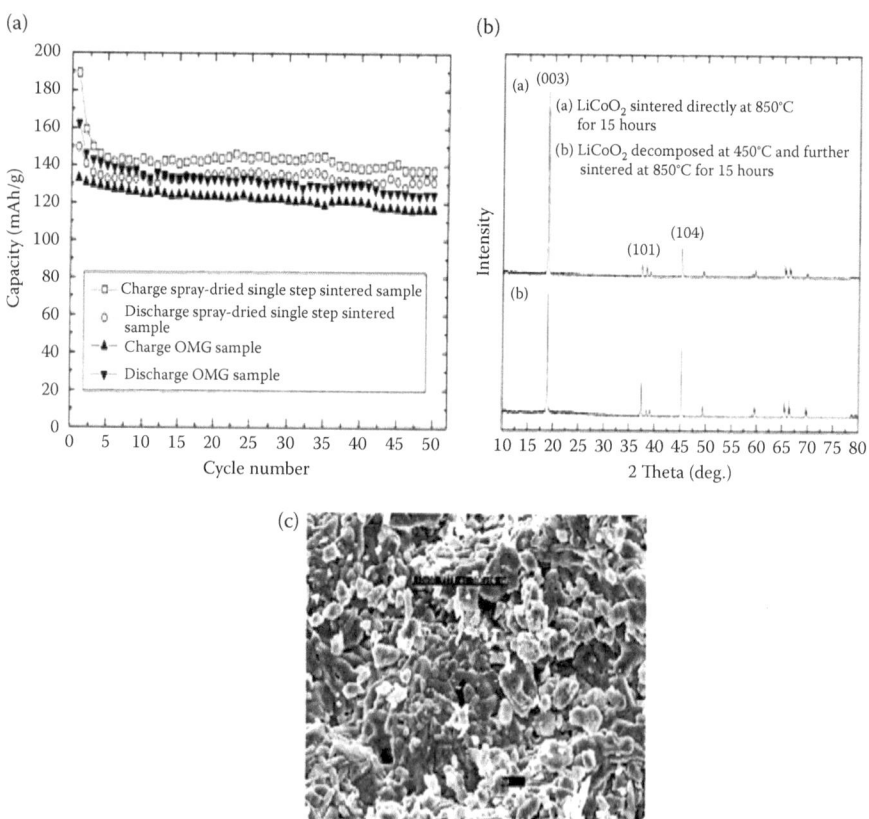

FIGURE 7.14
(a) Cycle life data, (b) powder XRD pattern and (c) SEM micrographs of LiCoO$_2$ prepared by spray drying method. (Adapted from W. Weppner and R.A. Huggins, *J. Electrochem. Soc.*, 124, 1977, 1569–1578; Reprinted with permission from *J. Power Sources*, 119–125, K. Konstantinov et al., Stoichiometry-controlled high-performance LiCoO$_2$ electrodematerials prepared by a spray solution technique, 195–200, Copyright 2003, Elsevier.)

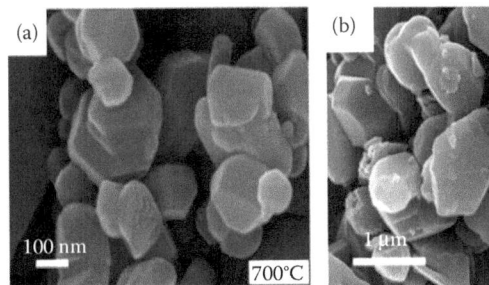

FIGURE 7.15
SEM micrograph of LiCoO$_2$ prepared at (a) 700°C and (b) 900°C. (Reprinted with permission from M.K. Jo et al., Effect of LiCoO$_2$ cathode nanoparticle size on high rate performance for Li-Ion batteries, *J. Electrochem. Soc.*, 156, 2009, A430–A434, Copyright 2002, The Electrochemical Society, Inc.)

particle size at 200°C is 50 nm, and increases to 100 nm, 300 nm and 1 μm at 500°C, 700°C and 900°C, respectively, while the BET surface area continues to decrease with increasing temperature, thus leading to a decrease in capacity.

$$LiCoO_2 \rightleftarrows Li_{1-x}CoO_2 + xLi^+ + e^- \tag{7.14}$$

Layered LiCoO$_2$ is a proven cathode material for LIBs. But it has certain limitations, such as (i) a limited availability, toxicity and high cost of cobalt, (ii) a low realisable reversible capacity of LiCoO$_2$ of ~165 mAh/g corresponding to ~0.6 equivalents of Li per mole of LiCoO$_2$; the complete removal of

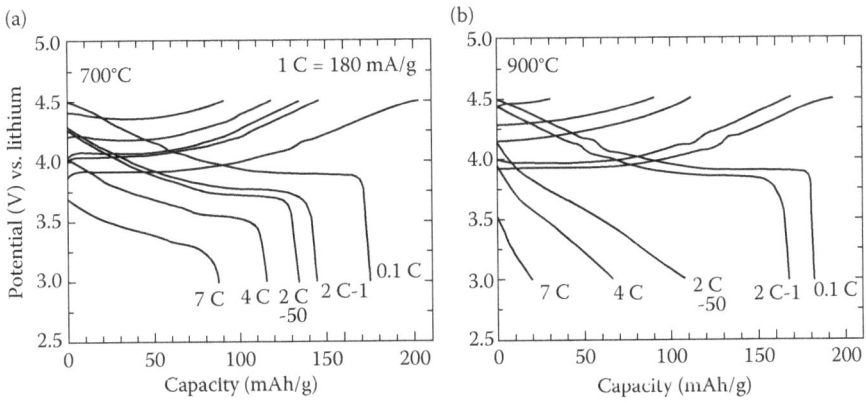

FIGURE 7.16
Galvanostatic profiles at various rates for LiCoO$_2$ prepared at (a) 700°C and (b) 900°C. (Reprinted with permission from M.K. Jo et al., Effect of LiCoO$_2$ cathode nanoparticle size on high rate performance for Li-Ion batteries, *J. Electrochem. Soc.* 156, 2009, A430–A434, Copyright 2002, The Electrochemical Society Inc.)

one equivalent Li per formula unit requires high potentials (>4.6 V vs. Li/Li$^+$), which promotes material degradation resulting from structural transformation, and (iii) safety problems associated with the release of oxygen from the material at highly de-intercalated state due to weak Co–O bonding. Accordingly, LiNiO$_2$, LiNi$_{1/3}$Co$_{1/3}$Mn$_{1/3}$O$_2$, spinel-LiMn$_2$O$_4$ and olivine-structured LiFePO$_4$, were attempted as alternatives.

The theoretical specific capacity for LiNiO$_2$ is 274 mAh/g, which corresponds to de-intercalation of one equivalent of lithium per formula unit of LiNiO$_2$, resulting in the formation of NiO$_2$. LiCoO$_2$ and LiNiO$_2$ have the same crystal structure. It has been difficult to synthesise LiNiO$_2$ with the ordered α-NaFeO$_2$ phase owing to the close similarity of ionic radii (0.74 Å) of Li$^+$ and Ni^{3+}. This makes the lithium and nickel ions occupy random positions in the LiNiO$_2$ crystalline lattice between the 3a and 3b sites. Li$^+$ ions occupying 3b Wycoff positions are strongly bound in the oxygen sub-lattice and, hence, not available for reversible intercalation/de-intercalation.

Any moderately synthesised LiNiO$_2$ sample that can intercalate/de-intercalate ~0.7~0.8 mol of Li per formula unit are potential cathodes, even if 20~30% of lithium ions are occupying the 3b positions. Chang et al. have synthesised LiNiO$_2$ with a particle size of 0.3 μm possessing minimum defect concentration through rotary evaporation. The electrodes were cycled at a load-current density of 100 μA/cm^2 (~C/100 rate) to estimate the capacity fade. After the first discharge at which a capacity of 210 mAh/g was achieved, a decline of 1.4 mAh/g per cycle was observed in subsequent cycles [30]. The intercalation/de-intercalation curve shown in Figure 7.17 is characterised by roughly three plateaus—4.18, 4.03 and 3.65 V—which correspond to sharp peaks observed in dQ/dV versus V plot and are assigned to the coexistence of several hexagonal phase (H$_1$, H$_2$, H$_3$) with monoclinic phase (M).

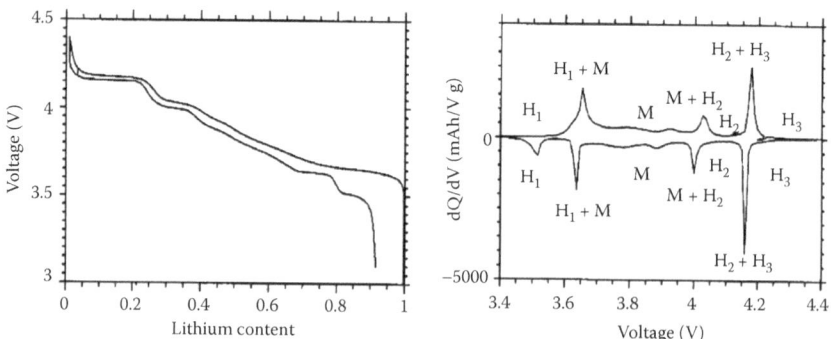

FIGURE 7.17

First cycle voltage profile versus Li content (left) and its differential capacity dQ/dV versus V (right) for LiNiO$_2$. (Reprinted with permission from C.C. Chang, J.Y. Kim and P.N. Kumta, Influence of crystallite size on the electrochemical properties of chemically synthesized stoichiometric LiNiO$_2$, *J. Electrochem. Soc.* 149, 2002, A1114–A1120, Copyright 2002, The Electrochemical Society, Inc.)

Lee et al. synthesised doubly substituted derivatives of LiNiO$_2$ [31]—Li$_{0.962}$Ni$_{0.893}$Co$_{0.099}$Al$_{0.046}$O$_2$ (NCA) and Li$_{0.951}$Ni$_{0.895}$Co$_{0.101}$Fe$_{0.053}$O$_2$ (NCF)—and compared it with Li$_{0.951}$Ni$_{1.049}$O$_2$ (LNO). The lattice parameters were found to be a = 2.8813 Å and c = 14.201 Å for LNO, a = 2.8721 Å and c = 14.186 Å for NCA and a = 2.876 Å and c = 14.203 Å for NCF indexed on a pseudo-hexagonal lattice of the rhombohedral R$_{3m}$ space group. More importantly, upon structural refinement, they found that the additionally substituted second element Co and third elements Al or Fe in LNO occupy the entire 3b Wycoff sites. The lattice-site-specific composition are given by [Li$_{0.95}$Ni$_{0.049}$]$_{3a}$[Ni]$_{3b}$O$_2$ for LNO, [Li$_{0.962}$Ni$_{0.038}$]$_{3a}$[Ni$_{0.855}$Co$_{0.099}$Al$_{0.046}$]$_{3b}$O$_2$ for NCA and [Li$_{0.951}$Ni$_{0.049}$]$_{3a}$[Ni$_{0.846}$Co$_{0.101}$Fe$_{0.053}$]$_{3b}$O$_2$ for NCF with average oxidation states of total metal ions as 2.907, 2.927 and 2.907, respectively. At a discharge rate of C/20, the cycle life performance of NCA was found to be much superior to NCF and LNO, though the first-cycle irreversible capacity of NCA is much lower in relation to LNO and NCF (Figure 7.18). NCA is able to preserve a stable single phase up to a highly oxidised region, which facilitates good cycle life and large reversible capacity [30,31].

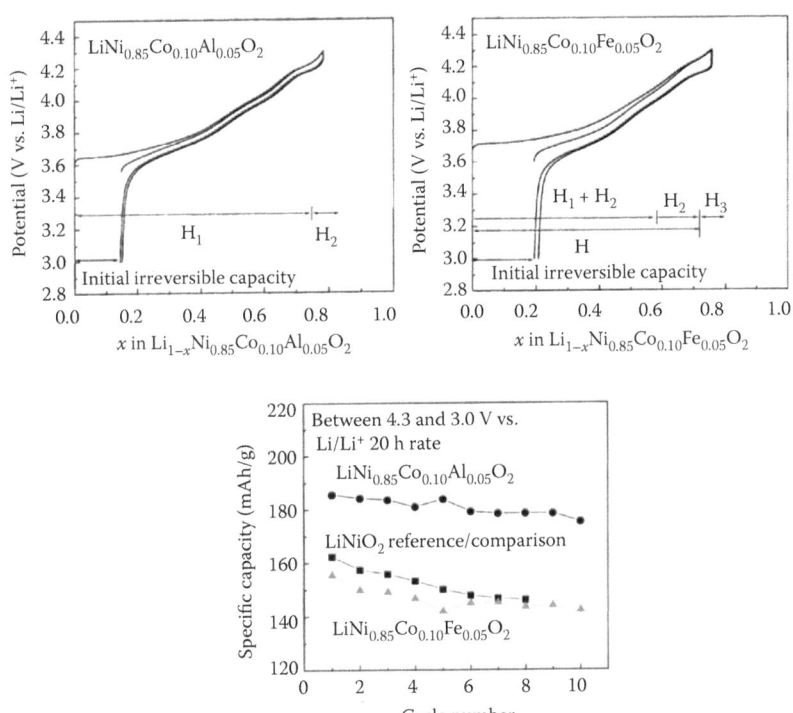

FIGURE 7.18
Galvanostatic lithium intercalation/de-intercalation profile and cycle life data (C/20 rate) of LiNi$_{0.85}$Co$_{0.10}$Al$_{0.05}$O$_2$ and LiNi$_{0.85}$Co$_{0.10}$Fe$_{0.05}$O$_2$. (Reprinted with permission from *J. Power Sources*, 97–98, K.K. Lee et al., Characterization of LiNi$_{0.85}$Co$_{0.10}$M$_{0.05}$O$_2$ (M = Al, Fe) as a cathode material for lithium secondary batteries, 308–312, Copyright 2001, Elsevier.)

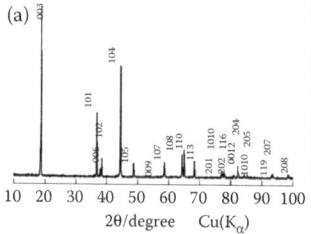 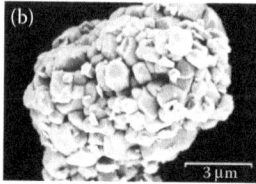 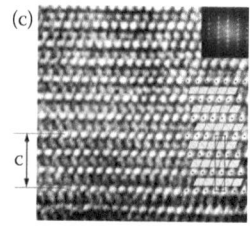

FIGURE 7.19
(a) XRD pattern, (b) SEM image and (c) HRTEM image of $LiNi_{1/3}Co_{1/3}Mn_{1/3}O_2$. (Reprinted with permission from N. Yabuuchi et al., Solid-state chemistry and electrochemistry of $LiNi_{1/3}Co_{1/3}Mn_{1/3}O_2$ for advanced lithium-ion batteries II. Preparation and characterization, *J. Electrochem. Soc.*, 152, 2005, A1434–A1440, Copyright 2005, The Electrochemical Society, Inc.)

MnO_2-based materials are both easily available and environmentally friendly [32]. Among various manganese dioxides, electrochemical manganese dioxide (EMD) is a widely commercialised material due to its use in dry cells. Studies on EMD reflect them to be 3 V-materials and, hence, their application in modern LIBs remains limited.

$LiNi_{1/3}Co_{1/3}Mn_{1/3}O_2$ is also referred as 'NCM' by LIB battery technologists and cathode manufacturers [33,34]. NCM was first synthesised by Ohzuku et al., with the aim to synthesise a cathode material bearing the advantages of cheap manganese transition metal with a reduced cobalt content and the inclusion of nickel for its superior electrochemical capacity. The properly synthesised NCM crystallises in an ordered super structure. The XRD pattern, SEM image and HRTEM image of NCM reported by Ohzuku et al. are shown in Figure 7.19a,b and c, respectively. The electron diffraction pattern is shown in Figure 7.20. The XRD pattern shows sharp and strong diffraction

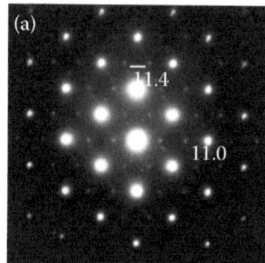 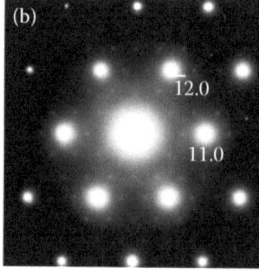 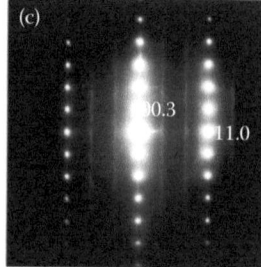

FIGURE 7.20
Electron diffraction patterns of $LiCo_{1/3}Ni_{1/3}Mn_{1/3}O_2$; (a) [22.1], (b) [00.1] and (c) [11.0] zone where the reflection indices are based on the α-NaFeO2-type $R\bar{3}m$ lattice in the hexagonal setting. (Reprinted with permission from N. Yabuuchi et al., Solid-state chemistry and electrochemistry of $LiNi_{1/3}Co_{1/3}Mn_{1/3}O_2$ for advanced lithium-ion batteries II. Preparation and characterization, *J. Electrochem. Soc.* 152, 2005, A1434–A1440, Copyright 2005, The Electrochemical Society, Inc.)

lines that are characteristic of layered structures and the SEM image shows agglomerates of particle sizes from 0.5 to 1.0 µm. The HRTEM image clearly reveals a layered stacking of units indicating that NCM consists of a transition metal sheet and lithium metal sheet piled up alternatively in a cubic close-packing oxygen array. The redox couples were identified by Ozhuku et al. as Ni^{2+}/Ni^{4+}, Co^{2+}/Co^{3+} with Mn remaining unchanged in its valency. $LiNi_{1/3}Co_{1/3}Mn_{1/3}O_2$ can also be prepared from LiOH and $(Ni_{1/3}Co_{1/3}Mn_{1/3})(OH)_2$ by calcination at 1000°C [35]. The galvanostatic charge–discharge curves and cyclability in the higher cut-off voltage studied by Lee et al. is shown in Figure 7.21. The discharge capacity is found to increase gradually with the increase of an upper cut-off voltage. Also, the stable cyclability shows that the Co dissolution is greatly suppressed, indicating that NCM is a more reliable and promising material than $LiCoO_2$.

Patoux et al. have prepared NCM by mixed metal nitrate evoporation followed by calcination at temperatures ranging between 600°C and 1000°C [36]. The samples prepared by this method crystallise in the typical layered α-$NaFeO_2$ structure with lattice parameters: a = 2.86 Å and c = 14.24 Å. The particle size increases progressively from 9 Å for the sample synthesised at 600°C to 70 Å at 1000°C. The first discharge curves from different upper cut-off votlage and their cyclablity is shown in Figure 7.22.

Although several compositions of $Li_xMn_yO_z$ are possible with spinel structures, the AB_2O_4 spinel is the most desired spinel phase in the Li–Mn–O system due to its neat and flat de-intercalation to form a fully delithiated phase of λ-MnO_2. Gummow et al. used ternary phase diagrams shown in Figure 7.23 to review the utilisation of various Li–Mn–O phases in LIBs [37]. Improved capacity retention is demonstrated when a small amount of Mn in the spinel structure is replaced with Li, Mg or Zn. The improved cyclability

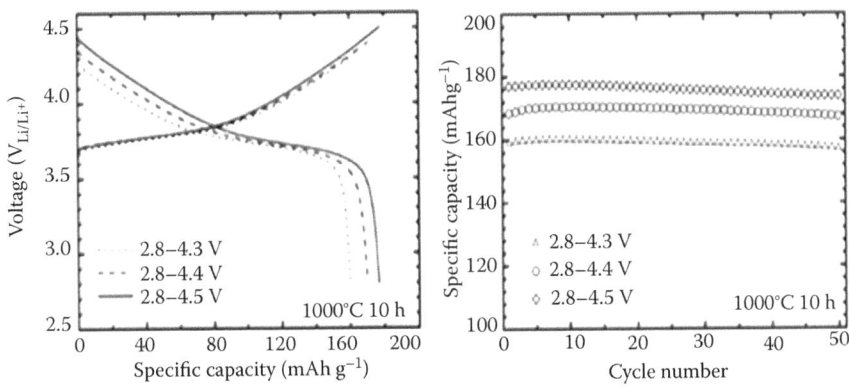

FIGURE 7.21
Galvanostatic charge–discharge curves and cycle life data for $LiNi_{1/3}Co_{1/3}Mn_{1/3}$ prepared at 1000°C. (Reprinted with permission from *Electrochim. Acta*, 50, M.H. Lee et al., Synthetic optimization of $Li[Ni_{1/3}Co_{1/3}Mn_{1/3}]O_2$ via co-precipitation, 939–948, Copyright 2004, Elsevier.)

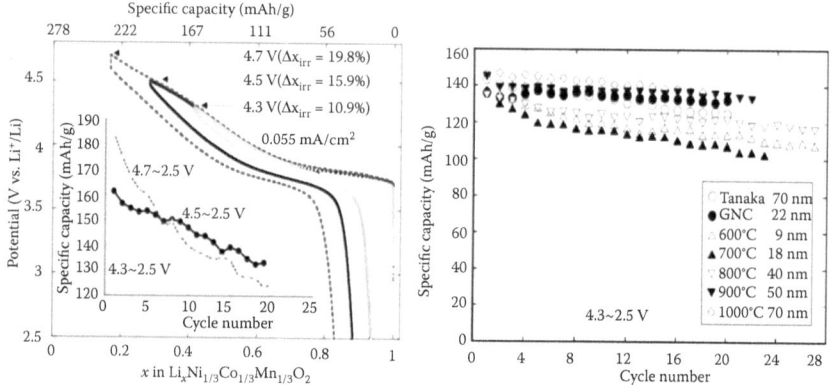

FIGURE 7.22
Typical charge–discharge curves for LiNi$_{1/3}$Co$_{1/3}$Mn$_{1/3}$ sample synthesised at 1000°C charged to varying cut-off potentials with cycle life data for LiNi$_{1/3}$Co$_{1/3}$Mn$_{1/3}$ samples synthesised by at varying temperatures along with the data from Tanaka Inc. and GNC Inc. (Reprinted with permission from *Electrochem. Commun.*, 6, S. Patoux and M.M. Doeff, Direct synthesis of LiNi$_{1/3}$Co$_{1/3}$Mn$_{1/3}$O$_2$ from nitrate precursors. 767–772, Copyright 2004, Elsevier.)

is primarily attributed to the suppression of JT effect on deep discharge of the doped spinel cathodes.

Guan et al. have synthesised nanoparticles of spinel LiMn$_2$O$_4$ and measured the total (electronic and ionic) conductivity for better understanding of energy storage in the material. The spinel LiMn$_2$O$_4$ with ~500–700 nm particles exhibits total conductivity of 1×10^{-4} Ω^{-1}/cm as against the conductivity value of 10^{-6} Ω^{-1}/cm for micron-sized (2 µm) particles sample [38].

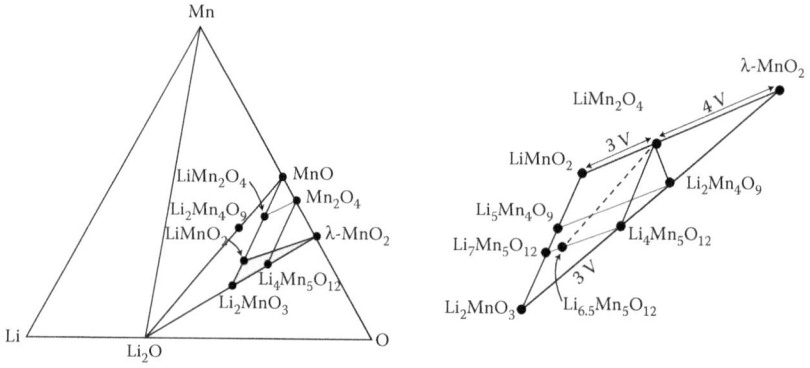

FIGURE 7.23
Triangular phase diagram for Li–Mn–O system with expanded emphasis on Li$_x$MnO$_2$ subsystem. (Reprinted with permission from *Solid State Ionics*, 69, R.J. Gummow et al., Improved capacity retention in rechargeable 4 V lithium/lithium manganese oxide (spinel) cells, 59–67, Copyright 1994, Elsevier.)

The total conductivity was separated into electronic and ionic contributions by blocking electrode measurements. Both electronic and ionic conductivity increase by two orders of magnitude upon reduction of the particle size from 2 µm to 750 nm. Spinel-LiMn$_2$O$_4$ with an average size of ~500–700 nm (sample XE800S) and 2 µm (sample SS800A) demonstrated that increasing the synthesis temperature increases the electronic and ionic conductivities (Figure 7.24).

Mn^{3+} is a d^4-JT cation with high-spin configuration and tends to stabilise in electronic energy through a cubic to tetragonal distortion of MnO$_6$ octahedron. Discharging from the fully charged state of Mn$_2$O$_4$, the oxidation state of the Mn ion progressively decreases from 4.0 to 3.5 for LiMn$_2$O$_4$. The material comprises 50% of Mn^{3+} and 50% of Mn^{4+} ions with an average Mn oxidation state of 3.5. This is the stage at which the 4 V capacities are realised. Any further lithiation insertion is possible given additional capacity, but at the cost of JT distortion and 3 V capacity. Utilising the same cathode material in 4 V, the 3 V regions induces severe strain in the particles, decreasing the cycle life substantially. Thus, any attempt to increase the initial Mn oxidation state to reduce the effect of JT distortion leads to the sacrifice of 4 V capacity. Even the substitution of divalent ions, such as Zn or Mg, would increase the Mn oxidation state, but lower the capacity [37].

Curtis et al. reported the synthesis of nanoparticles of LiMn$_2$O$_4$ spinels by a sol–gel method and studied the effects of annealing LiMn$_2$O$_4$ in a temperature range between 350°C and 950°C. The initial discharge profiles of LiMn$_2$O$_4$ annealed at various temperatures and their cyclability are shown in Figure 7.25. Though the nanoparticles reduce the capacities in the 4 V discharge region, both the capacity and rechargability improved in the 3 V regions [39].

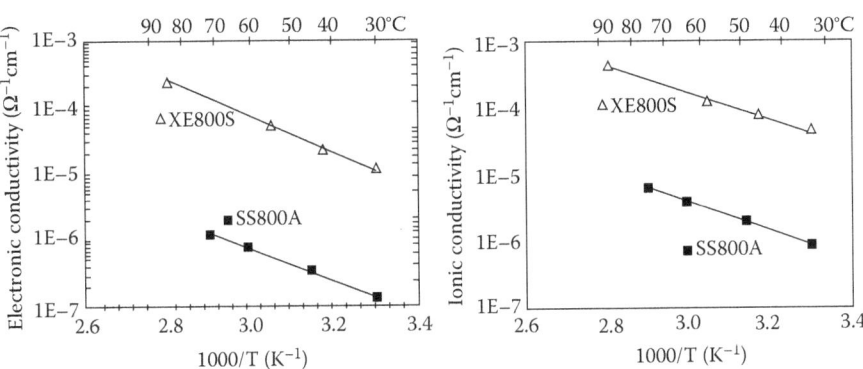

FIGURE 7.24
Electrical conductivity data of LiMn$_2$O$_4$ obtained at varying temperatures resolved into electronic and ionic conductivity data. (Reprinted with permission from *Solid State Ionics*, 110, J. Guan and M. Liu, Transport properties of LiMn$_2$O$_4$ electrode materials for lithium-ion batteries, 21–28, Copyright 1998, Elsevier.)

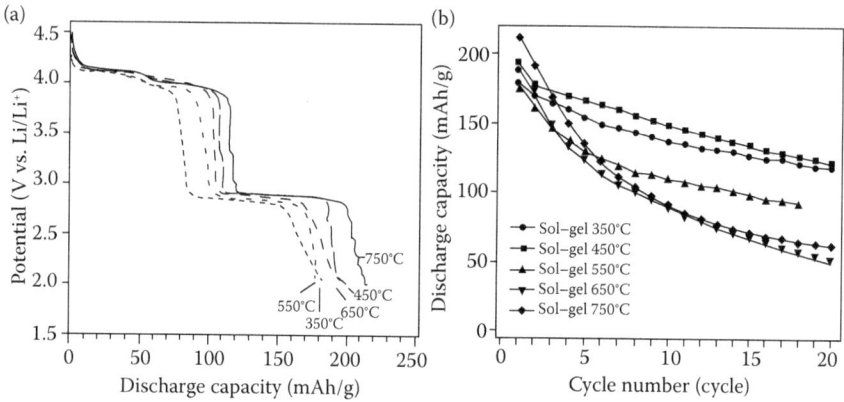

FIGURE 7.25
(a) Galvanostatic discharge profiles and (b) cycle life data of LiMn$_2$O$_4$ samples prepared at 350°C, 450°C, 550°C and 750°C. (Reprinted with permission from *J. Electrochem. Soc.*, 151, C.J. Curtis, J. Wang and D.L. Schulz, Preparation and characterization of LiMn$_2$O$_4$ spinel nanoparticles as cathode materials in secondary Li batteries, 2004, A590–A598m, Copyright 2004, The Electrochemical Society, Inc.)

Hosono et al. have applied a novel strategy for the synthesis of highly oriented spinel-LiMn$_2$O$_4$ nanorods [40]. The ion-exchange of Li source with Na$_{0.44}$MnO$_2$ nanorods results in LiMn$_2$O$_4$. This LiMn$_2$O$_4$ is composed of highly oriented single crystalline nanorods in the growth direction <001> and the perpendicular direction in <100>, illustrated in Figures 7.26 and 7.27. Figure 7.28 shows single crystalline LiMn$_2$O$_4$ nanowires with a high aspect ratio of about several 100s and a diameter of around 50–100 nm.

Figure 7.28 shows the second discharge curves of single crystalline spinel LiMn$_2$O$_4$ nanowires prepared by Hosono et al. [40] and the commercial

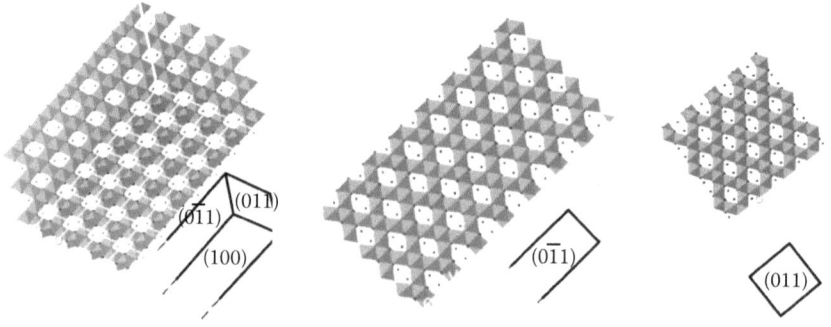

FIGURE 7.26
Pictorial three-dimensional representation of the lattice planes for single crystalline LiMn$_2$O$_4$ nanorods. (Reprinted with permission from E. Hosono et al., Synthesis of single crystalline spinel LiMn$_2$O$_4$ nanowires for a lithium ion battery with high power density, *Nano Lett.* 9, 2009, 1045–1051, Copyright 2009, American Chemical Society.)

FIGURE 7.27
SEM micrograph of nanofibres of $LiMn_2O_4$, TEM micrograph and HRTEM and SAED for nanofibres of $LiMn_2O_4$. (Reprinted with permission from E. Hosono et al., Synthesis of single crystalline spinel $LiMn_2O_4$ nanowires for a lithium ion battery with high power density, *Nano Lett.* 9, 2009, 1045–1051, Copyright 2009, American Chemical Society.)

$LiMn_2O_4$ obtained at various rates of 0.1, 5, 10 and 20 A/g. The discharge capacities at these rates are found to be 118, 108, 102 and 88 mA/g, respectively, indicating that above 90%, 85% and 75% capacities at 0.1 A/g can be retained at higher discharge rates of 5, 10 and 20 A/g. The charge–discharge cycle life data shows the performance of a single crystalline $LiMn_2O_4$ after 100 cycles at a 5 A/g rate, the capacity is still retained without any degradation [40].

Kim et al. [41] synthesised crystalline nanorods of length 1.2 µm and diameter 130 nm with the rod direction <110> in the spinel (Figure 7.29). Although the electrochemical performance is better when compared to micron-sized particles, the specific capacity obtained at a discharge rate of 1 C was just 100 mAh/g.

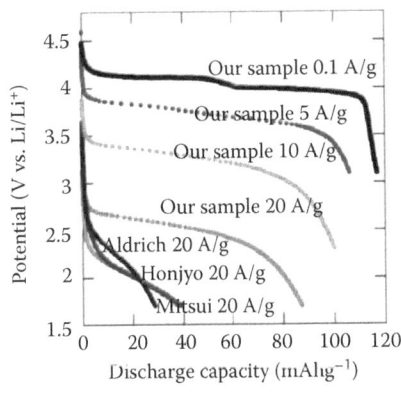
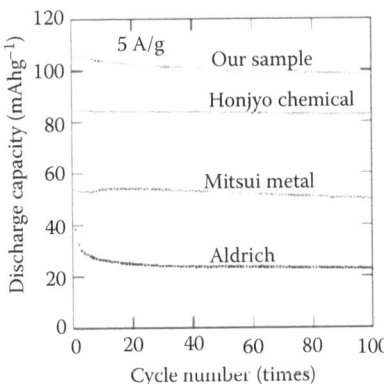

FIGURE 7.28
Discharge curves for various $LiMn_2O_4$ samples at different rates and cycle life data. (Reprinted with permission from E. Hosono et al., Synthesis of single crystalline spinel $LiMn_2O_4$ nanowires for a lithium ion battery with high power density, *Nano Lett.* 9, 2009, 1045–1051, Copyright 2009, American Chemical Society.)

FIGURE 7.29
SEM micrographs at low and high magnifications for $LiMn_2O_4$ nanorods. (Reprinted with permission from D.K. Kim et al., Spinel $LiMn_2O_4$ nanorods as lithium ion battery cathodes, *Nano Lett.* 8, 2008, 3948–3952, Copyright 2008, American Chemical Society.)

$LiFePO_4$ was conceptualised and designed as the cathode material for LIBs by Goodenough et al. in 1996 [42] (Figure 7.30). Figure 7.30 shows the crystal structure of $LiFePO_4$ olivine, which is an ideal hcp model. In this structure, lithium moieties occupy M1 sites and Fe occupy M2 sites. Since M1 sites form linear chains of edge-shared octahedra, Li undergoes reversible extraction/insertion from/into these chains. The chemically de-lithiated $FePO_4$ is found to have an olivine isostructure with the same space group. However, the lattice parameters in $FePO_4$ are slightly increased. A schematic representation of the interface shown in Figure 7.30c illustrates that Li insertion proceeds from the surface of the particle moving inward behind a two-phase interface. As

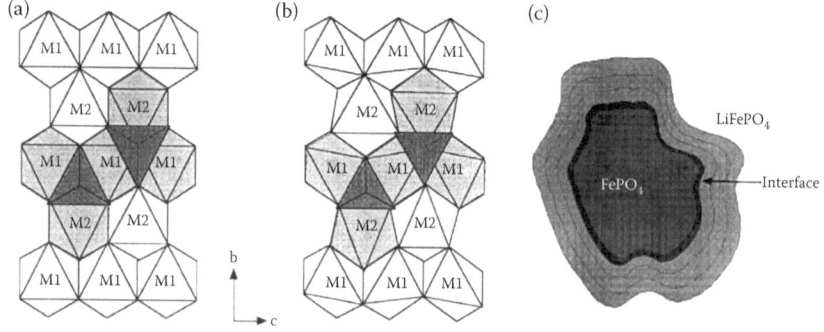

FIGURE 7.30
(a) Ideal crystal structure of olivine, (b) Actual structure of $LiFePO_4$ and (c) Schematic representation of interface between $LiFePO_4$ and $FePO_4$. (Reprinted with permission from A.K. Padhi, K.S. Nanjundaswamy and J.B. Goodenough, Phospho-olivines as positive-electrode materials for rechargeable lithium batteries, *J. Electrochem. Soc.* 144, 1997, 1188–1194, Copyright 1997, The Electrochemical Society, Inc.)

Li insertion proceeds, the surface area decreases and a critical surface area is reached where the Li transportation no longer sustains the current and the cell performance becomes diffusion-limited. Electrochemical intercalation was limited to ~0.6 Li/formula unit, and a specific capacity of 100–110 mAh/g was achieved. Wang et al. [43] reported the synthesis of 1D nanowires of $LiFePO_4$ by hydrothermal methods using ferrous sulphate, ammonium dihydrogen phosphate and lithium hydroxide. These 1D nanowires had a diameter in the range of few hundred nanometres and lengths extending a few tens of micrometres. The first cycle capacity of these nanoparticles was ~150 mAh/g and remained stable at 140 mAh/g for up to 70 cycles.

Kim et al. [44] synthesised $LiFePO_4$ nanoparticles of about 20–40 nm width and length in size by a polyol medium of Fe acetate and Li acetate by refluxing them at 335°C for 16 h. The crystallites obtained in this synthesis possessed a large area of crystallographic <010> planes, which contribute to the facile one-dimensional Li-ion diffusion in the orthorhombic lattice (Figure 7.31). These $LiFePO_4$ nanoparticles showed a reversible capacity of 166 mAh/g (Figure 7.32).

$LiFePO_4$ has inherent low electronic conductivity that results in its poor rate capability. Hwang et al. [45] attempted to improve the electronic conductivity of $LiFePO_4$ by synthesising it as a carbon composite by sol–gel method. The electronic (electrical) conductivity of $LiFePO_4$–carbon composite comprising around 25–30 nm crystallites increases progressively over the synthesis temperature range of 650–950°C (Table 7.3). At 850°C, the sample reaches optimum conductivity delivering a capacity of 148 mAh/g with the retention of carbon coating.

Single crystalline $LiFePO_4$ nanorods of length 1 μm and width 40 nm with orthorhombic crystal structure: a = 10.320 Å, b = 6.000 Å, c = 4.697 Å was synthesised by Murugan et al. [46]. Since $LiFePO_4$ have poor electronic and ionic conductivities, these nanoparticles were subsequently coated by a mixed ionic–electronic conducting polymer, PEDOT (p-toluenesulphonic acid doped

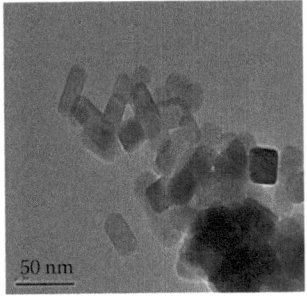
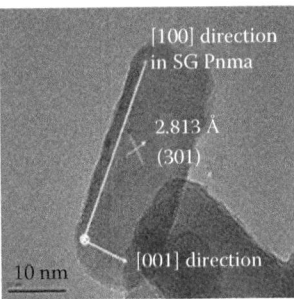

FIGURE 7.31
TEM micrographs and HRTEM images of $LiFePO_4$ nanoparticles [44]. (Reprinted with permission from D.H. Kim and J.K. Kim, Synthesis of $LiFePO_4$ nanoparticles in polyol medium and their electrochemical properties, *Electrochem. Solid State Lett.* 9, 2006, A439–A442, Copyright 2006, The Electrochemical Society, Inc.)

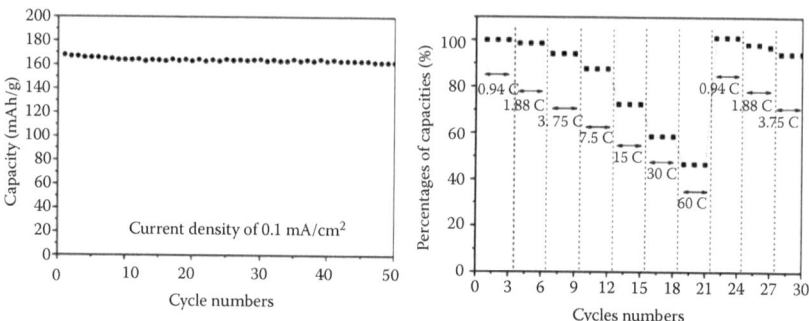

FIGURE 7.32
Cycle life data and rate capability for LiFePO$_4$ nanoparticles at varying temperatures. (Reprinted with permission from D.H. Kim and J.K. Kim, Synthesis of LiFePO$_4$ nanoparticles in polyol medium and their electrochemical properties, *Electrochem. Solid State Lett.* 9, 2006, A439–A442, Copyright 2006, The Electrochemical Society, Inc.)

TABLE 7.3
Electronic Conductivity Data for Carbon-Coated LiFePO$_4$ Samples Synthesised at Varying Temperatures

Synthesis Temperature (°C)	Conductivity (S/cm)	H/C Ratio
400	1.63×10^{-7}	0.090
500	7.19×10^{-7}	0.073
550	2.51×10^{-6}	0.055
650	3.24×10^{-4}	0.041
750	9.99×10^{-4}	0.032
850	2.46×10^{-3}	0.025

Source: Reprinted with permission from K.F. Hsu et al., Synthesis and characterization of nano-sized LiFePO$_4$ cathode materials prepared by a citric acid-based sol–gel route, *J. Mater. Chem.* 14, 2004, 2690–2695. Copyright 2004 Royal Society of Chemistry.

poly(3,4-ethylene dioxythiophene) (Figure 7.33). The as-prepared LiFePO$_4$ nanorods, which are about 40 nm in size, exhibited a reversible capacity of 135 mAh/g capacity. The PEDOT-coated nanorods of LiFePO$_4$ exhibited a reversible capacity of 166 mAh/g, very close to the theoretical capacity of 170 mAh/g, and better rate capability than as-prepared LiFePO$_4$ (Figure 7.34).

7.4.2 Anode Materials

7.4.2.1 Lithium Metal

Several studies have been conducted on rechargeable non-aqueous batteries with Li as the negative plate. Anodes with metallic lithium can store about a 10 times higher energy per unit weight than lithiated carbon. Although

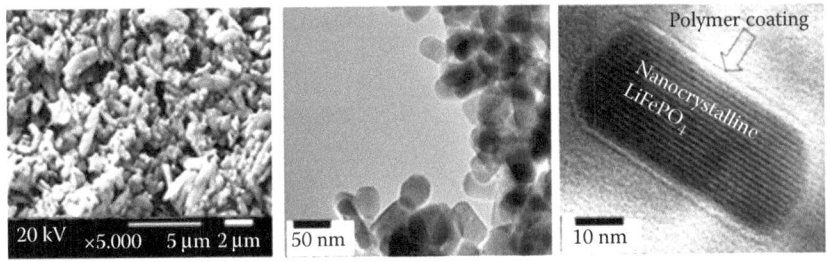

FIGURE 7.33
SEM micrograph of LiFePO$_4$, TEM micrograph of LiFePO$_4$ and HRTEM images of PEDOT-coated LiFePO$_4$. (Reprinted with permission from *Elecrochem. Commun.*, 10, M.A. Vadivel, T. Muraliganth and A. Manthiram, Rapid microwave-solvothermal synthesis of phospho-olivine nanorods and their coating with a mixed conducting polymer for lithium ion batteries, 903–906, Copyright 2008, Elsevier.)

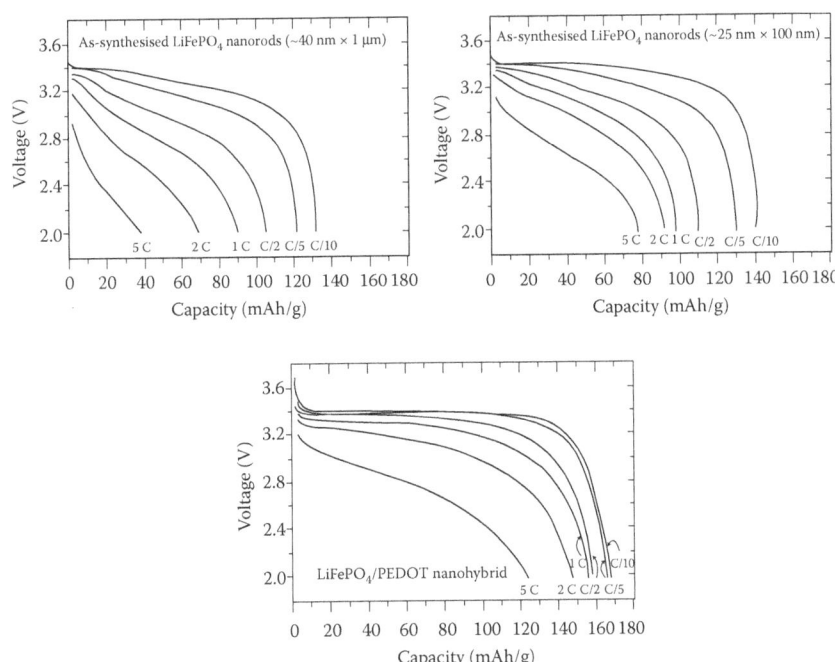

FIGURE 7.34
Typical discharge profiles collected at various rates for the samples: LiFePO$_4$ consisting of 40 nm × 1 μm nanorods, LiFePO$_4$ consisting of 25 nm × 100 nm nanorods and PEDOT-coated LiFePO$_4$. (Reprinted with permission from *Elecrochem. Commun.*, 10, M.A. Vadivel, T. Muraliganth and A. Manthiram, Rapid microwave-solvothermal synthesis of phospho-olivine nanorods and their coating with a mixed conducting polymer for lithium ion batteries, 903–906, Copyright 2008, Elsevier.)

primary cells have been produced commercially, secondary cells employing Li metal as anodes are plagued with problems such as surface passivation, dendritic growth of Li, internal shorting and build-up of cell pressure during repeated cycling, electrical isolation of active Li, increase of internal resistance, thermal instability, cell explosion and so on. As a consequence, R&D efforts have been initiated on alternate anode materials, such as carbon–silicon composites, that will act as a host for reversible insertion–extraction of Li^+ ions.

7.4.2.2 Carbonaceous Materials

In the last two decades, a variety of carbonaceous materials, both natural and synthetic, have been examined for their lithium intercalation properties as anodes in LIBs. Although several carbonaceous materials, especially disordered carbons, have been demonstrated to possess lithium insertion capacities, their capacity fade upon repeated cycling has impeded their implementation in LIBs.

Carbon has become the material of choice for anodes in the present-generation lithium-ion cells. Coke-based anodes offer a stable specific capacity of 180 mAh/g. In contrast, hard carbon materials offer specific capacities over 1000 mAh/g but have not achieved broad acceptance due to their instability and irreversibility on cycling. Indeed, the commercial and industrial goal is to reach specific capacities exceeding 600 mAh/g for hundreds of cycles.

Graphitic carbon is considered to be an important material for various batteries. The crystalline-layered structure allows the individual graphene layers to freely move independent of each other under shearing conditions of stress. This arises as carbon graphene layers are separated by a perpendicular distance of 3.4 Å and held coplanar to each other by weak van der Waals forces. Small external forces such as mechanical shear are sufficient to move the layers apart, giving other physical properties such as high electrical conductivity and high thermal conductivity.

Graphite exists in two polymorphs—hexagonally packed form (2H) and rhombohedrally packed form (3R)—which differ by 0.6 kJ/mol with the 2H being thermodynamically more stable than 3R [47]. In a 2H graphite structure, the hexagonal networks of carbon layers are arranged in the sequence of …ABABAB… as shown in Figure 7.35a, where the B layers are shifted to a registered position with respect to A layers. In the 3R structure, the stacking sequence is …ABCABC… as shown in Figure 7.35b, where the C layers are shifted by the same distance with respect to the B layers as the B layers are shifted with respect to the A layers. Figure 7.35c shows the XRD patterns of pure 2H and mixture of 2H and 3R in different contents. The two phases are identifiable from XRD patterns. The 2H form can be converted to the 3R form through grinding and the reverse is achieved by heating to higher temperatures. From the application point of view, as anodes of LIBs, the 2H and 3R

Advanced Batteries and Improvements in Electrode Materials 289

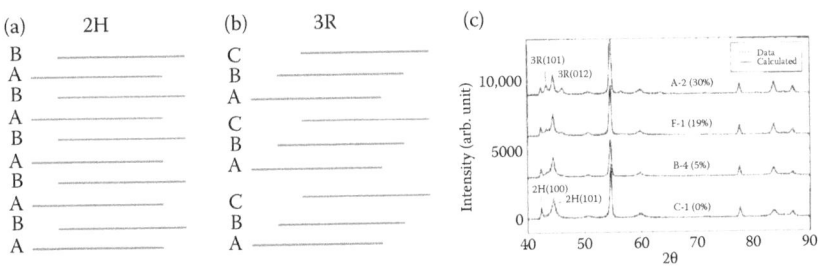

FIGURE 7.35
A schematic representation of (a) 2H, (b) 3R graphitic crystallographic forms and (c) powder XRD for pure 2H and the mixture of 2H and 3R. (Reprinted with permission from *J. Power Sources*, 68, H. Shi et al., Graphite structure and lithium intercalation, 291–295, Copyright 1997, Elsevier.)

forms differ by a voltage of 6 mV at the plateau of reversible insertion/de-insertion. Any attempted phase conversions of graphite induces turbostratic disorder that tilts the individual graphene layers in a random fashion like 'an unorganised pack of cards', although the structural difference is detected by powder XRD. The inter-layer distance is the same in both the forms being 3.36 Å and the C–C distance within the layer is 1.42 Å.

Crystallographic studies indicate that lithium de-intercalation from pre-lithiated graphite proceeds through specified stages of phases, popularly called 'staging' [48]. A typical example of a discharge curve is shown in Figure 7.36. The fully lithiated phase is called stage-1 corresponding to $Li_{1.0}C_6$, followed by stage-2 corresponding to $Li_{0.5}C_6$, followed by C_6. During lithium intercalation, the graphene layers that are shifted by half the value of the 'a'

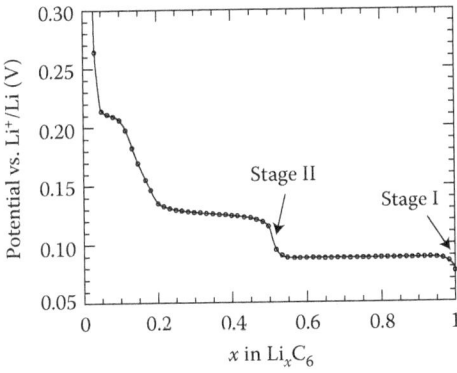

FIGURE 7.36
Voltage as function of Lithium composition in Li_xC_6 during discharge. (Reprinted with permission from J.-S. Filhol et al., Phase diagrams for systems with low free enrgy variation: A coupled theory/experiments method applied to Li-graphite, *J. Phys. Chem.*, 112, 2008, 3982–3988, American Chemical Society.)

axis in the 'a and b' direction are brought, coincidentally, exactly one carbon on top of other, which results in the lithium atom occupying the middle position between the top six and bottom six carbon atoms. This gives the coordination configuration LiC_{12}. In the stage-1 material ($Li_{1.0}C_6$), every middle position is occupied by carbon, whereas in the stage-2 material ($Li_{0.5}C_6$), every alternate lithium position is occupied. During stage-2 of lithiation, the inter-layer distance increases by a small factor, and again during stage-1 of lithiation, the distance increases by another small factor. These stages are identified with the help of slow scan cyclic voltammetry or by differentiating the slow-scan galvanostatic curve of lithium insertion–de-insertion. The voltammogram exhibits cathodic and anodic peaks corresponding to the transition from one stage of lithium intercalation to other. Additionally, different stages of lithiated graphite are identified by using a high-resolution XRD.

7.4.2.2.1 Natural Graphite

Utilisation of natural graphite mined from the earth's crust is considered to be the best source of graphitic carbon for anodes in LIBs. Natural graphite is available in several locations, such as Sri Lanka, China, India and Brazil.

Natural graphites typically behave like soft carbons with respect to the shape of the intercalation–de-intercalation profiles and the plateau potential, but are little inferior in reversible capacity values [49]. Natural graphites are naturally formed mineral-grade graphite, usually in large particles (millimetres). Reduction of particle size enhances BET surface area and edge surface area, thereby improving the lithium intercalation/de-intercalation property, and allowing capacities in the range of ~250 mAh/g to be attained (Figure 7.37).

Natural graphite performs unsatisfactorily when used as an LIB anode. There are several methods reported to improve the performance of natural graphites such as thermal modification or thermal treatment, and coating the natural graphite with graphitised carbon [50–52]. Menachem et al. studied the effect of burning graphite to remove volatile and related impurities [50]. From their studies, it is observed that the density of graphite is retained to the same extent of 2.20 (to 2.18) even after burning 11% of the contents. The lithium intercalation capacity increased from 350 mAh/g for $Li_{0.94}C_6$ to 391 mAh/g for $Li_{1.05}C_6$. The irreversible capacity is reduced from 156 to 141 mAh/g, and the degradation during cycling is also decreased from 1.3% to 0.6% per cycle. Both studies indicate that natural graphites contribute to the undesirable irreversible capacity. Yoshio et al. have varied the amount of carbon coating on natural graphite by the thermal vapour deposition method [51], and observed that increasing the carbon coating amount increases the average particle size with a decrease in BET surface area (as summarised in Table 7.4). The electrochemical performance of these carbon-coated natural samples was investigated in propylene carbonate (PC)-based electrolyte and ethylene carbonate (EC) mixed with dimethyl carbonate (DMC) in different ratios.

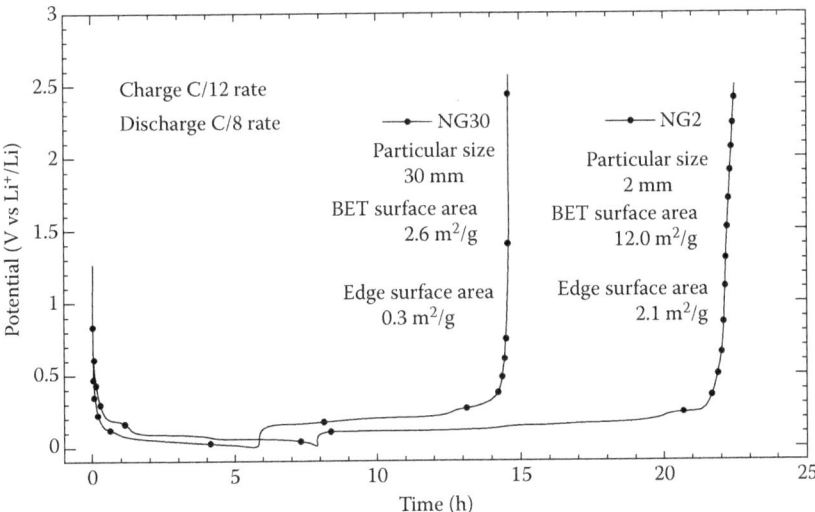

FIGURE 7.37
Lithium intercalation/de-intercalation profiles of two samples of natural graphites as cathodes in LIBs. (Reprinted with permission from *J. Power Sources*, 103, K. Zaghib et al., Effect of particle size on lithium intercalation rates in natural graphite, 2001, 140–146, Copyright 2001, Elsevier.)

The first discharge and charge capacity of these samples are very dependent on the amount of carbon coating, as presented in Table 7.5. As the amount of carbon coating on natural graphite increases, the capacity loss decreases, thus increasing the charge-to-discharge coulombic efficiency.

Zhang et al. attempted to coat natural graphite particles with a polymer, polyvinyl chloride (PVC), and further carbonised the polymer layer by thermal treatment at 900°C [52]. Thus, prepared carbon-coated natural graphite samples are found to possess superior properties such as higher intercalation

TABLE 7.4
Physical Parameters for Carbon-Coated Natural Graphite Samples

Carbon Coating on Graphite (wt.%)	BET Surface Area (m²/g)	Particle Size (μm)	d_{002} (nm)	$L_{c(002)}$ (nm)
0	5.6	12.6	0.33544	102.4
8.6	4.5	18.5	0.33555	83.8
13.4	3.1	20.4	0.33553	135.2
17.6	2.5	22.3	0.33553	93.3

Source: Reprinted with permission from M. Yoshio et al., Effect of carbon coating on electrochemical performance of treated natural graphite as lithium-ion battery anode material, *J. Electrochem. Soc.*, 147(4), 2000, 1245–1250, Copyright 2000, The Electrochemical Society, Inc.

TABLE 7.5

Electrochemical Performance for Carbon-Coated Natural Graphite Samples in Various Electrolyte Solutions

Amount of Carbon Coating (wt.%)	EC:DMC (1:2) 1 M LiPF$_6$ (mAh/g)			PC:DMC (1:2) 1 M LiPF$_6$ (mAh/g)		
	1st Discharge Capacity	1st Charge Capacity	Irreversible Capacity	1st Discharge Capacity	1st Charge Capacity	Irreversible Capacity
0	392	341	50	583	308	275
8.6	412	348	64	434	319	116
13.4	381	335	46	404	309	95
17.6	357	338	20	397	317	81

Source: Reprinted with permission from M. Yoshio et al., Effect of carbon coating on electrochemical performance of treated natural graphite as lithium-ion battery anode material, *J. Electrochem. Soc.*, 147(4) 2000, 1245–1250, Copyright 2000, The Electrochemical Society, Inc.

capacity, better cycle life and moderately higher rate capability than mesocarbon micro-beads (MCMB), reaching 325 mAh/g at 0.1 C as shown in Figure 7.38. It is also found that the sequence of heating has a pronounced effect on the type of carbon coating, which in turn brings the improvement in intercalation/de-intercalation capacity [52].

7.4.2.2.2 Synthetic Graphite

The anodes of LIBs may consist of either soft or hard carbon [53]. Soft carbon may be considered as typical (ordered) graphite. A typical example for soft carbon is the carbonisation of petroleum pitch at 2850°C for 1 h under the flow of inert gas. Dahn et al. have reported a study on different hard and soft carbon samples from various sources, including petroleum pitch from a Japanese source. Figure 7.39 shows the voltage versus composition profiles during first charge and discharge cycle and second discharge for Li/soft carbon cells where soft carbon is heat-treated to ~2850°C. The inset shows a change in composition with respect to voltage, that is, dx/dV against voltage. The soft carbon samples are capable of intercalating 1 mol of Li per 6 mol of carbon giving a lithiated phase with compositional formula: Li$_{1.0}$C$_6$. A fairly flat discharge curve at a voltage close to 0.1 V vs. Li/Li$^+$ was observed. Upon de-insertion, almost 0.9 equivalent of lithium is de-intercalated from the pre-lithiated material at a flat potential just few mV above the intercalation potential. The remainder 0.1 equivalent lithium formula units with 6 mol of carbon (Li$_{0.1}$C$_6$) is termed as 'irreversible capacity'. Irreversible capacity depends on many factors ranging from particle size of the graphite crystals to electrolyte effects, solid electrolyte interface (SEI) layer and so on. Further cycling yields a reversible capacity of ~340 mAh/g with Li$_{0.9}$C$_6$ (Figure 7.40).

The theoretical as well as the reversible capacities are found to increase with the H/C ratio [54]. Porous carbons were prepared from a mesophase

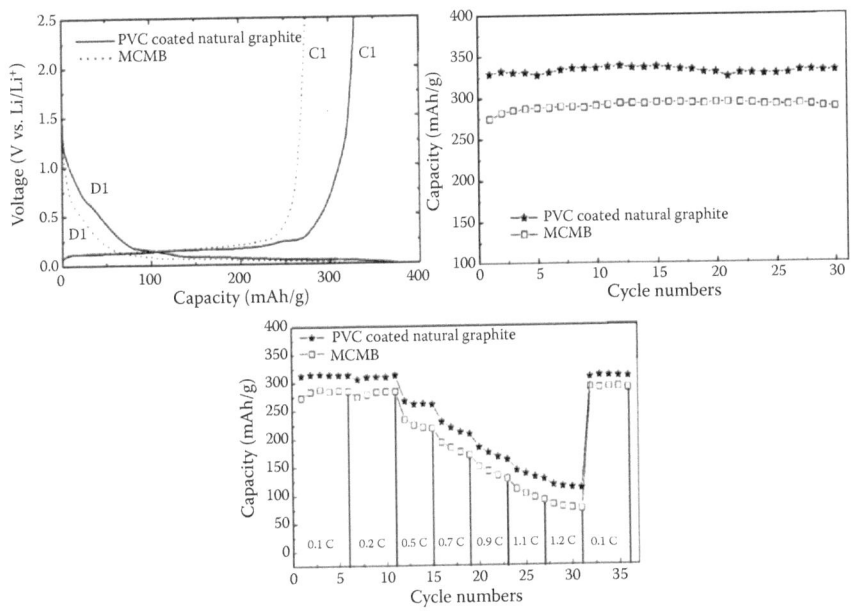

FIGURE 7.38
Galvanostatic lithium intercalation/de-intercalation profiles, cycle life data and rate capability for PVC coated natural graphite and MCMB samples. (Reprinted with permission from H.L. Zhang et al., Poly(vinyl chloride) (PVC) coated idea revisited: Influence of carbonization procedures on PVC-coated natural graphite as anode materials for lithium ion batteries. *J. Phys. Chem. C*, 122, 2008, 7767–7772, Copyright 2008, American Chemical Society.)

pitch by template method (replica method). When the carbonisation temperature is increased from 700°C to 2500°C, the intercalation capacity decreases drastically from 1600 to 500 mAh/g, due to the progressive loss of hydrogen content. Fujimoto et al. have tested MCMB samples heat-treated at varying temperatures between 700°C and 1000°C [55]. Upon heat treatment, the H, O and N contents decrease, making the samples carbon rich. But the samples prepared at lower heat treatment temperatures (~700°C) exhibit higher initial intercalation/de-intercalation capacity as well as a higher associated irreversible capacity. Heat treatment up to 1000°C is conducted to observe an intercalation capacity of 531 mAh/g and de-intercalation capacity of 325 mAh/g. The irreversible capacity is ~206 mAh/g, which is too high for the carbons to be employed as anodes for LIBs (Figure 7.40).

Hard carbons can be produced from several carbon sources, such as polymers, petroleum products, carbohydrates and so on. Hard carbons are characterised by high H/C ratio. Hard carbon produced from sugar pyrolysis at a relatively low temperature (~1050°C) is found to have a hydrogen content of around 0.5% with high initial (first cycle) capacity of around 750 mAh/g of which ~250 mAh/g of capacity comes in the region between 1.0 and 0.1 V

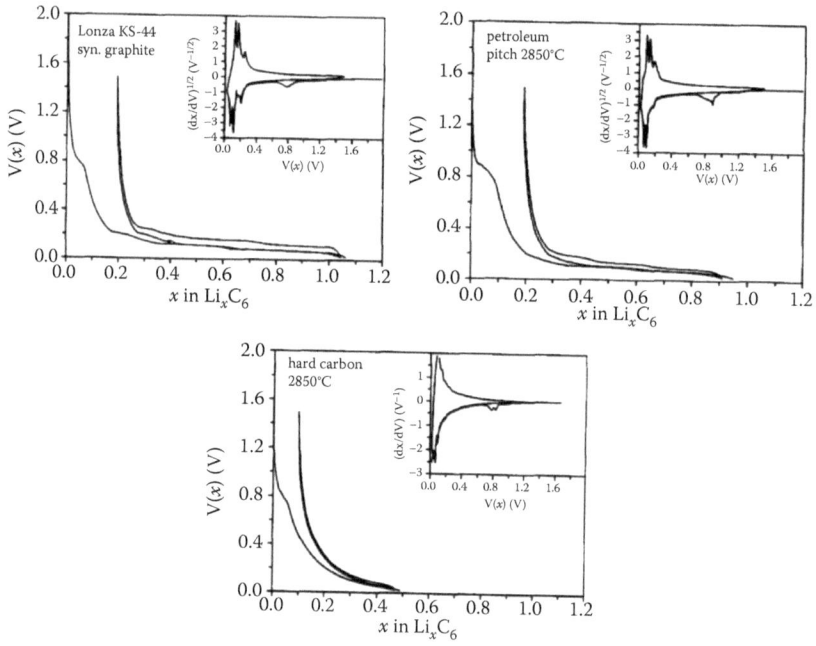

FIGURE 7.39
Lithium intercalation/de-intercalation profiles of synthetic graphite (Lonza KS-44), petroleum pitch (2850°C) and hard carbon (2850°C). (Reprinted with permission from J.R. Dahn et al., Dependence of the electrochemical intercalation of lithium in carbons on the crystal structure of the carbon, *Electrochim. Acta*, 38, 1993, 1179–1191, Copyright 1993, Elsevier.)

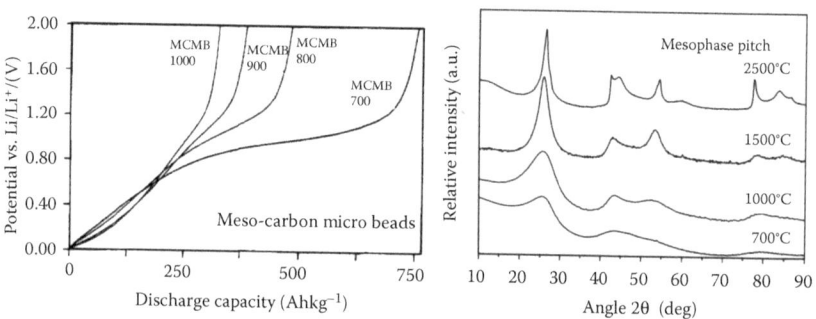

FIGURE 7.40
Lithium de-intercalation profiles of MCMB carbons heat-treated at 700, 800, 900 and 1000°C, respectively. Powder XRD patterns of mesophase pitch-derived carbon pyrolysed at varying temperatures. (Reprinted with permission from *J. Power Sources*, 54, H. Fujimoto et al. Irreversible capacity of lithium secondary battery using meso-carbon micro beads as anode material, 1995, 440–443, Copyright 2007. With permission from Elsevier, and Y.S Hu et al., Synthesis of hierarchically porous carbon monoliths with highly ordered microporous and their application in rechargeable lithium batteries with high rate capability, *Adv. Funct. Mater.*, 2007, 17, 1873–1878, Copyright 2007, Wiley-VCH Verlag GmBH & Co.)

and the remaining 500 mAh/g comes from the flat plateau at around 0.1 V vs. Li/Li$^+$.

Graphitic carbons (mostly soft carbons) can take up different morphologies depending on the method of preparation and carbon source [56]. Lampe-Onnerud et al. have reported carbons of different morphologies: beads, fibres, flakes and potatoes (as shown in the SEM images in Figure 7.41, and the results summarised in Table 7.6 and Figure 7.42). The crystallite size of beads and potatoes are around ~20 nm in the direction perpendicular to 'graphene planes', where the flakes are extended to ~100 nm in the 'c-direction'. Graphite flakes show very high reversible capacities, close to 370 mAh/g for the initial 30 cycles, due to the higher degree of graphitisation, after which the capacity falls rather steeply. Fibres show no advantages in terms of storage capacity or cycle life. The beads and potatoes have morphologies similar to graphite edges (i.e. closed tapered and blunt edges), exhibit specific capacity of ~350 mAh/g and long cycle life, due to limited or minimised passivation of the edges, leading to thinner SEI formation (Figures 7.41 and 7.42 and Table 7.6). The capacity loss is reduced by one or more of the following

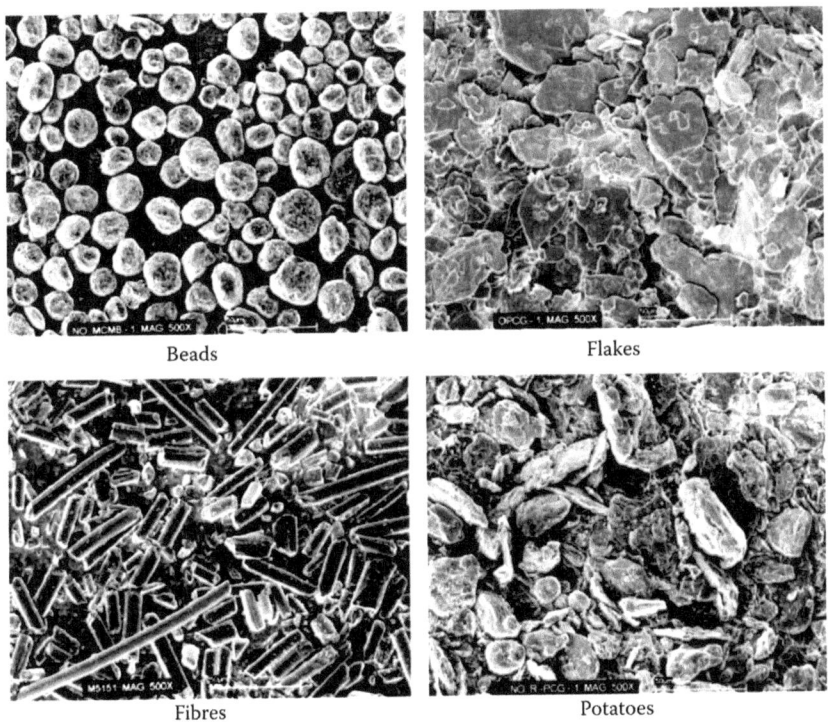

FIGURE 7.41
SEM micrographs of four different morphologies of graphite samples. (Reprinted with permission from *J. Power Sources*, 97–98, C. Lampe-Onnerud et al., Benchmark study on high performing carbon anode materials, 2001, 133–136, Copyright 2001, Elsevier.)

TABLE 7.6
Physical, Chemical and Electrochemical Properties of Graphites Having Different Morphologies: Beads, Flakes, Fibres and Potatoes

Graphite Sample	L_c (nm)	L_a (nm)	Particle Size 50% Diameter (μm)	Surface Area (m²/g)	Electrode Density (g/cc)	1st Cycle Capacity (mAh/g)	1st Cycle Capacity in Cell (mAh/g)	1st Cycle Capacity in Cell (mAh/cc)	Capacity Retention after 20 Cycles
Beads	21	76	21.6	1.08	1.65	346	311	513	99%
Fibres	17	66	17.9	1.30	1.61	325	293	471	99%
Flakes	>100	>100	22.0	2.97	1.91	370	333	637	96%
Potatoes	16	73	19.1	2.42	1.43	350	315	451	99%

Source: Reprinted with permission from *J. Power Sources*, 97–98, C. Lampe-Onnerud et al., Benchmark study on high performing carbon anode materials, 2001, 133–136. Copyright 2001, Elsevier.

methods: attaining high degree of graphitisation, minimising the surface oxygen concentration, attaining a large crystal size and use of PVDF instead of Teflon as binder [57].

Pyrolysis of aromatic hydrocarbons is one of the effective processes to form graphene-like sheets and subsequently stacking them to form graphitic carbons. Most of these carbons are hard carbons, that is, poorly ordered carbons, like a randomly shuffled pack of cards.

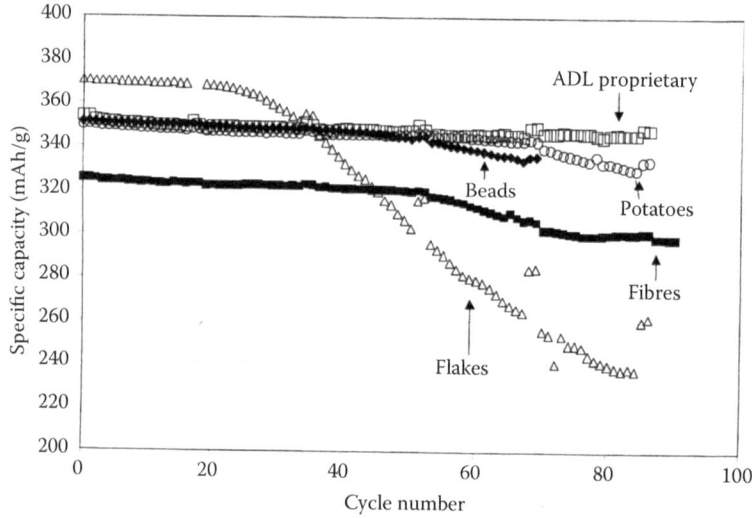

FIGURE 7.42
Cycle life data for graphite samples with different morphologies: beads, flakes, fibres and potatoes. (Reprinted with permission from *J. Power Sources*, 97–98, C. Lampe-Onnerud et al., Benchmark study on high performing carbon anode materials, 2001, 133–136, Copyright 2001, Elsevier.)

Hara et al. have prepared high-capacity carbons through the pyrolysis of 3,4,9,10-peryleneterta-3,4,9,10-carboxylic di-anhydride (PTCDA) at temperatures ranging between 650°C and 1000°C [58], as demonstrated in Figure 7.43a. The PTCDA upon pyrolysis undergo a three-phase stepwise formation of small graphite layers, as illustrated in Figure 7.43b. In phase (1), the initial carbonisation begins as the polymerisation and aromatisation of the PTCDA; this is followed by phase (2)—the growth of small condensed aromatics to large condensed aromatics as small graphene-like layers, and subsequently phase (3)—three dimensional growth of these small graphene-like layers into buckled and small stacked graphite layers. Carbon samples prepared at different pyrolysis temperature thus differentiated by their hydrogen-to-carbon (H/C) ratio. These XRD patterns (Figure 7.44a) correspond to the layer-by-layer structure of carbon indicating pyrolysed PTCDAs are not in amorphous state but disordered carbon comprising graphite layers. It is found that the H/C ratio decreases drastically from 0.16 to as low as 0.01. The first galvanostatic charge–discharge cycle for

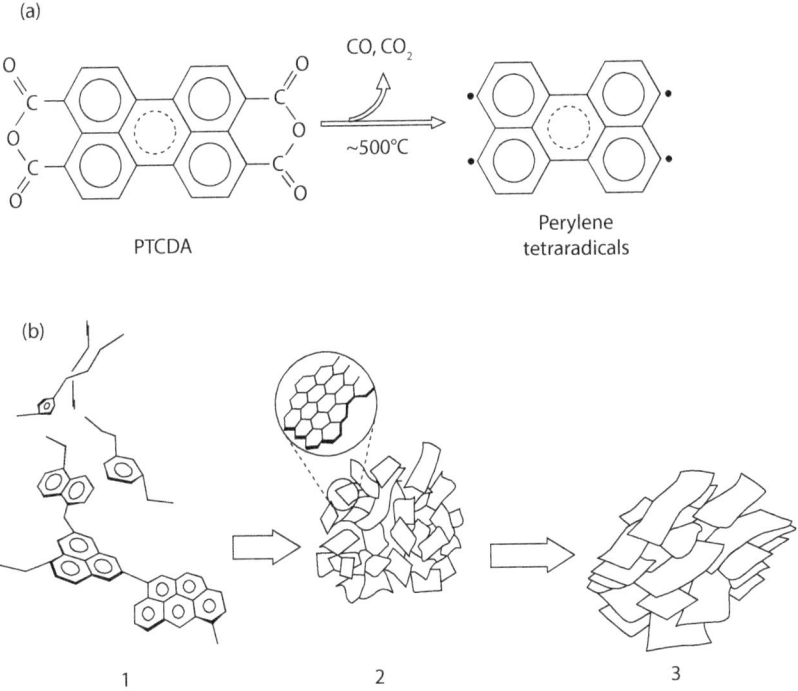

FIGURE 7.43
(a) Chemical reaction process involving the formation of perylene tetraradicals from PTCDA and (b) step-wise carbonisation process of perylene tetraradicals to small layered graphite. (Reprinted with permission from M. Hara et. al., Structural and electrochemical properties of lithiated polymerized aromatics. Anodes for Lithium-ion cells, *J. Phys. Chem.*, 1995, 99, 16338–16343, Copyright 1995, American Chemical Society.)

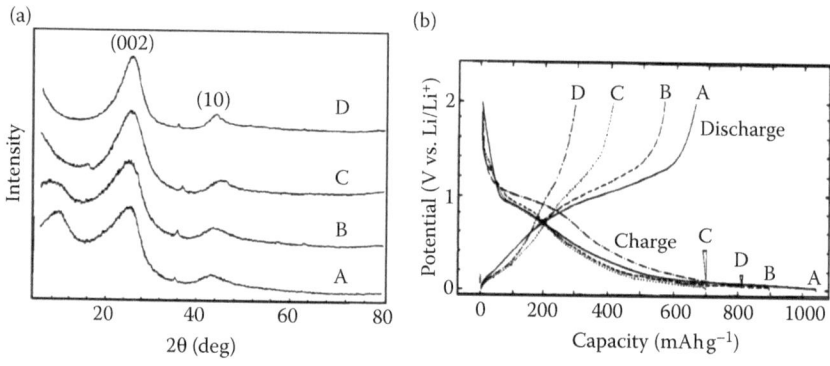

FIGURE 7.44
(a) Powder XRD patterns and (b) first-cycle galvanostatic charge–discharge curves for samples A, B, C and D, where sample A was prepared at 550°C with H/C ratio of 0.26; sample B at 650°C with H/C ratio of 0.16; sample C at 800°C with H/C ratio of 0.05 and sample D at 1000°C with H/C ratio of <0.01. (Reprinted with permission from M. Hara et. al., Structural and electrochemical properties of lithiated polymerized aromatics. Anodes for Lithium-ion cells, *J. Phys. Chem.*, 1995, 99, 16338–16343, Copyright 1995, American Chemical Society.)

all these samples is shown in Figure 7.44b. During the first cycle, carbon sample prepared at 650°C with high H/C ratio yields a superior discharge capacity of ~1020 mAh/g followed by a charge capacity of ~620 mAh/g, which amounts to an irreversible capacity of ~400 mAh/g. The detailed IR spectral studies revealed that the reversible capacity was attributed to doping–undoping of the ionic complex as lithium naphthalene for samples pyrolysed below 650°C. With progressive heat treatments, carbons lose their hydrogen content and perform poorly. It is found that the H/C ratio is critical for good performance of pyrolysed carbons. The reversible capacities decrease with increasing pyrolysis temperature. The discharge profiles of the disordered graphite of lower H/C ratios (≤0.05) (samples C and D) differ from those of higher H/C ratios (samples A and B). These features indicate that the structural shift from the polymer to the disordered carbon with small graphite layer results in the decrease of the reversible capacities and the change in the lithium doping–undoping mechanism.

Renouard et al. have prepared 'hard carbon' by the pyrolysis of hexa phenyl benzene (HPB) at 600°C for 108 h [59]. The samples comprised two types of morphologies: (a) bundles of nanofibres of approximately 100 nm in diameter and ~3–5 μm in length (Figure 7.45a) and (b) spherical particles with diameters ranging from a few hundred nanometres to 4 μm (Figure 7.45b). The chemical nature of the carbon consists of 'all-benzanoid polycyclic aromatic hydrocarbon (PAH)'. The heat treatment at 600°C gave rise to five fully fused benzene rings. Fusing of five benzene rings gives a slightly curved corrolenulene structure, which may be responsible for the formation of spherical particles with increased curvature. The cyclic voltammogram

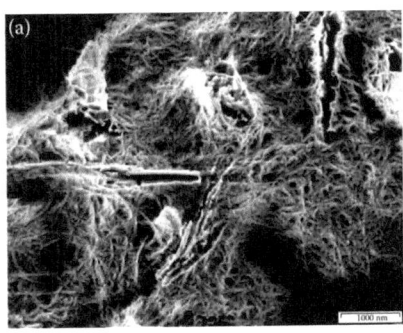
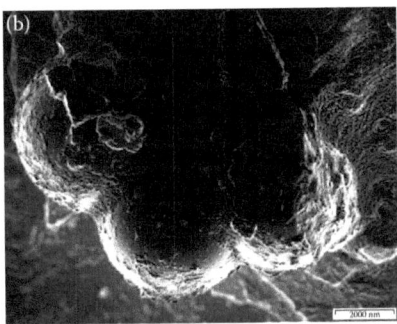

FIGURE 7.45
SEM micrograph images of the products obtained by pyrolysis of HPB at 600°C for 108 h showing presence of (a) bunch of nanofibres and (b) agglomeration of spherical balls. (Reprinted with permission from *J. Power Sources*, 139, T. Renouard et al., Pyrolysis of hexa(phenyl)benzene derivatives: A molecular approach toward carbonaceous materials for Li-ion storage, 2005, 242–249, Copyright 2005, Elsevier.)

for the HPB-pyrolysed carbon reflects the behaviour much similar to hard carbons with oxidation (lithium de-intercalation peak) at ~0.3 V and reduction (lithium intercalation) at ~0.2 V (Figure 7.46). This behaviour is much akin to natural carbons. One remarkable feature of the HPB-pyrolysed carbons is the stable intercalation/de-intercalation capacity of ~500 mAh/g over prolonged cycles, after the first cycle capacity of ~1000 mAh/g.

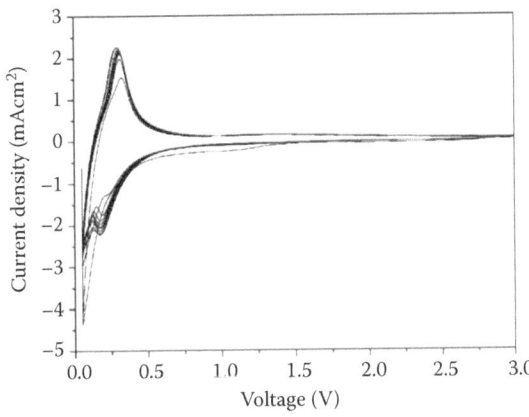

FIGURE 7.46
Cyclic voltammograms at 0.2 mV/s scan rate for the electrode made from products obtained by pyrolysis of HPB at 600°C for 108 h. (Reprinted with permission from *J. Power Sources*, 139, T. Renouard et al., Pyrolysis of hexa(phenyl)benzene derivatives: A molecular approach toward carbonaceous materials for Li-ion storage, 2005, 242–249, Copyright 2005, Elsevier.)

As an extension of this study, Bonino et al. have studied the effect of grinding on the performance of the HPB-pyrolysed carbon samples [60]. The SEM images of grinded HPB-pyrolysed carbon samples are shown in Figure 7.47a and b. From their studies, it is found that 10 min of grinding is effective for the capacity of about 2.3 Ah/g albeit with a gradual fall in the initial cycles' capacity to reach a roughly stable value of ~700 mAh/g (Figure 7.47c).

Carbon fibres are a class of materials that are either graphitisable or non-graphitisable. Yoon et al. have prepared carbon fibres at varying temperatures—600°C, 2000°C and 2800°C—and reported its structure as shown in the TEM image in Figure 7.48 [61]. The fibres prepared at 600°C comprise graphite layers stacked in the direction of the fibres. Upon heat treatment at 2000°C and 2800°C, the graphite sheets exfoliate and fold at the edges

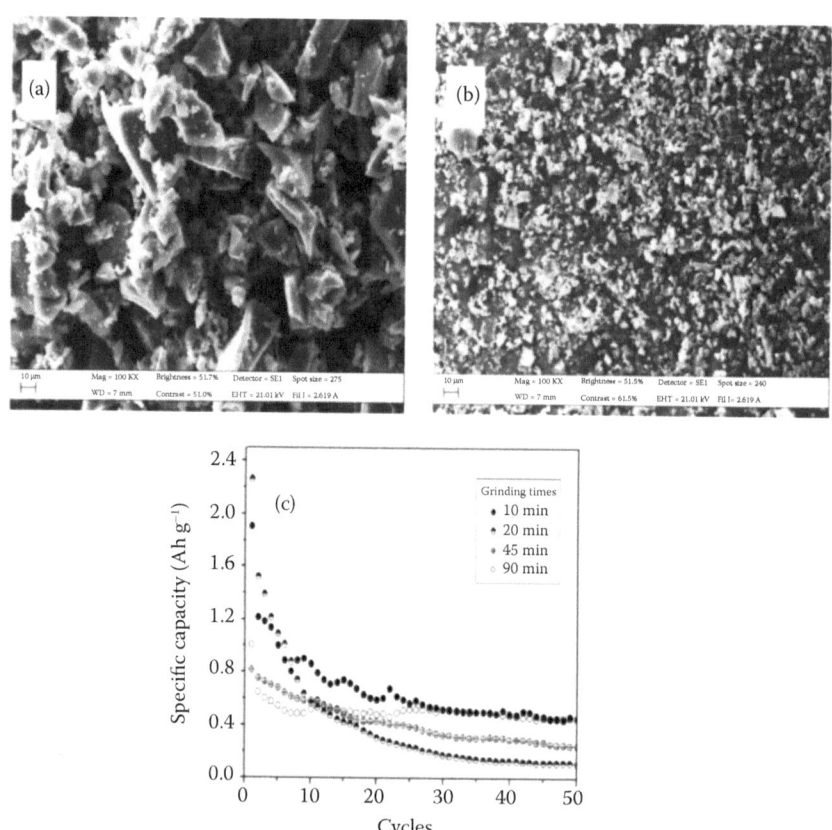

FIGURE 7.47
SEM micrographs of the grinded HPB samples for (a) 10 min, (b) 90 min and (c) cycle life data. (Reprinted with permission from *Electrochimica Acta.*, 51, F. Bonino et al., Structural and electrochemical studies of a hexaphenylbenzene pyrolysed soft carbon as anode material in lithium batteries, 2006, 3407–3412, Copyright 2006, Elsevier.)

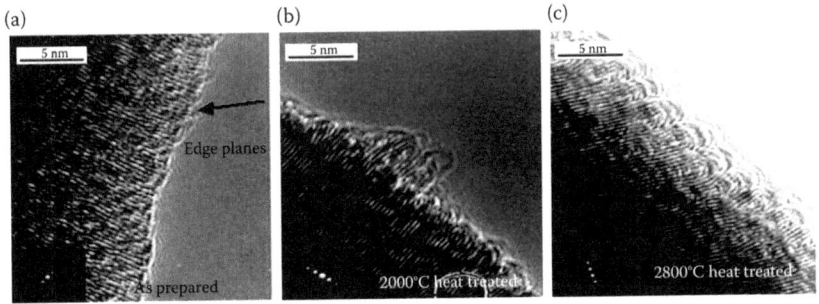

FIGURE 7.48
HRTEM images of graphitic nanofibres prepared at (a) 600°C, (b) 2000°C and (c) 2800°C. (Reprinted with permission from *Carbon*, 42, S.H. Yoon et al., Novel carbon nanofibres of high graphitization as anodic materials for lithium ion secondary batteries, 2004, 21–32, Copyright 2004, Elsevier.)

along the direction of the fibre growth. These specially designed fibres possess similar behaviour of typically ordered graphite, providing an intercalation/de-intercalation capacity of ~350 mAh/g, with potentials close to 0.1 V vs. Li/Li$^+$ (Figure 7.49).

Non-graphitisible carbon fibres synthesised by Tatsumi et al. exhibit high irreversible capacities [62]. Upon heat treatment, the reversible capacity reduces substantially, making these fibres unattractive for anodes in LIBs. The de-intercalation capacity of the carbon fibres is classified into two regions: (1) the high-voltage sloping region and (2) low-voltage plateau region. Although the total capacity of the sample prepared at 1000°C exceeds 400 mAh/g, the plateau region capacity is as low as 200 mAh/g.

Beguin et al. have prepared disordered carbons from viscose [63]. Carbon cloth was fabricated through the carbonisation of viscose fibres at

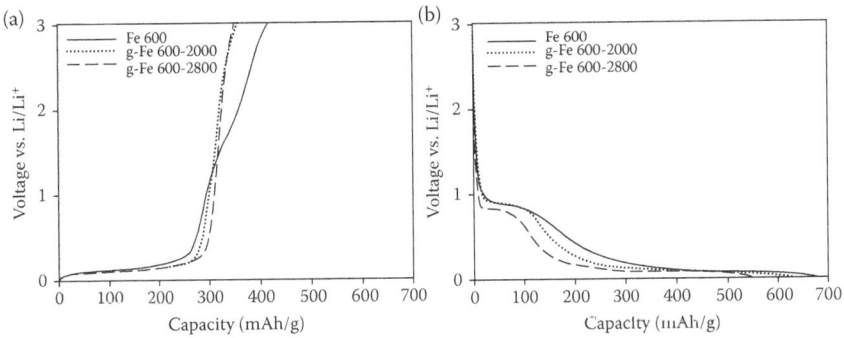

FIGURE 7.49
(a) Lithium de-intercalation and (b) intercalation profiles for graphitic nanofibres prepared at 600°C (solid line), 2000°C (dotted line) and 2800°C (dashed line). (Reprinted with permission from *Carbon*, 42, S.H. Yoon et. al., Novel carbon nanofibres of high graphitization as anodic materials for lithium ion secondary batteries, 2004, 21–32, Copyright 2004, Elsevier.)

temperatures between 400°C and 1000°C for 15 min. The carbon cloth acts as a self-standing electrode without the use of binder and conductive additive. The carbon cloth is coated with pyrolytic carbon by CVD of propene. Irrespective of the pyrolytic carbon coating, the BET surface area of the carbon cloth is lower than 1 m²/g, but this could be increased to 356 m²/g after mild oxidation by air at 450°C for 10 min. Initially, the carbon cloth with pyrolytic carbon surface has narrow micropores, which were widened after heat treatment. These findings were confirmed by the HRTEM image shown in Figure 7.50a for the 002 lattice fringes image of a transverse section of a fibre after propene pyrolysis. The inter-layer d002 distance was found to be below 0.7 nm by image analysis.

From the galvanostatic charge/discharge profiles, it is inferred that the carbon cloth electrode without pyrolytic carbon coating could offer 1.34 eq. of reversible Li intercalation/de-intercalation capacity and 0.71 eq. of irreversible capacity. With the lamellar pyrolytic carbon coating (using propene as

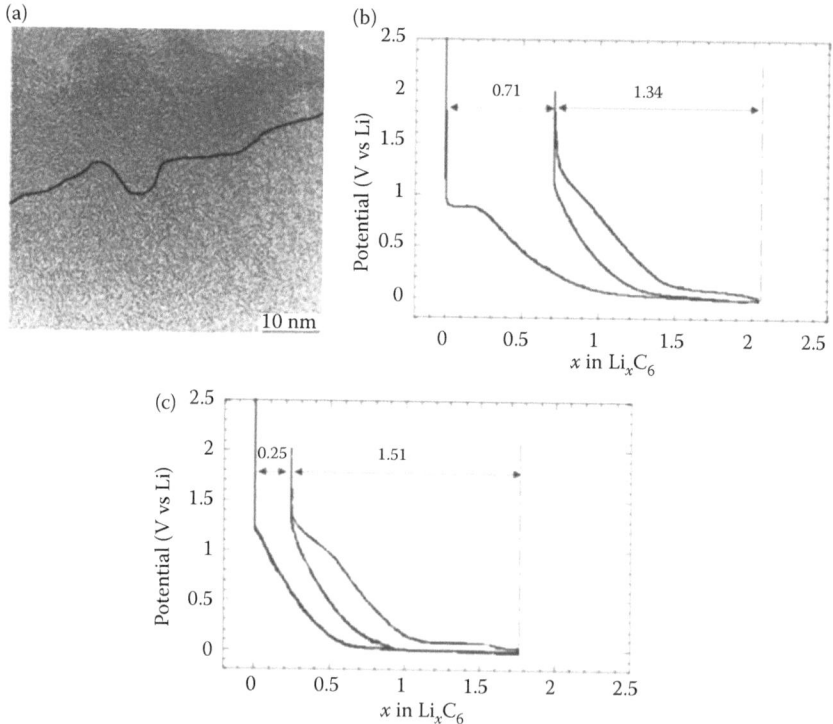

FIGURE 7.50
(a) HRTEM image of a transverse section of propene pyrolised carbon fibre, (b) galvanostatic charge/discharge of the carbon cloth outgassed at 150°C without carbon coating and (c) 900°C with carbon coating. (Reprinted with permission from *J. Physics and Chemistry of Solids*, 65, F. Beguin et al., A better understanding of the irreversible lithium insertion mechanisms in disordered carbons, 2004, 211–217, Copyright 2004, Elsevier.)

the carbon source), the modified carbon cloth provided 1.51 eq. of reversible capacity, and the irreversible capacity reduced to 0.25 eq., that is, a 50% improvement (Figure 7.50b and c). The same carbon cloth, when coated with carbon using propylene as the hydrocarbon source gave rise to 1.57 eq. of reversible capacity and 0.26 eq. of irreversible capacity [64]. In another study, Beguin et al. [65] prepared carbons coated using cellulose to reduce the irreversible capacity to 0.18 eq., and increased the reversible capacity to 1.62 eq. These data are shown in Figure 7.51a and b.

Tokumitsu et al. have prepared synthetic carbons by the pyrolysis of aromatic precursors, which are related to coronene structures [66]. The basic unit of coronene consists of five benzene rings fused together to enclose a five-membered ring in the centre. Three organic precursors—coronene, phenolphthalein and acenapthelene—were selected and subjected to heat treatment at the temperatures of 800°C, 1000°C and 2780°C. In all these cases, the initial discharge capacity is found to decrease upon heat treatment due to graphitisation. In the disordered state, the material provides large charge/discharge capacity associated with a large irreversible capacity. Upon heat treatment, the irreversible capacity reduces and the discharge capacity also falls. The effect is more pronounced in the case of coronene, which upon heat treatment at 800°C gives a capacity of 675 mAh/g that reduces drastically to 290 mAh/g upon heat treatment at temperatures between 1000°C and 2780°C (Table 7.7).

Kim et al. have treated the carbons obtained by the pyrolysis of polyacrylonitrile (PAN) by sulphuric acid to study the effect of heat and acid treatment on electrochemical anodic performance [67]. The PAN-based hard carbon samples are first heat-treated at 900°C and subsequently treated with sulphuric acid for various time durations ranging from 4 to 20 h—(sample TAN9S(h), where S represents sulphuric acid and h represents the duration

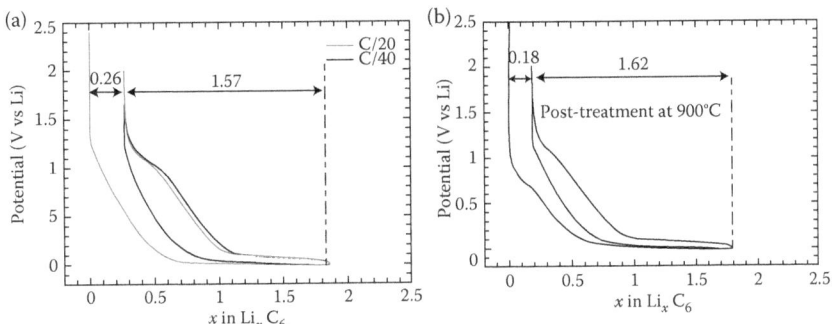

FIGURE 7.51
Galvanostatic charge/discharge profiles for carbon cloth electrodes coated with carbon by (a) propylene, (b) cellulose. (Reprinted with permission from *Carbon*, 43, F. Beguin et al., Correlation of the irreversible lithium capacity with the active surface area of modified carbons, 2005, 2160–2167, Copyright 2005, Elsevier.)

TABLE 7.7

Structural, Physical and Electrochemical Data of Carbons from Various Hydrocarbon Sources at Different Temperatures and Pressures

Source Compound	Heat Treatment Temperature (°C)	d_{002} (nm)	L_c (nm)	L_a (nm)	Density (g/cc)	Cavity Index (vol%)	Discharge Capacity (mAh/g)
Coronene	800	0.343	1.86	1.82	1.69	51.6	675
	1000	0.343	2.08	2.02	1.84	43.5	289
	2780	0.336	55.9	69.6	2.23	2.9	292
Phenolphthalein	800	0.355	1.11	1.97	1.54	57.0	386
	1000	0.352	2.00	2.00	1.43	55.5	282
	2780	0.344	2.58	4.00	1.42	49.8	134
Acenaphthylene	800	0.339	2.17	1.97	1.65	49.7	497
	1000	0.343	2.18	2.01	1.91	41.0	264
	2780	0.337	82.3	108	2.22	2.6	317

Source: Reprinted with permission from *J. Power Sources*, 54, K. Tokumitsu et al., Charge/discharge characteristics of synthetic carbon anode for lithium secondary battery, 1995, 444–447, Copyright 1995, Elsevier.

of acid treatment in hours). The sample convention is applied to TAN11 in which the heat treatment temperature is 1100°C. Figure 7.52 shows the cycle life data of these two series of carbon samples. Moderate capacities (~250 mAh/g) were obtained for the carbons without acid treatment due to the high irreversible capacities (Figure 7.52). Following sulphuric acid

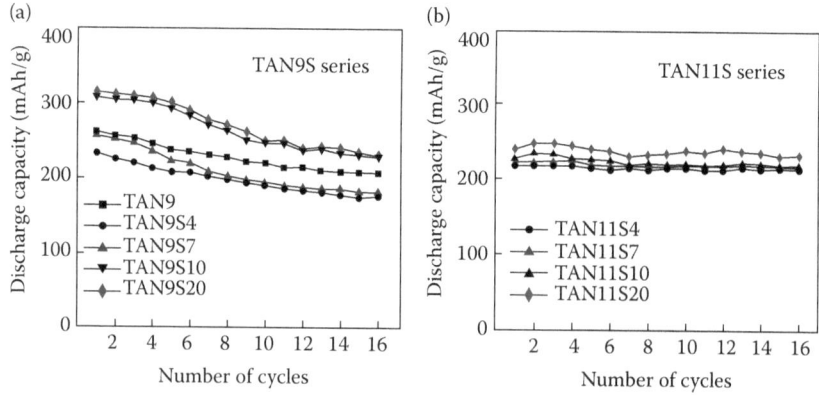

FIGURE 7.52

Cycle life data for (a) TANS9S series and (b) TAN11S series of graphitic carbons heat-treated at 900°C and 1100°C respectively, with sulphuric acid treatment denoted by S followed by the duration of the treatment in hours. (Reprinted with permission from *Carbon*, 43, Y.J. Kim et al., Effects of sulfuric acid treatment on the microstructure and electrochemical performance of a polyacrylonitrile (PAN)-based carbon anode, 2005, 163–169, Copyright 2005, Elsevier.)

treatments, TAN9 exhibits an increased reversible capacity of up to ~20% and a decreased irreversible capacity of up to ~8%. TAN11 experienced smaller changes compared to TAN9, but displayed better cycling behaviour. It was suggested that the acidity/basicity of the surface groups may have an influence on improving the electrochemical performance of hard carbon anode materials for LIBs, but the effect of prolonged acid treatment was insignificant.

Cho et al. have prepared crystalline nanofibrillar carbon from carbon cotton (CC), which was obtained by heating (PAN)/FeCl$_3$ hybrid precursor by thermal processing [68]. In this method, the CC hybrid precursor was carbonised at different temperatures of 1100°C and 1500°C in nitrogen atmosphere, and labelled as CC1100 and CC1500, respectively. To increase the crystallinity, CC1500 was further annealed at 800°C in the presence of hydrogen (CC1500A). HRTEM images of these three samples are shown in Figure 7.53. CC1100 consists of poorly packed, highly twisted carbon layers (Figure 7.53a). Well-packed and less twisted carbon layers with structural defects were formed after increasing the annealing temperature to 1500°C (Figure 7.53b). Surprisingly, Figure 7.53c shows that the structural defects disappeared upon further heat treatment in hydrogen at 800°C. Figure 7.54 shows the first constant current charge and discharge curves and cycle life data for these three carbon samples. Upon heating, the nanofibres between 1100°C and 1500°C, the charge/discharge capacity increases with individual graphene sheets ordered in the direction perpendicular to the fibres, and a flat intercalation profile. On the other hand, the discharge profile was not flat. The materials were suitably tested at different potential levels but nanofibres were found highly crimped (Figure 7.54). The carbon nanofibrils CC1500A is found to have a better rechargeable capacity of ~630 mAh/g, which is attributed to its micro-textural features, that is, the web-like open framework structure consisting of highly crystalline and highly twisted carbon nanofibrils.

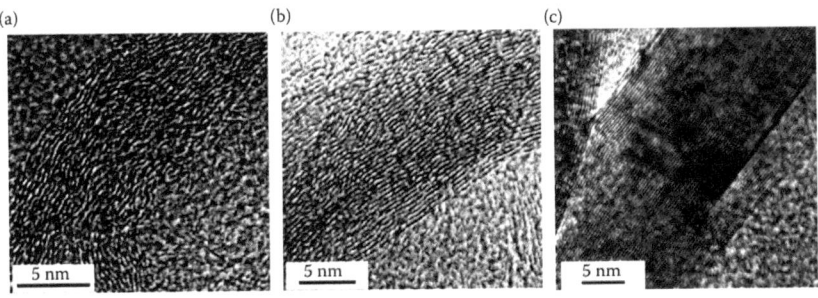

FIGURE 7.53
SEM micrograph images for three types of carbon cottons carbonised at different temperatures (a) CC1100, (b) CC1500 and (c) CC1500A. (Reprinted with permission from *Electrochemica Acta*, 53, H.G. Cho et al., The enhanced anodic performance of highly crimped and crystalline nanofibrillar carbon in lithium-ion batteries, 2007, 944–950, Copyright 2007, Elsevier.)

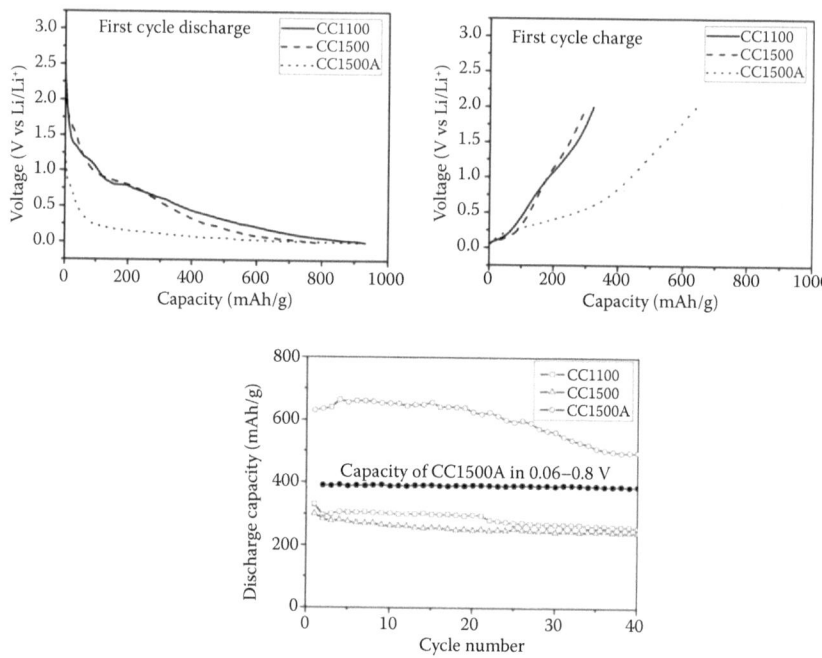

FIGURE 7.54
First cycle galvanostatic lithium intercalation/de-intercalation profiles and cycle life data for CC1100, CC1500 and CC1500A carbons. (Reprinted with permission from *Electrochemica Acta*, 53, H.G. Cho et al., The enhanced anodic performance of highly crimped and crystalline nanofibrillar carbon in lithium-ion batteries, 2007, 944–950, Copyright 2007, Elsevier.)

Subramanian et al. have prepared carbon nanofibres by vapour growth method [69] and obtained a new nanostructure with graphene sheets arranged in a slanted position with stacking in two parallel strands that form a cup-like structure as shown in Figure 7.55. CNFs showed a narrow pore size distribution of 1–3 nm with an average BET surface area of 100 m²/g.

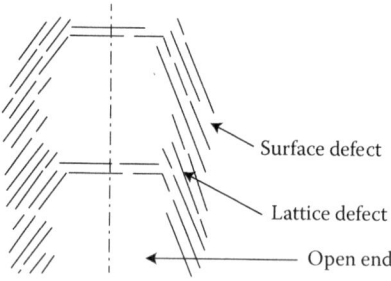

FIGURE 7.55
Structural model of carbon nanofibers (CNFs). (Reprinted with permission from V. Subramanian et al., High rate of reversibility anode materials of lithium batteries from vapor-grown carbon nanofibres, *J. Phys. Chem. B*, 110, 2006, 7178–7183, Copyright 2006, American Chemical Society.)

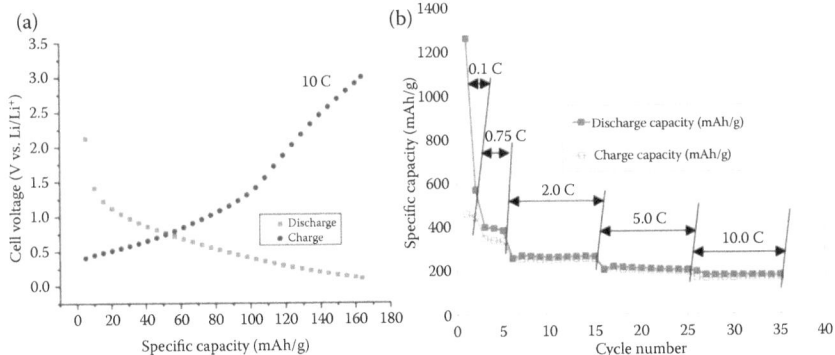

FIGURE 7.56
(a) Galvanostatic profile for lithium intercalation/de-intercalation at 10C rate and (b) rate capability for vapour grown carbon nanofibres. (Reprinted with permission from V. Subramanian et al., High rate of reversibility anode materials of lithium batteries from vapor-grown carbon nanofibres, *J. Phys. Chem.*, B, 110, 2006, 7178–7183, Copyright 2006, American Chemical Society.)

The charge–discharge curves of these CNFs at 10 C rate and cycling performance are presented in Figure 7.56a. The Li-ion insertion capacity is 195 mAh/g while the de-insertion capacity is 166 mAh/g, indicating a high reversible capacity at a high rate of 10 C. The high-rate capability of these CNFs (as illustrated in Figure 7.56b) can be attributed to the inherent cup-like structure, where the CNFs have a large number of lattice, surface defects and open ends. These lead to enhanced intercalation and accessibility of Li ions to the interiors of the CNFs.

Monolithic macroporous carbon have been prepared with a perfect spherical morphology [70], by template method using poly(methyl methacrylate) (PMMA) colloidal crystal template and resorcinol–formaldehyde (RF) resin as the source of carbon. Thermal processing of the PMMA/RF composite in an inert atmosphere produced three-dimensional macroporous ordered carbon monoliths (3DOM). The dimensions of 3DOM carbon monoliths depend on the size of the template. The SEM image of the spherical carbon presented in Figure 7.57a shows, when not using a template, the bulk of the carbon monoliths prepared consists of RF carbon spheres of ~104 nm average diameter. The diameter of carbon spherical particle prepared with a template (shown in Figure 7.57b) is ~300 nm with 10 nm of wall thickness occupied by carbon atoms and the inner hollow space is a macroporous sphere of a diameter ~285 nm. Templated 3DOM carbon spheres with diameters of 408 and 469 nm are labelled as C1 and C2 in SEM. Both reported a specific discharge capacity of ~300 mAh/g and a specific current of 15.2 mA/g (Figure 7.58), which is a 10-fold increase from the non-templated carbon. The key advantage of using 3DOM as an electrode is that a binder is not required for electrode preparation. Both template carbons have similar rate capabilities, their initial coulombic efficiencies were low (43–46%) but increased to >90% in subsequent cycles.

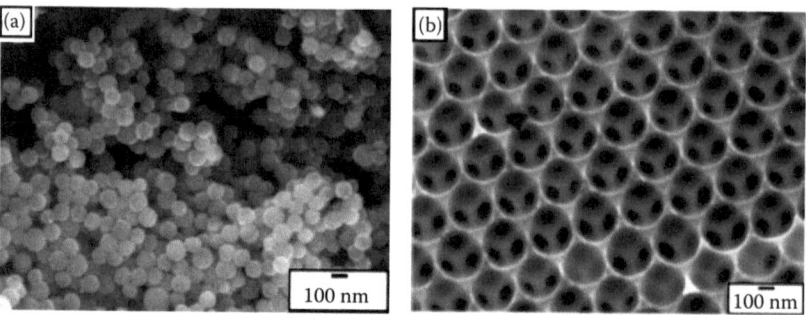

FIGURE 7.57

SEM micrographs for carbon sphere prepared (a) without template and (b) with template. (Reprinted with permission K.T. Lee et. al., Synthesis and rate performance of monolithic macroporous carbon electrodes for lithium-ion secondary batteries, *Adv. Funct. Mater.*, 15, 2005, 547–556, Copyright 2005, Wiley-VCH GMbH & Co.)

Hu et al. have synthesised ordered mesoporous and macroporous carbon monoliths by template method using silica monoliths as template. In this preparation, the mesophase pitch was used as the carbon precursor [54]. The silica monolith templates were filled with a 10 wt.% solution of mesophase pitch in tetrahydrofuran solvent and then carbonised at different temperatures. The physical characteristics of these various mesoporous and macroporous carbon are presented in Table 7.8, which demonstrates these carbons have more prominent graphitic structures as the carbonisation temperature increases. SEM images of the carbon monoliths obtained at carbonisation temperature of 700°C and 2500°C are shown in Figure 7.59.

The first discharge and charge curves of these different carbon samples are shown in Figure 7.60a and their rate capabilities up to 60 C in Figure 7.60b.

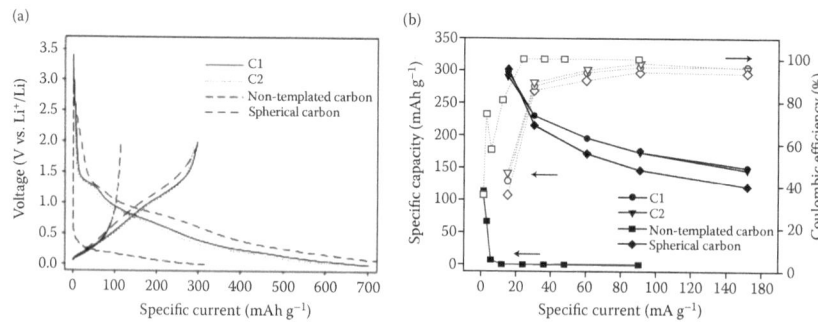

FIGURE 7.58

(a) Charge–discharge profiles of templated 3DOM carbon spheres with diameters of 408 nm (C1) and 469 nm (C2), spherical carbon, and non-templated carbon samples and (b) rate capabilities of these carbon electrodes. (Reprinted with permission K.T. Lee et. al., Synthesis and rate performance of monolithic macroporous carbon electrodes for lithium-ion secondary batteries, *Adv. Funct. Mater.*, 2005, 15, 547–556, Copyright 2005, Wiley-VCH GMbH & Co.)

TABLE 7.8
Physical Characteristics of Mesoporous and Macroporous Carbons Synthesised at Varying Temperatures

Heat Treatment Temperature (°C)	Surface Area (m²/g)	Pore Volume (cm³/g)	Mesopore Diameter (nm)	Macropore Diameter (μm)	Stacking Height (nm)	Layer Size (nm)	H/C Atomic Ratio
700	330	0.55	7.3	1.4	1.3	1.1	0.257
850	327	0.59	7.3	1.4	3.4	1.4	0.094
1000	277	0.47	7.3	1.4	4.8	1.9	0.018
1500	150	0.20	7.3	1.4	9.4	3.8	—
2500	61	0.13	5~15	1.4	—	—	—

Source: Reprinted with permission from Y.-S. Hu et al., Synthesis of hierarchially porous carbon monoliths with highly ordered microstructure and their application in rechargeable lithium batteries with high rate capability, *Adv. Funct. Mater.*, 2007, 17, 1873–1878. Copyright 2007, Wiley-VCH Verlag GmbH & Co.

The plateau between 0.8 and 0.7 V is ascribed to solid electrolyte interface formation, and is found to decrease with an increase in carbonisation temperature and, thus, decreases with surface area. These carbon samples also show good high-rate capabilities. For the material carbonised at 700°C, which had 540 mAh/g capacity at 1 C rate, the capacities were 260 mAh/g at 10 C; 145 mAh/g at 30 C and 70 mAh/g at 60 C. The cycling performance displayed better stability at high current rates.

7.4.2.3 Other Carbonaceous Materials

Chabre et al. have studied the electrochemical intercalation/de-intercalation on fullerene samples (C60) [71]. The first redox cycle for pure C60 shown in

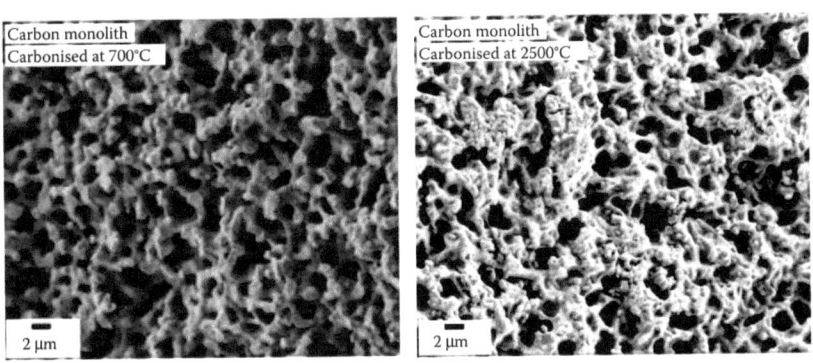

FIGURE 7.59
SEM micrograph for carbon monolith carbonised at 700°C and 2500°C. (Reprinted with permission from Y.-S. Hu et al., Synthesis of hierarchially porous carbon monoliths with highly ordered microstructure and their application in rechargeable lithium batteries with high rate capability, *Adv. Funct. Mater.*, 17, 2007, 1873–1878. Copyright 2007, Wiley-VCH Verlag GmbH & Co.)

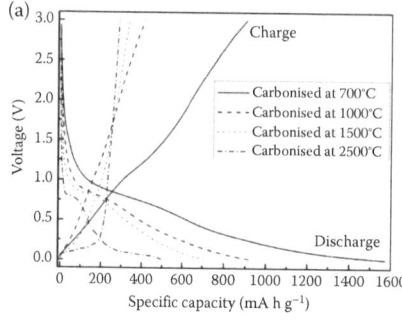
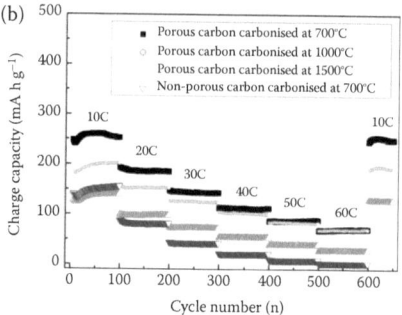

FIGURE 7.60
(a) Galvanostatic discharge and charge curves of carbon samples carbonised at different temperatures cycled at C/5 rate and (b) rate capability of carbon samples carbonised at different temperatures along carbon samples carbonised without the template at 700°C. (Reprinted with permission from Y-S. Hu et al., Synthesis of hierarchically porous carbon monoliths with highly ordered microstructure and their application in rechargeable lithium batteries with high rate capability, *Adv. Funct. Mater.*, 2007, 17, 1873–1878. Copyright 2007, Wiley-VCH Verlag GmbH & Co.)

Figure 7.61a shows three well-defined reduction peaks labelled as A, B and C; and fourth and fifth peaks labelled as D and E are observed as shoulder at 1.0 V, and as a very large wave, respectively. Integrating current against time, the same sequence of redox reactions is demonstrated in Figure 7.61b as voltage versus Li_xC_{60}. This shows the domain limits for the successive electron plus the ion insertion process appears at $x = 0.5, 2, 3, 4$ and 12. From Figure 7.61c, it is evident that the Li intercalation into the system is reversible between 1.3 and 3.0 V with three sets of redox reactions (or) intercalation/de-intercalation steps. The three steps are marked by a high degree of faradiac efficiency with each corresponding to ~1.1/~1.2 eq. of Li for every C_{60} molecule. The final step from 1.3 to 0.2 V corresponds to intercalation of 8 eq. of Li in total forming $Li_{12}C_{60}$.

Yoo et al. have synthesised families of graphene nanosheets (GNS) and proved that when GNS are assembled in a random manner, the sheets are capable of storing (or absorbing lithium) on both sides of the graphene sheets [72]. Hence, an intercalation capacity of 1.5 times greater than the intercalation capacity of graphite is acheivable for graphenes, namely 372 mAh/g for LiC_6. The lithiated graphene behaves as a molecular compound rather than a crystalline material where lithium occupies specific positions of the structure involving stages of intercalation and de-intercalation. The initial capacity is further increased by the synthesis of nanocomposites of (i) graphene nanosheets (GNS), (ii) carbon nanotubes (CNT) and (iii) graphene and fullerene C_{60}. The higher initial capacities of the composites are attributed to the fact that inter-layer distance is maintained as high as 4.0 Å over large graphene sheet assembly range.

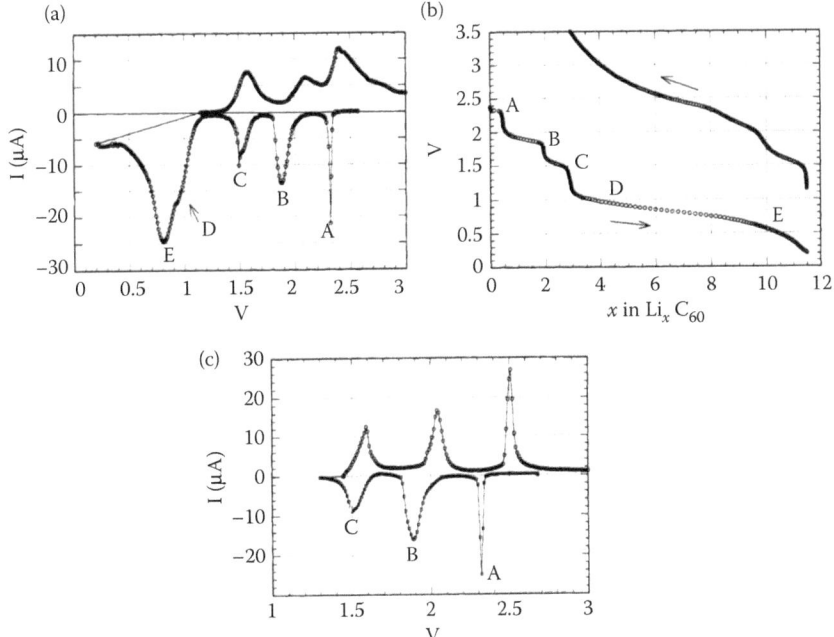

FIGURE 7.61
(a) Voltammograms of the first reduction/oxidation cycle of pure C60, (b) voltage versus composition obtained by integrating the voltammograms in (a), (c) first reduction/oxidation voltammogram for C_{60} in the range 0.1~3.0 V and 1.3~3.0 V vs. Li/Li$^+$. (Reprinted with permission from Y. Chabre et al., Electrochemical intercalation of lithium into solid C_{60}, *J. Am.Chem. Soc.*, 114, 1992, 764–766, Copyright 1992, American Chemical Society.)

GNS are produced by layer-by-layer exfoliation of graphite into oxidised graphene sheets (OGS) followed by a reduction. The thickness of nanoplatelets* comprising a stacked assembly of 10~20 layers of GNS lies between 3 and 7 nm (Figure 7.62a). The nanocomposite of (i) GNS and CNT, and (ii) GNS and C_{60} were prepared by reaction with the addition of single-walled carbon nanotubes and C_{60} to GNS, respectively. The first cycle capacity of the reaction sequence samples—graphite, GNS, GNS + CNT and GNS + C_{60}—are 360, 540, 730 and 784 mAh/g, respectively (Figure 7.62b). More importantly, the reversible capacity values observed after 20 cycles of intercalation/de-intercalation of lithium are 240, 290, 480 and 600 mAh/g, respectively (Figure 7.62c).

Lithium storage in carbonaceous materials depends upon the structure of the host nanomaterial. It is observed that the d-spacing of GNS changes with stacking layers accumulated to form the nanosheet; the d-spacing of GNS increased with a decrease in the number of graphene layers, as illustrated in Figure 7.63a. Cross-sectional TEM images can be used to measure

---

* Nanoplatelets are disc shaped or graphene sheets of nanometre-sized platelets.

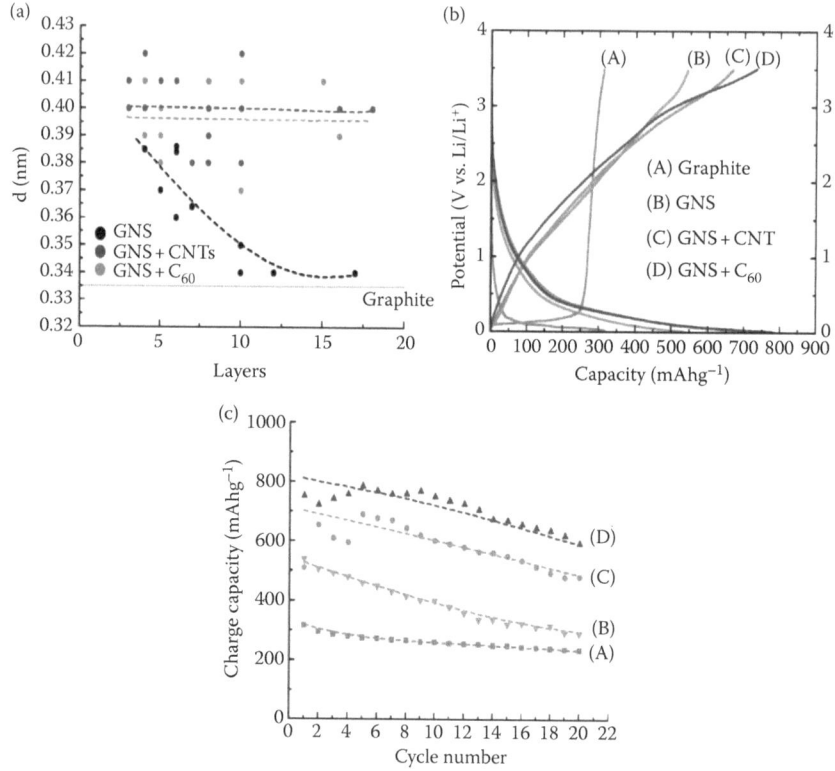

FIGURE 7.62
(a) Relationship between the number of layers of graphene sheets and the average inter-layer distance of graphene material. (b) Li intercalation/de-intercalation profiles and (c) cycle life data for GNS, GNS + CNT, GNS + C_{60} composites. (Reprinted with permission from E.J. Yoo et al., Large reversible Li storage of graphene nanosheet families for use in rechargeable lithium ion batteries, *Nano Lett.*, 8, 2008, 2277–2282, Copyright 2008, American Chemical Society.)

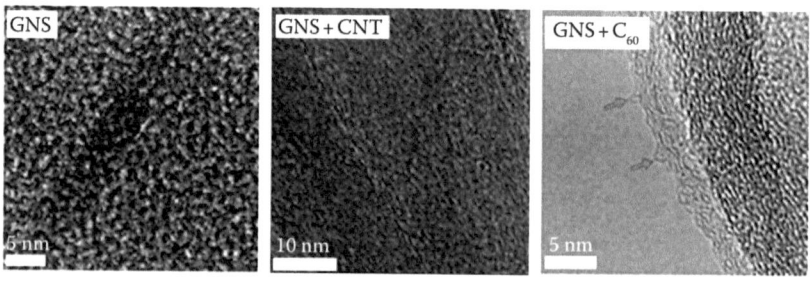

FIGURE 7.63
TEM micrograph for graphene nanosheets: GNS, GNS + CNT and GNS + C_{60} composites. (Reprinted with permission from E.J. Yoo et al., Large reversible Li storage of graphene nanosheet families for use in rechargeable lithium ion batteries, *Nano Lett.*, 8, 2008, 2277–2282, Copyright 2008, American Chemical Society.)

layer-to-layer distance. The discharge capacity data along with the TEM micrographs suggest that expansion in the d-spacing of the graphene layers cause additional sites for the accommodation of lithium ions, resulting in relatively higher storage and retrieval of lithium in the presence of C_{60}.

7.4.3 Other Anode Materials

Nano silicon/Si–C composites and inter-metallic phases are being envisaged as high-capacity anodes. Si has the highest theoretical specific capacity of ~4800 mAh/g. However, the reactivity of Si towards Li is poor. Recently, it has been demonstrated that Si and Si–C nanocomposites do show considerable reactivity with Li and are very promising. Efforts are needed to reduce the capacity fade and improve the performance. Various compositions of inter-metallic composites consisting of nanoparticles of the constituent metals need to be studied. The constituent metals will be those that have well-separated thermodynamic potentials for lithium insertion such that the active–inactive concept will come into operation to ensure long cyclability. For example, inter-metallic phases of Sn and Fe such as Sn_2Fe, $SnFe$, Sn_2Fe_3 and Sn_3Fe_5 have been studied. Lithiation of the phases yields Li_xSn ($x \leq 4.4$) in a matrix of inactive Fe (~10 nm).

Oxides as anode materials have also shown promise due to their ability to provide a stable matrix in which the metals are embedded in nanoform. The poor electronic conductivity of SnO_2 matrix results in large irreversible capacity losses. Research on novel oxide matrices consisting of spectator ions capable of providing the much-needed conductivity and resilience to absorb the strain associated with the intercalation/de-intercalation reactions is the need of the hour [73]. Recently, several binary and ternary oxides have shown promise, which includes the zero-strain material, $Li_4Ti_5O_{12}$ [74]. Studies need to be focused on such novel oxide materials in nanoform to achieve high capacities with low capacity fade.

7.5 Electrolytes for Lithium Batteries

Progress in LIBs relies as much on improvements in the electrolyte as it does on the electrodes. The most desirable polymer electrolytes are those formed by solvent-free membranes, but the poor ionic conductivity of these materials at room temperatures has prevented them from realisation. Efforts have been expended in the literature in dispersing nanoscale inorganic fillers in solvent-free, polyether-based electrolytes to increase their conductivity by several fold [75]. Other avenues are also being explored to achieve high-conductivity polymer electrolytes employing ionic liquids (ILs) and polymer-in-salt nanostructures. Although nanostructured materials offer a variety of advantages, the challenge lies in mitigating undesirable electrode/electrolyte reactions and finding easy and economically viable synthetic routes. For example, Zaghib et al. have evaluated the effect of different lithium salts used in a non-aqueous electrolyte [76] and found that the performance of graphite whiskers anode decreases with the electrolyte as $LiClO_4 > LiCF_3SO_3 > LiAsF_6 > LiPF_6 > LiBF_3$.

Electrolytes have a strong effect on the performance of carbon anodes [77], due to the reaction at low potentials and the formation of decomposition products. These decomposition products form a film on the particles of the electrode material, resulting in a solid electrolyte interface (SEI). The SEI formation has more detrimental effects on the anode material than the cathode material. It is well known that ethylene carbonate (EC) and propylene carbonate (PC) are two important solvents for LIBs. Fujimoto et al. [77] have studied the influence of EC and PC on the natural graphite and found EC to be better in terms of passivation (low SEI formation) and retention of crystallinity of the natural graphite particles. The plateau voltage of natural graphite voltage lies flat and is located at a very low voltage (~0.1 V vs. Li/Li$^+$) for the EC-based electrolyte as compared to a higher voltage observed for the PC electrolyte due to passivation. Loss of crystallinity in natural graphite particles is observed with an increased proportion of PC in a mixture of EC and PC (Figure 7.64) [77].

The current target of electrolytes in LIB is to possess (i) high ionic conductivity (>10^{-2} S/cm^2), (ii) an extra ordinary higher electrochemical window up to 5.8 V (suitable for even Li–F$_2$ system of 6 V) and (iii) excellent cycling stability beyond 1000 cycles with better reversibility. Ionic liquids (ILs), organic salts with low melting points (100°C), have been the focus of many investigations as one of the most promising candidates as a liquid electrolyte because

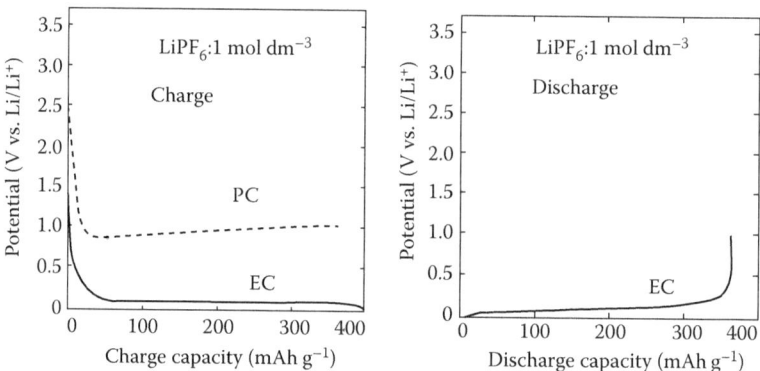

FIGURE 7.64
Galvanostatic lithium intercalation/de-intercalation profiles for graphitic anodic materials in EC and PC electrolytes. (Reprinted with permission from *J. Power Sources*, 72, M. Fujimoto et al., Influence of solvent species on the charge–discharge characteristics of anatural graphite electrode, 1998, 226–230, Copyright 1998, Elsevier.)

of their chemical stability, low inflammability, negligible vapour pressure, high ionic conductivity and wide electrochemical window. It has been demonstrated that ILs can radically improve the performance, cyclability and long-term stability of the electrochemical energy devices. Also, the durability of such devices with IL electrolytes is reported to be significantly better than those of traditional organic and aqueous electrolytes.

Two main strategies have been pursued so far with a view to translate the benefits of ILs to polymer electrolytes. The first strategy involves the design of polymer electrolytes comprising conventional polymer matrices and ionic liquids. The second strategy consists of designing functional polymers with some of the characteristics of ILs. The combination of the two strategies would develop a novel family of tailor-made polymer electrolytes. Recently, gel electrolytes such as ionic gel electrolytes containing silica/TiO_2 nanoparticles have been developed [78]. These ionic gel electrolytes are physically cross-linked gels wherein gelation is brought about by hydrophobic and hydrophilic interaction of gelators. Chemically cross-linked ionic gel electrolytes are an upcoming category of materials for enhancing the energy storage characteristics that need to be developed.

7.6 Outlook

Given their critical role in energy security and in reducing greenhouse gas emissions, electrochemical energy should be a prime objective for policy makers. Large-scale economical storage of electrical energy for applications ranging from portable gadgets to transportation, power grid and beyond remains a great challenge and our weakest link to the future. Electrochemical storage technologies provide solutions for both de-centralised units as well as stationary use. While their portability guarantees a niche market for them, competition can arise from other energy storage technologies for large-scale, stationary applications. It is thus necessary to reduce capital cost, and enhance the service life and reliability of electrochemical energy storage systems. Mature technologies such as those of lead–acid and nickel–metal hydride batteries may soon be replaced by those of advanced lead–acid and LIBs in the near term. Second-generation nickel–iron, lead–carbon hybrid and flow batteries should be able to meet the storage requirements in the medium term. Emerging technologies such as lithium–metal polymer batteries, lithium–sulphur and lithium–air systems are expected to blossom only in the long term. In order to ensure that such technologies measure up to the demands, cross-cutting research must be undertaken with special emphasis on the synthesis and characterisation of multi-functional and nanostructured materials, and high-performance electrolytes, backed by a sound theory of the physico-chemical phenomena and process that occur at the molecular scale in these systems. Experience tells that uninterrupted power supply systems and power trains supported by battery technology have not proven to be complete solutions mainly because of the restrictions on chemical power sources with relatively low power density and cycle lives. Coupling of supercapacitors with batteries could address this long-standing issue.

References

1. D. Linden and T.B. Reddy, *Handbook of Batteries*, 3rd edn. New York: McGraw-Hill, 2002.
2. R.V. Kumar and T. Sarakonsri, *High Energy Density Lithium Batteries*, Eds. K.E. Aifantis, S.A. Hackney and R.V. Kumar. Wiley-VCH, 2010.
3. D. Pavlov, *Essentials of Lead-Acid Batteries*, Eds. B. Haripraksh, T. Premkumar and A.K. Shukla. India: SAEST, 2006.
4. L. Antropov, *Theoretical Electrochemistry*. Moscow: Mir, 1972.
5. T.R. Crompton, *Battery Reference Book*, 3rd edn. Oxford: Newnes, 2000.
6. H.A. Catherino, F.F. Feres and F. Trinadad, Sulfation in lead–acid batteries. *J. Power Sources* 129, 2004, 113–120.
7. K. Sawai, T. Funato, M. Watanabe, H. Wada, K. Nakamura, M. Shiomi and S. Osumi, Development of additives in negative active-material to suppress sulfation during high-rate partial-state-of-charge operation of lead–acid batteries, *J. Power Sources* 158, 2006, 1084–1090.
8. M.L. Soria, J.C. Hernandez, J. Valenciano, A. Sanchez and F. Trinidad, New developments on valve-regulated lead–acid batteries for advanced automotive electrical systems, *J. Power Sources* 144, 2005, 473–485.
9. M.L. Soria, F. Trinidad, J.M. Lacadena, A. Sanchez and J. Valenciano, Advanced valve-regulated lead-acid batteries for hybrid vehicle applications, *J. Power Sources* 168, 2007, 12–21.
10. J. Morales, G. Petkova, M. Cruz and A. Caballero, Nanostructured lead dioxide thin electrode, *Electrochem. Solid State Lett.* 7, 2004, A75–A77.
11. J. Morales, G. Petkova, M. Cruz and A. Caballero, Synthesis and characterization of lead dioxide active material for lead-acid batteries, *J. Power Sources* 158, 2006, 831–836.
12. A.K. Shukla, S. Venugopalan and B. Hariprakash, Nickel-cadmium: Overview. *Encyclopedia of Electrochemical Power Sources*, J. Garche, C. Dyer, P. Moseley, Z. Ogumi, D. Rand and B. Scrosati, Eds. Vol. 4. Amsterdam: Elsevier, 2009, pp. 452–458.
13. H. Bode, K. Dehmelt and J. Witte, Zur kenntnis der nickelhydroxidelektrode—I. Über das nickel (II)-hydroxidhydrat, *Electrochim. Acta* 11, 1966, 1079–1087.
14. P. Gao, Y. Wang, Q. Zhang, Y. Chen, D. Bao, L. Wang, Y. Sun, G. Li and M. Zhang, Cadmium hydroxide nanowires—new high capacity Ni-Cd battery anode materials without memory effect, *J. Mater. Chem.* 22, 2012, 13922.
15. D. Ohms, M. Kohlhase, G. Benczur-Urmossy and G. Schadlich, New developments on high power alkaline batteries for industrial applications, *J. Power Sources* 105, 2002, 127–133.

16. A.K. Shukla and B. Hariprakash, Nickel-Iron, *Encyclopedia of Electrochemical Power Sources*, J. Garche, C. Dyer, P. Moseley, Z. Ogumi, D. Rand and B. Scrosati, Eds. Vol. 4. Amsterdam: Elsevier, 2009, pp. 522–527.
17. A.K. Shukla, M.K. Ravikumar and T.S. Balasubramanian Nickel/iron batteries, *J. Power Sources* 51, 1994, 29–36.
18. B. Hariprakash, S.K. Martha, M.S. Hegde and A.K. Shukla, A sealed, starved-electrolyte nickel–iron battery, *J. Appl. Electrochem.* 35, 2005, 27–32.
19. B. Hariprakash, A.K. Shukla, S. Venugoplan, Nickel-Metal hydride: Overview. *Encyclopedia of Electrochemical Power Sources*, J. Garche, C. Dyer, P. Moseley, Z. Ogumi, D. Rand and B. Scrosati, Eds. Vol. 4. Amsterdam: Elsevier, 2009, pp. 494–501.
20. S.R. Ovshinsky, M.A. Fetcenko and J.A. Ross, Nickel metal hydride battery for electric vehicle, *Science* 260, 1993, 176–181.
21. N. Furukawa, Development and commercialization of nickel-metal hydride secondary batteries, *J. Power Sources* 51, 1994, 45–59.
22. K. Shinyama, Y. Magari, H. Akita et al. Investigation into the deterioration in storage characteristics of nickel-metal hydride batteries during cycling, *J. Power Sources* 143, 2005, 265–269.
23. L. Sun, H.K. Liu, D. Bradhurst, S.X. Dou, *Electrochem. Solid State Lett.* 2, 1999, 164.
24. M. Winter, J.O. Besenhard, M.E. Spahr and P. Novak, Insertion electrode materials for rechargeable lithium batteries, *Adv. Mater.* 10, 1998, 725–763.
25. M.S. Whittingham, Lithium batteries and cathode materials, *Chem. Rev.* 104, 2004, 4271–4301.
26. K. Mizushima, P.C. Jones, P.J. Wiseman and J.B. Goodenough, Li_xCoO_2 ($0 < x \sim 1$): A new cathode material for batteries of high energy density, *Mater. Res. Bull.* 15, 1980, 783–789.
27. W. Weppner and R.A. Huggins, Determination of the kinetic parameters of mixed-conducting electrodes and application to the system Li_3Sb, *J. Electrochem. Soc.* 124, 1977, 1569–1578.
28. K. Konstantinov, G.X. Wang, Y. Yao, H.K. Liu and S.X. Dou, Stoichiometry-controlled high-performance $LiCoO_2$ electrodematerials prepared by a spray solution technique, *J. Power Sources* 119–125, 2003, 195–200.
29. M.K. Jo, Y.S. Hong, J.B. Choo and J.P. Cho, Effect of $LiCoO_2$ cathode nanoparticle size on high rate performance for Li-Ion batteries, *J. Electrochem. Soc.* 156, 2009, A430–A434.
30. C.C. Chang, J.Y. Kim and P.N. Kumta, Influence of crystallite size on the electrochemical properties of chemically synthesized stoichiometric $LiNiO_2$, *J. Electrochem. Soc.* 149, 2002, A1114–A1120.
31. K.K. Lee, W.S. Yoon, K.B. Kim, K.Y. Lee and S.T. Hong, Characterization of $LiNi_{0.85}Co_{0.10}M_{0.05}O_2$ (M = Al, Fe) as a cathode material for lithium secondary batteries, *J. Power Sources*, 97–98, 2001, 308–312.
32. T. Ohzuku, M. Kitagawa and T. Hirai, Electrochemistry of manganese dioxide in lithium nonaqueous cell. II). X-Ray diffractional and electrochemical characterization on deep discharge products of electrolytic manganese dioxide, *J. Electrochem. Soc.* 137, 1990, 40–46.
33. Y. Koyama, N. Yabuuchi, I. Tanaka, H. Adachi and T. Ohzuku, Solid-state chemistry and electrochemistry of $LiNi_{1/3}Co_{1/3}Mn_{1/3}O_2$ for advanced lithium-ion batteries. I) First-principles calculation on the crystal and electronic structures, *J. Electrochem. Soc.* 151, 2004, A1545–A1551.

34. N. Yabuuchi, Y. Koyama, N. Nakayama and T. Ohzuku, Solid-state chemistry and electrochemistry of LiNi$_{1/3}$Co$_{1/3}$Mn$_{1/3}$O$_2$ for advanced lithium-ion batteries. II) Preparation and characterization, *J. Electrochem. Soc.* 152, 2005, A1434–A1440.
35. M.H. Lee, Y.J. Kang, S.T. Myung and Y.K. Sun, Synthetic optimization of Li[Ni$_{1/3}$Co$_{1/3}$Mn$_{1/3}$]O$_2$ via co-precipitation, *Electrochim. Acta* 50, 2004, 939–948.
36. S. Patoux and M.M. Doeff, Direct synthesis of LiNi$_{1/3}$Co$_{1/3}$Mn$_{1/3}$O$_2$ from nitrate precursors. *Electrochem. Commun.* 6, 2004, 767–772.
37. R.J. Gummow, A. de-Kock, M.M. Thackeray, Improved capacity retention in rechargeable 4 V lithium/lithium manganese oxide (spinel) cells, *Solid State Ionics* 69, 1994, 59–67.
38. J. Guan and M. Liu, Transport properties of LiMn$_2$O$_4$ electrode materials for lithium-ion batteries, *Solid State Ionics* 110, 1998, 21–28.
39. C.J. Curtis, J. Wang and D.L. Schulz, Preparation and characterization of LiMn$_2$O$_4$ spinel nanoparticles as cathode materials in secondary Li batteries, *J. Electrochem. Soc.* 151, 2004, A590–A598.
40. E. Hosono, T. Kudo, I. Honma, H. Matsuda and H. Zhau, Synthesis of single crystalline spinel LiMn$_2$O$_4$ nanowires for a lithium ion battery with high power density, *Nano Lett.* 9, 2009, 1045–1051.
41. D.K. Kim, P. Muralidharan, H.W. Lee, R. Ruffo, Y. Yang, C.K. Chan, H. Peng, R.A. Huggins and Y. Cui, Spinel LiMn$_2$O$_4$ nanorods as lithium ion battery cathodes, *Nano Lett.* 8, 2008, 3948–3952.
42. A.K. Padhi, K.S. Nanjundaswamy and J.B. Goodenough, Phospho-olivines as positive-electrode materials for rechargeable lithium batteries, *J. Electrochem. Soc.* 144, 1997, 1188–1194.
43. G. Wang, S. Shen and J. Yao One-dimensional nanostructures as electrode materials for lithium-ion batteries with improved electrochemical performance, *J. Power Sources* 189, 2009, 543–549.
44. D.H. Kim and J.K. Kim, Synthesis of LiFePO$_4$ nanoparticles in polyol medium and their electrochemical properties, *Electrochem. Solid State Lett.* 9, 2006, A439–A442.
45. K.F. Hsu, S.Y. Tsay and B.J. Hwang, Synthesis and characterization of nano-sized LiFePO$_4$ cathode materials prepared by a citric acid-based sol–gel route, *J. Mater. Chem.* 14, 2004, 2690–2695.
46. M.A. Vadivel, T. Muraliganth and A. Manthiram, Rapid microwave-solvothermal synthesis of phospho-olivine nanorods and their coating with a mixed conducting polymer for lithium ion batteries, *Elecrochem. Commun.* 10, 2008, 903–906.
47. H. Shi, J. Barker, M.Y. Saidi, R. Koksbang and L. Morris, Graphite structure and lithium intercalation, *J. Power Sources* 68, 1997, 291–295.
48. J.-S. Filhol, C. Combelles, R. Yazami and M.-L. Doublet, Phase diagrams for systems with low free enrgy variation: A coupled theory/experiments method applied to Li-graphite, *J. Phys. Chem.* 112, 2008, 3982–3988.
49. K. Zaghib, F. Brochu, A. Guerfi and K. Kinoshita, Effect of particle size on lithium intercalation rates in natural graphite, *J. Power Sources* 103, 2001, 140–146.
50. C. Menachem, E. Peled, L. Burslein and Y. Rosenberg, Characterization of modified NG7 graphite as an improved anode for lithium-ion batteries, *J. Power Sources* 68, 1997, 277–282.

51. M. Yoshio, H. Wang, K. Fukuda, Y. Hara and Y. Adachi, Effect of carbon coating on electrochemical performance of treated natural graphite as lithium-ion battery anode material, *J. Electrochem. Soc.* 147, 2000, 1245–1250.
52. H.L. Zhang, F. Li, C. Liu and H.M. Cheng, Poly(vinyl chloride) (PVC) coated idea revisited: Influence of carbonization procedures on PVC-coated natural graphite as anode materials for lithium ion batteries. *J. Phys. Chem. C* 122, 2008, 7767–7772.
53. J.R. Dahn, A.K. Sleigh, H. Shi, J.N. Reimers, Q. Zhong and B.M. Way, Dependence of the electrochemical intercalation of lithium in carbons on the crystal structure of the carbon, *Electrochim. Acta* 38, 1993, 1179–1191.
54. Y.S. Hu, P. Adelhelm, B.M. Smarsly, S. Hore, M. Antoneitti and J. Maier, Synthesis of hierarchically porous carbon monoliths with highly ordered microstructure and their application in rechargeable lithium batteries with high-rate capability, *Adv. Funct. Mater.* 17, 2007, 1873–1878.
55. H. Fujimoto, A. Mabuchi, K. Tokomitsu and T. Kishoh, Irreversible capacity of lithium secondary battery using meso-carbon micro beads as anode material, *J. Power Sources* 54, 1995, 440–443.
56. C. Lampe-Onnerud, J. Shi, P. Onnerud, R. Chamberlian and B. Barnett, Benchmark study on high performing carbon anode materials, *J. Power Sources* 97–98, 2001, 133–136.
57. W. Lu and D.D.L. Chung, Effect of the pitch-based carbon anode on the capacity loss of lithium-ion secondary battery, *Carbon* 41, 2003, 945–950.
58. M. Hara, A. Satoh, N. Takami and T. Ohsaki, Structural and electrochemical properties of lithiated polymerized aromatics, anodes for lithium-ion cells. *J. Phys. Chem.* 99, 1995, 16338–16343.
59. T. Renourard, L. Gherghel, M. Wachlter, F. Bonino, B. Scrosati, R. Nuffer, L. Mathis and K. Mullen, Pyrolysis of hexa(phenyl)benzene derivatives: A molecular approach toward carbonaceous materials for Li-ion storage, *J. Power Sources* 139, 2005, 242–249.
60. F. Bonino, S. Brutti, M. Piana, S. Natale, B. Scrosati, L. Gherghel and K. Mullen, Structural and electrochemical studies of a hexaphenylbenzene pyrolysed soft carbon as anode material in lithium batteries, *Elecrochim. Acta* 51, 2006, 3407–3412.
61. S.H. Yoon, C.W. Park, H.J. Yang, Y. Karai, I. Mochida, R.T.K. Baker and N.M. Rodrigues, Novel carbon nanofibers of high graphitization as anodic materials for lithium ion secondary batteries, *Carbon* 42, 2004, 21–32.
62. K. Tatsumi, J. Kawamura, S. Higuchi, Y. Hoshibu, H. Nakayama and Y. Sawada, Anode characteristics of non-graphitizable carbon fibers for rechargeable lithium-ion batteries, *J. Power Sources* 68, 1997, 263–266.
63. F. Beguin, F. Chevallier, C. Vix, S. Saadallah, J.N. Rouszaud and E. Frackowiak, A better understanding of the irreversible lithium insertion mechanisms in disordered carbons, *J. Phys. Chem. Solids* 65, 2004, 211–217.
64. F. Beguin, F. Chevallier, C. Vix, S. Saadallah, V. Bertagna, J.N. Rouzand and E. Frackowiak, Correlation of the irreversible lithium capacity with the active surface area of modified carbons, *Carbon* 43, 2005, 2160–2167.
65. F. Beguin, F. Chevallier, S. Gautier, J.P. Salvetat, C. Clinard, E. Fraikowiak and H.N. Rauzand, Effects of post-treatments on the performance of hard carbons in lithium cells, *J. Power Sources* 97–98, 2001, 143–145.

66. K. Tokumitsu, A. Mabuchi, H. Fujimoto and T. Daoh, Charge/discharge characteristics of synthetic carbon anode for lithium secondary battery, *J. Power Sources* 54, 1995, 444–447.
67. Y.J. Kim, H.J. Lee, S.W. Lee, S.W. Cho and C.R. Park, Effects of sulfuric acid treatment on the microstructure and electrochemical performance of a polyacrylonitrile (PAN)-based carbon anode, *Carbon* 43, 2005, 163–169.
68. H.G. Cho, Y.J. Kim, E.U. Sung and C.R. Park, The enhanced anodic performance of highly crimped and crystalline nanofibrillar carbon in lithium-ion batteries, *Elecrochim. Acta* 53, 2007, 944–950.
69. V. Subramanian, H. Zhu and B. Wei, High rate reversibility anode materials of lithium batteries from vapor-grown carbon nanofibers, *J. Phys. Chem. B* 110, 2006, 7178–7183.
70. K.T. Lee, J.C. Lytle, N.S. Ergang, S.M. Oh and A. Stein, Synthesis and rate performance of monolithic macroporous carbon electrodes for lithium-ion secondary batteries, *Adv. Funct. Mater.* 15, 2005, 547–556.
71. Y. Chabre, D. Djurado, M. Armand, W.R. Ramanow, N. Coustel, J.P. MacCauly, J.E. Fischer and A.B. Smith, Electrochemical intercalation of lithium into solid C_{60}, *J. Am.Chem. Soc.* 114, 1992, 764–766.
72. E.J. Yoo, J.D. Kim, E. Hosono, H.S. Zhou, T. Kudo and I. Honma, Large reversible Li storage of graphene nanosheet families for use in rechargeable lithium ion batteries, *Nano Lett.* 8, 2008, 2277–2282.
73. J. Lin, Z. Peng, C. Xiang, G. Ruan, Z. Yan, D. Natelson and J.M. Tour, Graphene nanoribbon and nanostructured SnO_2 composite anodes for lithium ion batteries, *ACS Nano* 7(7), 2013, 6001.
74. G. Liu, H. Wang, G. Liu, Z. Yang, B. Jin and Q. Jiang, Synthesis and electrochemical performance of high-rate dual-phase $Li_4Ti_5O_{12}$–TiO_2 nanocrystallines for Li-ion batteries, *Electrochim. Acta* 87, 2013, 218.
75. A.S. Arico, P. Bruce, B. Scrosati, J.M. Tarascon and W.V. Schalkwijk, Nanostructured materials for advanced energy conversion and storage devices, *Nat. Mater.* 4, 2005, 366–377.
76. K. Zaghib, K. Tatsumi, H. Abe, T. Ohsahi, Y. Sawada and S. Higachi, Electrochemical behavior of an advanced graphite whisker anodic electrode for lithium-ion rechargeable batteries, *J. Power Sources* 54, 1995, 435–439.
77. M. Fujimoto, Y. Shoji, Y. Kida, R. Ohishita, T. Nohma and K. Nishio, Influence of solvent species on the charge–discharge characteristics of a natural graphite electrode, *J. Power Sources* 72, 1998, 226–230.
78. P. Wang, S.M. Shaik, P. Comte, I. Exnar and M. Gratzel, Gelation of ionic liquid-based electrolytes with silica nanoparticles for quasi-solid-state dye-sensitized solar cells, *J. Am. Chem. Soc.* 125, 2003, 1166–1167.

8

Lead–Carbon Hybrid Ultracapacitors and Their Applications

Ashok Kumar Shukla, Anjan Banerjee
and Musuwathi Krishnamoorthy Ravikumar

CONTENTS

8.1	Introduction	321
8.2	Operating Principle of Lead–Carbon HUCs	325
8.3	Assembly of 12 V/kF-Range Lead–Carbon HUCs	329
	8.3.1 Performance Data for 12 V/kF-Range Lead–Carbon HUCs	329
	8.3.2 Performance Curves for 12 V/kF-Range Lead–Carbon HUCs at Varying Temperatures	333
	8.3.3 Coulombic Efficiency for 12 V/kF-Range Lead–Carbon HUCs	333
	8.3.4 Constant-Power Discharge Data for 12 V/kF-Range Lead–Carbon HUCs	333
	8.3.5 Impedance Data for 12 V/kF-Range Lead–Carbon HUCs	335
	8.3.6 Energy Density and Power Density Data for 12 V/kF-Range Lead–Carbon HUCs	337
	8.3.7 Self-Discharge, Leakage Current and Parallel Resistance Data for 12 V/kF-Range Lead–Carbon HUCs	339
	8.3.8 Pulse-Cycling Data for 12 V/kF-Range Lead–Carbon HUCs	341
8.4	Applications	342
Acknowledgements		345
References		345

8.1 Introduction

In his April 2009 address to the National Academy of Sciences, U.S. President Obama stated, '…. in no area will innovation be more important than in the development of new technologies to produce, use and save energy' [1]. Truly, energy is both enabling and pervasive. Of all forms of energy, electrical energy is the most useful, convenient and cleanest. Electricity has helped shape our history for over two centuries. Today, electricity is one of our prime power sources. We depend on electricity for heating, cooling, lighting and cooking, work and entertainment, transportation and communication, farming and industry, and health care. But we need to use it wisely to conserve our limited resources, protect the environment, reduce pollution and promote the economic growth. Our standard of living truly depends on electric power and how we use it.

Electricity is a form of energy that begins with the atom. Atoms are made up of protons that carry positive charge, neutrons that bear no charge and electrons that are negatively charged. Electricity is produced when an external force called voltage causes electrons to move from atom to atom. The product of the voltage and current is called power. The flow of electrons is called electric current. There are two types of electricity: static electricity and current electricity. For more than 2000 years, scientists have known about and experimented with static electricity. But it is only in the last 200 years that we have figured out ways to generate electric current and use it to do work. Static electricity occurs when there is an imbalance of positively and negatively charged atoms. Electrons then jump from atom to atom, releasing energy. For example, lightning is a result of the static form of electricity. Condensers, familiarly known as capacitors, are devices that store energy in the form of static electricity. By contrast, current electricity is the constant flow of electrons. There are two types of current electricity: direct current (DC) and alternating current (AC). DC means that the electrons move in one direction while AC means that the electrons flow in both the directions. There are two basic ways of producing electricity: through generators and batteries. Power plants make AC through generators while batteries make DC where chemical reactions force electrons to flow. The excess electricity generated by the power plants can be stored in the batteries to meet the peak power demand. Storage of electrical energy as an idea is not new. Fossil fuel is nothing but energy stored in the form of chemicals, and on demand, can be converted to electrical energy through thermal route. Storage makes resource management efficient, and this applies to electrical energy as well. In recent years, storage of electrical energy is attaining new importance. Among the various types of energy storage technologies, electrochemical devices such as batteries and electrochemical capacitors or supercapacitors or ultracapacitors have seen most significant developments.

The important parameters by which the energy storage technologies can be characterised are: (a) *Extent*, which is the amount of energy that the said device can store. This can be specified as gravimetric energy density (Wh/kg) and this emphasises the weight of the device. Alternatively, volumetric energy density (Wh/L) can also be a measure, and this emphasises the size of the device. This will be an important specification in dealing with applications where available space is restricted; (b) The *delivery rate*, which is a certain rate at which the energy needs to be made available. With end-use applications in mind, availability of power can be rated as W/kg and or W/L; (c) The *life* for which the energy storage devices will be used to store and withdraw energy. The device then cycles between storage and supply or charge and discharge. No device is ever perfectly reversible, even if mechanical wear and tear is ignored. It has a finite life and is measured in terms of number of cycles of charge and discharge it can withstand before replacement is needed; (d) The *response time is* a critical factor; when the need for additional power arises a storage system has to respond quickly. Large systems that are used in energy management, are the ones that take time to respond. The quality power systems are generally electrochemical in nature and are used along with electronic control systems that have been designed to make these respond rapidly. An example is a UPS commonly used with a PC. The electronics however do add to the cost; (e) The *safety* of the device during its use is a critical factor. The commercial success of a device depends on safety during its use; (f) *Recyclability* of the materials employed in the device is an important factor as it controls the environmental pollution; and (g) Last but not least is the all important factor of *cost*, which directly determines consumer interest and marketability.

It is possible to think of energy stored or power delivered. Thus, we can have capital cost per unit of energy stored or unit of power delivered. Operating costs can also be evaluated in similar terms. Other specifications based additionally on per unit weight or volume basis are also used. It should be kept in mind that the value of energy in different applications is different and costing is more profitably done with the end use in mind, rather than in a general manner. However, although storage batteries have the adequate energy densities, there are disadvantages in power applications owing to their limited power delivery. By contrast, electrochemical capacitors have less energy density but fast response time. Accordingly, electrochemical capacitors that can exhibit high power densities have attracted the attention in these power applications.

Based on the current R&D trend, electrochemical capacitors can be divided into three classes: (1) electrical double-layer capacitors (EDLCs), (2) pseudo-capacitors and (3) hybrid ultracapacitors or hybrid supercapacitors [2]. EDLCs rely on carbon-based structures utilising an electrical double-layer capacitance effect based on non-faradaic accumulation of electrostatic charge at the carbon surface/electrolyte through reversible ion adsorption onto the carbon surface of the electrodes. By contrast, fast faradaic charge transfer brought about by the charging of an electrical double layer at the electrode/electrolyte interface determines the working of pseudo-capacitors. A combination of faradaic and non-faradaic components would generate electrochemical capacitors that exhibit high capacitance for pulse power as well as sustained energy. These electrochemical capacitors are referred to as hybrid ultra capacitors (HUCs).

Among the aforesaid electrochemical capacitors, HUCs are attractive as these could optimize the pulse power characteristics of an EDLC and the sustained-energy characteristics of a battery. Accordingly, the charge storage mechanism in a HUC combines the mechanism of energy storage in a battery and an EDLC. During the charge and discharge of an HUC, the battery-type electrode undergoes charge transfer or faradaic reactions—oxidation and reduction—while the double-layer-type electrode undergoes non-faradaic reactions such as double-layer charge and discharge. HUCs have the benefits of both the batteries and EDLC: the high energy density of batteries and high power density with the high cycle-life of EDLC. Considering the technology need, availability, safety, recyclability and cost, lead–carbon HUCs are set to play a seminal role in future energy storage and management.

Historically, the lead–acid battery was invented by Gaston Planté in 1859. It is noteworthy that every time a demand for a new type of portable or reserve energy source emerges, lead–acid batteries face challenges that require technology and/or design solutions that have been met quite appreciably. The problems faced with lead–acid batteries for newer applications—in hybrid electric vehicles (HEVs) and similar duties requiring high-rate partial state-of-charge applications—are mostly related to negative plates, which cannot accept high charging currents generated during regenerative braking and the like. Lead–acid batteries operating at partial state-of-charge face sulphation of the negative plates resulting in battery failure.

To circumvent the problem, lead–carbon HUCs have been developed either by employing pasted positive plates of conventional lead–acid batteries or by employing electrochemically deposited PbO_2 on suitable substrates, such as titanium as positive plates in conjunction with activated carbon as negative plates [3–12]. The conventional pasted-type lead–acid battery electrodes require slow charge (C/10 rate) and discharge (C/5 rate) schedules for sustaining the desired cycle-life. Electrochemically deposited PbO_2 electrodes on a substrate such as titanium can be charged and discharged at much faster rates. However, these electrodes can store only limited amounts of energy and, hence, exhibit lower energy and power density values besides being expensive. Furthermore, activated carbon electrodes used for such lead–carbon HUCs are reportedly made of metal as a substrate that, albeit having corrosion-protective coatings, could be prone to corrosion, which could limit their cycle life.

In a cost-effective approach to circumvent the aforesaid problems, substrate-integrated (SI) PbO_2 positive plates have been used to replace the conventional plates [13–17]. Lead foils are used as substrate for positive plates that are subjected to electrochemical formation cycles during which a thin layer of PbO_2 is formed on the lead substrate. Since the substrate lead and electrochemically formed PbO_2 are integral part of the same electrode, they are termed as substrate-integrated PbO_2 (SI–PbO_2) positive plates. Such electrochemically formed SI–PbO_2 positive plates have several advantages over the conventional positive plates; in particular, SI–PbO_2 plates can be charged and discharged quickly with high coulombic efficiency and exhibit high cycle-life in addition to their high power density. SI–PbO_2 plates are easy to prepare and hence are cost effective. Negative plates in lead–carbon HUCs comprise EDLC-type electrodes obtained from activated carbon-coated graphite sheets. Since graphite is known to be a corrosion-resistant material, such lead–carbon HUCs have long cycle-life. Among lead–carbon HUCs with flooded, absorbent glass mat (AGM) and polymeric silica gelled configurations, lead–carbon HUCs with gelled configuration exhibit the best long-term performance. This chapter highlights operating principles, electrochemical features and performance parameters of the 12 V/kF-range lead–carbon HUCs in AGM and gelled configurations.

8.2 Operating Principle of Lead–Carbon HUCs

The operating principle of the lead–carbon HUC is shown in Figure 8.1. The PbO_2 electrode acts as battery electrode with the charge–discharge reactions similar to the positive plate (cathode) in a lead acid battery. Accordingly, positive plate voltage in a lead–carbon HUC is akin to the cathode of a lead–acid battery. The charge and discharge reactions for PbO_2 electrode and its voltage in lead–carbon HUCs can be written as follows:

$$½ PbSO_4 + H_2O \underset{discharge}{\overset{charge}{\rightleftharpoons}} ½ PbO_2 + ½ H_2SO_4 + H^+ + e^-$$

$$V^o_{cathode} = 1.69 \text{ V vs. SHE} \quad (8.1)$$

$V^o_{cathode}$ is the standard electrode potential and changes with the concentration of the electrolyte. Accordingly, it is referred as $V_{cathode}$ in the following text.

An activated carbon anode is an EDLC-electrode. EDLCs are governed by the same physics as parallel-plate electrolytic capacitors, but since EDLCs use much thinner dielectric medium and high surface-area electrodes,

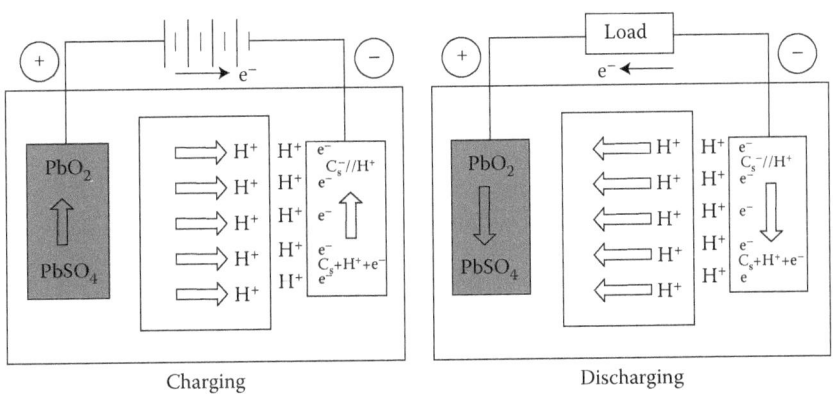

FIGURE 8.1
Working principle of the lead–carbon HUC.

EDLCs tend to store relatively larger charges. The total charge (q) stored in the double-layer capacitor is proportional to its potential (V) with its capacitance (C) as proportionality constant; that is q α V or q = C × V. Accordingly, V = q/C. The charge and discharge reactions on the activated carbon anode and its potential can be expressed as follows:

$$C_s + H^+ + e^- \underset{\text{discharge}}{\overset{\text{charge}}{\rightleftharpoons}} (C_s^- // H^+); \quad V_{anode} - V_{discharged\text{-}anode} = q_{anode}/C_{anode} \quad (8.2)$$

In Equation 8.2, C_s represents the carbon atoms at the electrode surface, // represents the double layer where charges are accumulated on either side, V_{anode} and $V_{discharged\ anode}$ refer to the potential of the electrode (vs. SHE) in its charged and discharged states, respectively, q_{anode} is the charge on the carbon anode and C_{anode} is its capacitance in sulphuric acid electrolyte. The nature of interaction between C_s^- and H^+ is not yet established and is a subject of further study. However, the double-layer behaviour of $C_s^- // H^+$ could be inferred from AC impedance, and cyclic voltammetry.

The net cell reaction for lead–carbon HUC is as follows:

$$\tfrac{1}{2} PbSO_4 + H_2O + C_s \underset{\text{discharge}}{\overset{\text{charge}}{\rightleftharpoons}} \tfrac{1}{2} PbO_2 + \tfrac{1}{2} H_2SO_4 + (C_s^- // H^+) \quad (8.3)$$

The cell voltage is expressed as

$$V_{cell} = V_{cathode} - (q_{anode}/C_{anode}) \quad (8.4)$$

Accordingly, the cell voltage for lead–carbon HUC depends on the anode capacitance and total charge on the carbon anode in sulphuric acid electrolyte.

The total quantity of energy stored in an ideal capacitor can be estimated as follows. The electrical energy (dE) stored in a charged capacitor is

$$dE = \int \frac{q}{C} dq \quad (8.5)$$

When energy is stored from 0 to a certain amount of energy E, then the voltage of the capacitor increases from 0 to V and the charge accumulated on the electrode increases from 0 to q. Accordingly,

$$E(Ws) = \frac{1}{2C} \times q^2 \quad (8.6)$$

The electrical energy stored during charging of a battery is related to its capacity (Ah) as

$$E = I \times t \times V \tag{8.7}$$

where V is the nominal charging voltage in volts (V), I is the charging current in amperes (A) and t is the charing time in hours (h). The charging voltage for the battery depends on the nature of mechanism of electrode reactions. In the case of battery electrodes that operate on heterogeneous mechanisms through dissolution–precipitation, the voltages during charge and discharge remain nearly constant as in the case of a lead–acid battery. In such a case, the energy stored during charging of a battery is

$$E = I \times t \times (V_{cathode} - V_{anode}) \tag{8.8}$$

Equation 8.8 can be written as

$$E = (I \times t \times V_{cathode}) - (I \times t \times V_{anode}) \tag{8.9}$$

where the first term on the right-hand side of Equation 8.9 is the energy stored in the cathode while the second term is for the anode. In an HUC, either anode or cathode is replaced with capacitor electrode. In a lead–carbon HUC, the anode is acting as a capacitor-type electrode and, hence, according to Equation 8.6, the energy stored while charging such an HUC can be obtained as

$$E = (I \times t \times V_{cathode}) - (½ \times C_{anode} \times V_{anode}^2) \tag{8.10}$$

This can be written as

$$E = I \times t \times \left(V_{cathode} - \frac{V_{anode}}{2} \right) \tag{8.11}$$

The specific energy can be obtained by dividing Equation 8.11 with the weight of the device.

When single cells comprising a battery with equivalent capacity (Ah) are connected in series, their voltage (V_b) increases to ($n \times V_b$), where n is the number of cells connected in series with capacity (Ah) remaining the same as that for a single cell. When the single cells of a battery are connected in parallel, the voltage (V) remains the same as that of a single cell, but capacity increases to ($n \times$ Ah).

For fully charged capacitors of voltage (V_c) and equivalent capacitance (C) connected in series, the net voltage of the capacitor bank increases to ($n \times V_c$), but the capacitance of the capacitor bank reduces to C/n, where n is the number of equivalent capacitors connected in series in the capacitor bank. When fully charged capacitors of equivalent capacitances are connected in parallel, then the voltage of the capacitor bank remains the same as that of a single capacitor (V_c), but the capacitance increases to ($n \times C$).

It is noteworthy that when the single cells are assembled with one electrode as a battery-type electrode and the other as a capacitor-type electrode, as in the case of lead–carbon HUCs, its electrical circuit characteristics is similar to that of conventional electrical capacitors. Accordingly, n_p number of lead–carbon HUCs connected in parallel have a voltage similar to that of a single capacitor voltage and the total capacitance is expressed as

$$C_t = C_1 + C_2 + C_3 + C_4 \cdots \tag{8.12}$$

When all capacitors have equivalent capacitance (C), Equation 8.12 could be expressed as

$$C_t = n_p \times C \tag{8.13}$$

Similarly, when n_s number of lead–carbon HUCs is connected in series, their voltage increases to $n_s \times V$ and the total capacitance is expressed as

$$1/C_t = 1/C_1 + 1/C_2 + 1/C_3 + 1/C_4 \cdots \tag{8.14}$$

When all capacitors have equivalent capacitance (C), Equation 8.14 is given as

$$C_t = C/n_s \tag{8.15}$$

Since the lead–carbon HUCs have the same electrical circuit characteristics as that of conventional capacitors, the total capacitance of a lead–carbon HUC bank, with each cell having equivalent capacitance and connected in both series and parallel as shown in Figure 8.2, can be expressed as

$$C_t = (n_p/n_s) \times C \tag{8.16}$$

The desired size of a lead–carbon HUC capacitor bank can be calculated using Equations 8.12 through 8.16 for any specific application. The total resistance of the capacitor bank is expressed as

$$R_{total} = R_{cell} \times (n_s/n_p) \tag{8.17}$$

It is very imperative that capacitor bank sizing is critical for any applications. At lower voltages, a constant power requires higher current as the voltage decreases. This is often overlooked during the initial analysis, and can result in undersizing a solution. For the sizing calculations, should the

Lead–Carbon Hybrid Ultracapacitors and Their Applications

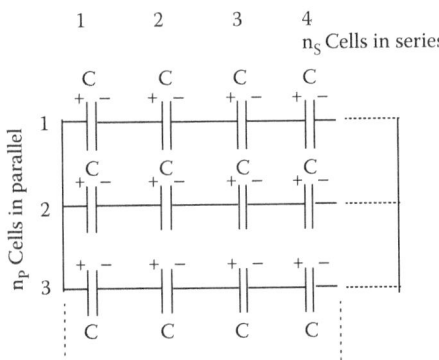

FIGURE 8.2
Equivalent circuit for similar capacitors connected in series and parallel.

total voltage drop ($dV_{total} = I \times R_{esr} + I \times dt/C$) be greater than the application specified limit, then one should either move to the next size-up cell or place two series stacks in parallel.

8.3 Assembly of 12 V/kF-Range Lead–Carbon HUCs

The 12 V lead–carbon HUCs with activated carbon electrodes sandwiched between SI–PbO$_2$ electrodes were assembled in an AGM configuration in 6 M aq. sulphuric acid or silica gel with 1.4 g/cc aq. sulphuric acid using commercial six-compartment 12 V/42 Ah lead–acid battery containers. A typical 12 V/kF-range lead–carbon HUC is shown in Figure 8.3. Each compartment contained seven positive plates and six negative plates in parallel configuration.

8.3.1 Performance Data for 12 V/kF-Range Lead–Carbon HUCs

12 V lead–carbon HUCs were performance tested by a constant-current charge up to a cut-off voltage followed by an hour constant voltage charge and discharge at various currents. Two ranges of load current values were selected (lower range values: 3, 6, 9, 12, 15 A and higher range values: 25, 50, 75, 100 A). Capacitance values (C) were calculated from the negative slopes of discharge profiles [18]. Since, $q = C \times V$, where q is the charge on the electrodes and V is the voltage difference between the electrodes, and as $q = I \times t$, where I and t represent current and time, respectively, we can express, $V = (I \times t)/C$. By differentiating with respect to time at constant current we get $dV/dt = I/C$. During the constant-current charge–discharge of lead–carbon HUCs, the

FIGURE 8.3
The 12 V/kF-range lead–carbon HUC.

voltage varies almost linearly with time. The charge profiles have a positive slope of I/C while the discharge profiles have a negative slope of –I/C. In practical situations, during the discharge of lead–carbon HUCs, there will be an initial voltage drop due to the internal ohmic resistance or IR drop, where R is the ohmic resistance or equivalent series resistance (ESR). However, the capacitance values were calculated excluding ESR.

HUCs were subjected to a constant-current charge to a cut-off voltage of 13.8 V followed by a potentiostatic charge at 13.8 V for 1 h. Subsequently, the lead–carbon HUCs were discharged galvanostatically up to 6 V. The capacitance values at a current of 3 A were found to be 1.38 and 2.45 kF, and decreased to 1.25 and 2.31 kF at a discharge current of 15 A for AGM and silica gelled electrolyte configurations, respectively. At higher currents—25, 50, 75, 100 A—capacitance further decreases. Performance degradation happens at higher discharge-loads due to the non-uniform distribution of currents through the porous electrodes. All discharge profiles for lead–carbon HUCs at constant-currents are shown in Figures 8.4a,b and 8.5a,b. The capacitance for the lead–carbon HUC in the silica gelled electrolyte configuration are found to be higher than in AGM configurations, owing to the enhancement of active surface area of positive plates and the effect of polymeric structure of silica gel on negative plate [17].

Between the AGM and gel configurations, appreciable increase in the surface area of the active material for the SI–PbO_2 electrode is observed for gelled configurations providing higher capacitance for the latter. Silica particles appear to have a distinct role in controlling the nucleation of PbO_2 on positive plates during electrochemical formation. Silica particles affect the nucleation process of PbO_2 and control its particle size. Accordingly, in the gelled electrolyte, the particle size of PbO_2 is smaller

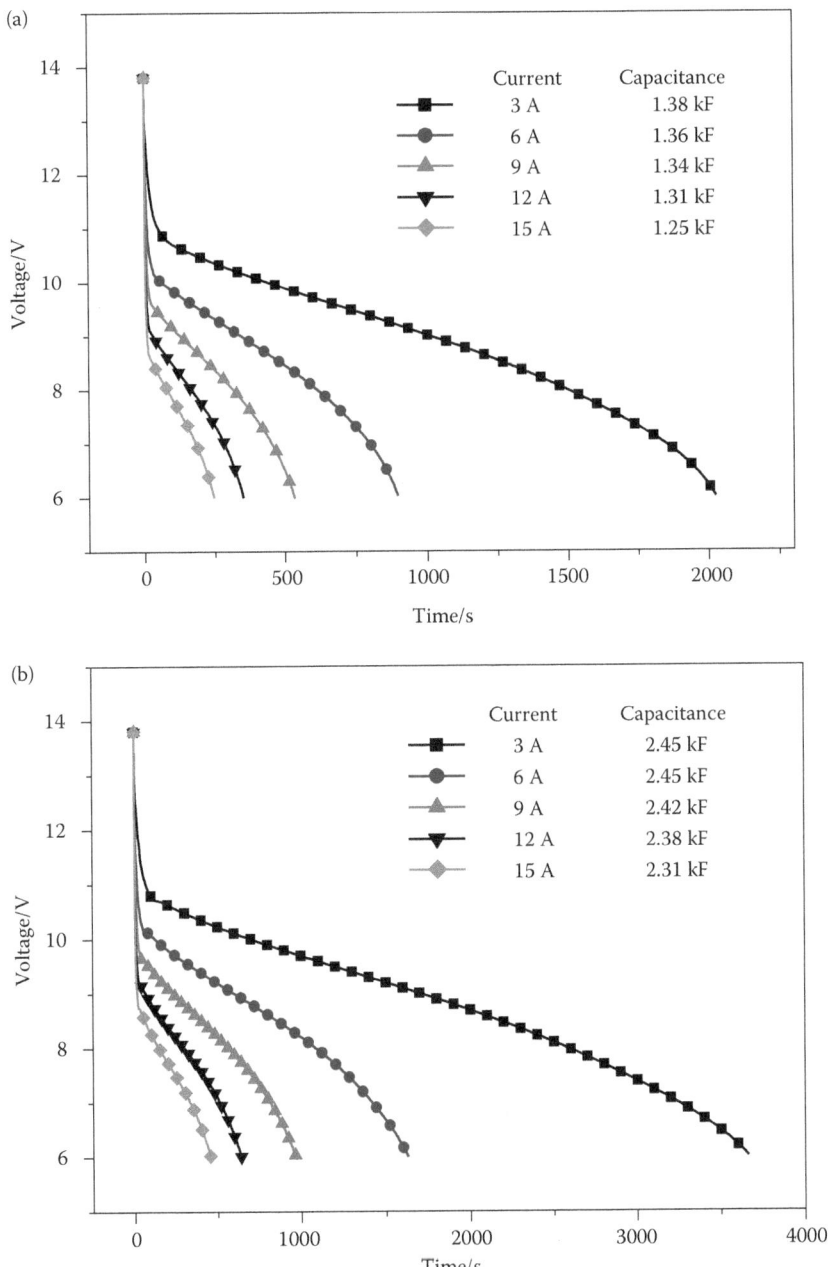

FIGURE 8.4
Constant-current discharge profiles at varying current loads: 3, 6, 9, 12, 15 A for (a) AGM and (b) gelled 12 V/kF-range lead–carbon HUCs.

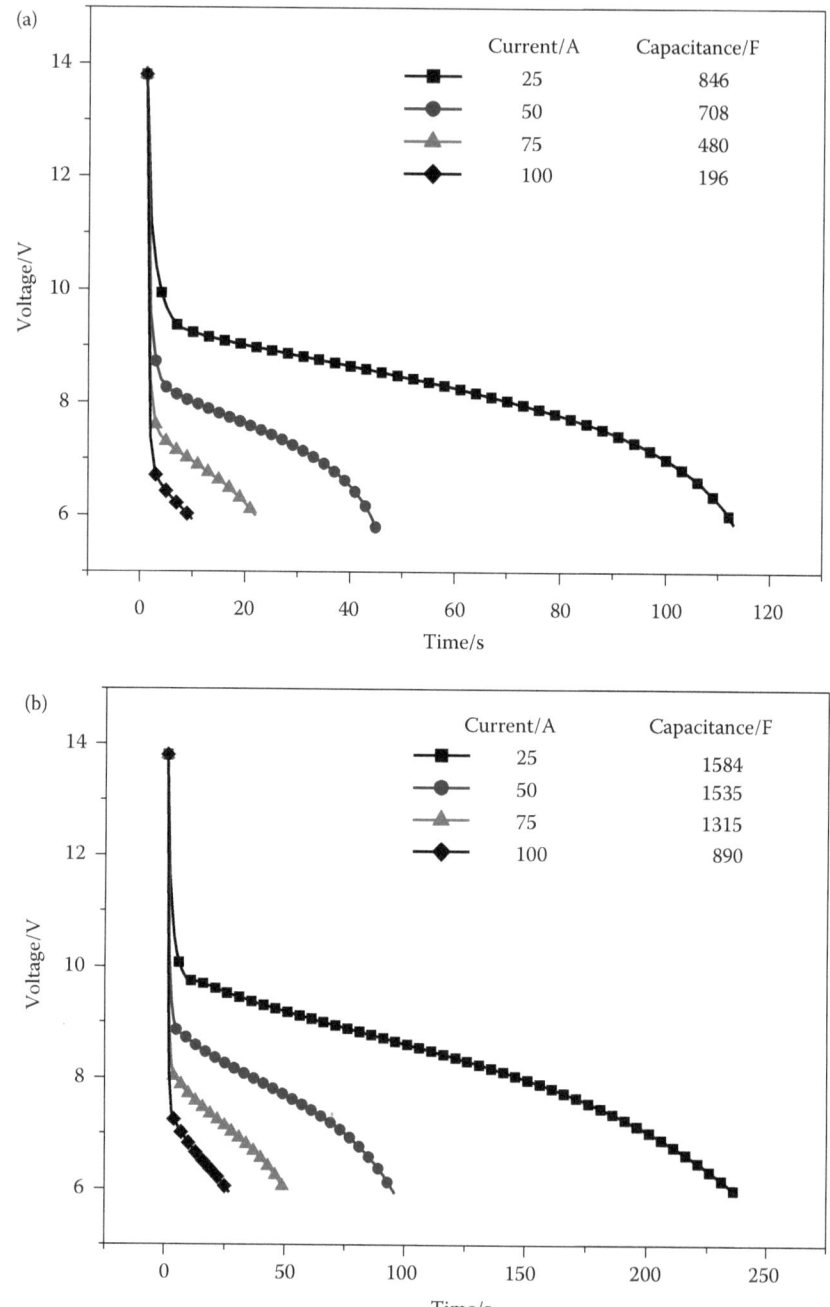

FIGURE 8.5
Constant-current discharge profiles at varying current loads: 25, 50, 75, 100 A for (a) AGM and (b) gelled 12 V/kF-range lead–carbon HUCs.

than AGM yielding higher surface area and porosity. Higher surface area and porosity helps the electrolyte to enter the pores of electrode active materials and reduce the electrode/electrolyte contact resistance as well as more utilisation of active material. Performance enhancement has been observed due to the very good electrolyte distribution in the pores of the active material of the positive plate in gelled configuration compared to AGM configuration.

8.3.2 Performance Curves for 12 V/kF-Range Lead–Carbon HUCs at Varying Temperatures

The change in the capacitance values at varying temperatures differ for AGM and silica gelled lead–carbon HUCs as shown in Figure 8.6a and b. Capacitance increases with increasing operating temperature because elevated temperatures facilitate the improvement of the conductivity and the diffusion rate of the ions. Similarly, capacitance decreases with drop in temperature due to lower electrolyte conductivity and the freezing of electrolytes at sub-zero temperatures. The charge transfer resistance is also a function of temperature. When the operating temperature increases from –20°C to 50°C, large amounts of charge can cross the activation barrier of an electrochemical reaction very easily. Accordingly, the working temperature range for lead–carbon HUCs is kept between –20°C and 50°C.

8.3.3 Coulombic Efficiency for 12 V/kF-Range Lead–Carbon HUCs

Coulombic efficiencies for lead–carbon HUCs with AGM and silica gelled electrolyte configurations were calculated from constant-current charge–discharge at 3 A current. From the data in Figure 8.7, the efficiency values are estimated to be 93% and 92% for AGM and gelled electrolyte lead–carbon HUCs, respectively. The high coulombic efficiency is achieved mainly by using activated carbon capacitor electrodes as negative plates.

8.3.4 Constant-Power Discharge Data for 12 V/kF-Range Lead–Carbon HUCs

Constant-power discharge data at 10, 20, 30, 40 and 50 W for 12 V lead–carbon HUCs are shown in Figure 8.8a and b. Lead–carbon HUCs were first charged galvanostatically with 3 A up to a cut-off voltage of 13.8 V followed by 1 h potentiostatic charge at 13.8 V. The constant-power discharge curves show that lead–carbon HUCs have the capability to deliver energy as well as power for specified duration. The lead–carbon HUC is a hybrid device with a lead–acid battery and a carbon-based EDLC. Eventually this kind of device is limited by capacitor behaviour. To calculate energy density using the $(1/2 \times C \times V^2)$ formula is an overestimation for this device

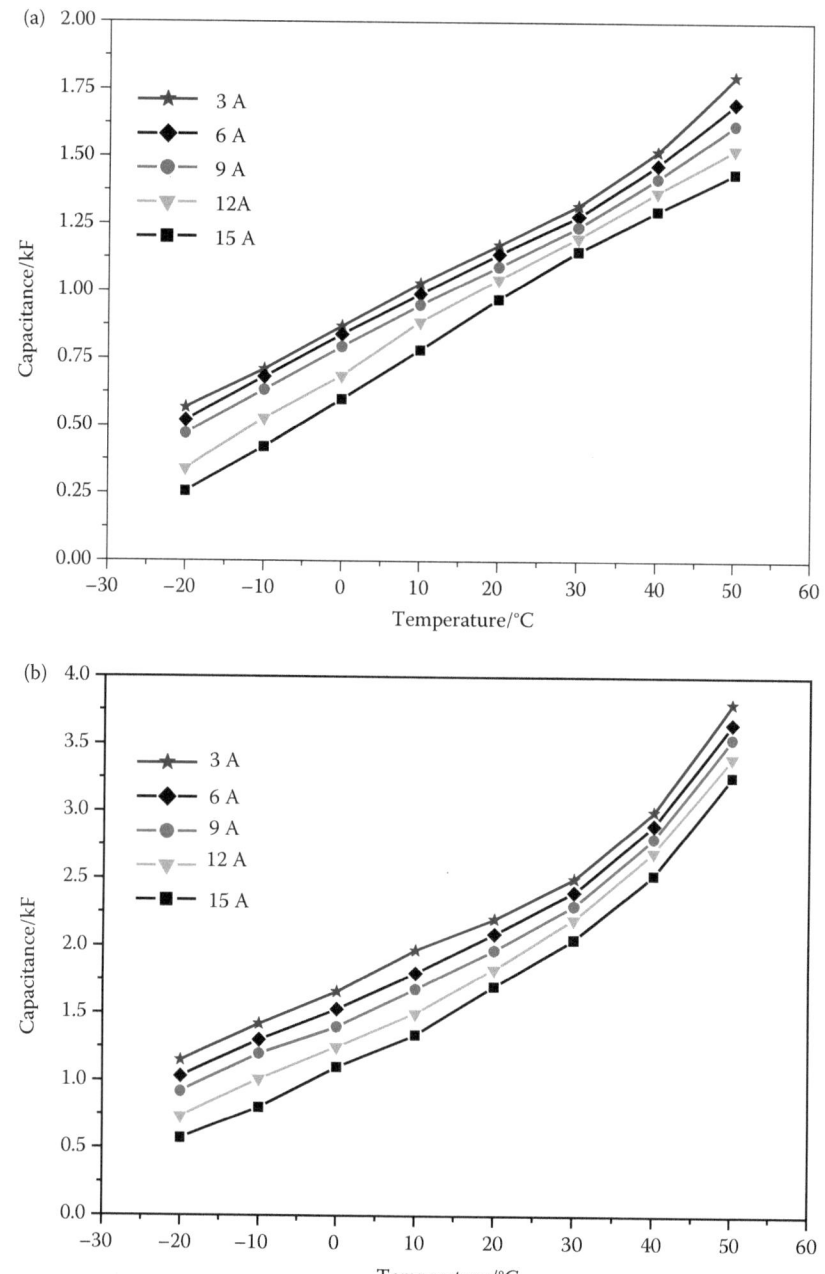

FIGURE 8.6
Capacitance data at varying temperatures for (a) AGM and (b) gelled 12 V/kF-range lead–carbon HUCs.

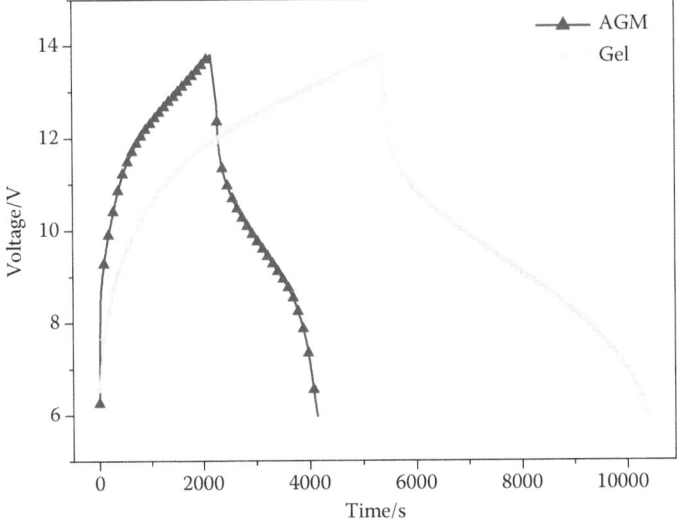

FIGURE 8.7
Constant-current charge–discharge curves of 12 V/kF-range lead–carbon HUCs.

because it is not an ideal capacitor. In this regard, constant-power experiments are useful for calculating the energy density and power density. Different constant-power loads have been chosen and the device has been discharged through those constant-power loads over a definite voltage limit. The multiplication of the discharge time and the corresponding power gives the energy value accordingly in Equation 8.7 and it is normalised by the weight represented as energy density. Different constant-power values are chosen for reproducibility of data. The energy density values at different power loads are almost similar, but at higher power loads, the values slightly reduce due to fast discharge rate. The power density has been calculated from the aforementioned energy density data, divided by response time of the device obtained from impedance spectroscopy analysis.

8.3.5 Impedance Data for 12 V/kF-Range Lead–Carbon HUCs

Internal resistance measurements had been carried out using AC impedance spectroscopy at different state-of-charge values of the lead–carbon HUCs. The intercept at the abscissa at high frequency values in the Nyquist plot indicates the resistance value as shown in Figure 8.9. Response time is another important parameter of HUCs that can be estimated from a Bode plot (phase vs. frequency). In the Bode plot, the resistance and capacitive reactance are equal at −45° phase angle; the corresponding frequency is referred

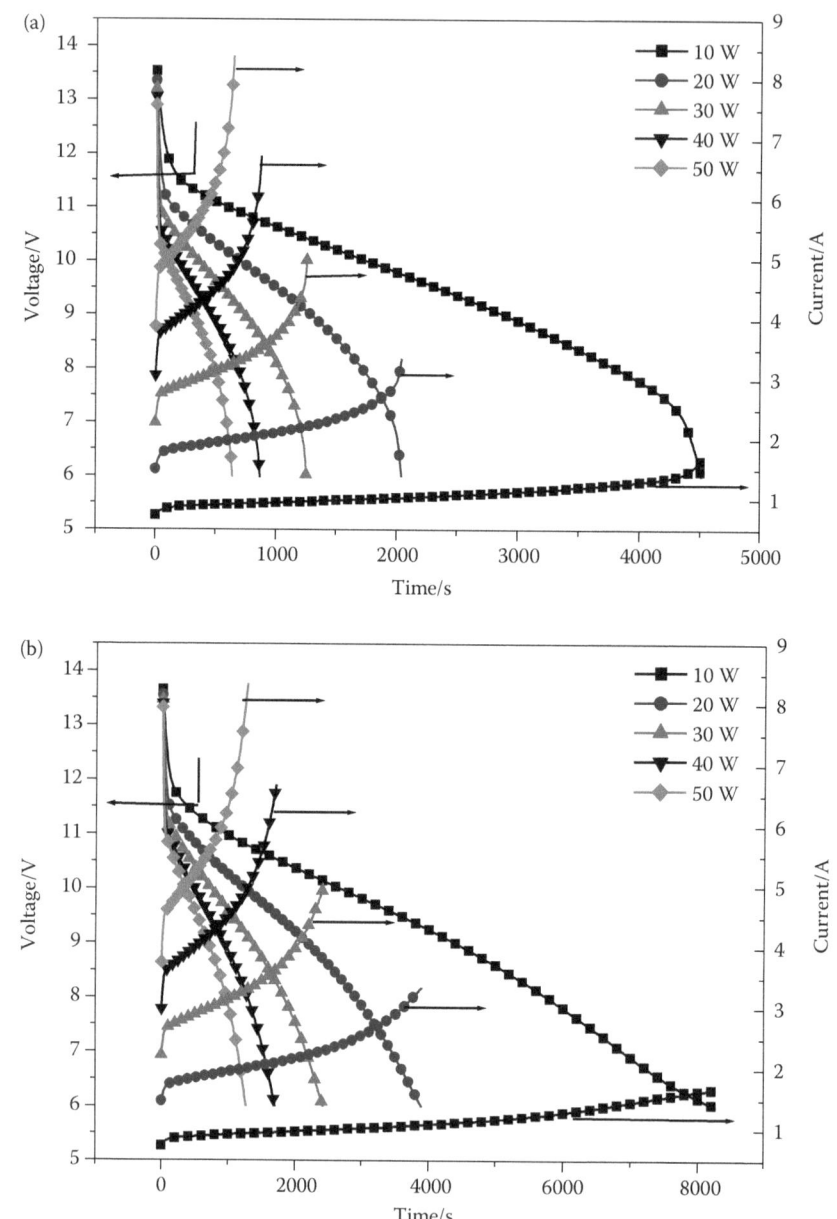

FIGURE 8.8
Constant power discharge profiles at varying power loads. (a) AGM and (b) gelled 12 V/kF-range lead–carbon HUCs.

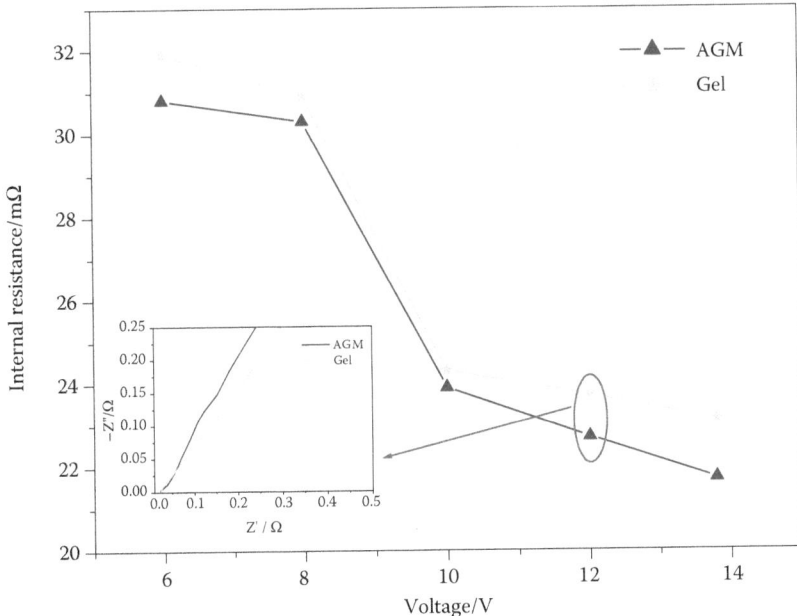

FIGURE 8.9
Internal resistance *vs.* potential profiles for Lead-Carbon HUCs. Inset: Nyquist plots of AGM and gelled 12 V/kF-range Lead-Carbon HUCs.

to as the breaking frequency. Beyond this frequency, capacitive reactance becomes larger than resistance. The time corresponding to the breaking frequency is represented as response time for the capacitor. The response times for the lead–carbon HUCs are estimated to be 1.66 and 4.68 s for the AGM and gelled electrolyte configurations, respectively, as shown in Figure 8.10a and b. The lower response times indicate the effectiveness of lead–carbon HUCs in pulse power applications.

8.3.6 Energy Density and Power Density Data for 12 V/kF-Range Lead–Carbon HUCs

The battery is recognised as an energy device and the capacitor as a power device. But lead–carbon HUC devices can optimized both these characteristics. The energy density and power density values are estimated from the constant-power discharge profiles of lead–carbon HUCs that was discussed in Section 8.3.4. The energy and power density data for lead–carbon HUCs are summarised in Table 8.1. Energy density of the device is estimated from the constant power loads from 13.8 to 6 V window at 50 W discharge rates.

The energy density of a lead–carbon HUC is higher than the conventional electrochemical capacitor because the lead–carbon HUC has a battery component in it, which gives the enhancement of energy density. But these

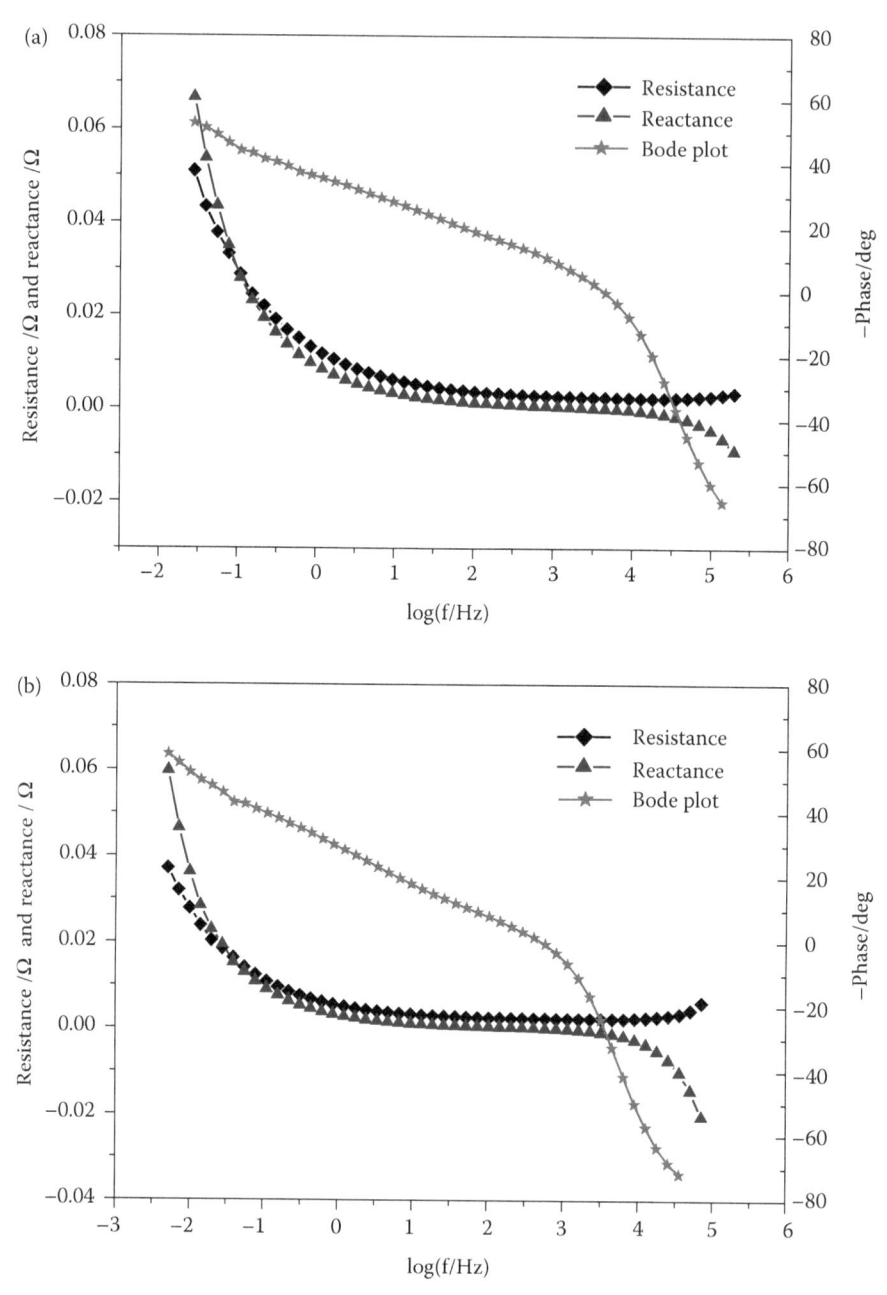

FIGURE 8.10
Response time data for (a) AGM and (b) gelled 12 V/kF-range lead–carbon HUCs.

TABLE 8.1
Energy Density and Power Density Data for AGM and gelled 12 V/kF-range Lead–Carbon HUCs

	Response Time (s)	Energy Density (Wh/kg)	Power Density (W/kg)
AGM	1.66	0.9	1937
Gel	4.86	1.8	1333

energy density data are poor in comparison with lead–acid battery. By contrast, the power density is very high, and as good as a conventional electrochemical capacitor. Actually, the advantage of the device is that it is mainly a power device with appreciably stored energy. A smart technology is required to explore the advantages of the device in an efficient manner by coupling it with an energy device like a battery. The lead–carbon HUC is a power device so it can both charge and discharge very rapidly. The lead–carbon HUC can be charged very fast by any means, either conventional or non-conventional, such as solar, wind, mechanical dynamo, electrical grid power and so on, and this charge transfers to a battery for long-range storage. Then this combined system becomes very useful for energy harvesting, storage and management. The battery by itself is unable to do this job because it cannot charge in a fast manner; that is why the battery is not efficient to store non-conventional energy such as solar, wind and so on.

8.3.7 Self-Discharge, Leakage Current and Parallel Resistance Data for 12 V/kF-Range Lead–Carbon HUCs

Self-discharge for the lead–carbon HUC was measured after charging for 3 h at 13.8 V followed by 24 h in an open-circuit condition at room temperature (28°C) and 50°C. As shown in Figure 8.11a and b, at higher temperatures, the self-discharge is only marginally increased. Self-discharge data were also obtained for 7 days at room temperature as shown in Figure 8.11c with around 25% voltage loss. The self-discharge energy loss factor (SDLF) was calculated from the voltage loss data with respect to time according to the following equation:

$$(SDLF) = 1 - (V/V_w)^2 \qquad (8.18)$$

In Equation 8.18, V and V_w are the voltage after the self-discharge period and its respective working voltage. Leakage current was measured during 24 h after fixing the voltage at 13.8 V. Parallel resistance after 24 h was obtained as a ratio of working voltage, that is, 13.8 V, and leakage current after 24 h. Self-discharge, leakage current and parallel resistance values for AGM and gelled lead–carbon HUCs are given in Table 8.2. By nature, the pure electrochemical capacitor is a leaky system, so its self-discharge is more significant than an energy device like the battery. To overcome this high self-discharge, the battery plate is introduced into the HUC. So, in the case of the lead–carbon HUC,

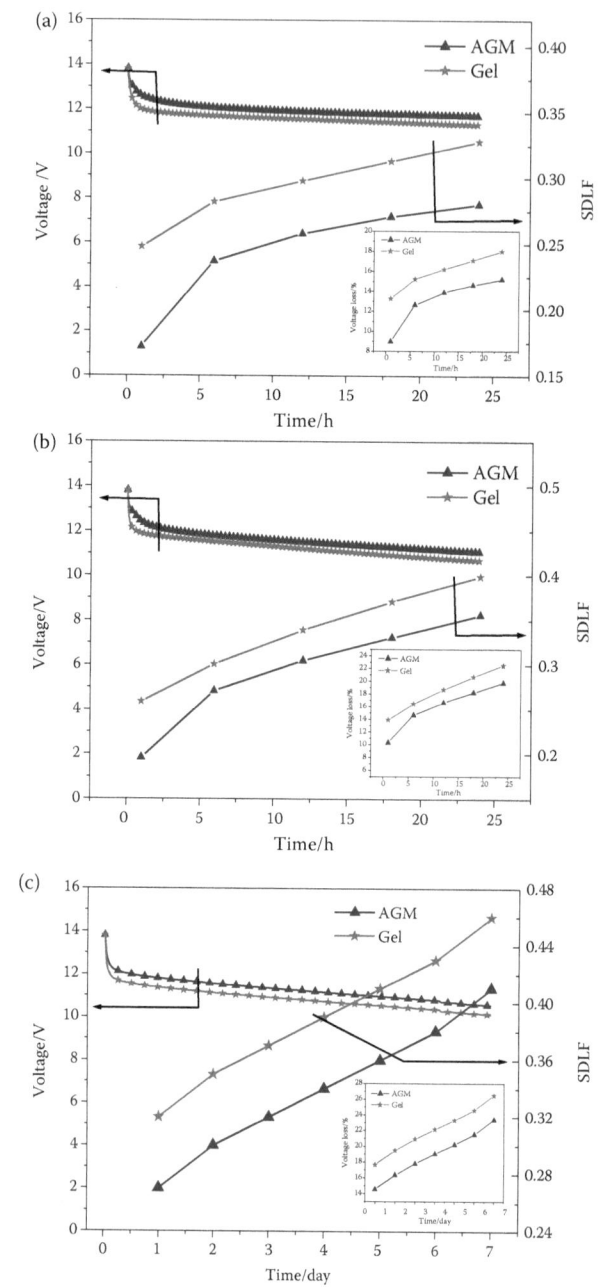

FIGURE 8.11
Self-discharge profiles of 12 V/kF-range lead–carbon HUCs (a) 1 day at room temperature, (b) 1 day at 50°C and (c) 7 days at room temperature. Inset: Percent voltage loss versus time plot.

TABLE 8.2

Self-Discharge, Leakage Current and Parallel Resistance Data for AGM and gelled 12 V/kF-range Lead–Carbon HUCs

Parameter	AGM	Gel
Self-discharge at room temperature for 1 day (28°C) (%)	15	18
Self-discharge at 50°C for 1 day (%)	19	22
Self-discharge at room temperature for 7 days (28°C) (%)	23	26
Leakage current (mA)	80	180
Parallel resistance (Ω)	173	77

both the self-discharge rate and the leakage current are reduced. It is accepted that with an open circuit condition, the voltage loss (namely self-discharge) arises due to a leakage current through parallel resistance (as per Ohm's law). Higher parallel resistance would lead to lower leakage current flow, thus reducing self-discharge rates. In the case of AGM and gelled lead–carbon HUCs, the gelled configuration exhibits more self-discharge than AGM, thus giving higher leakage current and lower parallel resistance accordingly as evident from Table 8.2. The difference in self-discharge percentages between AGM and gelled lead–carbon HUCs is not drastic, at both room temperature and the elevated temperature of 50°C.

The small difference in self-discharge values between AGM and gelled configurations may be attributed to the surface area of positive active materials. Owing to the use of silica gel electrolytes, the surface area of a positive active material increases and this large surface area may be responsible for some parasitic side reactions at the open circuit condition for gelled configurations. In general, increasing the surface area enhances the rate of reaction and that may be reflected in this case, but a thorough study is needed to elucidate the proper reason.

8.3.8 Pulse-Cycling Data for 12 V/kF-Range Lead–Carbon HUCs

The 12 V lead–carbon HUCs were subjected to pulsed cycle life tests under a charge–discharge current load of 30 A, first for 50,000 cycles for a duration of 1 s followed by a rest for 5 s. The effect of high-rate pulsed cycling on the capacitance of these lead–carbon HUCs was also studied. This was followed by 50,000 cycling with 60 A charge–discharge pulse rates for 1 s followed by a 10 s rest. The capacitance measurements were carried out at 3 A current discharges after each 10,000 cycles; the data are shown in Figure 8.12. After 100,000 cycles, the capacitance loss was calculated as 36% and 23% for AGM and gelled lead–carbon HUCs, respectively. Initially, smaller particle size as well as higher surface area and porosity of positive active material also contributed to the long cycle life of gelled configurations compared to AGM.

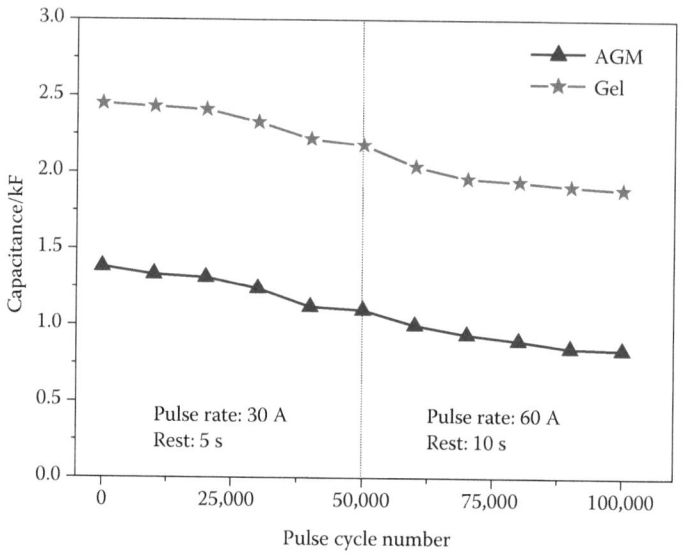

FIGURE 8.12
Pulse cycle life test data for 12 V/kF-range lead–carbon HUCs.

Agglomeration of the particle of positive active material has been observed during the cycling of the device and the particle size of PbO_2 increases. This is also responsible for the capacitance loss. But this agglomeration process is less vigorous in a silica-gelled environment and that gives less capacitance loss compared to AGM configuration [17]. The overall enhanced cycle life of the lead–carbon HUCs is mainly attributed for the coupling of the capacitor electrode with the battery-type electrode. These data are truly attractive for the long-term applications of lead–carbon HUCs.

8.4 Applications

Major applications of electrochemical capacitors appear to be in high-pulse power and short-term power holds [19]. Many appliances now incorporate digital components with memory where even a very brief interruption in the power supply could cause the loss of stored information. In such applications, the capacitor can act as a back-up supply for short periods. Batteries are the alternative to the capacitors for these applications, but batteries have a limited lifetime and, therefore, need to be replaced regularly. Instead, electrochemical capacitors are a preferred choice as back-up power supply due to their long lifetime. At present, battery-powered electric vehicles have the limitations of low power density, limited charge/discharge cycles, high-temperature dependence and long charging time. Electrochemical capacitors are bereft of these limitations albeit they are faced with other limitations such as low energy density and high costs.

Peak load requirements that result from accelerating or climbing uphills could be met by the high-power device, namely the electrochemical capacitors bank. Utilisation of electrochemical capacitors also facilitates harvesting energy from regenerative braking. An electrochemical capacitor is presently the power supply in hybrid cars for start/stop applications. When a hybrid vehicle stops, its internal combustion engine (ICE) shuts down; this is restarted by the electrical system powered by an electrochemical capacitor that is recharged when the ICE resumes powering the vehicle. This helps to reduce fuel consumption. Electrochemical capacitors are ideally suited for city transit buses with stop-and-go driving, in trash trucks that can experience as many as a thousand start/stop cycles during a day, and in delivery vans that operate on similar drive cycles. The primary challenges for any energy storage unit used in heavy-duty hybrid vehicles are long cycle life and the need to dissipate the heat generated due to charge/discharge losses. Electrochemical capacitors are highly efficient and have limited heat dissipation owing to their low energy content.

These capacitors could also help in improving the power quality as a static-synchronous-compensator system that injects or absorbs power from a distribution line to compensate for any voltage fluctuations. Such a system requires a DC energy storage device of some sort from which energy could be drawn or stored. Since the majority of voltage perturbations on the distribution bus are shortlived, usually not lasting for more than 10 cycles, electrochemical capacitors are an attractive option for energy storage and delivery to improve the power quality.

Presently, batteries are being used widely in the portable power appliances, such as UPS, laptops and mobile phones. Many such devices draw high-power pulsed currents, which result in the reduction of battery performance. Batteries in parallel with electrochemical capacitors could be an effective alternative for these applications. Capacitors also hold promise in solar energy storage. In solar photovoltaic applications, batteries need to be replaced every 1–3 years because of continuous cycling that has a detrimental effect on batteries. But electrochemical capacitors can be charged and discharged quickly for a large number of cycles, and need to be replaced every 20 years only, which is similar to the life span of the photovoltaic panels. Life-cycle costs are therefore reduced by eliminating frequent maintenance requirements.

Energy efficiency is always of primary concern in renewable power generation. In this regard, electrochemical capacitors are attractive as they exhibit a much higher charging efficiency than batteries. Besides, the pitch of wind turbine three-rotor blades can be adjusted to respond to current conditions. Pitch adjustments allow wind turbines to maximise energy generation.

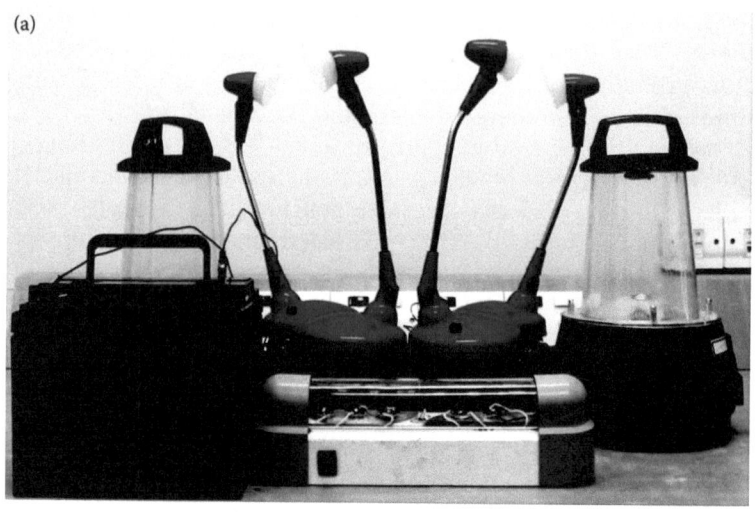

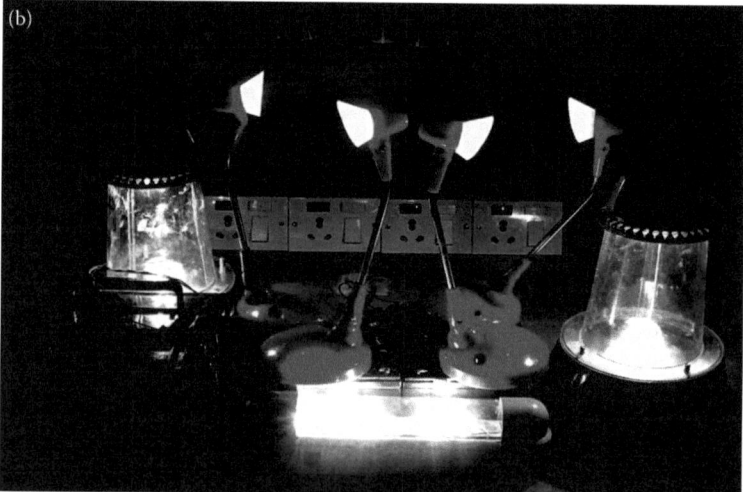

FIGURE 8.13
(See colour insert.) Lead–carbon hybrid ultracapacitor setup with LED lights in (a) OFF and (b) ON positions.

In order to adjust the blades, turbines need power. To date, wind turbine manufacturers have relied on battery-driven energy storage systems for that power. In a wind turbine, pitch control of each of the blades ensures optimum positioning for efficient use of wind speed for both performance and safety. Though wind turbine blades are long and awkward to reposition, it is vital to adjust them quickly to grab useful winds. It is even more vital to quickly get blades turned out of potentially damaging winds. Engineers accomplish the pitch control either mechanically or electrically, but electrical control systems are more reliable. However, when electrical control systems rely on battery-based back-up systems, the potential maintenance advantage over hydraulic systems is not necessarily realised. Electrochemical capacitors are becoming the power source of choice in wind turbine nacelles to power blade-pitch control.

Lead–carbon HUCs are attractive for the aforesaid applications. Modern energy services are crucial to human well-being and to a country's economic development and yet globally over 1.3 billion people do not have access to electricity and 2.6 billion people are without clean cooking facilities. More than 95% of these people are either in sub-Saharan Africa or developing Asia and 84% are in rural areas [20]. Today, the global population is 7 billion and is estimated to grow to 9 billion by 2050, and about 10 billion by 2100. It is predicted that most of this population growth will be in Asia and Africa, where rapidly rising economic growth will place additional demands on energy supply. The popularisation of ultracapacitors in rural lighting looks to be an effective route. Lead–carbon HUCs could be attractive options for rural lighting as depicted in Figure 8.13a and b that show several variants (1–5 W) of LED lamps employing 12 V lead–carbon HUCs.

Acknowledgements

Financial support from the Department of Science & Technology, Government of India, and Indian Institute of Science, Bangalore is gratefully acknowledged.

References

1. The White House, Office of the Press Secretary, "Remarks by the President at the National Academy of Sciences Meeting," April 27, 2009. http://www.whitehouse.gov/the-press-office/remarks-president-national-academy-sciences-annual-meeting. Accessed April 15, 2014.
2. A. K. Shukla, A. Banarjee, M. K. Ravikumar and A. Jalajakshi, Electrochemical capacitors: Technical challenges and prognosis for future markets, *Electrochim. Acta* 84, 2012, 165–173.
3. W. G. Pell and B. E. Conway, Peculiarities and requirements of asymmetric capacitor devices based on combination of capacitor and battery-type electrodes, *J. Power Sources* 136, 2004, 334–345.
4. D. Zhou and L. Gao, Effect of electrochemical preparation methods on structure and properties of PbO_2 anodic layer, *Electrochim. Acta* 53, 2007, 2060–2064.

5. N. Yu and L. Gao, Electrodeposited PbO_2 thin film on Ti electrode for application in hybrid supercapacitor, *Electrochem. Commun.* 11, 2009, 220–222.
6. N. Yu, L. Gao, Z. S. Hao and Z. Wang, Electrodeposited PbO_2 thin film as positive electrode in PbO_2/AC hybrid capacitor, *Electrochim. Acta* 54, 2009, 3835–3841.
7. I. Gyuk, Supercapacitors for electric storage: scope and projects, *Proceedings of Advanced Capacitor World Summit*, July 14–16, 2004, Washington, DC.
8. E. Buiel, Development of lead-carbon hybrid battery/super capacitors, *Proceedings of Advanced Capacitor World Summit*, July 17–19, 2006, San Diego, CA.
9. S. Kazaryan, Characteristics of the $PbO_2 | H_2SO_4 | C$ ECs by universal supercapacitor, LLC, *Proceedings of Advanced Capacitor World Summit*, July 23–25, 2007, San Diego, CA.
10. A. I. Belyakov, Asymmetric electrochemical supercapacitors with aqueous electrolytes, *ESSCAP*, November 6–7, 2008, Rome, Italy.
11. S. A. Kazaryan, S. V. Litvinenko and G. G. Kharisov, Self-discharge of heterogeneous electrochemical supercapacitor of $PbO_2 | H_2SO_4 | C$ related to manganese and titanium ions, *J. Electrochem. Soc.* 155, 2008, A464–A473.
12. Axion Pb-C, Technology, http://www.axionpower.com.
13. A. K. Shukla, M. K. Ravikumar and S. A. Gaffoor, International Patent No. WO 2011/161686 A1.
14. A. K. Shukla, A. Banerjee, M. K. Ravikumar and S. A. Gaffoor, International Patent No. WO 2013/011464 A1.
15. A. K. Shukla, A. Banerjee, M. K. Ravikumar and S. A. Gaffoor, US Patent No. 2012/0327560 A1.
16. A. Banerjee, M. K. Ravikumar, A. Jalajakshi, S. A. Gaffoor and A. K. Shukla, A 12 V substrate-integrated PbO_2-activated carbon asymmetric hybrid ultracapacitor with silica-gel based inorganic-polymer electrolyte, *ECS Trans.* 41, 2012 101–113.
17. A. Banerjee, M. K. Ravikumar, A. Jalajakshi, P. Suresh Kumar, S. A. Gaffoor and A. K. Shukla, Substrate-integrated lead-carbon hybrid ultracapacitor with flooded, absorbent glass mat and silica-gel electrolyte configurations, *J. Chem. Sci.* 124, 2012, 747–762.
18. B. E. Conway, *Electrochemical Supercapacitor: Scientific Fundamentals and Technological Applications*, New York: Kluwer Academic/Plenum Publishers, 1999.
19. J. M. Miller, *Ultracapacitor Applications*, London, UK: The Institution of Engineering and Technology, 2011.
20. S. Chu and A. Majumdar, Opportunities and challenges for a sustainable energy future, *Nature* 488, 2012, 294–303.

9

Vanadium Flow Batteries: From Materials to Large-Scale Prototypes

Huamin Zhang

CONTENTS

9.1 Introduction 347
9.2 Vanadium Flow Battery: Principle and Constructions 349
9.3 Key Materials Development of Vanadium Flow Battery 351
    9.3.1 Membranes Separators..351
        9.3.1.1 Perfluorinated Membranes 351
        9.3.1.2 Non-Fluorinated Membranes.......................................355
        9.3.1.3 Anion Exchange Membranes 357
    9.3.2 Electrode Materials...361
        9.3.2.1 Carbon-Based Materials..366
        9.3.2.2 Electrode Materials Modification 367
    9.3.3 Electrolytes..372
    9.3.4 Stacks and System..375
        9.3.4.1 Stack Structure 375
        9.3.4.2 Design of Cell Stack 377
        9.3.4.3 Sealing of the Stack...379
9.4 Application and Demonstration 380
9.5 Challenges and Prospective 384
Acknowledgements 387
References..387

9.1 Introduction

Renewable energies from sources such as solar and wind attracts significant attention due to growing issues of energy shortage and air pollution. However, the random nature of these intermittent renewable sources makes it quite challenging for its use and dispatch through the grid. One effective solution is to connect the power station and the grid with electrical energy storage (EES) devices [1]. Furthermore, the energy storage technique plays a critical role in future 'smart grid and electric vehicle' applications. To date, various EES technologies have been developed for specific applications, based on their own merits (Figure 9.1) [2]. Considering the performance requirements and cost, electrochemical systems are superior to other physical systems, which are usually restricted by the relatively long response times and specific geographical locations. The electrochemical process or batteries are ideal candidates for EES due to large control window for power rate and capacity, fast response time, low environmental footprint [3] and portability (no longer restricted by geographical location).

Put it simply, batteries are devices that can convert chemical energy into electricity by the oxidation or reduction of active species. This process is reversible such that the system can be used for both energy storage and electricity delivery. Normally, batteries from different active species could meet specific EES applications such as UPS (uninterrupted power supply), grid support and energy management, according to their range requirement of power rate and capacity (Figure 9.1). For large-scale energy storage systems (grid support), it could be realised via lead–acid batteries, sodium–sulphur, lithium ions and flow batteries. Although lead–acid battery is a mature technology, its short lifetime and low deep discharge ability limited its usage in large-scale energy storage.

Flow battery is an emerging technique effective for large-scale energy storage due to its attractive features such as long life time, active thermal

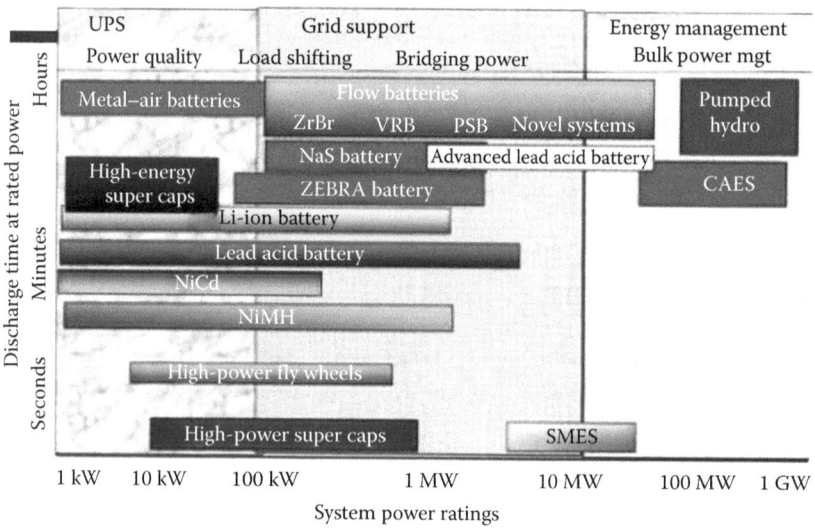

FIGURE 9.1
The basic characteristics of different electrical energy storage techniques. (From D. Rastler, *Electricity Energy Storage Technology Options: A White Paper Primer on Applications, Costs and Benefits*. EPRI, Palo Alto, CA, 2010.)

management and independence of energy and power rating [4]. A flow battery system (FBS) is an EES approach that was originally initiated by NASA in the 1970s. It employs various soluble species with different valances to realise charge and discharge processes [5]. Compared with traditional batteries, the reactants of flow battery were stored in external tanks instead of within electrode. A stack including series single cells acts as the reactor, the electrochemical electron-transfer reactions occur in single cells as their reactants flow through them. This special configuration offers flow batteries some key advantages compared to other battery systems. The most important aspect could be the independent design of power and energy. The power of flow batteries is determined by stacks (electrode area, cell numbers), whereas the energy is dependent on the volume and concentration of active species.

Similar to other batteries, there are various flow battery chemistries, including zinc–bromine batteries, polysulphide–bromide batteries, iron–chrome batteries, soluble lead–acid batteries, zinc/nickel flow batteries and vanadium flow batteries (VFBs) [3,6]. Among these, zinc–bromine is a hybrid flow battery, where the deposit/dissolution of zinc occurs at the positive side during charge and discharge. During charge, zinc is deposited onto the negative electrode, while bromine is formed at the positive electrode. The highly corrosive bromine and the electroplating of zinc at the anode remain major limitations for its further application. Unlike zinc–bromide batteries, polysulphide–bromide and iron–chrome batteries are considered as the true FBSs since both anolyte and catholyte redox couples remain dissolved in the electrolyte during charge and discharge. Despite this advantage, polysulphide–bromide batteries still suffer from durability issues due to the corrosive nature of bromine, and both batteries suffer from the electrolytes crossing over from the cathode to the anode during charge and discharge as well. Among the existing flow battery technologies, VFBs are considered one of the most attractive techniques for large-scale energy storage, with the flowing anolyte and catholyte solutions being the same species, crossover of active species could be avoided [7,8].

9.2 Vanadium Flow Battery: Principle and Constructions

As shown in Figure 9.2, a VFB utilises vanadium ions with different valences (positive side: VO^{2+} [V(IV)] and VO_2^+ [V(V)], negative side: V^{2+} [V(II)] and V^{3+} [V(III)]) as active species; the charge and discharge process is realised via oxidation and reduction of different vanadium ions. During charge, the V^{3+} is reduced to V^{2+}, while VO^{2+} is oxidised to VO_2^+ in the presence of external electrical sources; the reverse reaction will occur during discharge. The

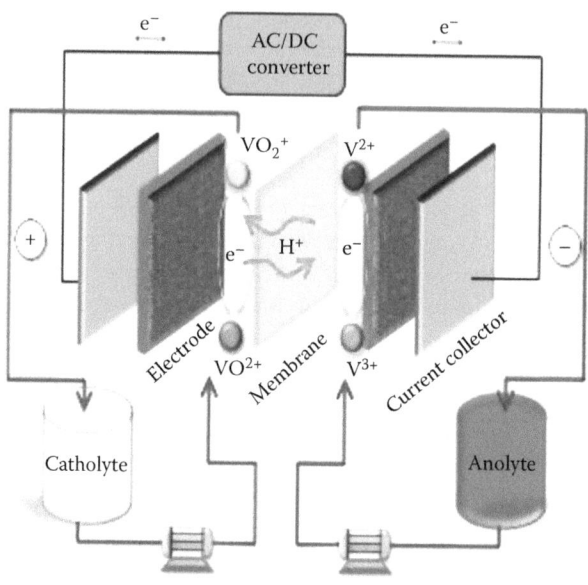

FIGURE 9.2
(See colour insert.) The principle of a VFB. (Reprinted with permission from C. Ding et al., Vanadium flow battery for energy storage: Prospects and challenges, *J. Phys. Chem. Lett.* 4, 2013, 1281–1294. Copyright 2013 American Chemical Society.)

overall electrochemical reaction gives a cell voltage of 1.26 V at 25°C and unit activities (i.e. standard voltage) [9]. All active species are dissolved in supporting electrolytes (normally in sulphuric acid), the concentration of vanadium ions and sulphuric acid is in between 1–2 M and 1–3 M, respectively. VFBs have a unique advantage over other flow batteries, in which the electrolytes at the positive and negative sides are identical in their discharged states [10], making electrolyte shipment and storage even simpler and less expensive. Furthermore, the migration of ions from one electrolyte to the other through the membrane causes minimal contamination, as the same elements of electrolytes are used at both sides.

VFBs are normally constructed by assembling several cells together in a series to form a battery stack [10] (Figure 9.3). Electrodes at either side of a bipolar plate play the role of separating each cell from the other (Figure 9.3a), where the bipolar plate acts as the current-conducting medium between the negative electrode of one cell and the positive electrode of the next (Figure 9.3b). The cells in the battery are electrically connected in a series and the electrolyte flows through the cells in parallel (Figure 9.3c). The electrolytes at the positive and negative sides are isolated by a membrane. The key components of a VFB are the electrolyte, the carbon felt electrode, the ion exchange membrane and the bipolar plate. Each of these components will be discussed in detail in the following section.

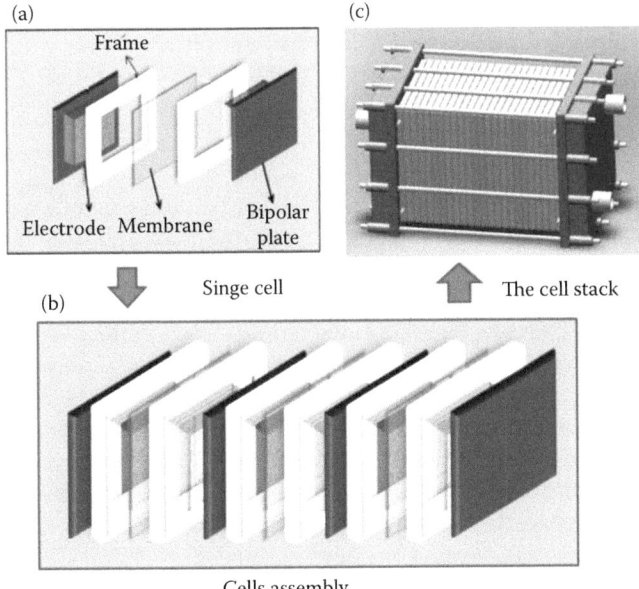

FIGURE 9.3
The assembly of (a) single cell, (b) multiple cells and (c) cell stack. (X.F. Li et al., Ion exchange membranes for vanadium redox flow battery (VRB) applications, *Energy Environ. Sci.* 4, 2011, 1147–1160. Copyright 2011. Reprinted with permission from Royal Society of Chemistry.)

9.3 Key Materials Development of Vanadium Flow Battery

9.3.1 Membranes Separators

9.3.1.1 Perfluorinated Membranes

In a VFB, a membrane plays the role of isolating electrolytes and transporting ions to complete the current circuit [11] (Figure 9.4). The ideal membranes should carry the following characteristics: (a) Good ion conductivity: the

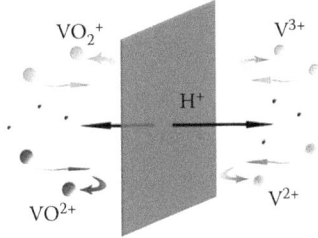

FIGURE 9.4
The membrane separator of a VFB.

membranes need to complete the current circuit by transferring ions. A high ion conductivity is required to minimise the ohmic losses and increase/maintain the voltage efficiency (VE); (b) High ion selectivity: these membranes should possess low permeation rates of active species to minimise self-discharge induced by side reactions; and (c) Good chemical stability or sustainability to ensure the cycle life of a battery.

The membranes traditionally used in VFBs are perfluorinated sulphonic acid polymers such as Dupont's Nafion®, which consists of a flexible hydrophobic perfluorinated backbone and a side chain with highly hydrophilic terminal sulphonic acid groups ($-SO_3H$) (Figure 9.5a). In this structure, the hydrophobic Teflon backbone provides excellent mechanical and chemical stability, while the hydrophilic zone, originating from the assembled sulphonated groups, provides the membranes with high ion conductivity. The flexible long ionic side chain and perfluorinated backbone of Nafion allow continuous networks of ionic channels through the Nafion membrane [12]. As observed in Figure 9.5b, the classic cluster-network model describes the transport mechanism of ions in Nafion. In this structure, an equal distribution of sulphonated ion clusters (also described as 'inverted micelles') with a 40 Å (4 nm) diameter is held within a continuous fluorocarbon lattice, where these clusters are interconnected by ~10 Å (1 nm) channels and formed proton transport channels.

Although Nafion membranes show both high proton conductivity and chemical stability in VFB, the extremely high cost and low ion selectivity

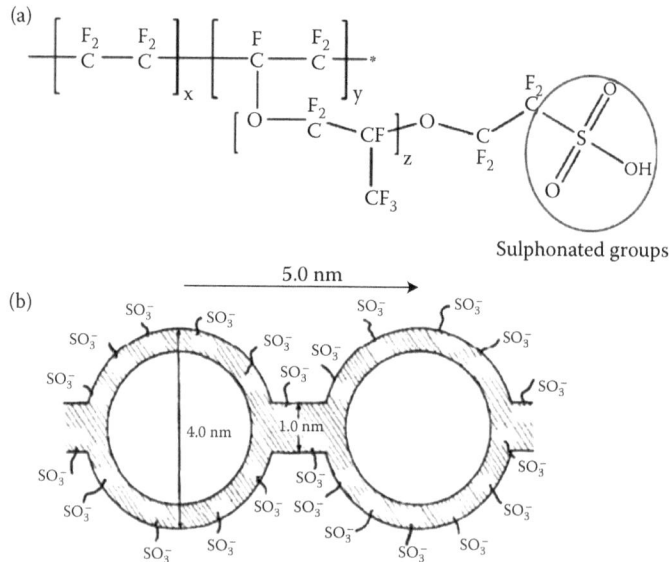

FIGURE 9.5
(a) The proposed structure of Nafion. (b) A cluster-network model for the morphology of hydrated Nafion. (Reprinted from *J. Membr. Sci.*, 13, W.Y. Hsu and T.D. Gierke, Ion transport and clustering in Nafion perfluorinated membranes, 307–326. Copyright 1983, with permission from Elsevier.)

of these membranes (high vanadium ion crossover) prevented further VFB commercialisation. Various approaches to improve ion selectivity were carried out, including modifying the structures of Nafion and using new class of membranes. Modified perfluorinated membranes mainly focused on the improvement of their ion selectivity in VFBs. The general way is to block the hydrophilic clusters by allowing special interaction between inorganic or organic components, and the sulphonic acid groups of Nafion. For example, inorganic nanoparticles such as SiO_2, TiO_2 and ZrP can be introduced into Nafion to enhance the membrane selectivity via blocking the hydrophilic clusters [14–16], as shown in Figure 9.6. SiO_2 nanoparticles were successfully introduced in Nafion by sol–gel method. The modified SiO_2–Nafion membranes can effectively reduce the crossover of vanadium, as the polar clusters (pores) of the original Nafion were filled with SiO_2 nanoparticles during the *in situ* sol–gel reaction of tetraethylorthosilicate (TEOS). The VFB assembled with a Nafion/SiO_2 hybrid membrane shows higher coulombic efficiency (CE, the ratio of electrical charge used during discharge compared to that used during charge) and energy efficiency (EE, the ratio of energy between the discharge and charge processes) than that with a pristine Nafion. At a current density of 20 mA cm^{-2}, maximum EE of 79.9% and 73.8% were obtained for the modified membrane and the pristine Nafion, respectively.

Organic components can also be introduced in perfluorinated membranes, to increase the membrane selectivity. Polyvinylidene (PVDF) can effectively suppress the swelling of Nafion due to its high crystallite nature and further improve the selectivity [17]. By simple blending of PVDF with Nafion, the CE of the battery can be improved due to its higher ion selectivity, and the EE increases as well. The use of acid–base membranes such as SPEEK (sulphonated poly(ether ether ketone)), and PSf-ABIm (polysulphone-2-amide-benzimidazole) is another way to improve membrane selectivity via the interactions between sulphonated acid and amide basic groups [18,19].

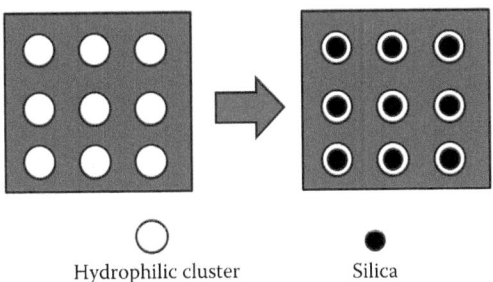

FIGURE 9.6
Inorganic–organic composite membranes. (X.F. Li et al., Ion exchange membranes for vanadium redox flow battery (VRB) applications, *Energy Environ. Sci.* 4, 2011, 1147–1160. Copyright 2011. Reprinted with permission from Royal Society of Chemistry.)

Apart from the blend membranes, introducing an ultrathin vanadium-resistant layer on membrane support could be another choice. For example, a positively charged polymer layer such as polyethylenimine (PEI), polypyrole and polyaniline can be deposited on the surface of a Nafion membrane to increase the ion selectivity [20–22]. Even though the modification of Nafion is an effective way to improve its selectivity, the extremely high cost of the membrane matrix still limited its further application.

Furthermore, fluorinated polymers with a short side chain could be another choice. Perfluorinated membranes with different side chains were investigated in VFB, including their morphology and performance (Figure 9.7) [23]. Nafion 115 with a long side chain, and Aquivion-E87-12S with a short side chain (SSC-M2) were selected as samples (Figure 9.7a and b). The SAXS and TEM results indicated that the membranes with a long side chain can be more easily aggregated to form ion clusters, which further leads to a higher degree of phase separation in the structure, as illustrated in the schematic diagram in Figure 9.7c. On the contrary, the membranes with a short side chain show less pronounced hydrophilic–hydrophobic phase-separated structures. In a VFB, the membranes with short side chains show better selectivity or higher

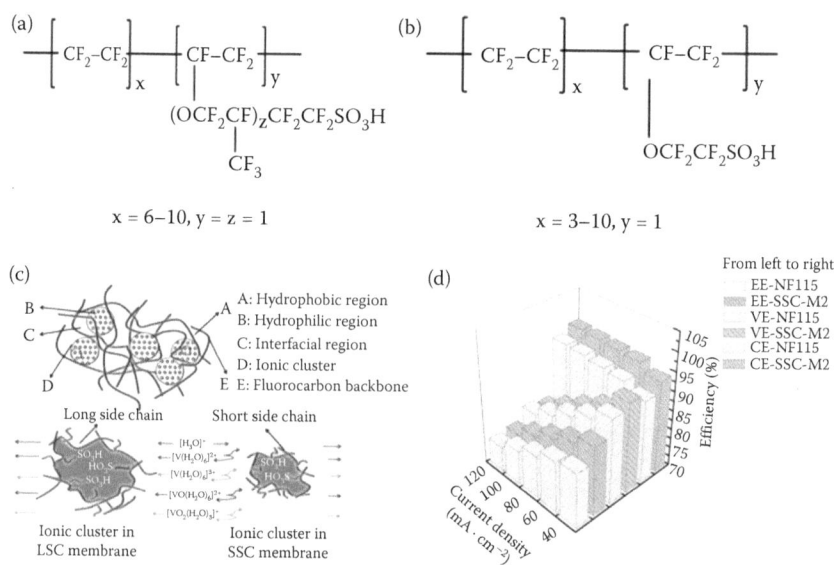

FIGURE 9.7
A comparison between fluorinated polymers with different side chains: (a) Nafion 115 (long side chain); (b) SSC-M2 (Aquivion-E87-12S with short side chain); (c) schematic principles of perfluorosulphonic acid membranes with different side chain lengths employed in VFBs; (d) VFB performance of Nafion 115 and SSC-M2. (From C. Ding et al., Morphology and electrochemical properties of perfluorosulfonic acid ionomers for vanadium flow battery applications: Effect of side-chain length, *ChemSusChem* 6, 2013, 1262–1269. Copyright Wiley–VCH Verlag GmbH & Co. KGaA. Reprinted with permission.)

CE and accordingly lower capacity decay rate (dark bars in Figure 9.7d). Membranes with a shorter side chain could solve the problem of low selectivity of Nafion.

9.3.1.2 Non-Fluorinated Membranes

As an alternative to perfluorinated membrane, non-fluorinated hydrocarbon-based membranes have been investigated due to their low cost and tunable conductivity. These membranes are mainly divided into cation exchange membranes (CEM) and anion exchange membranes, according to their fixed ion exchange groups of polymers. The CEM were first investigated by Skyllas-Kazacos et al., where they introduced commercial membranes such as sulphonated polyethylene (PE) and sulphonated polystyrene in VFB [24]. The battery assembled with these membranes showed a high circuit voltage; however, their VE is rather low due to the high ohmic polarisation. Later on, inspired by the extensive research on fuel cell membranes, sulphonated aromatic polymers such as polyether ether ketone and polyaryl ether sulphone were applied in VFB. All these membranes possess an aromatic rigid polymer main chain, which provides high mechanical and chemical stability; while the randomly distributed sulphonated groups in the polymer chains provide membranes with high ion conductivity.

Compared with perfluorinated membranes, the rigid and less hydrophobic nature of these polymers chain together with the less acidic sulphonic groups of these new membranes induce a lower degree of hydrophilic/hydrophobic phase separation, resulting in a higher selectivity in VFBs. For example, CEM with functional groups such as sulphonated poly(arylenethioether ketone), sulphonated poly(fluorenyl ether ketone) and poly(arylene ether sulphone) were investigated by Chen et al. in VFBs (Figure 9.8) [25–27]. Similar to other ion exchange membranes, the sulphonic acid groups ensure membrane

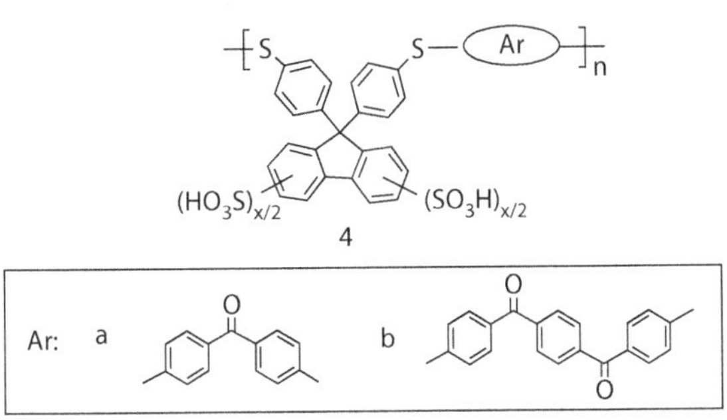

FIGURE 9.8
The structure of sulphonated poly(arylenethioether ketone). (D.Y. Chen et al., Synthesis and characterization of novel sulfonated poly(arylene thioether) ionomers for vanadium redox flow battery applications, *Energy Environ. Sci.* 3, 2010, 622–628. Copyright 2010. Reprinted with permission from Royal Society of Chemistry.)

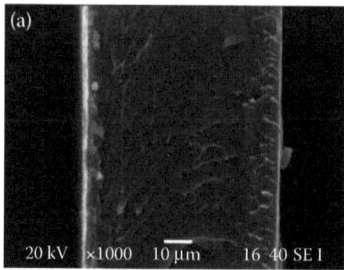 

FIGURE 9.9
A crosssection of SPEEK membranes (a) before and (b) after 80 cycles. (Reprinted from *J. Power Sources*, 196, Z.S. Mai et al., Sulfonated poly (tetramethydiphenyl ether ether ketone) membranes for vanadium redox flow battery application, 482–487. Copyright 2011, with permission from Elsevier.)

conductivity, while the aromatic backbone provides membranes with good mechanical and chemical stability. Most of the membranes showed much lower VO^{2+} permeability and a higher CE than that of Nafion.

SPEEK membranes were prepared and first used in VFBs by our group recently [28]. The single cell assembled with SPEEK membranes exhibited a better performance than that of Nafion at the same operating condition. VFB single cells with SPEEK membranes showed a very high EE (>84%), comparable to that of Nafion, but a much higher CE (>97%) at the current density of 50 mA cm^{-2}. The membranes remained stable after being continuously run for 80 cycles (Figure 9.9).

Gary et al. investigated S-Radel (sulphonated polysulphone) membranes in VFB; even though these membranes show excellent performance, their stability need to be further improved [29]. Introducing some cross-linked groups such as aryl groups (SDPEEK) could be an effective way to improve membrane stability [30] (Figure 9.10). The single cell assembled with SDPEEK membranes retained a stable performance after being run continuously for 900 h,, which showed superior cycle stability compared with other SPEEK membranes (Figure 9.11). Moreover, the crosslinking process can effectively improve membrane selectivity and dimension stability.

FIGURE 9.10
The structure of SDPEEK. (Reprinted from *J. Power Sources*, 217, H.Z. Zhang et al., Crosslinkable sulfonated poly (diallyl-bisphenol ether ether ketone) membranes for vanadium redox flow battery application, 309–315. Copyright 2012, with permission from Elsevier.)

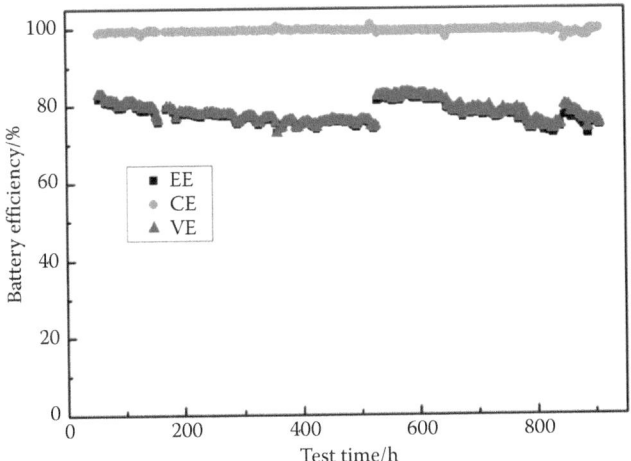

FIGURE 9.11
The cycle performance of SDPEEK. (Reprinted from *J. Power Sources*, 217, H.Z. Zhang et al., Crosslinkable sulfonated poly (diallyl-bisphenol ether ether ketone) membranes for vanadium redox flow battery application, 309–315. Copyright 2012, with permission from Elsevier.)

9.3.1.3 Anion Exchange Membranes

Anion exchange membranes attract quite a lot of attention due to their repellent effect to positive vanadium ions, and appears to be beneficial to minimise the vanadium ion cross-mixing [31,32]. Jian et al. prepared quaternised poly(phthalazinone ether ketone/sulphone) membranes and investigated the effect of the amination agent on these polymers [33,34]. The positive-charged quaternised groups in membranes can effectively repel vanadium ions and, thus, show much better ion selectivity. The VFB performance of these membranes is comparable to Nafion 117 and their permeability is around two orders of magnitude lower than Nafion 115. Recently, anion conductive membranes based on quaternised poly (tetramethydiphenyl ether sulphone) were designed and fabricated for VFB applications (Figure 9.12) [35]. Four substituted methyl groups in polymer chains provide the membranes with a much wider window for controlling the degree of quaternisation. (Figure 9.12). The recorded vanadium permeability of prepared membranes (9×10^{-9} cm^2 min^{-1}) is around two orders of magnitude lower than that of Nafion 115 (11×10^{-7} cm^2 min^{-1}). VFBs assembled with these membranes show a much higher CE and EE at different current densities than that of Nafion 115 (N115) (Table 9.1) [35].

Even though most of the above-mentioned polymer membranes show very good ion selectivity and ion conductivity, their chemical stability remains one of the most critical issues. The introduction of ion exchange groups in hydrocarbon membranes could especially accelerate the membrane degradation. To solve these problems, porous membranes were first introduced

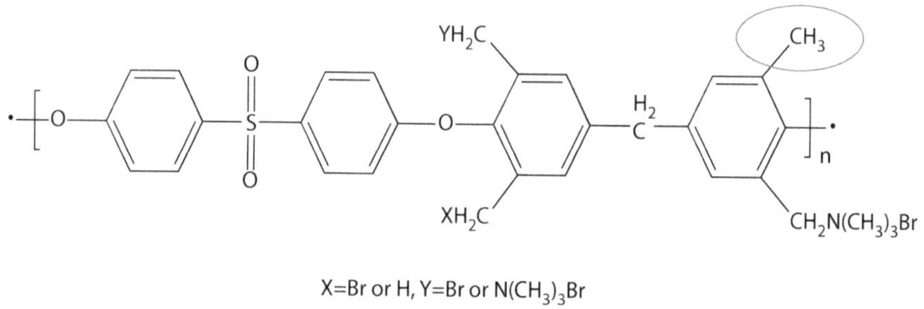

FIGURE 9.12
The structure of quaternised poly (tetramethydiphenyl ether sulphone). (Z.S. Mai et al., Anionconductive membranes with ultralow vanadium permeability and excellent performance in vanadium flow batteries, *ChemSusChem*, 2013, 6, 328–335. Copyright Wiley-VCH Verlag GmbH & Co. KGaA. Reprinted with permission.)

in VFBs based on the idea of separating vanadium ions from protons via pore size exclusion, where the restriction caused by ion exchange groups from traditional ion exchange membranes can be overcome [36] (Figure 9.13). Inspired by the principle of nanofiltration (NF) membranes, which is a membrane process that can separate ions with different size and charge densities by adjusting the pore size distribution of membranes; the vanadium ions and protons in VFBs could be possibly separated by porous membranes as well as by control of the membrane morphology.

The first reported porous NF membrane is made from hydrolysed polyacronitrile (PAN-H). The pore size and pore size distribution were well controlled via changing the cast parameters. The results indicated that the PAN-H membrane can successfully separate vanadium ions from protons due to their difference in radius, charge density and specific interactions with electrolyte and membranes (Figure 9.14). The PAN-H membranes show very good potential in VFBs where they achieve a CE up to 95% and an EE of 76% at the operating condition of 80 mA cm^{-2}.

TABLE 9.1
Battery Performance of Quaternised Polyether Sulphone and Nafion

Membrane[a]	Efficiencies [%] Coulombic	Voltage	Energy
QAPES1-40	98.8	89.3	88.3
QAPES1-60	98.7	87.1	86.0
GAPES2-60	98.9	84.0	83.1
N115	94.0	89.8	84.4

Source: Z.S. Mai et al., Anion-conductive membranes with ultralow vanadium permeability and excellent performance in vanadium flow batteries, *ChemSusChem* 2013, 6, 328–335. Copyright Wiley-VCH Verlag GmbH & Co. KGaA. Reprinted with permission.

[a] QAPES1-40, QAPES1 with thickness of 40 μm; QAPES1-60, QAPES1 with thickness of 60 μm; QAPES2-60, QAPES2 with thickness of 60 μm; where the IEC of QAPES1 > QAPES2.

Vanadium Flow Batteries

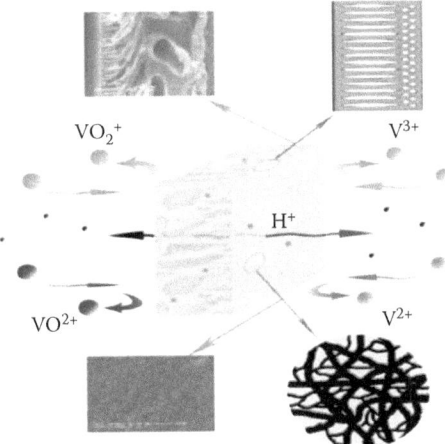

FIGURE 9.13
(See colour insert.) The porous nanofiltration (NF) membranes for VFB separator. (H.Z. Zhang et al., Nanofiltration (NF) membranes: The next generation separators for all vanadium redox flow batteries (VRBs)?, *Energy Environ. Sci.* 4, 2011, 1676. Copyright 2011. Reprinted with permission from Royal Society of Chemistry.)

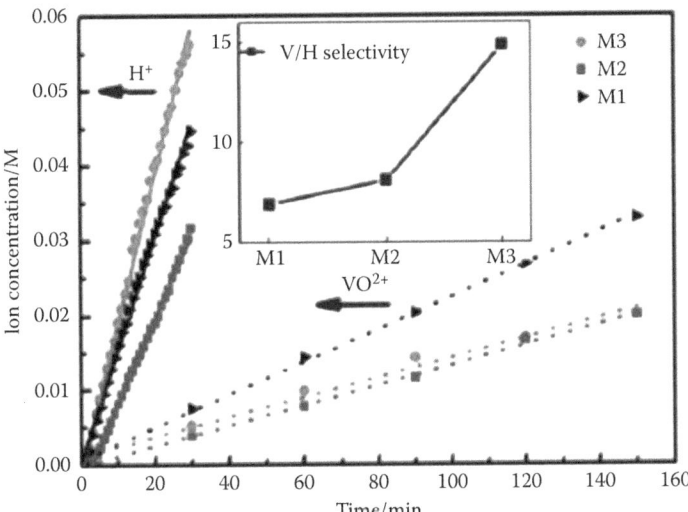

FIGURE 9.14
The V/H selectivity of porous membranes (M1 to M3, pore size decreasing). (H.Z. Zhang et al., Nanofiltration (NF) membranes: The next generation separators for all vanadium redox flow batteries (VRBs)?, *Energy Environ. Sci.* 4, 2011, 1676. Copyright 2011. Reprinted with permission from Royal Society of Chemistry.)

A hydrophobic porous PVDF membrane was investigated in VFBs and showed a very impressive performance and stability [37]. The battery assembled with the PVDF membrane is very stable and able to continuously run for more than 1000 cycles, reaching nearly 80% EE and 95% CE. Polyether sulphone (PES) membranes with different morphologies could be obtained by adding poly (vinyl pyrrolidone) (PVP) in the casting solutions [38]. With increasing PVP in the cast solutions, their morphologies changed from finger-like pores to sponge-like pores, the pore size becomes larger and more continuous (Figure 9.15). The EE of the cell assembled with these membranes first increases then decreases with higher PVP content in cast solutions.

A compromise between membrane selectivity and ion conductivity is one of the most important challenges. Normally, large pores are preferred in membranes to ensure ion conductivity, while membranes with smaller pores are preferred to suppress the vanadium ions permeability. Thus, improving the ion selectivity while preserving the conductivity becomes one of the most urgent tasks in this field. For example, silica nanoparticles could be introduced in the pores and on the surface of porous membranes to improve the membrane selectivities by blocking the pores [39]. The high acid adsorption of silica nanoparticles can maintain the membrane conductivity, since the VFB is under a strong acid medium. The results showed that the V/H (vanadium to proton permeability ratio) selectivity increased more than 3 times after blocking the membrane pores with silica. As a result, the CE of the membranes increases from 92% to 98% after modification.

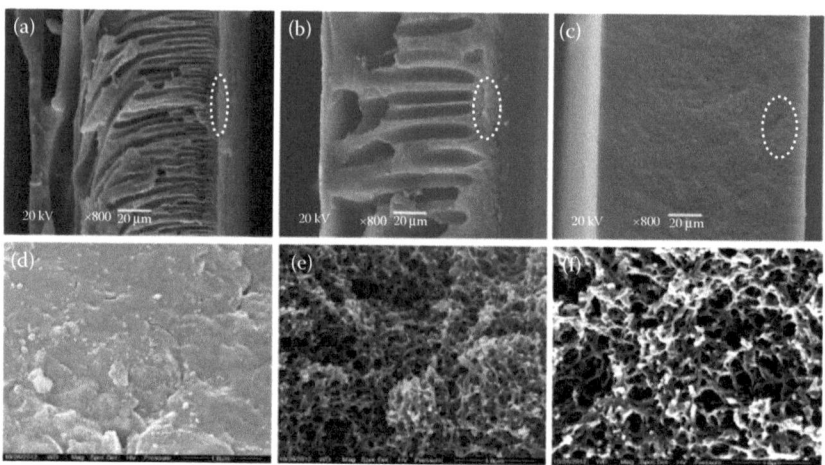

FIGURE 9.15
Morphology changes with the addition of PVP in the cast solutions 15 wt%, 25 wt% to 35 wt%, respectively, (a)–(c) under low magnification, (d)–(f) high magnification view of the region marked in white. (Reprinted from *J. Power Sources*, 233, Y. Li et al., Porous poly (ether sulfone) membranes with tunable morphology: Fabrication and their application for vanadium flow battery, 202–208. Copyright 2013, with permission from Elsevier.)

Vanadium Flow Batteries

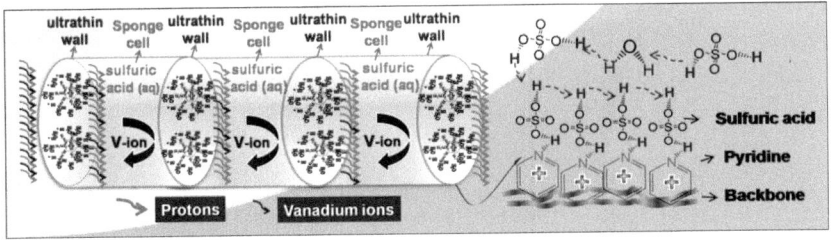

FIGURE 9.16
(See colour insert.) The schematic diagram of selective transfer of vanadium ions and protons. (H.Z. Zhang et al., Advanced charged membranes with highly symmetric spongy structures for vanadium flow battery application, *Energy Environ. Sci.* 6, 2013, 776–781. Copyright 2013. Reprinted with permission from Royal Society of Chemistry.)

Another effective way is to design and fabricate membranes with a specific morphology. For example, highly symmetric, sponge-like porous membranes with positive-charged groups were fabricated [40]. The membranes perfectly combined excellent V-ion rejection with high proton conductivity (Figure 9.16), which shows better overall performance than commercial Nafion membranes [40].

In summary, the membrane plays an important role in reducing the cost of a VFB system. The properties of membranes explored are summarised in Table 9.2. Even though the modification of perfluorinated polymers were investigated extensively, high cost remains to be the biggest obstacle. Non-fluorinated ion exchange membranes show very good selectivity due to their less continuous phase-separated structure; however, their instability is still a critical issue to be overcome. Therefore, porous membrane separators could be a better option; since ion exchange groups are not essential, the separation of vanadium ions and protons can be realised by tuning pore size distribution. Although the membrane properties have significantly improved, issues such as the transport mechanism of different ions in porous membranes, and the conflict between the membrane selectivity and ion conductivity would still require further research to resolve.

9.3.2 Electrode Materials

VFB electrode materials currently adapted can be divided into two types: (1) metal and (2) carbon materials. A range of metal electrode materials such as Pb, Au, Pt, platinised titanium (Pt-Ti) and iridium oxide dimensionally stable electrodes (DSAs) have been evaluated for their suitability as positive electrodes in VFBs [41]. It was found that the electrochemical reversibility for the VO^{2+}/VO_2^+ redox couple was not sufficient on the Au electrode. The Pb and Ti electrodes were easily passivated in the potential range, where the VO^{2+}/VO_2^+ redox couple reactions occur. The passivation film formed on the surface would increase the electric resistance. The growth of the passivation

TABLE 9.2
Properties and Performance of All the Membranes Tested in Vanadium-Redox Flow Batteries

Membrane	Structure	IEC (mmol g^{-1})	Conductivity (mS cm^{-1})	Permeability of VO^{2+} (cm^2 min^{-1})	Current Density (mA cm^{-2})	CE (%)	VE (%)	EE (%)
Nafion 115 [28]	—	0.91	—	7.95×10^{-7}	50	91.7	92.3	84.7
Nafion/SiO$_2$ [14]	—	0.96	56.2	—	—	> Nafion 117	> Nafion 117	79.9
Nafion/organic silica/TiO$_2$ [15]	C$_2$H$_5$-O-Si-O-Ti-OC$_4$H$_9$ (with CH$_3$, OC$_4$H$_9$, CH$_3$, OC$_4$H$_9$ substituents)	0.95	—	4.3×10^{-7}	30	94.8	82.2	77.9
Nafion/PVDF [17]	—	—	—	—	80	> r-Nafion	≈ r-Nafion	85
Nafion/SPEEK [18]	PEEK–SO$_2$–NH-R-NH–SO$_2$–Nafion	1.67	—	1.928×10^{-7}	50	97.6	85.3	83.3
SPEEK/PSf-ABIm [19]	—	1.26	—	1.12×10^{-7}	20	96.7	90.6	87.7
PEI/Nafion [20]	—	0.89	—	5.23×10^{-7}	50	96.2	88.4	85.1

Vanadium Flow Batteries

Membrane	Structure	(value)		(diffusion)	Temp			
Nafion/PANI [22]	—	1.29	—	2.63×10^{-7}	25		97.9	83.3
SPTK [25]		1.91	10.5	1.2×10^{-13} m^2 s^{-1}	50	79.6	—	—
SPTKK [25]		1.89	13.6	3.1×10^{-13} m^2 s^{-1}	50	81.8	—	—
SPFETKs [26]		1.67	40% higher than Nafion 117	4.4×10^{-12} m^2 s^{-1}	50	84.2	—	—
Sulphonated poly(arylene ether sulphone) [27]		1.45	17.0	8.9×10^{-14} m^2 s^{-1}	50	69.0	—	—
SPEEK [28]		1.2	—	0.36×10^{-7}	50	98.5	88.8	87.5
S-Radel [29]			—	2.07×10^{-7}	50/100	97.3	77	75.2
SDPEEK [30]			—	2.3×10^{-6} cm^2 h^{-1}	50	97.8	89.9	87.9

continued

TABLE 9.2 (continued)
Properties and Performance of All the Membranes Tested in Vanadium-Redox Flow Batteries

Membrane	Structure	IEC (mmol g^{-1})	Conductivity (mS cm^{-1})	Permeability of VO^{2+} (cm^2 min^{-1})	Current Density (mA cm^{-2})	CE (%)	VE (%)	EE (%)
TEA-treated CMPSf [31]		1.37	18.2	1.6×10^{-8}	80	99	—	81
QAPPESK [33]		—	—	—	40	96.4	91.6	88.3
QAPPEK [34]		1.64	—	—	40	98.4	87.3	85.9

Membrane	Structure			Permeability	Current density (mA cm^{-2})	CE (%)	VE (%)	EE (%)
QAPES [35]	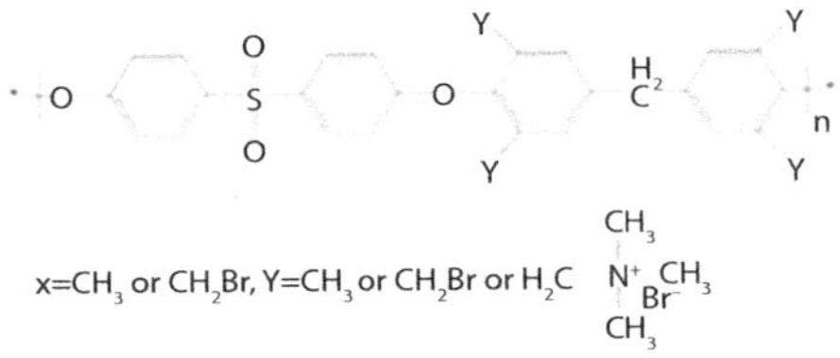 (structure shown)	—	—	0.09×10^{-7}	60	98.7	87.1	86.0
PVDF [37]	—	—	—	8.7×10^{-7}	80	90.0	85.4	76.8
Silica/PAN [39]	—	—	—	—	80	98	—	79
	—	—	—	—	40	—	—	>88
CPSF-Py [40]	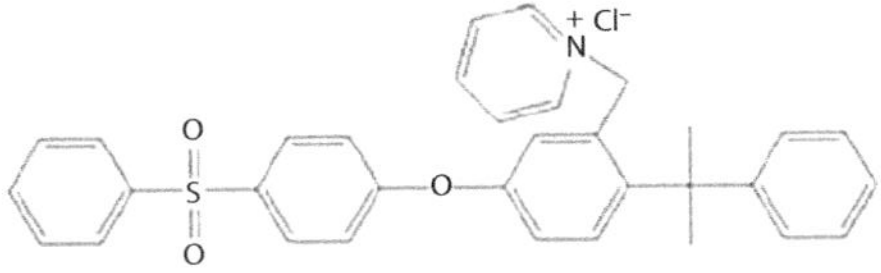 (structure shown)				120	—	—	>81

film can be prevented by using Pt-Ti and DSA electrodes, but its widespread application is limited by high cost.

9.3.2.1 Carbon-Based Materials

Carbon materials are ideal VFB electrode materials due to the excellent stability with respect to the acidic electrolyte, and reasonable cost [42]. Various carbon materials such as carbon felt, graphite felt, carbon paper and graphite powder have been investigated as the electrodes for VFBs [43–45]. Among them, carbon felt or graphite felt are the preferable electrode materials due to their three-dimensional network structure, providing the channels for electrolyte, large specific surface area, high conductivity and better chemical and electrochemical stability. Zhong et al. [43] compared two typical graphite felts based on rayon or polyacrylonitrile (PAN) precursors, and found that the electrical conductivity of the PAN-based felt is superior to that of its rayon-based counterpart. Furthermore, XPS analysis reveals that the rayon-based felt reacts more easily with oxygen and forms C=O groups due to its microcrystalline structure, while the PAN-based felt is more resistant to oxidation and preferentially forms C–O groups.

In recent years, carbon nanomaterials such as carbon nanotubes [46], graphite oxide [47,48] and graphene oxide (GO) nanosheets [49,50] have become other alternatives. Han et al. [49] treated GO nanosheets at various temperatures, which demonstrated excellent electrocatalytic activity towards VO^{2+}/VO_2^+ and V^{2+}/VO^{3+} redox couples. This was attributed to the relatively large active surface area, and the presence of a large amount of hydroxyl and carboxyl acid active groups on basal planes and sheet edges, as shown in Figure 9.17. However, electrodes based on single GO nanosheets suffered from low rate capability due to poor electrical conductivity; thus

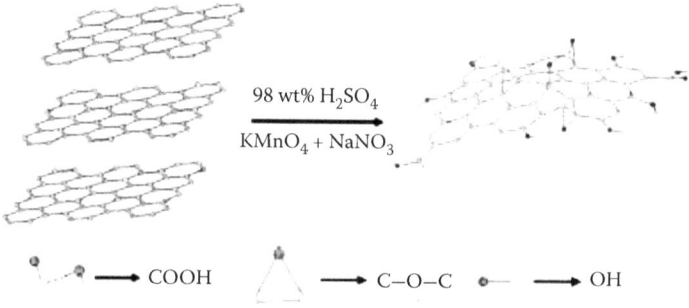

FIGURE 9.17
The procedure for preparation of graphene oxide (GO) nanosheets. (Reprinted from *Carbon*, 49, P.X. Han et al., Graphene oxide nanoplatelets as excellent electrochemical active materials for VO^{2+}/VO_2^+ and V^{2+}/VO^{3+} redox couples for a vanadium redox flow battery, 693–700. Copyright 2011, with permission from Elsevier.)

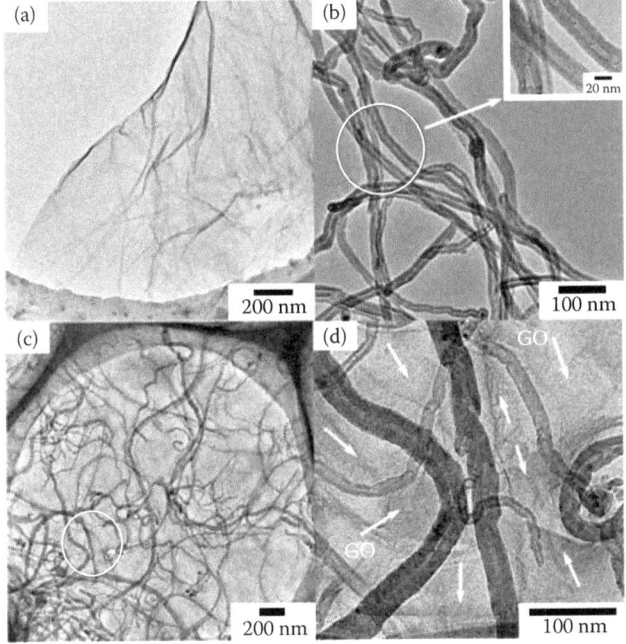

FIGURE 9.18
TEM images of (a) GO nanosheets, (b) MWCNTs and (c) GO/MWCNTs. (d) A detailed view of the white circle area in (c). (P.X. Han et al., Graphene oxide nanosheets/multi-walled carbon nanotubes hybrid as an excellent electrocatalytic material towards VO^{2+}/VO_2^+ redox couples for vanadium redox flow batteries, *Energy Environ. Sci.* 4, 2011, 4710–4717. Copyright 2011. Reprinted with permission from Royal Society of Chemistry.)

the EE and power capability were severely limited [49]. Therefore, in order to improve the electron and ion transport capability, they prepared a GO nanosheets/multi-walled carbon nanotubes (GO/MWCNTs) hybrid, as shown in Figure 9.18c and d, by an electrostatic spray technique on a glassy carbon electrode (GCE) after efficient ultrasonic treatment [50]. The results showed that after the introduction of GO nanosheets (Figure 9.18a), MWCNTS (Figure 9.18b) formed an electrocatalytic hybrid with a mixed conducting network. Compared with the pure GO nanosheets and MWCNTs, GO/MWCNTs hybrid electrode exhibited much better electrocatalytic redox reversibility towards VO^{2+}/VO_2^+ redox couples. The value of I_{pa}/I_{pc} was almost constant at about 1.20 in the range between 5 and 100 mV s^{-1}, which is close to the value of unity for reversibility (Figure 9.19).

9.3.2.2 Electrode Materials Modification

In order to enhance the reversibility of the electrode reaction and the electrocatalytic activity, modification of the electrode materials is necessary by

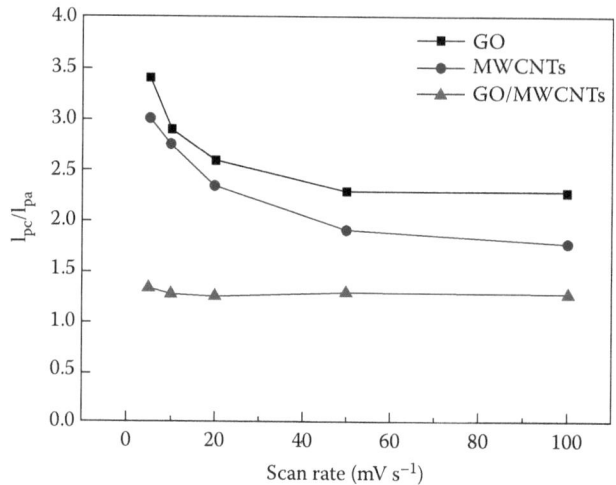

FIGURE 9.19
Curves of I_{pa}/I_{pc} as a function of scan rate for VO^{2+}/VO_2^+ redox couples for GO, MWCNTs, and hybrid GO/MWCNTs electrodes. (P.X. Han et al., Graphene oxide nanosheets/multi-walled carbon nanotubes hybrid as an excellent electrocatalytic material towards VO^{2+}/VO_2^+ redox couples for vanadium redox flow batteries, *Energy Environ. Sci.* 4, 2011, 4710–4717. Copyright 2011. Reprinted with permission from Royal Society of Chemistry.)

adjusting the material hydrophilicity and pore structure. A variety of modification methods have been investigated, which can be classified as oxidation treatment, nitrogenisation treatment and introduction of electrocatalyst.

The carbon atoms on the graphene edge plane or carbon atoms with lattice defects on the graphene basal plane have high surface energy, which leads to them being easily oxidised to hydroxyl, carbonyl, carboxyl and other functional groups in an oxidation condition. The oxygen-containing functional groups can facilitate electrocatalytic activity of carbon electrode materials for the vanadium redox reactions. The mechanism on the improved activity of carbon electrode materials towards V^{2+}/V^{3+} and VO^{2+}/VO_2^+ redox couples by –OH groups can be expressed in Figure 9.20. It involves the transport of vanadium ions V^{2+}, V^{3+}, VO^{2+} or VO_2^+) from the bulk of the solution to the electrode surface, their ion exchange with hydrogen ions in the –OH groups on carbon and the subsequent electron transfer reaction [51].

Skyllas-Kazacos [52,53] first studied the effect of thermal and acid treatments of carbon and graphite felts on the electrochemical activity of these materials as electrodes for VFBs. They prepared the graphite felt with a large amount of chemisorbed oxygen on the surface by treating graphite felt in 400°C air or boiling concentrated sulphuric acid for 5 h. Significant improvement from 78% to 88% in EE was observed in cell charge–discharge cycling tests after thermal or acid treatments of the graphite felt; the cell resistance dropped after treatment. The increased electrocatalytic activity of the treated

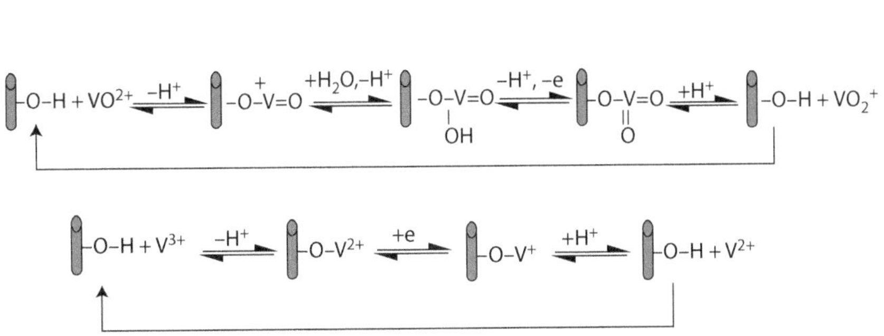

Vanadium Flow Batteries

$$\text{-O-H} + VO^{2+} \underset{+H^+}{\overset{-H^+}{\rightleftharpoons}} \text{-O-V=O}^+ \xrightarrow{+H_2O, -H^+} \underset{\text{OH}}{\text{-O-V=O}} \underset{+e}{\overset{-H^+, -e}{\rightleftharpoons}} \underset{\text{O}}{\overset{\parallel}{\text{-O-V=O}}} \overset{+H^+}{\rightleftharpoons} \text{-O-H} + VO_2^+$$

$$\text{-O-H} + V^{3+} \overset{-H^+}{\rightleftharpoons} \text{-O-V}^{2+} \overset{+e}{\rightleftharpoons} \text{-O-V}^+ \overset{+H^+}{\rightleftharpoons} \text{-O-H} + V^{2+}$$

FIGURE 9.20
The catalytic mechanism of –OH or –COOH functional group towards V^{2+}/V^{3+} and VO^{2+}/VO_2^+ couples. (Reprinted with permission from *Carbon*, 48, L. Yue et al., Highly hydroxylated carbon fibres as electrode materials of all-vanadium redox flow battery, 3079–3090. Copyright 2010, with permission from Elsevier.)

graphite felt was attributed to the increased surface concentration of the C–O and C=O functional groups.

Subsequently, various oxidation methods such as electrochemical oxidation, mild oxidation, plasma treatment and gamma-ray irradiation were developed [54,55] and the battery performance improved due to the introduction of comparing oxygen-containing functional groups with the untreated carbon materials.

Apart from oxygen-containing functional groups, nitrogenous groups also can improve the electrocatalytic activity of carbon electrode materials for the vanadium redox reactions. Carbon atoms adjacent to nitrogen dopants possess a substantially high positive charge density to counterbalance the strong electronic affinity of the nitrogen atom. Thus, the 'positively charged' carbon atoms can work as the active sites for the oxidation reaction. At the same time, the five valence electrons of nitrogen atoms contribute the extra charge to the bond in graphene layers, which enhances the basicity of the carbon surface and the electrical conductivity of the nitrogen-doped carbon, beneficial for the reduction process [56].

Shao et al. [56] synthesised nitrogen-doped mesoporous carbon materials (N-MPC) with the pore size of 6–10 nm by heat-treated mesoporous carbon prepared with a soft-template method in NH_3 as shown in Figure 9.21a. The electrocatalytic kinetics and the reversibility of the redox couple VO^{2+}/VO_2^+ were significantly enhanced on N-MPC electrodes compared with MPC and graphite electrodes (Figure 9.21b). It was suggested that nitrogen doping could facilitate the electron transfer on electrode/electrolyte interface for both oxidation and reduction processes [56]. Furthermore, these nitrogen-doped carbon materials are more hydrophilic with increased electrochemically active sites. The experimental results of Wu et al. [57] and Wang et al. [58] also confirmed this phenomenon.

The introduction of electrocatalysts on the surface of carbon felt to reduce the reaction overpotential of vanadium ion redox couples is another effective method. Sun and Skyllas-Kazacos [59] prepared the metallised graphite

 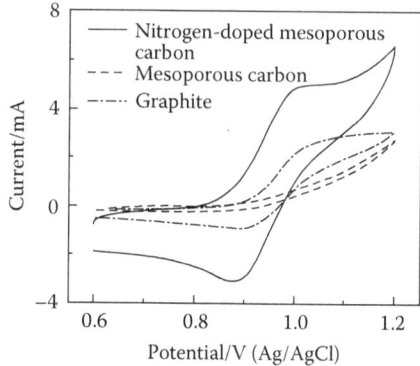

FIGURE 9.21
TEM image of nitrogen-doped mesoporous carbon and cyclic voltammograms (CV) on different electrodes in 3.0 M H_2SO_4 + 1.0 M $VOSO_4$ (50 mV s^{-1}). (Reprinted from *J. Power Sources*, 195, Y.Y. Shao et al., Nitrogendoped mesoporous carbon for energy storage in vanadium redox flow batteries, 4375–4379. Copyright 2010, with permission from Elsevier.)

fibre electrode by impregnation or ion exchange with solutions containing Pt^{4+}, Pd^{2+}, Au^{4+}, Mn^{2+}, Te^{4+}, In^{3+} and Ir^{3+}. The cyclic voltammetry measurement results showed that the electrode modified with Ir^{3+} exhibited the best electrochemical behaviour for the various vanadium redox species, while electrodes modified by Pt^{4+}, Pd^{2+} and Au^{4+} were unsuitable due to the high hydrogen evolution rates. The electrodes treated with Mn^{2+}, Te^{4+} and In^{3+} also exhibited significantly improved electrochemical activity compared with the untreated fibre electrode. Wang et al. studied the performance of the graphite felt modified by pyrolysis of Ir reduced from H_2IrCl_6 [60]. AC impedance and steady-state polarisation measurements showed that the Ir-modified materials have a higher anodic peak—the cathodic peak currents of the VO^{2+}/VO_2^+ couple—and lower overpotential of the desired VO^{2+}/VO_2^+ redox process. Furthermore, Ir modification of graphite felt enhanced the electroconductivity of electrode materials because the resistivity of Ir is 5.1×10^{-6} Ω cm, while the resistivity of PAN carbon felt is 8×10^{-2} Ω cm. A decrease of 25% in cell resistance was observed for a cell with Ir-modified graphite felt as the electrodes compared to the cell using non-modified felt.

Flox et al. [61] investigated graphene-supported monometallic (Pt) and bimetallic ($CuPt_3$) cubic nanocatalysts as positive electrode materials for improving the VO^{2+}/VO_2^+ redox process in the VFB (Figure 9.22). The presence of the $CuPt_3$ nanocubes on grapheme conferred higher electrocatalytic activity due to the much higher electroactive area compared to that obtained with the Pt nanoparticles. The electrochemical surface area (as derived from the CV curve in Figure 9.22g) of the nano-($CuPt_3$)-decorated graphene electrode (3.13 cm^2) was 105% higher than that of non-decorated graphene (1.53 cm^2). It is hypothesised that the $CuPt_3$ nanoparticles exhibit a highly

Vanadium Flow Batteries

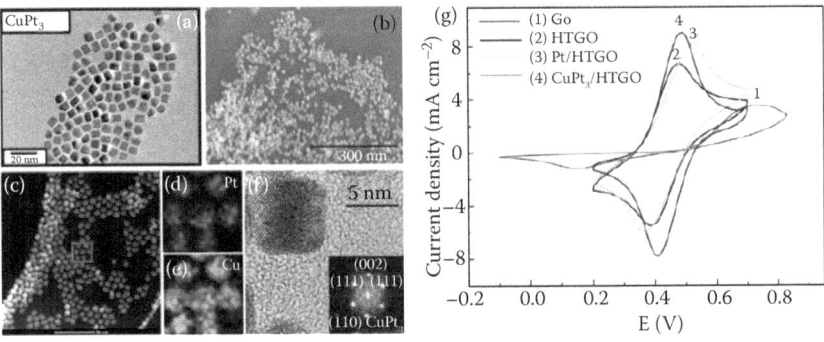

FIGURE 9.22
(See colour insert.) (a) TEM of the CuPt$_3$. (b) SEM images of the CuPt$_3$/HTGO. (c) High-angle annular dark-field scanning transmission electron microscopy image of the CuPt$_3$ nanocubes and EDX mapping for Pt (d) and Cu (e) obtained on the squared region in (c). (f) HRTEM of a CuPt$_3$ nanotube and corresponding fast Fourier transform. (g) CV of GO (line 1), HTGO (line 2), Pt/HTGO (line 3) and CuPt$_3$/HTGO (line 4). (Reprinted from *Carbon* 50, C. Flox et al., Active nano-Cupt3 electrocatalyst supported on graphene for enhancing reactions at the cathode in all-vanadium redox flow batteries, 2372–2374. Copyright 2012, with permission from Elsevier.)

active Pt–Cu bicomponent catalytic activity, as surface Pt and Cu atoms are primarily responsible for positive reaction kinetics.

The high cost of noble metal catalysts limited their commercial application. Advanced catalysts of both low cost and high catalytic activity, such as WO$_3$ [62] and Mn$_3$O$_4$ [63], are studied. Yao et al. [62] prepared a carbon paper electrode coated with supported tungsten trioxide via an impregnation method, which resulted in improved electrocatalytic activity towards both VO^{2+}/VO$_2^+$ and V^{2+}/V^{3+} redox reactions. As shown in Figure 9.23a, for the pristine carbon paper, the shapes of the oxidation peak and the reduction peak are unsymmetrical, and the corresponding currents I_{pa} and I_{pc} are small, indicating that the electrochemical activity is relatively low. When the carbon paper is coated with super active carbon (SAC), both oxidation and reduction peak currents increased remarkably, and the peak potential separation decreased from 311 to 151 mV, suggesting that the electrochemical activity and the kinetic reversibility on this electrode have improved compared to the pristine. When the carbon paper is coated with WO$_3$/SAC, the peak currents are further increased, the magnitudes of which are almost the same (I_{pa}/I_{pc} is close to 1.03). The peak potential separation is only 106 mV, meaning that the kinetic reversibility is further improved as well. A similar result can also be observed for V^{2+}/V^{3+} redox reactions (Figure 9.23b). As a result, VE and EE of the VFB assembled with WO$_3$/SAC-coated carbon paper at 50 mA cm^{-2} are 85.2% and 80.5%, respectively, which are much higher than that of the cell assembled with pristine carbon paper electrodes.

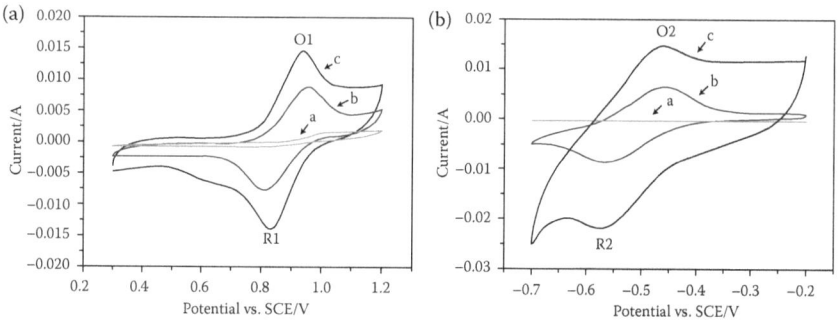

FIGURE 9.23
CV curves of (a) carbon paper, (b) carbon paper coated with SAC and (c) carbon paper coated with WO$_3$/SAC. Left image: 0.05 M V(IV) + 0.05 M V(V) + 3 M H$_2$SO$_4$ solution. Right image: 0.05 M V(II) + 0.05 M V(III) + 3 M H$_2$SO$_4$ solution. (Reprinted from *J. Power Sources*, 218, C. Yao et al., Carbon paper coated with supported tungsten trioxide as novel electrode for all-vanadium flow battery, 455–461. Copyright 2012, with permission from Elsevier.)

As mentioned above, advanced catalysts with both low cost and high catalytic activity towards both VO^{2+}/VO_2^+ and V^{3+}/V^{2+} redox reactions should be investigated further for commercial application. Furthermore, the effects on ohmic and concentration polarisation should not be ignored. For example, nanocarbon electrode delivers high electrocatalytic activity due to the large surface area, while the pore channels in this electrode are almost on the nanometer size scale, which is inconvenient for the transfer of electrolytes and leads to the increase of concentration polarisation. Therefore, besides the electrocatalytic activity, conductivity and the pore structure should also be considered in developing a high-performance electrode, which enables high power density in order to decrease the size and cost of a VFB. In addition, the reaction mechanism of both VO^{2+}/VO_2^+ and V^{3+}/V^{2+} redox reactions on the different electrode materials in the VFB should be clarified further.

9.3.3 Electrolytes

In a typical setup, a flow battery consists of two electrolyte reservoirs from which the electrolytes are circulated by pumps through an electrochemical cell stack [4]. The electrolytes in VFB serve as the medium to store and release energy, whereas the volume and the concentration of electrolyte determine the energy of battery.

In a VFB, the electrolyte is composed of active species (vanadium ions with different valence states) and supporting electrolytes (e.g. sulphuric acid, hydrochloric acid, methane sulphonic acid and a mixed acid of the above, or organic electrolyte). The V(II)/V(III) redox couple serves as anolyte, while the V(IV)/V(V) redox couple serves as catholyte [64]. The conversions between electrical energy and chemical (or electrochemical) energy are realised via changes in vanadium valence states through electrode reactions, when the electrolytes are pumped from storage tanks to flow-through electrodes in a cell stack [65]. The liquid electrolyte and dissolved redox species in a VFB are in intimate proximity with the electrodes, making a quick response (on the order of sub-seconds) possible for utility applications [9]. While the nominal change in the oxidation state of vanadium on either side of the membrane that stores and releases charges, there is a certain change in the *pH* (1 ~ 3.5) of the solution over the course of a charge and discharge cycle [66]. As such, the VFB eliminates cross-contamination and electrolyte maintenance problems faced by all other flow batteries [7]. Furthermore, VFB cells can be overcharged and overdischarged within the limits of the capacity of the electrolytes, and can be cycled from any state of charge or discharge without permanent damage to the cells or electrolytes [67].

In particular, the performance of VFBs would be determined by the following factors: (a) concentration of vanadium ions, (b) vanadium thermal stability toward ambient temperatures, (c) the concentration of sulphonic acid, (d) density, (e) viscosity, (f) state of charge (SOC), (g) conductivity and (h) electrochemical behaviour of electrolytes [68], among which the concentration and thermal stability are of greatest importance [66,69]. Since the VFB uses soluble redox couples as active materials, the concentration of the electrolytes will determine the system-specific energy density. Consequently, a higher solubility of all redox agents is always desired. Of equal importance is the thermal stability of the electrolyte, which defines the operating temperature window of a VFB. It is worth noting that the electrolyte needs to be stable over the entire SOC, especially for nearly 0% SOC and 100% SOC, which usually represent the most adverse conditions during cell operation and electrolyte storage. The electrolyte also needs to possess adequate ionic conductivity to provide a decent rate capability. Viscosity usually is not an issue in the aqueous system, but one need to take into consideration the high pressure drop due to high viscosity when designing a non-aqueous VFB system [68].

Traditional sulphuric acid-based VFB technology is significantly limited by the vanadium ion solubility and stability in electrolyte solutions over a certain temperature window, which limits the system not only to a low energy density of <25 Wh L^{-1} (<1.7 M of active vanadium concentration), but also to a narrow operational temperature range (10–40°C) [9,66]. Restricted by the electrolyte thermal stability, VFBs would require additional temperature management, which inadvertently lowers the EE by parasitic losses. To extend the operational temperature window of VFB and reduce the cost of the system, great efforts have been spent to increase the stability of vanadium species in sulphuric acid solutions in the last few decades [69,70]. Although the thermal stability of V(V) can be dramatically enhanced by increasing the sulphuric acid concentration to more than 5 M [69], this would have adverse impact on the solubility of the V(II), V(III) and V(IV) ions due to the common ions effect [68].

Other approaches to inhibit the precipitation of V(V) under high temperature were studied, for example, addition of some organic or inorganic chemicals as stabilising agents, or additives to be used as precipitation inhibitors [71–77]. Among all the studied chemicals, K_2SO_4, sodium hexametaphosphate (SHMP), polyacrylic acid (PA) and its mixture with CH_3SO_3H, Coulter IIIA dispersant and inositol have proven to be promising in stabilising the positive electrolyte (Figure 9.24) [71–74]. However, the exact mechanism of stabilisation by these additives has not yet been clarified [71]. In addition to this, it seems that most additives could hardly improve the *in situ* capacity retention of VFBs under relatively high temperature. It is difficult for all four types of vanadium oxidation states to be stabilised at their corresponding high levels with the same additives.

The supporting electrolyte is also found to be critical in determining the stability of the different vanadium ions produced during charge–discharge

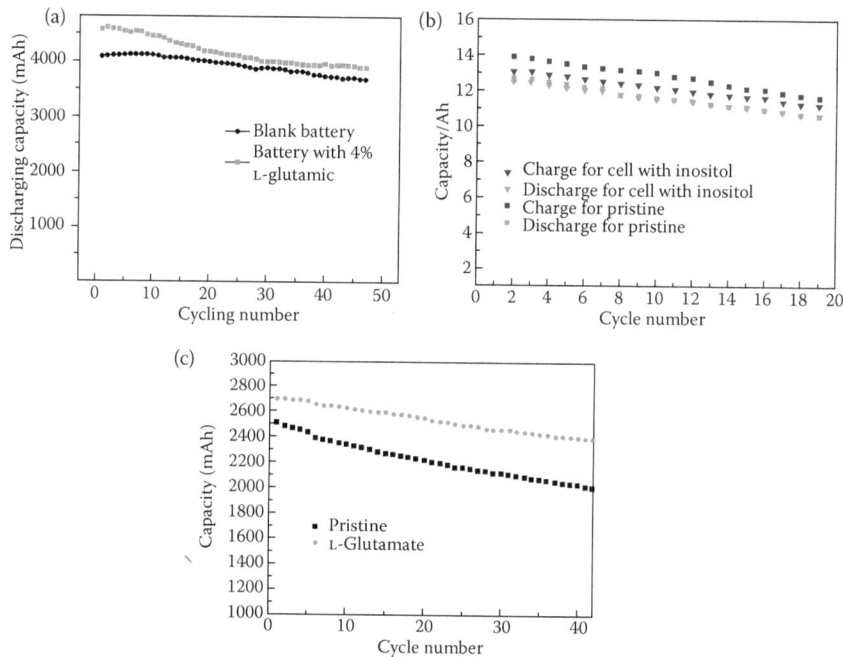

FIGURE 9.24
The capacity retention profiles of a battery with different additives and the pristine sample. (a) With 4% L-glutamic acid (Reprinted with permission from *Electrochim. Acta*, 95, X.X. Liang et al., Effect of L-glutamic acid on the positive electrolyte for all-vanadium redox flow battery, 80–86. Copyright 2012, with permission from Elsevier.); (b) with inositol (Reprinted with permission from *Electrochim. Acta*, 78, X. Wu et al., Influence of organic additives on electrochemical properties of the positive electrolyte for all-vanadium redox flow battery, 475–482. Copyright 2013, with permission from Elsevier.); (c) with L-glutamate (Reprinted with permission from Y. Lei et al., Effect of amino acid additives on the positive electrolyte of vanadium redox flow batteries, *J. Electrochem. Soc.* 160, 2013, A722–A727. Copyright 2010. The Electrochemical Society.)

cycling of the VFB. An ideal supporting electrolyte would possess: (1) fast electrochemical kinetics of the active species at the electrode–electrolyte interface and (2) good electrolyte solubility range (>2 M) and (3) minimal cross-contamination of active electrolytes. Sulphuric acid is among the best supporting electrolytes, providing sufficient solubility for each of the vanadium oxidation states, though V_2O_5 still have limited solubility [78]. In order to increase the solubility and stability of the electrolytes further, another new system of supporting electrolyte was proposed by Yang's group [79,80], by employing hydrochloric acid or sulphate–chloride mixed acid system instead of pure sulphonic acid solutions. The improved stability may be attributed to the formation of a vanadium dinuclear $[V_2O_3 \cdot 4H_2O]^{4+}$ or a dinuclear-chloro complex $[V_2O_3Cl \cdot 3H_2O]^{3+}$. Except for the introduction of chloride as a supporting electrolyte, several organic electrolytes have also been proposed. The vanadium concentration can be increased up to 2 M by using 1.5 M CH_3SO_3H and 1.5 M H_2SO_4 mixed acid solution [81], the VFB showed a much higher energy density of 39.87 Wh L^{-1} than that of using H_2SO_4 as the sole electrolyte (36.27 Wh L^{-1}). Table 9.3 summarized the characteristics and performance of VFBs employing electrolytes with different additives or different supporting electrolytes.

To improve the durability of the system, some fundamental research, such as the mechanism of mass imbalance between catholyte and anolyte, and the control mechanism need to be further investigated.

9.3.4 Stacks and System

9.3.4.1 Stack Structure

9.3.4.1.1 Construction of Cell Stack

In real applications, a cell stack is composed of several single cells as mentioned in Section 9.2. The cell size can be designed to achieve the desired power rate, but in practice other factors also need to be taken into consideration, such as thermal stability, mechanical strength, shunt current, dimension of the components and so forth. Figure 9.25 provides an overview of the construction of a cell stack. Generally, the cell stack is constructed in a parallel-feed design, in which the cells are connected electrically in a series and the electrolytes are pumped into the cells in parallel.

9.3.4.1.2 Flow Pattern of the Electrolyte in Public Pipeline

As shown in Figure 9.26, the flow pattern of the electrolyte in the public pipeline can be classified into 'U' profile and 'Z' profile. Generally, the former type is more commonly adopted. The equivalent fluid mechanics schematic is given in Figure 9.27. The electrolytes flow into all cells individually along the public pipeline, and the uniformity of the electrolyte in each cell can significantly affect the stack performance, such as EE, life-cycle and so on.

TABLE 9.3

Comparison of Characteristics and Performance of VRFs Employing Electrolytes with Different Additives or Different Supporting Electrolytes

Supporting Electrolyte	Additives	Solubility	Temperature Range	Stability	Energy Density (Wh L^{-1})	Reference
3 M H$_2$SO$_4$	None	<1.7 VOSO$_4$	50°C	5.2 h	<25 Wh L^{-1}	[79]
3 M H$_2$SO$_4$	2–5% K$_2$SO$_4$	4 M VOSO$_4$	4°C	Stable for 80 days	—	[71]
3 M H$_2$SO$_4$	3% SHMP 7% wt (PA)	4 M VOSO$_4$	4°C	Stable for 80 days	—	[71]
5.0 M H$_2$SO$_4$	0.4% wt CH$_3$SO$_3$H 0.05–0.1 w/w	1.8 M VO$_2^+$	40°C	860 h	—	[72]
5.0 M H$_2$SO$_4$	Coulter HIA dispersant	1.8–2.2 VOSO$_4$	50°C	4.8 days	—	[73]
3.0 M H$_2$SO$_4$	—	1.8 M VO$_2^+$	40°C	109 h		[74]
10 M HCl	None	2.3 M V^{n+}	−5 to 40°C	15 days	35.5 (100 mA cm^{-2})	[79]
2.5 M H$_2$SO$_4$ + 6 M HCl	None	3 M V^{n+}	−5 to 50°C	>10 days	39.5 (100 mA cm^{-2}) 39.87	[80]
1.5 M H$_2$SO$_4$ + 1.5 M CH$_3$SO$_3$H	None	2 M VOSO$_4$	No data	No data	(120 mA cm^{-2})	[81]

Vanadium Flow Batteries

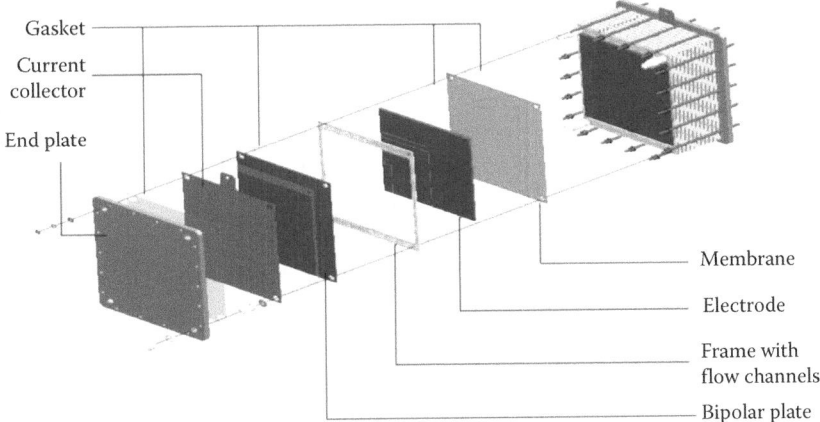

FIGURE 9.25
The construction of a cell stack.

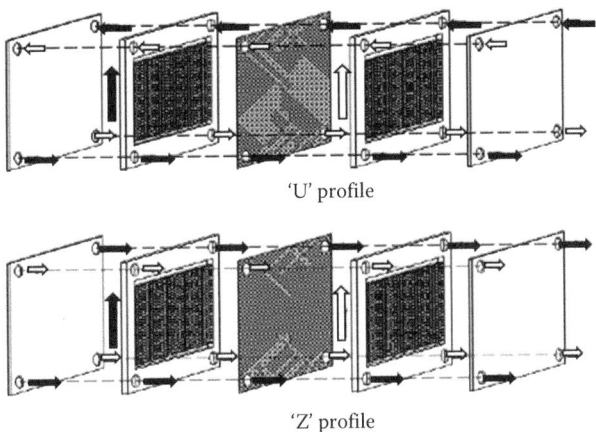

FIGURE 9.26
The flow patterns of the electrolyte in a public pipeline.

9.3.4.2 Design of Cell Stack

9.3.4.2.1 Power Output and Energy Efficiency

There are two important technical parameters that need to be considered in the design of the cell stack: power output and EE, which can be expressed as

$$P_{out} = I_{app} \times A \times V_{ave} \times N \qquad (9.1)$$

$$EE = CE \times VE \qquad (9.2)$$

where P_{out} is the power output, I_{app} and P_{ave} represent the applied current density and average voltage of single cell, respectively, and A and N are the

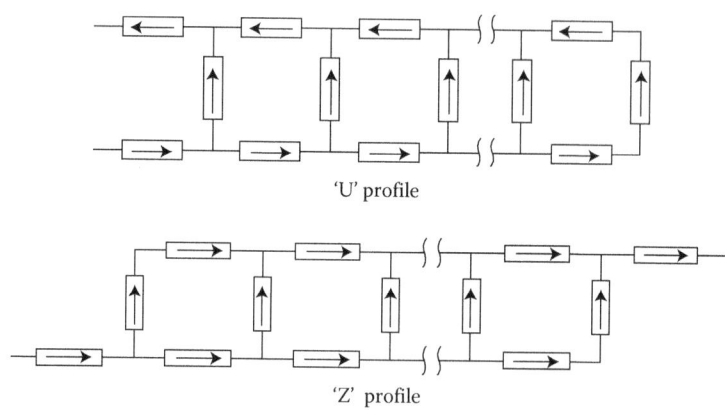

FIGURE 9.27
Equivalent fluid mechanics schematic of the two flow patterns.

electrode surface area and the cell number in the stack, respectively. The energy efficiency EE is the product of coulombic efficiency CE and voltage efficiency VE.

The battery performance can be accurately defined according to the two technical parameters and the applied current density. In general, the larger the value for these two parameters at a certain applied current density, the better the battery performance for a given cell stack. The increase in power output corresponds to the decrease in EE. Strategies such as reducing the battery resistance, improvement in the selectivity and conductivity of the membrane and optimisation of the electrolyte flow in the cell can be effective in improving the power output at the same EE.

9.3.4.2.2 Flow Field

The flow field is another key factor in the design of a cell stack, which governs the uniformity of the electrolyte in the electrode, and then affects the distribution of the current density and overpotential across the cell [82].

An uneven distribution of the current density or overpotential causes a decline in stack performance directly. As a result, it is important to optimise the flow field according to the uneven distribution of the electrolyte flow in the electrode, in order to improve the performance of the stack.

9.3.4.2.3 Shunt Current

As mentioned above, the single cells of a stack are connected via their channels with the public pipeline. Each cell shares the positive/negative electrolytes with the corresponding positive/negative electrolyte public pipeline. Because of the conductivity of the electrolyte and non-zero electrical field potential gradient, the shunt current will incur. The equivalent circuit of shunt current is shown in Figure 9.28. There are two problems introduced by the presence of shunt current: (1) the corrosion of key materials or components of the cell stack,

Vanadium Flow Batteries

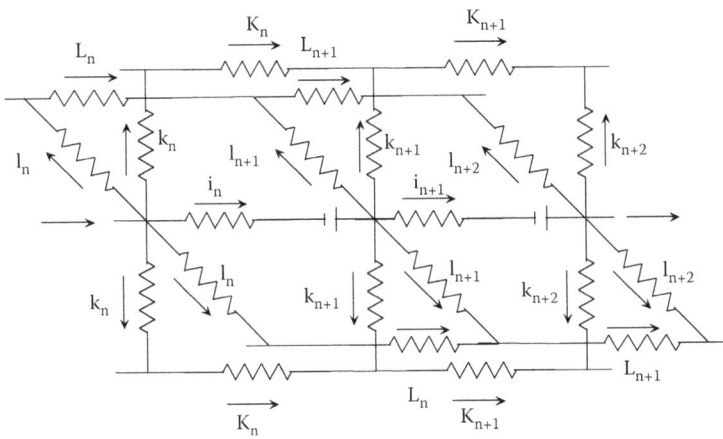

FIGURE 9.28
An equivalent circuit of shunt current. (Reprinted from *J. Power Sources*, 196, F. Xing, H. Zhang and X. Ma, Shunt current loss of the vanadium redox flow battery, 10753–10757. Copyright 2011, with permission from Elsevier.)

which will shorten the cycle life, and (2) capacity loss, which will decrease the EE. A lot of strategies have been formulated to decrease shunt current, such as reducing the number of single cells in a series, lowering the overpotential, charging/discharging under a higher operating current, adding clapboards into the cells and optimising the structure of the electrode frame [83–85].

In Figure 9.28, 'k' and 'K' are the channel current and manifold current for anodic electrolyte, respectively, 'l' and 'L' are the channel and manifold current for cathodic electrolyte, respectively, 'i_n' is the current passing through the nth single cell and 'I_t' is the applied current.

9.3.4.3 Sealing of the Stack

A proper sealing must be applied to avoid the leakage of the electrolyte in the cell stack [86]. The battery may fail when the electrolyte leakage takes place at the outer casing of the stack. This form of physical leakage can be perceived and handled easily. However, the inner leakage as a result of the mutual mixture of the positive/negative electrolyte in the public pipeline will be difficult to detect, and impacts the battery performance greatly.

There are two main types of seals shown in Figure 9.29. One is the face seal, which combines with a self-locking device. Compared with the face seal, the line seal relates to a smaller packing force and, hence, does not contain any self-locking devices.

The internal structure of stacks is mainly designed via modelling, which will not be discussed here. More detailed information can be obtained from our recently published review [7].

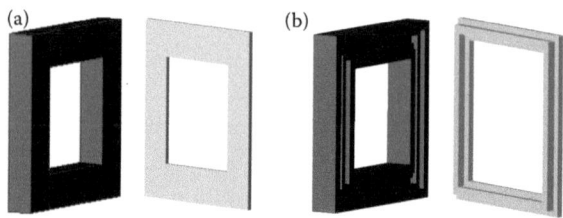

FIGURE 9.29
Two different types of stack seals: (a) face seal and (b) line seal.

9.4 Application and Demonstration

The current application of VFB is mainly applied in areas such as renewable energy; peak shaving and load levelling of grid, micro grid and battery backup power. In the field of renewable energy, a VFB system ensures smooth output and improves the reliability of the renewable powers. Up to now, several demonstrations already confirmed the possibility of VFBs in the field of renewable energies. For example, a 200 kW/800 kWh VFB system was installed to combine with a 2.45 MW wind farm in King Island in Australia. The VFB system plays the role of stabilising the transient power transport changes, and loading fluctuations in wind power generation.

Another typical VFB for renewable energy was installed by SEI in 2005, a 4 MW/6 MWh VFB system (Figure 9.30) was installed in Hokkaido and combined with a 30.6 MW wind farm. The setup is composed of four 1 MW/1.5 MWh systems and has continuously run for more than 3 years, finishing more than 270,000 cycles, thereby demonstrating the excellent reliability of VFBs [87].

Quite recently, Sumitomo Electric Industries Ltd. has completed a megawatt-class electric power generation/storage system. The system consists of a VFB (1 MW/5 MWh) and Japan's largest concentrated photovoltaic (CPV) field (200 kW) [88]. In this system, the VFB functions as a storage facility of electric power generated by the CPV units. The system is able to connect to external commercial power networks, and store electricity provided by power companies during the night. This system started running in 2012.

Development and demonstration of VFBs have also started recently in China, among which Dalian Institute of Physics, Chinese Academy of Sciences (DICP) and its spin-off Rongke Power Ltd* collaboratively devel-

---

* Rongke Power is one of the leading producers of integrated vanadium flow battery energy storage systems. The company is currently involved in VFB materials and components development and production, and can provide effective solutions from engineering to finished turnkey systems. Rongke Power has the technical and manufacturing capacity to provide integrated solutions and support services to customers around the world (http://www.rongkepower.com/).

Vanadium Flow Batteries

FIGURE 9.30
(**See colour insert.**) An overview of 4 MW/6 MWh VFB system in Japan. (Reprinted with permission from S. Eckroad, Vanadium redox flow batteries: An in-depth analysis, EPRI: California, 2007. Copyright 2007, Electric Power Research Institute.)

oped this technology (Figure 9.31). DICP started VFB research in 2000, from materials to system integration, where they successfully developed VFB stacks with power ratings from 500 W to 22 kW. VFB systems with scales from 10 kWs to MWs were explored to meet different kinds of applications (Figure 9.32).

In recent years, more than 10 demonstrations ranging from kWs to MW systems were successfully installed. In the application of renewable energies, a 5 MW/10 MWh VFB system was recently installed to combine with a 50 MW wind farm in Liaoning province in China by Rongke Power Ltd. (Figure 9.33). This is by far the largest VFB system installed in the world.

The demand of power increased dramatically recently; however, the demand differs greatly between night and day, therefore affecting the final efficiency of power generation and the stability of the grid. VFB could offer an effective solution to increase the power generation efficiency by storing off-peak electricity for use in peak hours [84] (Figure 9.34). Meanwhile, compared with electricity in peak hours, the electricity in off-peak hours is more cost effective and can be used for factory production to further cut the operational cost of manufacturing.

In 1997, a 200 kW/800 kWh VFB system was installed by Kashima-Kita for the application of load levelling and peak shaving [87]. This system has successfully completed 150 cycles in a year, and the EE is around 80%

FIGURE 9.31
The megawatt-class power generation/storage system. (Reprinted with permission from N. Chouhan and R.S. Liu, *Electrochemical Technologies for Energy Storage and Conversion*, Wiley-VCH Verlag GmbH & Co. KGaA, 2012, pp. 1–43. http://global-sei.com/qss/servlet/qss.)

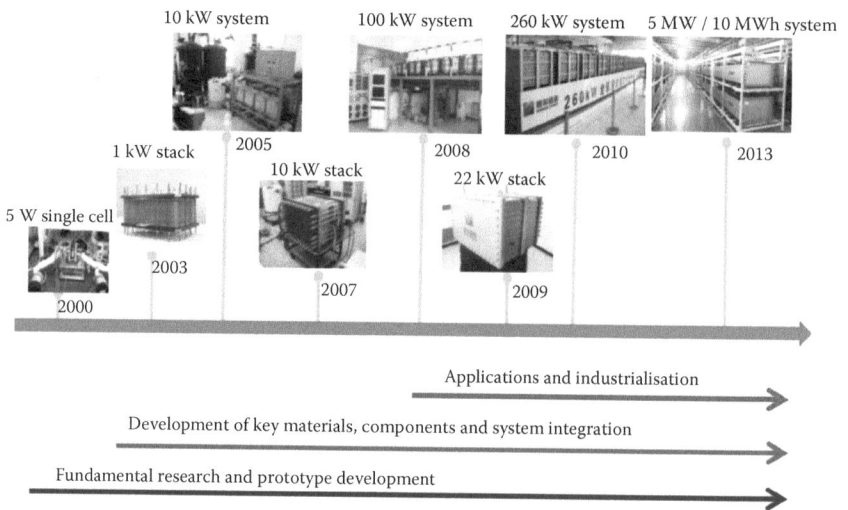

FIGURE 9.32
The development of VFB technology in Dalian Institute of Chemical Physics, Chinese Academy of Sciences (DICP).

under a current density between 80 and 100 mA cm^{-2}. A series of VFB systems with different power rate/capacity (200 kW/1.6 MWh, 500 kW/5 MWh, 1.5 MW/1.5 MWh, 500 kW/2 MWh) was installed to improve grid safety and stability [90]. More recently, a 600 kW/3600 kWh VFB system was successfully installed in Gills Onions Company and brought into operation under the permission of Southern California Edison power company. This system plays the role of storing electricity at night and releasing electricity during the day; hence, releasing the burden from the grid during peak demand and thereby increasing the power quality.

Vanadium Flow Batteries

FIGURE 9.33
5 MW/10 MWh VFB system installed by Rongke Power. (a) An overview of the wind farm, (b) VFB system, (c) electrical control system and (d) electrolyte tanks.

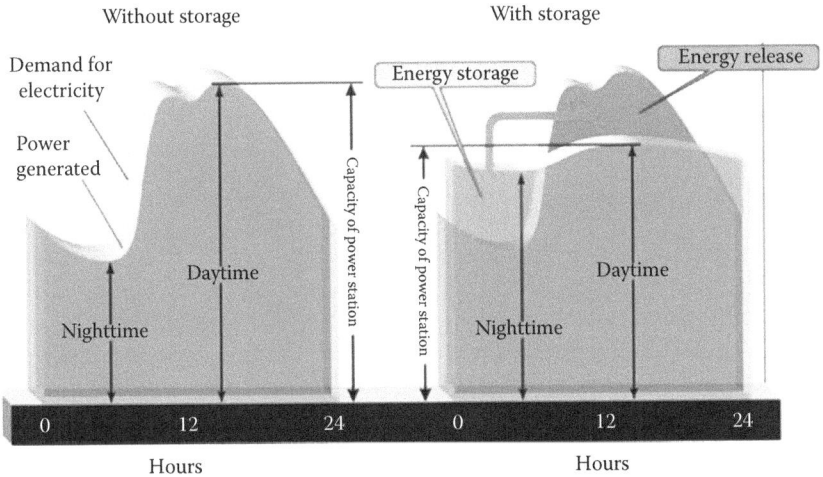

FIGURE 9.34
The principle of peak shaving and load levelling achieved through energy storage. (Reprinted with permission from Z. Yang et al., Electrochemical energy storage for green grid, *Chem. Rev.* 111, 2011, 3577–3613. Copyright 2011, American Chemical Society.)

Microgrids are modern, small-scale versions of the centralised electricity system. They achieve specific local goals, such as reliability, carbon emission reduction, diversification of energy sources and cost reduction, established by the community being served. A microgrid is normally an independent network including a distributed power generation system, energy storage and electrical load. A typical micro smart grid was built in Osaka in 2011 [87], a total of 8.2 kW renewable power generator, including CPV, and a 2 kW/10 kWh VFB systems were combined (Figures 9.35 and 9.36). The VFB system plays the role of stabilising the renewable sources and ensuring high-quality power supply. This VFB system showed excellent performance with an EE of above 80% at a current density of 140 mA cm^{-2}.

In China, a 200 kW/800 kWh VFB system was successfully installed by Rongke Power in 2012 in Beijing (Figure 9.37). Different kinds of energy storages including VFB were installed to combine with a 2.5 MW wind power generation system and a 503 kW PV system. The system was installed in 2012 and continuously ran for more than 1 year, indicating good reliability.

Off-grid refers to a system that is isolated from the grid, that is, not being connected to the main or national transmission grid in electricity. In practice, an off-grid power supply system is mainly used to provide the essential electricity to a region without access to the grid. Currently, quite a lot of off-line system demonstrations focused on communication base stations in remote regions. Renewable power such as photovoltaic (PV) combined with energy storage system is normally used. The VFB system is normally in the power rate range of 5–10 kW. Examples include a 5 kW VFB system installed in Winafrique by Purdent [2], a 10 kW/100 kWh VFB system installed by Celstorm, which consists of 10 1 kW stacks and can store 4 days' electricity, saving 13,140 L of diesel per year [90].

9.5 Challenges and Prospective

The VFB, as an effective energy storage technique, is very attractive for the utilisation and stabilisation of renewable energy. This technology is currently in the demonstration phase, and will require much more effort to develop before commercialisation is possible. Although many challenges remain to be overcome, cost is still the most critical challenge. The current cost of a VFB is around $500–600 per kWh, which is too high to be commercially viable [2]. There are several possibilities to reduce the cost, among them is to improve the power density of VFB stacks and to explore cheaper materials. Design and fabrication of the optimised cell architecture could be another important way to increase the system's operating current density. New cell structures with low resistance, high electrochemical activity and low concentration polarisation are particularly important in improving

Vanadium Flow Batteries

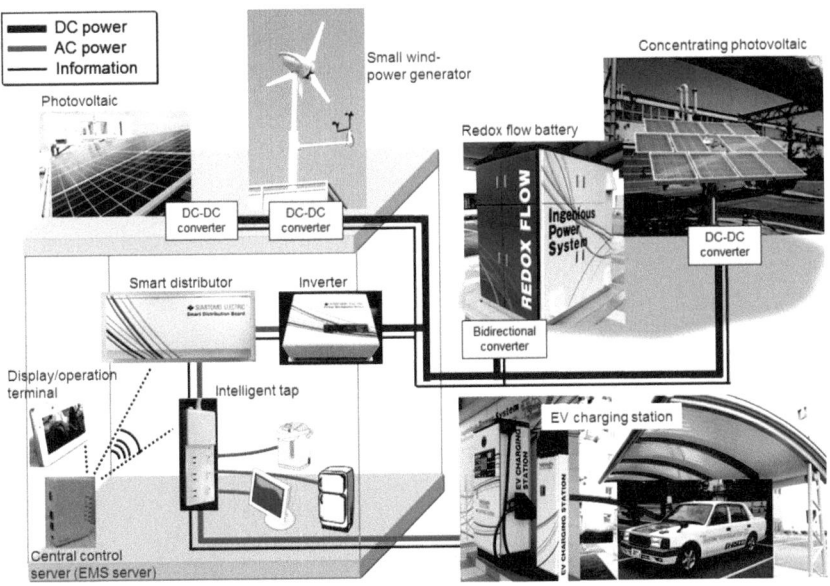

FIGURE 9.35
(**See colour insert.**) A schematic of a microgrid assembled in SEI. (Reprinted with permission from the website: http://global-sei.com/RandD/theme/power/index.html.)

the power density of VFB. In addition, the cost of the whole system can be decreased by material optimisation. For example, the energy density could be improved by improving the solubility of vanadium ions and the stability of the electrolytes. Novel membranes should be developed with low cost and high stability, achieved with advanced membrane materials or membrane

FIGURE 9.36
2 kW/10 kWh VFB systems. (Reprinted with permission from the website: http://global-sei.com/RandD/theme/power/index.html.)

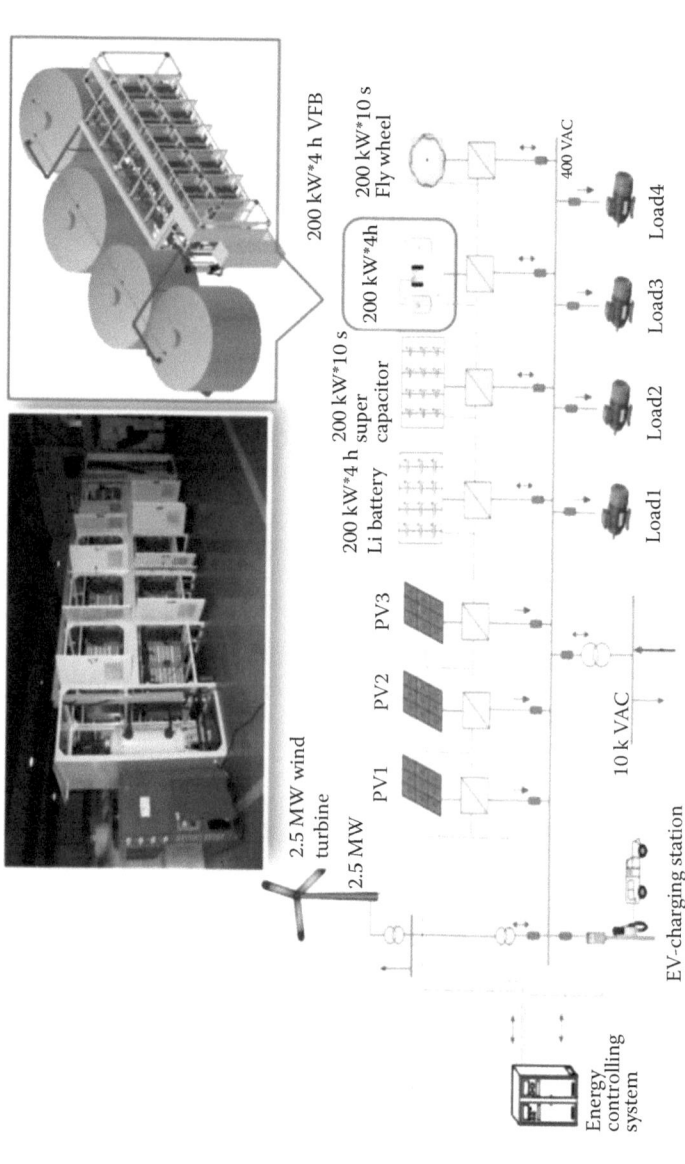

FIGURE 9.37
(See colour insert.) A 200 kW/800 kWh VFB system combined with wind power generation and PC system developed by Rongke Power Ltd. The call-outs are the schematic and actual photo of the 200 kW × 4 h VFB.

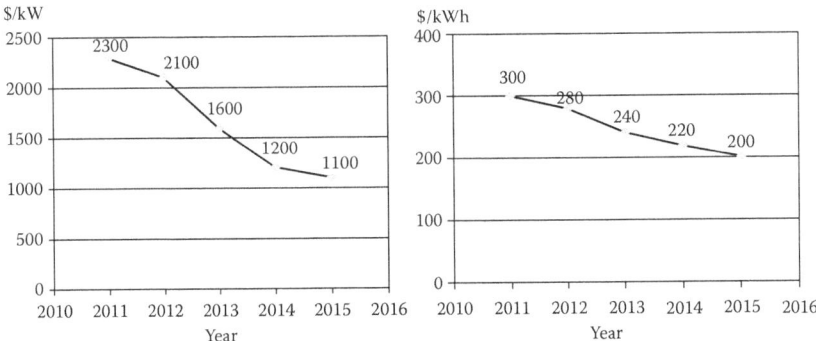

FIGURE 9.38
The expected VFB cost map of Rongke Power in the next few years.

structures, in order to efficiently decrease the battery cost and extend the operational life span. The capacity loss could also be decreased by (a) careful control of the charging SOC of VFB, (b) membranes developed with lower vanadium permeability and (c) electrodes with high electrochemical activity. With the above-mentioned modifications, the cost of VFB could be reduced dramatically. Figure 9.38 indicates the cost target of Rongke Power in the future; it is expected that the cost of a 1 MW/5 MWh VFB system can reach $400 per kWh in the very near future.

Acknowledgements

The authors acknowledge the financial support from National Basic Research Program of China (973 Program No. 2010CB227200) and China Natural Science Foundation (No. 21206158).

References

1. Z.G. Yang, J. Liu, S. Baskaran, C.H. Imhoff and J.D. Holladay, Enabling renewable energy and the future grid-with advanced electricity storage, *JOM-J. Min. Met. Mat. S.* 62, 2010, 14–23.
2. D. Rastler, *Electricity Energy Storage Technology Options: A White Paper Primer on Applications, Costs and Benefits.* EPRI, Palo Alto, CA, 2010.
3. M. Skyllas-Kazacos, M.H. Chakrabarti, S.A. Hajimolana, F.S. Mjalli and M. Saleem, Progress in flow battery research and development, *J. Electrochem. Soc.* 158, 2011, 55–59.

4. A.Z. Weber, M.M. Mench, J.P. Meyers, P.N. Ross, J.T. Gostick and Q. Liu, Redox flow batteries: A review, *J. Appl. Electrochem.* 41, 2011, 1137–1164.
5. L.H. Thaller, Electrically rechargeable redox flow cells, *9th Intersociety Energy Conversion Engineering Conference* 1, 1974, 924–928.
6. C. Ponce de Leóna, A. Frías-Ferrerb, J. González-Garcíab, D.A. Szántoc, F.C. Walsha, Redox flow cells for energy conversion, *J. Power Sources* 160, 2006, 716–732.
7. C. Ding, H.M. Zhang, X.F. Li, T. Liu and F. Xing, Vanadium flow battery for energy storage: Prospects and challenges, *J. Phys. Chem. Lett.* 4, 2013, 1281–1294.
8. M. Mohd Ruslim, A. Hamzah, M. Nizam and A. Seman, State-of-the-art of vanadium redox flow battery: A review on research prospects, *Int. Rev. Elec. Eng.*, 7, 2012, 5610–5622.
9. Z. Yang, J. Zhang, M.C. Kintner-Meyer, X. Lu, D. Choi, J.P. Lemmon and J. Liu, Electrochemical energy storage for green grid, *Chem. Rev.* 111, 2011, 3577–3613.
10. X.F. Li, H.M. Zhang, Z.S. Mai, H.Z. Zhang and I. Vankelecom, Ion exchange membranes for vanadium redox flow battery (VRB) applications, *Energy Environ. Sci.* 4, 2011, 1147–1160.
11. T. Mohammadia, S.C. Chiengb, M. Skyllas Kazacos, Water transport study across commercial ion exchange membranes in the vanadium redox flow battery, *J. Membr. Sci.* 133, 1997, 151–159.
12. K.A. Mauritz, and R.B. Moore, State of understanding of Nafion, *Chem. Rev.* 104, 2004, 4535–4585.
13. W.Y. Hsu and T.D. Gierke, Ion transport and clustering in Nafion perfluorinated membranes, *J. Membr. Sci.*13, 1983, 307–326.
14. J.Y. Xi, Z.H. Wu, X.P. Qiu and L.Q. Chen, Nafion/SiO_2 hybrid membrane for vanadium redox flow battery, *J. Power Sources* 166, 2007, 531–536.
15. X.G. Teng, Y.T. Zhao, J.Y. Xi, Z.H. Wu, X.P. Qiu and L.Q. Chen, Nafion/organic silica modified TiO_2 composite membrane for vanadium redox flow battery via in situ sol-gel reactions, *J. Membr. Sci.* 341, 2009, 149–154.
16. S.B. Sang, Q.M. Wu and K.L. Huang, Preparation of zirconium phosphate (Zrp)/ Nafion1135 composite membrane and H^+/VO^{2+} transfer property investigation, *J. Membr. Sci.* 305, 2007, 118–124.
17. Z.S. Mai, H.M. Zhang, X.F. Li, S.H. Xiao and H.Z. Zhang, Nafion/polyvinylidene fluoride blend membranes with improved ion selectivity for vanadium redox flow battery application, *J. Power Sources* 196, 2011, 5737–5741.
18. Q.T. Luo, H.M. Zhang, J. Chen, D.J. You, C.X. Sun And Y. Zhang, Preparation And Characterization Of Nafion/Speek Layered Composite Membrane And Its Application In Vanadium Redox Flow Battery, *J. Membr. Sci.* 325, 2008, 553–558.
19. X.S. Zhao, Y.Z. Fu, W.Li and A. Manthiram, Hydrocarbon blend membranes with suppressed chemical crossover for redox flow batteries, *RSC Adv.* 2, 2012, 5554–5556.
20. Q.T. Luo, H.M. Zhang, J. Chen, P. Qian and Y.F. Zhai, Modification of Nafion membrane using interfacial polymerization for vanadium redox flow battery applications, *J. Membr. Sci.* 311, 2008, 98–103.
21. J. Zeng, C.P. Jiang, Y.H. Wang, J.W. Chen, S.F. Zhu, B.J. Zhao and R. Wang, Studies on polypyrrole modified Nafion membrane for vanadium redox flow battery, *Electrochem. Commun.* 10, 2008, 372–375.

22. B. Schwenzer, S. Kim, M. Vijayakumar, Z. Yang and J. Liu, Correlation of structural differences between Nafion/polyaniline and Nafion/polypyrrole composite membranes and observed transport properties, *J. Membr. Sci.* 372, 2011, 11–19.
23. C. Ding, H.M. Zhang, X.F. Li, H.Z. Zhang, C. Yao and D.Q. Shi, Morphology and electrochemical properties of perfluorosulfonic acid ionomers for vanadium flow battery applications: Effect of side-chain length, *ChemSusChem* 6, 2013, 1262–1269.
24. T. Mohammadi and M. Skyllas-Kazacos, Preparation of sulfonated composite membrane for vanadium redox flow battery applications, *J. Membr. Sci.* 107, 1995, 35–45.
25. D.Y. Chen, S.J. Wang, M. Xiao and Y.Z. Meng, Synthesis and characterization of novel sulfonated poly(arylene thioether) ionomers for vanadium redox flow battery applications, *Energy Environ. Sci.* 3, 2010, 622–628.
26. D.Y. Chen, S.J. Wang, M. Xiao, D.M. Han and Y.Z. Meng, Synthesis of sulfonated poly(fluorenyl ether thioether ketone)swith bulky-block structure and its application in vanadium redox flow battery, *Polymer* 52, 2011, 5312–5319.
27. D.Y. Chen, S.J. Wang, M. Xiao and Y.Z. Meng, Synthesis and properties of novel sulfonated poly(arylene ether sulfone) ionomers for vanadium redox flow battery, *Energy Convers. Manage.* 51, 2010, 2816–2824.
28. Z.S. Mai, H.M. Zhang, X.F. Li, C. Bi and H. Dai, Sulfonated poly (tetramethydiphenyl ether ether ketone) membranes for vanadium redox flow battery application, *J. Power Sources* 196, 2011, 482–487.
29. S. Kim, J. Yan, B. Schwenzer, J. Zhang, L. Li, J. Liu, Z. Yang and M.A. Hickner, Cycling performance and efficiency of sulfonated poly(sulfone) membranes in vanadium redox flow batteries, *Electrochem. Commun.* 12, 2010, 1650–1653.
30. H.Z. Zhang, H.M. Zhang, X.F. Li, Z.S. Mai, W.P. Wei and Y. Li, Crosslinkable sulfonated poly (diallyl-bisphenol ether ether ketone) membranes for vanadium redox flow battery application, *J. Power Sources* 217, 2012, 309–315.
31. F.X. Zhang, H.M. Zhang and C. Qu, Influence of solvent on polymer prequaternization toward anion-conductive membrane fabrication for all-vanadium flow battery, *J. Phys. Chem. B*116, 2012, 9016–9022.
32. J. Qiu, M. Li, J. Ni, M. Zhai, J. Peng, L. Xu, H. Zhou, J. Li and G. Wei, Preparation of ETFE-based anion exchange membrane to reduce permeability of vanadium ions in vanadium redox battery, *J. Membr. Sci.* 297, 2007, 174–180.
33. X.G. Jian, C. Yan, H.M. Zhang, S.H. Zhang, C. Liu and P. Zhao, Synthesis and characterization of quaternized poly(phthalazinone ether sulfone ketone) for anion-exchange membrane, *Chin. Chem. Lett.* 18, 2007, 1269–1272.
34. S.H. Zhang, C.X. Yin, D.B. Xing, D.L. Yang and X.G. Jian, Preparation of chloromethylated/quaternized poly(phthalazinone ether ketone) anion exchange membrane materials for vanadium redox flow battery applications, *J. Membr. Sci.* 363, 2010, 243–249.
35. Z.S. Mai, H.M. Zhang, H.Z. Zhang, W.X. Xu, W.P. Wei, H. Na and X.F. Li, Anion-conductive membranes with ultralow vanadium permeability and excellent performance in vanadium flow batteries, *ChemSusChem* 6, 2013, 328–335.
36. H.Z. Zhang, H.M. Zhang, X.F. Li, Z.S. Mai and J.L. Zhang, Nanofiltration (NF) membranes: The next generation separators for all vanadium redox flow batteries (VRBs)?, *Energy Environ. Sci.* 4, 2011, 1676.
37. W.P. Wei, H.M. Zhang, X.F. Li, H.Z. Zhang, Y. Li and I. Vankelecom, Hydrophobic asymmetric ultrafiltration PVDF membranes: An alternative separator for VFB with excellent stability, *Phys. Chem. Chem. Phys.*15, 2013, 1766–1771.

38. Y. Li, H.M. Zhang, X.F. Li, H.Z. Zhang and W.P. Wei, Porous poly (ether sulfone) membranes with tunable morphology: Fabrication and their application for vanadium flow battery, *J. Power Sources* 233, 2013, 202–208.
39. H.Z. Zhang, H.M. Zhang, X.F. Li, Z.S. Mai and W.P. Wei, Silica modified nanofiltration membranes with improved selectivity for redox flow battery application, *Energy Environ. Sci.* 5, 2012, 6299–6303.
40. H.Z. Zhang, H.M. Zhang, F.X. Zhang, X.F. Li, Y. Li and I. Vankelecom, Advanced charged membranes with highly symmetric spongy structures for vanadium flow battery application, *Energy Environ. Sci.* 6, 2013, 776–781.
41. M. Rychcik and M. Skyllas-Kazacos, Evaluation of electrode materials for vanadium redox cell, *J. Power Sources* 19, 1987, 45–54.
42. S. Eckroad, Vanadium redox flow batteries: An in-depth analysis, EPRI: California, 2007, pp. 2–7.
43. S. Zhong, C. Padeste, M. Kazacos and M. Skyllas-Kazacos, Comparison of the physical, chemical and electrochemical properties of rayon-based and polyacrylonitrile-based graphite felt electrodes, *J. Power Sources* 45, 1993, 29–41.
44. H. Kaneko, K. Nozaki, Y. Wada, T. Aoki, A. Negishi and M. Kamimoto, Vanadium redox reactions and carbon electrodes for vanadium redox flow battery, *Electrochim. Acta* 36, 1991, 1191–1196.
45. S.Q. Liu, X.H. Shi, K.L. Huang, X.G. Li, Y.J. Li and X.W. Wu, Characteristics of carbon paper as electrode for vanadium redox flow battery, *Chin. J. Inorg. Chem.* 24, 2009, 798–802.
46. W.Y. Li, J.G. Liu and C.W. Yan, Multi-walled carbon nanotubes used as an electrode reaction catalyst for VO^{2+}/VO_2^+ for a vanadium redox flow battery, *Carbon* 49, 2011, 3463–3470.
47. Z. González, C. Botas, P. Álvarez, S. Roldán, C. Blanco, R. Santamaría, M. Granda and R. Menéndez, Thermally reduced graphite oxide as positive electrode in vanadium redox flow batteries, *Carbon* 50, 2011, 828–834.
48. W.Y. Li, J.G. Liu and C.W. Yan, Graphite-graphite oxide composite electrode for vanadium redox flow battery, *Electrochim. Acta* 56, 2011, 5290–5294.
49. P.X. Han, H.B. Wang, Z.H. Liu, X.A. Chen, W. Ma, J.H. Yao, Y.W. Zhuand and G.L. Cui, Graphene oxide nanoplatelets as excellent electrochemical active materials for VO^{2+}/VO_2^+ and V^{2+}/V^{3+} redox couples for a vanadium redox flow battery, *Carbon* 49, 2011, 693–700.
50. P.X. Han, Y.H. Yue, Z.H. Liu, W. Xu, L.X. Zhang, H.X. Xu, S.M. Dong and G.L. Cui, Graphene oxide nanosheets/multi-walled carbon nanotubes hybrid as an excellent electrocatalytic material towards VO^{2+}/VO_2^+ redox couples for vanadium redox flow batteries, *Energy Environ. Sci.* 4, 2011, 4710–4717.
51. L. Yue, W. Li, F. Sun, L. Zhao and L. Xing, Highly hydroxylated carbon fibres as electrode materials of all-vanadium redox flow battery, *Carbon* 48, 2010, 3079–3090.
52. B.T. Sun and M. Sykllas-kazacos, Modification of graphite electrode materials for vanadium redox flow battery application.1. thermal-treatment, *Electrochim. Acta* 37, 1992, 1253–1260.
53. B.T. Sun and M. Skyllas-Kazacos, Chemical modification of graphite electrode materials for vanadium redox flow battery application. 2. acid treatments, *Electrochim. Acta* 37, 1992, 2459–2465.

54. X.G. Li, K.L. Huang, N. Tan, S.Q. Liu and L.Q. Chen, Electrochemical modification of graphite felt electrode for vanadium redox flow battery, *J. Inorg. Mater.* 21, 2006, 1114–1120.
55. K.J. Kim, Y.J. Kim, J.H. Kim and M.S. Park, The effects of surface modification on carbon felt electrodes for use in vanadium redox flow batteries, *Mater. Chem. Phys.* 131, 2011, 547–553.
56. Y.Y. Shao, X.Q. Wang, M. Engelhard, C.M. Wang, S. Dai, J. Liu, Z.G. Yang and Y.H. Lin, Nitrogen-doped mesoporous carbon for energy storage in vanadium redox flow batteries, *J. Power Sources* 195, 2010, 4375–4379.
57. T. Wu, K.L. Huang, S.Q. Liu, S.X. Zhuang, D. Fang, S. Li, D. Lu and A. Q. Su, Hydrothermal ammoniated treatment of PAN-graphite felt for vanadium redox flow battery, *J. Solid State Electrochem.* 16, 2011, 579–585.
58. S.Y. Wang, X.S. Zhao, T. Cochell and A. Manthiram, Nitrogen-doped carbon nanotube/graphite felts as advanced electrode materials for vanadium redox flow batteries, *J. Phys. Chem. Lett.* 3, 2012, 2164–2167.
59. B.T. Sun and M. Skyllas-Kazacos, Chemical modification and electrochemical-behavior of graphite fiber in acidic vanadium solution, *Electrochim. Acta* 36, 1991, 513–517.
60. W.H. Wang and X.D. Wang, Investigation of Ir-modified carbon felt as the positive electrode of an all-vanadium redox flow battery, *Electrochim. Acta* 52, 2007, 6755–6762.
61. C. Flox, J. Rubio-Garcia, R. Nafria, R. Zamani, M. Skoumal, T. Andreu, J. Arbiol, A. Cabot and J.R. Morante, Active nano-$CuPt_3$ electrocatalyst supported on graphene for enhancing reactions at the cathode in all-vanadium redox flow batteries, *Carbon* 50, 2012, 2372–2374.
62. C. Yao, H.M. Zhang, T. Liu, X.F. Li and Z.H. Liu, Carbon paper coated with supported tungsten trioxide as novel electrode for all-vanadium flow battery, *J. Power Sources* 218, 2012, 455–461.
63. K.J. Kim, M.S. Park, J.H. Kim, U. Hwang, N.J. Lee, G. Jeong and Y.J. Kim, Novel catalytic effects of Mn_3O_4 for all vanadium redox flow batteries, *Chem. Commun.* 48, 2012, 5455–5457.
64. M. Skyllas-Kazacos and F. Grossmith, Efficient vanadium redox flow cell, *J. Electrochem. Soc.* 134, 1987, 2950–2953.
65. W. Wang, Q. Luo, B. Li, X. Wei, L. Li and Z. Yang, Recent progress in redox flow battery research and development, *Adv. Funct. Mater.* 23, 2013, 970–986.
66. M. Vijayakumar, L. Li, G. Graff, J. Liu, H. Zhang, Z. Yang and J.Z. Hu, Towards understanding the poor thermal stability of V^{5+} electrolyte solution in vanadium redox flow batteries, *J. Power Sources* 196, 2011, 3669–3672.
67. P. Modiba, *Electrolytes for Redox Flow Battery Systems*, University of Stellenbosch, 2010.
68. F. Rahman and M. Skyllas-Kazacos, Vanadium redox battery: Positive half-cell electrolyte studies, *J. Power Sources* 189, 2009, 1212–1219.
69. F. Rahman and M. Skyllas-Kazacos, Solubility of vanadyl sulfate in concentrated sulfuric acid solutions, *J. Power Sources* 72, 1998, 105–110.
70. X. Wu, J. Wang, S. Liu, X. Wu and S. Li, Study of vanadium(IV) species and corresponding electrochemical performance in concentrated sulfuric acid media, *Electrochim. Acta* 56, 2011, 10197–10203.

71. M. Skyllas-Kazacos, C. Peng and M. Cheng, Evaluation of precipitation inhibitors for supersaturated vanadyl electrolytes for the vanadium redox battery, *Electrochem. Solid-State Lett.* 2, 1999, 121–122.
72. J. Zhang, L. Li, Z. Nie, B. Chen, M. Vijayakumar, S. Kim, W. Wang, B. Schwenzer, J. Liu and Z. Yang, Effects of additives on the stability of electrolytes for all-vanadium redox flow batteries, *J. Appl. Electrochem* 41, 2011, 1215–1221.
73. F. Chang, C. Hu, X. Liu, L. Liu and J. Zhang, Coulter dispersant as positive electrolyte additive for the vanadium redox flow battery, *Electrochim. Acta* 60, 2012, 334–338.
74. X.X. Liang, S. Peng, Y. Lei, C. Gao, N.F. Wang, S.Q. Liu and D. Fang, Effect of L-glutamic acid on the positive electrolyte for all-vanadium redox flow battery, *Electrochim. Acta* 95, 2013, 80–86.
75. X. Wu, S. Liu, N. Wang, S. Peng and Z. He, Influence of organic additives on electrochemical properties of the positive electrolyte for all-vanadium redox flow battery, *Electrochim. Acta* 78, 2012, 475–482.
76. Y. Lei, S.Q. Liu, C. Gao, X. X. Liang, Z.X. He, Y.H. Deng and Z. He, Effect of amino acid additives on the positive electrolyte of vanadium redox flow batteries, *J. Electrochem. Soc.* 160, 2013, A722–A727.
77. S. Peng, N.F. Wang, C. Gao, Y. Lei, X.X. Liang, S. Q. Liu and Y. Liu, Stability of positive electrolyte containing trishydroxymethyl aminomethane additive for vanadium redox flow battery, *Int. J. Electrochem. Sci* 7, 2012, 4388–4396.
78. A. Parasuraman, T.M. Lim, C. Menictas and M.S. Kazacos, Review of material research and development for vanadium redox flow battery applications, *Electrochim. Acta* 101, 2013, 27–40.
79. S. Kim, M. Vijayakumar, W. Wang, J. Zhang, B. Chen, Z. Nie, F. Chen, J. Hu, L. Li and Z. Yang, Chloride supporting electrolytes for all-vanadium redox flow batteries, *Phys. Chem. Chem. Phys.* 13, 2011, 18186–18193.
80. L.Y. Li, S. Kim, W. Wang, M. Vijayakumar, Z.M. Nie, B.W. Chen, J.L. Zhang, G.G. Xia, J.Z. Hu, G.Graff, J. Liu and Z.G. Yang, A stable vanadium redox-flow battery with high energy density for large-scale energy storage, *Adv. Energy Mater.* 1, 2011, 394–400.
81. S. Peng, N.F. Wang, X.J. Wu, S.Q. Liu, D. Fang, Y.N. Liu and K.L. Huang, Vanadium species in CH_3SO_3H and H_2SO_4 mixed acid as the supporting electrolyte for vanadium redox flow battery, *Int. J. Electrochem. Sci* 7, 2012, 643–649.
82. D.S. Aaron, Q. Liu, Z. Tang, G.M. Grim, A.B. Papandrew, A. Turhan, T.A. Zawodzinski and M.M. Mench, Dramatic performance gains in vanadium redox flow batteries through modified cell architecture, *J. Power Sources* 206, 2012, 450–453.
83. L. Zhang, Y.S. Yang, Y.H. Wen, X.D. Wang and Z.L. Xie, Research progress of the shunt current in redox flow battery, *Chinese J. Power Sources* 33, 2009, 144–147.
84. B. LI, J.B. Guo, J.Z. Chen and D. Hui, Modelling and simulating of shunt current in redox flow battery, *Proceedings of the CSEE* 31, 2011, 27, 1–7.
85. F. Xing, H. Zhang and X. Ma, Shunt current loss of the vanadium redox flow battery, *J. Power Sources* 196, 2011, 10753–10757.
86. B.M. Broman and A. Zocchi, Membrane separated, bipolar multicell electrochemical reactor, US Patent 6,555,267, 2003.
87. M. Skyllas-Kazacos, G. Kazacos, G. Poon and H. Verseema, Recent advances with UNSW vanadium-based redox flow batteries, *Int. J. Energy Res.* 34, 2010, 182–189.

88. H. Nakahata, N. Ayai, T. Shibata, T. Shigematsu, R. Satoh, H. Nakaishi, T. Iwasaki, T. Hisada, K. Kitayama and H. Miyoshi, Development of smart grid demonstration systems, *SEI Tech. Rev.* 76, 2013, 8–13.
89. N. Chouhan and R.S. Liu, In: L. Zhang, R.S. Liu, H. Liu, A. Sun, J. Zhang. (Eds.), *Electrochemical Technologies for Energy Storage and Conversion*, Wiley-VCH Verlag GmbH & Co. KGaA, 2012, pp. 1–43.
90. a + f GMBH, Energy Storage Business Division, Intelligent Technology—A Clean Future, http://energy.gildemeister.com/de/.

10

Physical Properties of Negative Half-Cell Electrolytes in the Vanadium Redox Flow Battery

Asem Mousa and Maria Skyllas-Kazacos

CONTENTS

10.1 Introduction 396
10.2 Physical Properties of Vanadium (III) Sulphate Solutions 397
    10.2.1 Density of Vanadium (III) Sulphate Solutions 397
        10.2.1.1 Effects of Vanadium (III) Concentration 397
        10.2.1.2 Effect of Sulphuric Acid Concentration 399
        10.2.1.3 Temperature Effect 399
        10.2.1.4 Empirical Model for the Density of Vanadium Solutions 399
    10.2.2 Viscosity of Vanadium (III) Sulphate Solutions 401
        10.2.2.1 Effects of Vanadium (III) Sulphate Concentration 401
        10.2.2.2 Effects of Sulphuric Acid Concentration 403
        10.2.2.3 Temperature Effects 405
        10.2.2.4 Modelling the Viscosity of Vanadium (III) Sulphate Solutions 407
        10.2.2.5 Empirical Model for Vanadium (III) Sulphate Solutions 410
    10.2.3 Conductivity of Vanadium (III) Sulphate Solutions 412
        10.2.3.1 Effect of Vanadium (III) Sulphate Concentration 413
        10.2.3.2 Effect of Sulphuric Acid Concentration 418
        10.2.3.3 Temperature Effects 420
        10.2.3.4 Modelling the Conductivity Data 420
10.3 Physical Properties of Vanadium (II) Sulphate Solutions 422
10.4 Summary 425
References 426

Three physical properties of the supersaturated vanadium (II) and vanadium (III) sulphate solutions were investigated in this study: density, viscosity and conductivity. These solutions correspond to the charged and discharged negative half-cell electrolytes, respectively, of the vanadium redox flow battery (VRFB) and the vanadium/oxygen redox fuel cell. The effects of vanadium concentration, total sulphates concentration and temperature on these properties were determined. The measured values of each of the physical properties (density, viscosity and conductivity) were fitted to an empirical formula that describes the property as a function of vanadium (III) sulphate concentration, sulphuric acid concentration and temperature. The three properties showed good agreement between the measured and calculated values, indicating successful fitting. These empirical models will provide an important tool for use in the engineering design of vanadium flow battery systems for different applications and climatic conditions.

10.1 Introduction

The VRFB that was pioneered by Skyllas-Kazacos and co-workers at the University of New South Wales [1–5] is now regarded as one of the most promising technologies for large-scale energy storage and several companies are currently manufacturing systems for a wide range of applications [6–8]. One of the most important components of any flow battery is the electrolyte that stores energy in the form of fully soluble redox couple solutions. In the all-vanadium redox flow battery, V(II)/V(III) and V(IV)/V(V) redox couple solutions in a sulphuric acid-supporting electrolyte are employed in the negative and positive half-cells, respectively, as illustrated in Figure 10.1.

When designing a battery for a particular application, consideration must be given to the optimal electrolyte composition that will suit the specific ambient temperature range of the installation site and the operating conditions of the battery.

While the properties of the charged positive half-cell electrolyte dictate the performance of the VRFB at elevated temperatures due to the thermal precipitation of V(V) species, optimisation of the negative half-cell electrolyte is required for stable low-temperature operation. The V(II)/V(III) electrolyte is also employed in the vanadium/oxygen redox fuel cell (V/O_2) that is currently receiving considerable attention, especially for mobile applications. In this study, the properties of the V(II) and V(III) solutions were evaluated as a function of sulphuric acid concentration and temperature. Electrolyte properties such as density, viscosity and conductivity will vary with temperature and composition, and knowledge of these properties is critical for engineering design and optimisation of both the VRB and V/O_2 flow battery systems.

Physical Properties of Negative Half-Cell Electrolytes

FIGURE 10.1
(See colour insert.) Photograph of an all-vanadium redox flow battery showing the electrolyte reservoirs that store the two half-cell solutions that are pumped through the cell stack where energy is generated by the electrochemical reactions of the vanadium redox couples at inert electrodes.

10.2 Physical Properties of Vanadium (III) Sulphate Solutions

10.2.1 Density of Vanadium (III) Sulphate Solutions

This study was carried out on 0–1.5 M vanadium (III) sulphate in three sets of sulphuric acid solutions containing 1.0, 1.5 and 2.0 M acid. Measurements were made in the temperature range 20–40°C. The density of the vanadium solutions was calculated by measuring the weight of density bottles containing a known volume of the vanadium solution. The volumes of the density bottles were determined precisely at different temperatures using deionised water.

The calculated values of the density of the different vanadium (III) sulphate solutions at different temperatures are shown in Table 10.1.

10.2.1.1 Effects of Vanadium (III) Concentration

Figure 10.2 shows that the density of the V(III) solutions is directly proportional to the amount of vanadium (III) sulphate (present as $[V_2(SO_4)_3]$) in the solution. This figure shows that the solutions with high vanadium content almost have the same density at each acid concentration, indicating that the density of the solutions at these high vanadium concentrations is less affected by the acid concentration.

TABLE 10.1
Measured Density Values (g cm^{-3}) for Vanadium (III) Sulphate Solutions at Different Acid Concentrations and Temperatures

[H$_2$SO$_4$] (M)	Temperature (°C)	[V$_2$(SO$_4$)$_3$] (M)						
		0.0	0.3	0.5	0.8	1.0	1.3	1.5
2.0	15	1.121	1.205	1.283	1.358	1.427	1.492	1.553
	20	1.118	1.202	1.280	1.354	1.424	1.489	1.550
	25	1.116	1.198	1.277	1.351	1.421	1.486	1.547
	30	1.113	1.195	1.274	1.348	1.417	1.483	1.544
	35	1.110	1.192	1.270	1.344	1.414	1.479	1.541
	40	1.107	1.189	1.267	1.341	1.410	1.476	1.538
1.5	15	1.091	1.170	1.256	1.326	1.390	1.468	1.533
	20	1.089	1.168	1.253	1.322	1.387	1.465	1.529
	25	1.086	1.165	1.250	1.319	1.384	1.461	1.526
	30	1.084	1.162	1.247	1.315	1.381	1.458	1.522
	35	1.081	1.159	1.244	1.312	1.377	1.455	1.518
	40	1.078	1.155	1.241	1.308	1.373	1.451	1.515
1.0	15	1.058	1.151	1.238	1.321	1.400	1.470	1.545
	20	1.058	1.149	1.235	1.317	1.396	1.467	1.542
	25	1.057	1.147	1.232	1.314	1.391	1.463	1.539
	30	1.055	1.144	1.229	1.311	1.389	1.460	1.535
	35	1.052	1.141	1.227	1.307	1.387	1.457	1.532
	40	1.048	1.138	1.223	1.304	1.383	1.453	1.528

Note: [V(III)] = 2 × [V$_2$(SO$_4$)$_3$].

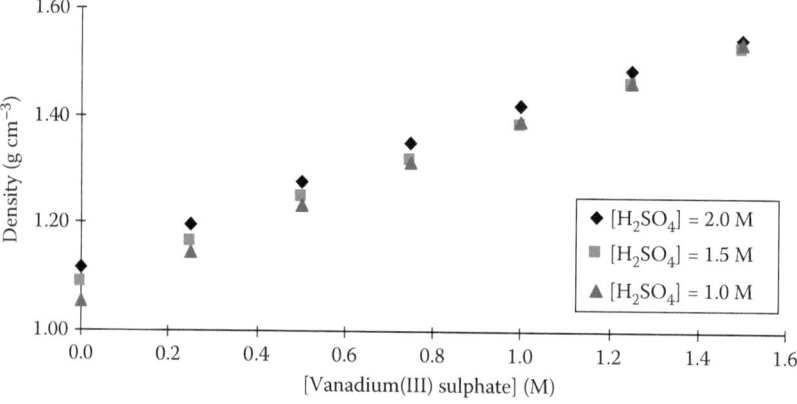

FIGURE 10.2
Variation in density of V$_2$(SO$_4$)$_3$ solutions with the increase in the salt concentration in different acid solutions at 25°C.

TABLE 10.2
Empirical Parameters and Regression Coefficients for Fitting of the Measured Density Values at 25°C

[H_2SO_4] (M)	A	B	C	Regression
1.0	−0.0353	0.3404	1.1156	1.0000
1.5	−0.0228	0.3263	1.0868	0.9995
2.0	−0.0288	0.3630	1.0573	0.9999

The data were fitted to a second-order polynomial (Equation 10.1) and the values of the second-, first- and zero-order coefficients were calculated for each solution as shown in Table 10.2. The second-order coefficient is reported as an indicator of the solute–solute interactions in the solution. The first-order coefficient is proportional to the variation in the solution density (molar volume) with the addition of vanadium sulphate and the last coefficient is the density of the acid–water mixture. The values of these coefficients at different acid concentrations do not show any trends. This might be due to the fact that the variations might be too small and lie within experimental error.

$$D = A[V_2(SO_4)_3]^2 + B[V_2(SO_4)_3] + C \tag{10.1}$$

where D is the solution density (g cm^{-3}), [$V_2(SO_4)_3$] is the concentration of vanadium (III) sulphate and A, B and C are the regression constants.

10.2.1.2 Effect of Sulphuric Acid Concentration

The data in Table 10.1 show that the change in the density of the solution is less affected by the changes in acid concentration compared to that of vanadium (III) sulphate. No attempt was therefore made to fit these data to a linear model.

10.2.1.3 Temperature Effect

From the experimental data, temperature was found to have a negligible effect on solution density. The data were fitted to a linear model but the values of the slopes of the four lines were close to zero, indicating that the extent of the thermal expansion of the solution was insignificant. A summary of the regression coefficient values is shown in Table 10.3.

10.2.1.4 Empirical Model for the Density of Vanadium Solutions

One of the main goals of this study was to find an empirical formula that allows the prediction of the density of vanadium (III) sulphate solutions under certain conditions. Hence, several models were tested to fit all the experimental data obtained in this study. These points covered a wide range of salt and

TABLE 10.3

Slope and Intercept of the Linear Variation in the $V_2(SO_4)_3$ Solution Density with Temperature

$[V_2(SO_4)_3]$ (M)	Slope	Intercept	Regression
0	−0.0006	1.1295	0.9999
0.5	−0.0007	1.2934	0.9998
1.0	−0.0007	1.4376	0.9998
1.5	−0.0006	1.5620	1.0000

Note: $[H_2SO_4] = 2.0$ M.

acid concentrations and temperatures. Several models that were reported in the literature to fit the data of similar systems were tested [9–13]. The best fit was obtained with a second-order model with respect to both the acid and the vanadium (III) sulphate concentration and first order with respect to temperature [9]. An exponential term was reported in some studies to describe the effect of the temperature on the density [11]. This was tested but did not make any significant improvement on the obtained data. Hence, Equation 10.2 was used to fit the data with an average error of 0.0046 g cm^{-3} ($R^2 = 0.9949$). The value of the coefficients A, B, C, D, E and F are shown in Table 10.4.

$$d = A + B \times m_{V_2(SO_4)_3} + C \times (m_{V_2(SO_4)_3})^2 + D \times m_{H_2SO_4} + E \times (m_{H_2SO_4})^2 + F \times T \tag{10.2}$$

where d is the solution density in g cm^{-3}, $m_{V_2(SO_4)_3}$ and $m_{H_2SO_4}$ are the molality of the salt and the acid, respectively, T is the solution temperature in °C and A, B, C, D, E and F are empirical regression coefficients.

The calculated density values used in the above model were based on the molality of vanadium (III) sulphate and excess sulphuric acid. Nevertheless, these concentration units are not commonly used to report the composition of vanadium solutions used in the vanadium redox battery. Hence, the data were re-fitted to the above model but using the molar concentration of the vanadium ions and the total molar concentration of the sulphate ions $[SO_4^{-1}]_T$ as shown in Equation 10.3. These concentration units are most frequently used to report the properties of the vanadium solutions. The calculated coefficients are shown in Table 10.4.

TABLE 10.4

Empirical Regression Coefficients Used in Equations 10.2 and 10.3

	A/A′	B/B′	C/C′	D/D′	E/E′	F/F′
Equation 10.2	1.04048	0.36368	−0.03338	0.03810	0.00042	−0.00068
Equation 10.3	1.00320	0.08822	0.00256	0.07363	−0.00491	−0.00061

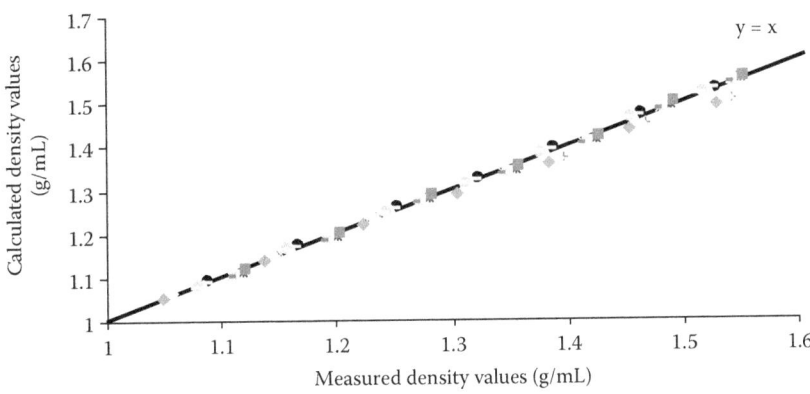

FIGURE 10.3
Calculated density values for $V_2(SO_4)_3$ solutions according to Equation 10.3 plotted against the measured values.

$$d = A' + B' \times [V] + C' \times [V]^2 + D' \times [SO_4^{-2}]_T + E' \times [SO_4^{-2}]_T^2 + F' \times T \qquad (10.3)$$

where d is the solution density in g cm^{-3}, [V] and $[SO_4^{-2}]_T$ are the molarity of vanadium (III) ion and the total sulphates, respectively, T is the solution temperature in °C and A', B', C', D', E' and F' are empirical regression coefficients. To further confirm the validity of the above model, the calculated density values for all the solutions were plotted against the measured values. Figure 10.3 shows that the calculated and measured values are similar within experimental errors ($R^2 = 0.9974$).

10.2.2 Viscosity of Vanadium (III) Sulphate Solutions

The viscosity of vanadium (III) sulphate solutions was determined using a U-shaped viscometer, details of which are described by Mousa [14]. The data covered a $V_2(SO_4)_3$ concentration range of 0–1.5 M (equivalent to 0–3.0 M total vanadium (III) concentration) and a sulphuric acid concentration between 1.0 and 2.0 M. The viscosity of these solutions was measured at temperatures between 15°C and 40°C. The experimental values are shown in Table 10.5.

10.2.2.1 Effects of Vanadium (III) Sulphate Concentration

The viscosity of the solutions was found to increase exponentially with the increase in the $V_2(SO_4)_3$ concentration at different temperatures as shown in Figure 10.4.

This behaviour has been reported for many concentrated solutions of both strong electrolytes and molecular solutes. Rahman observed a similar behaviour for vanadium (V) solution in sulphuric acid–water mixtures. His results

TABLE 10.5

Viscosity Values (in cP) for Different $V_2(SO_4)_3$ Solutions at Different Acid Concentrations and Temperatures

[H_2SO_4] (M)	Temperature (°C)	[$V_2(SO_4)_3$] (M)						
		0	0.3	0.5	0.8	1.0	1.3	1.5
2.0	15	1.5	2.0	2.8	3.7	5.9	8.5	13.0
	20	1.5	1.8	2.4	3.4	5.1	7.3	11.2
	25	1.4	1.7	2.1	3.0	4.4	6.2	9.6
	30	1.3	1.5	1.9	2.7	3.8	5.3	8.1
	35	1.2	1.4	1.7	2.4	3.3	4.5	6.9
	40	1.1	1.3	1.6	2.1	3.0	3.9	5.8
1.5	15	1.5	1.9	2.5	3.5	5.1	8.0	12.5
	20	1.4	1.8	2.3	3.1	4.3	6.9	10.8
	25	1.3	1.6	2.0	2.7	3.7	5.9	8.8
	30	1.3	1.5	1.8	2.4	3.3	5.1	7.4
	35	1.2	1.4	1.7	2.1	2.9	4.4	6.2
	40	1.1	1.3	1.6	1.9	2.6	3.8	5.3
1.0	15	1.4	1.7	2.3	3.2	4.2	7.5	11.7
	20	1.4	1.7	2.2	2.8	3.6	6.4	10.2
	25	1.3	1.5	1.9	2.4	3.1	5.6	7.9
	30	1.2	1.5	1.8	2.1	2.7	4.8	6.5
	35	1.2	1.3	1.6	1.9	2.4	4.1	5.5
	40	1.1	1.3	1.5	1.7	2.2	3.6	4.7

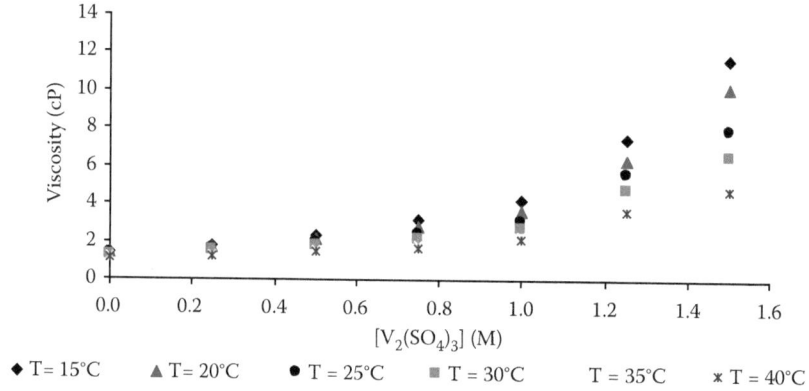

FIGURE 10.4
Variation in solution viscosity with the increase in $V_2(SO_4)_3$ concentration at different temperatures. [H_2SO_4] = 2.0 M.

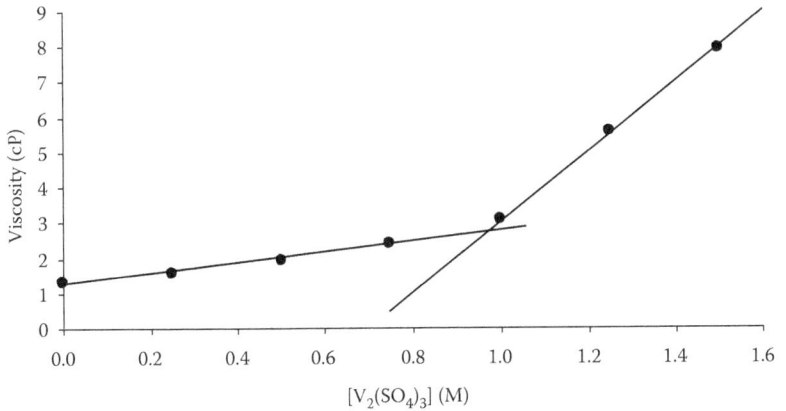

FIGURE 10.5
Illustration of the two regions of the viscosity–concentration curve of vanadium (III) sulphate solutions. [H_2SO_4] = 2.0 M, temperature = 25°C.

were in agreement with previous reports that proposed the formation of vanadium–vanadium dimers and polymers or vanadium–sulphate complexes [15].

The sharp increase in the viscosity is attributed to the stronger solute–solute attractions at the high concentration range. It is suggested that the structure of the solution undergoes significant changes as the solute concentration is increased above a certain critical concentration value. At this concentration, the number of water molecules in the solution becomes insufficient to hydrate the solute ions or molecules. This results in a change in the flow properties of the solution as it becomes less fluid and closer to the flow properties of molten glass, hence, the name 'glass transition concentration'.

The viscosity–concentration curve could be divided into two straight-line regions. The first line has a much lower slope value compared to that of the second region as shown in Figure 10.5. The 'glass transition' concentration could be estimated from the intersection point between the two lines [16–18]. The concentration of vanadium (III) sulphate at the intersection point of the two lines was estimated at different temperatures. Unfortunately, the results did not show any consistent trends as all the intersection points were around 1.0 M without any meaningful order. Another attempt was made to estimate the intersection point at different sulphuric acid concentrations but showed the same random behaviour. It is believed that the high error originates from the error in drawing the straight line in each region. Attempts were used to obtain the best line fitting for each region, but this was not successful due to the small number of points in each region.

10.2.2.2 Effects of Sulphuric Acid Concentration

The effect of acid concentration on the viscosity of the solution is illustrated in Figure 10.6, which shows the viscosity of a number of these solutions

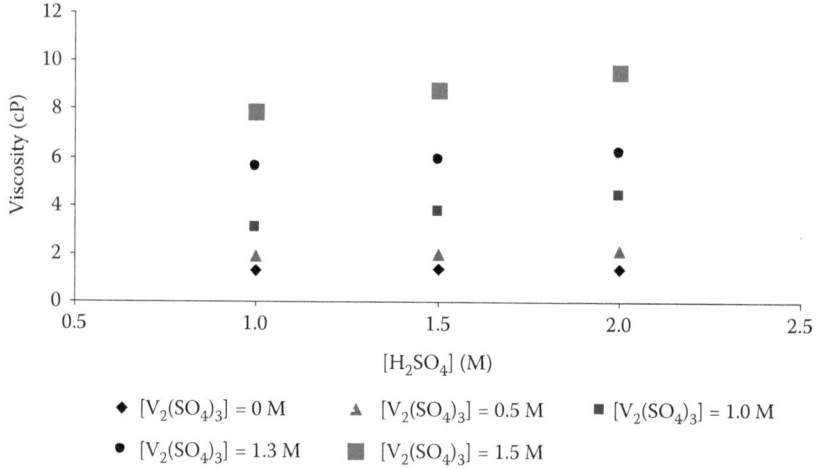

FIGURE 10.6
Viscosity of $V_2(SO_4)_3$ solutions as a function of acid concentration. Temperature = 25°C.

against the sulphuric acid concentration. This figure shows that the extent of the acid effect on the solution viscosity (indicated by the slope of the line) is proportional to the vanadium (III) sulphate concentration. The change of the value of the slope of these lines with the increase in the vanadium (III) sulphate concentration is shown in Figure 10.7. The figure should be considered

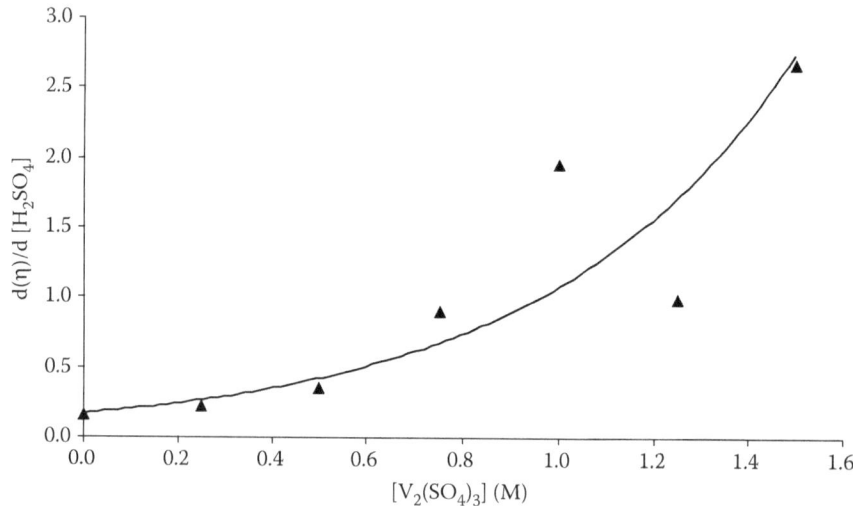

FIGURE 10.7
Variations in the slope of the viscosity–acid concentration lines with the increase in $V_2(SO_4)_3$ concentration.

qualitatively only due to the relatively high errors associated with obtaining the slopes values from the small number of points.

10.2.2.3 Temperature Effects

The change in the solution viscosity with the increase in the solution temperature for different solutions is shown in Figure 10.8. The viscosity of the solution decreases as the temperature of the solution is increased. This behaviour is expected as the increase in the temperature results in an increase in the kinetic energy of the species in the solution. This increase causes less solute–solvent and solute–solute interactions, resulting in a decrease in the resistance (friction) against the flow of the solution.

The resistance of the solution to flow (and hence the attraction forces in the solution) could be quantitatively measured from the 'activation energy of the flow'. The value of the activation energy can be obtained from the slope obtained according to Equation 10.4 [11].

$$\eta = A \exp(E_{vis}/RT) \quad (10.4)$$

where η is the viscosity of the solution at any temperature, A is a constant, E_{vis} is the 'viscosity activation energy', R is as constant and T is the temperature in kelvin.

A plot of $\ln(\eta)$ versus $1/T(K)$ is shown in Figure 10.9. Straight lines were obtained for all the solutions and the slope for these lines are shown in Table 10.6.

The change in the value of the activation energy with the increase in the $V_2(SO_4)_3$ concentrations in different sulphuric acid–water mixtures is shown in Figure 10.10. This figure shows an initial increase in the activation energy

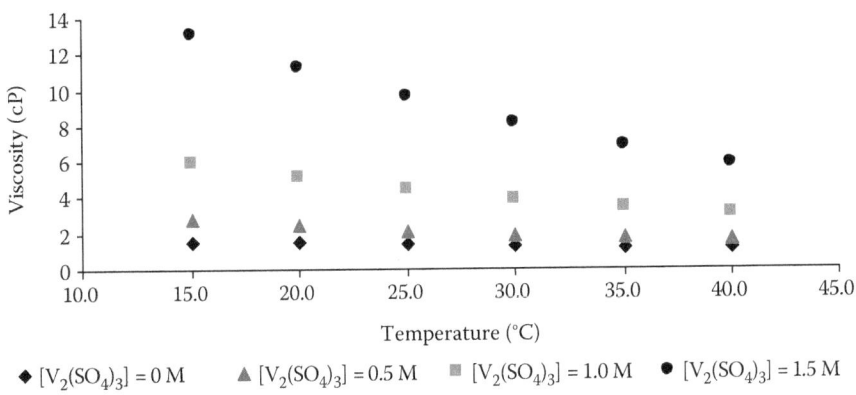

FIGURE 10.8
Viscosity of different $V_2(SO_4)_3$ solutions at different temperatures. $[H_2SO_4]$ = 2.0 M.

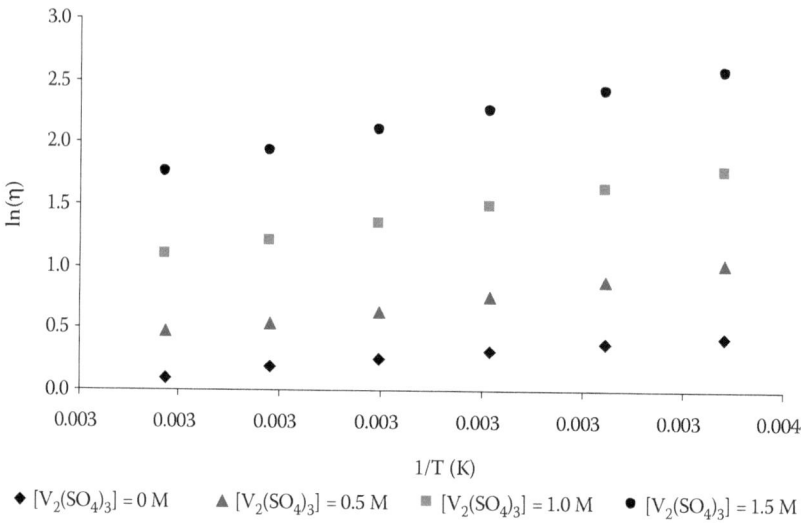

FIGURE 10.9
Arrhenius plot for the viscosity of $V_2(SO_4)_3$ solutions. $[H_2SO_4] = 2.0$ M.

TABLE 10.6

Activation Energy (kJ mol^{-1}) for Vanadium Solutions at Different Sulphuric Acid Concentrations

$[V_2(SO_4)_3]$ (M)	$[H_2SO_4]$ (M)		
	2.0	1.5	1.0
0	9.7	8.9	6.6
0.3	13.1	12.2	10.5
0.5	16.9	15.0	13.2
0.8	17.1	18.0	19.0
1.0	20.7	20.4	20.2
1.3	23.4	22.6	21.7
1.5	24.3	26.2	28.1

of the solution flow that is followed by a plateau up to a vanadium (III) sulphate concentration of 1.0 M (equivalent to 2.0 M total vanadium concentration). After this concentration, a sharp increase in the value of the activation energy is observed. This behaviour further supports the assumption of complex formation at higher concentrations of vanadium sulphate solutions.

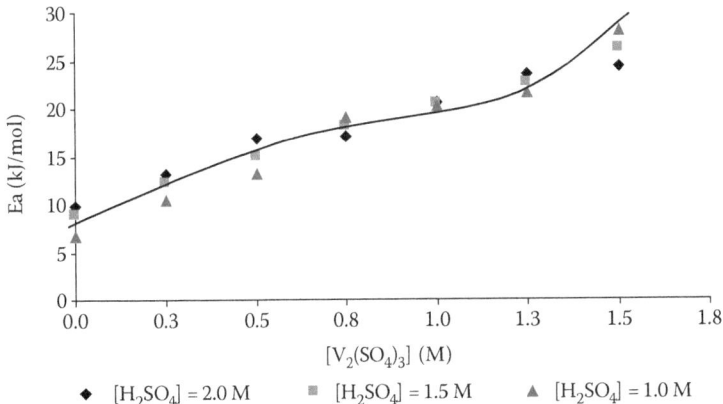

FIGURE 10.10
Change in the activation energy for the rate of flow with the increase in the $V_2(SO_4)_3$ concentration in different acid solution concentrations.

10.2.2.4 Modelling the Viscosity of Vanadium (III) Sulphate Solutions

Several models were tested to describe the variation of the solution viscosity with the increase in concentration of vanadium (III) sulphates. Unfortunately, the high degree of saturation of these solutions and the unavailability of much of the information needed to apply these theoretical models made these attempts unsuccessful except for the modified form of Jones–Doles equation (Equation 10.5).

$$\eta_r = \eta/\eta° = 1 + Ac^{1/2} + Bc + Dc^2 + Ec^{3.5} + Fc^7 \qquad (10.5)$$

where η_r, η and $\eta°$ are the relative viscosity, solution viscosity and solvent viscosity, respectively, c is the molar concentration of vanadium (III) sulphate and A, B, D, E and F are empirical coefficients.

To calculate the value of the 'measured relative viscosity' from experimental results, the viscosity of dilute sulphuric acid solutions was used as the viscosity of the solvent. The fact that the concentration of the acid was kept constant for each set of solutions allows it to be considered as the solvent. The validity of this assumption is based on the fact that the addition of the acid (up to 2.0 M) did not cause significant changes in the viscosity of water compared to the addition of vanadium sulphates. It is also assumed that the contribution of sulphuric acid to the increase in the viscosity of the solution is independent of the addition of vanadium (III) sulphate. Nevertheless, the accuracy of the parameters calculated below should always be treated with caution.

The values of the five coefficients in Equation 10.5 were determined for all the solutions using a linear regression function (Table 10.7). The calculated values were used to re-calculate the viscosity of each solution. A further confirmation was obtained by plotting these calculated viscosity values against

TABLE 10.7

Empirical Coefficients for the Modified Jones–Dole Equation (Equation 10.5) for $V_2(SO_4)_3$ Solutions at Different Acid Concentrations and Temperatures

[H_2SO_4] (M)	Temperature (°C)	A ($M^{-1/2}$)	B (M^{-1})	D (M^{-2})	E ($M^{-7/2}$)	F (M^{-7})
2.0	15	2.1223	−3.3988	3.6310	0.2146	0.0551
	20	1.1130	−2.2395	4.2635	−0.7909	0.1386
	25	1.1862	−2.6492	4.5201	−1.0967	0.1666
	30	1.2820	−2.8947	4.3581	−0.9206	0.1214
	35	1.1947	−2.7774	4.3619	−1.2146	0.1542
	40	0.3595	−0.3043	1.4081	0.0191	0.0593
1.5	15	1.9384	−3.3771	3.7454	0.0822	0.0676
	20	1.3628	−2.6054	3.5298	−0.1854	0.0947
	25	1.1057	−1.7938	2.2949	0.3238	0.0210
	30	0.1134	0.4546	0.1890	0.9777	−0.0281
	35	1.5933	−2.8544	2.6910	0.0950	0.0054
	40	1.3458	−2.8729	3.1514	−0.2482	0.0177
1.0	15	1.4089	−2.6801	3.7475	−0.0781	0.0792
	20	0.4575	−0.0215	0.4689	1.0547	0.0310
	25	−0.0827	0.9006	−0.1065	0.9796	−0.0017
	30	1.7704	−3.5489	3.4222	−0.1701	0.0237
	35	1.3375	−2.9978	3.1734	−0.2556	0.0228
	40	1.2108	−2.0180	1.3532	0.5563	−0.0411

the measured ones. A straight line passing through the origin and with a slope of 1 was obtained as shown in Figure 10.11. The first two coefficients in Equation 10.5 have similar physical meanings as those in the original Jones–Dole equation. The physical meaning of the other coefficients was not discussed in earlier reports. Hence, these parameters could be considered as empirical constants to fit the data to the Jones–Dole model.

The variations in the value of the A and B coefficient with the change in the acid concentration or the temperature of the solutions showed a high degree of scattering as shown in Figures 10.12 and 10.13. Nevertheless, it is possible to propose that the A-coefficient has a positive value of 1.5–2.0 $M^{-1/2}$ within the temperature and acid concentration ranges covered in this study. The B-coefficient seems to have a negative value between −2.8 and −3.8 M^{-1}. Although the data show a high degree of scattering, the B-coefficient curves seem to show a maximum at a temperature between 20°C and 25°C followed by a slow decrease. The positive A-values indicate strong solute–solute interaction in the solutions [19].

The negative B-coefficients indicate that the addition of $V_2(SO_4)_3$ has a 'structure-breaking' effect. The terms 'structure making' and 'structure breaking' are usually used to refer to the effect of the solute on the degree of 'structure or lattice' order of the water as discussed before. Hence, it is possible to extend

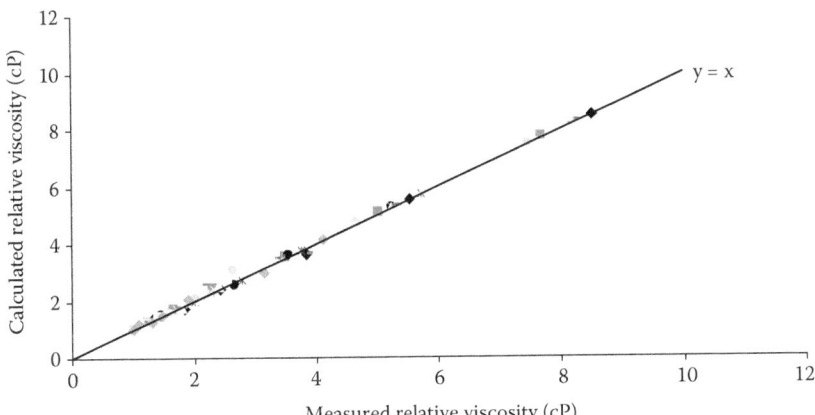

FIGURE 10.11
Correlation between measured viscosity values and those calculated from the modified Jones–Dole model (Equation 10.5).

these terms to the effect of the addition of vanadium sulphate on the degree of order in the sulphuric acid–water mixtures. Accordingly, the B-negative values might indicate that the addition of vanadium (III) sulphate (mainly vanadium (III) ions as a new component of the solution) results in breaking the acid–water structure of the solution. This effect could be attributed to two mechanisms: (i) the addition of vanadium sulphate increases the degree of saturation in the solution, resulting in a decrease in the ratio of free water

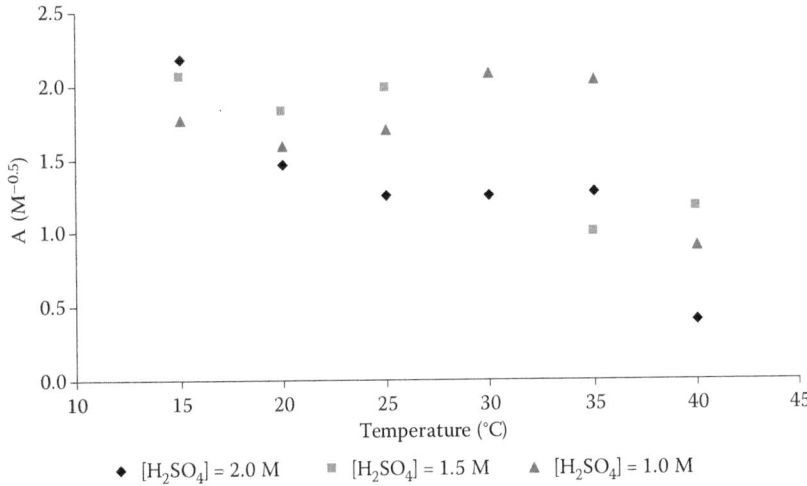

FIGURE 10.12
Variation of the A-coefficient value with the increase in the solution temperature at different acid concentrations.

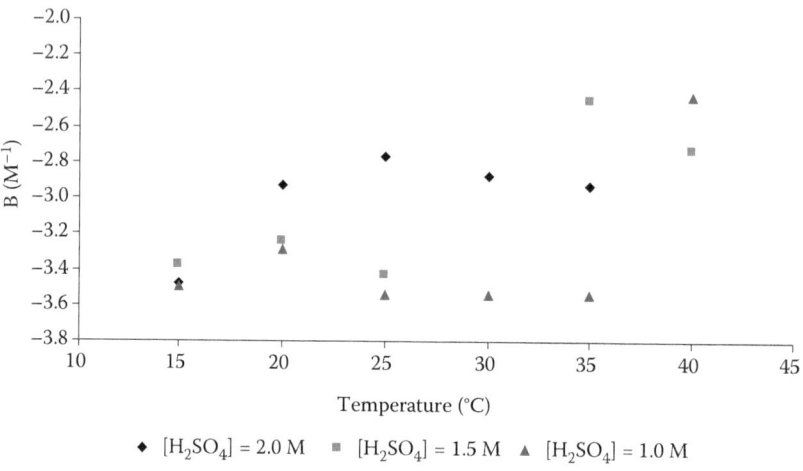

FIGURE 10.13
Variation of the B-coefficient value with the increase in the solution temperature at different acid concentrations.

molecules to the charged ions in the solutions. This results in the disruption of the hydration spheres around the ions and hence less order in the solution; (ii) this behaviour could be attributed to ion pairing between the vanadium (III) ions and the sulphate–bisulphate ions in the solution.

The variation in the B-value with the increase in solution temperature could be predicted from the first mechanism. Increasing the solution temperature should reduce the number of iceberg-like structures or the hydration spheres in the solution. Hence, this should reduce the effects of the salt on the solution order resulting in lower B-values at higher temperatures. Accordingly, the B-value should decrease with the increase in the temperature.

The effect of temperature on the degree of complex formation in the solution was not investigated in this study. Hence it is not possible to predict the B-coefficient value behaviour with the increase in temperature according to the second mechanism. The presence of the maxima in the B-value–temperature curves might indicate a competition between these two mechanisms. Nevertheless, it is not safe to make these assumptions due to the significant errors in these values.

10.2.2.5 Empirical Model for Vanadium (III) Sulphate Solutions

As mentioned in the previous section, fitting the experimental data obtained in this study to an empirical formula is of significant practical value for this study. Such a formula will allow calculation of the viscosity of the solution at different vanadium and acid concentrations and temperature. Hence, several models that are reported in the literature [9–11, 13] were tested before arriv-

TABLE 10.8
Values of the Empirical Coefficients for Equations 10.6 and 10.7

	A/A'	B/B'	C/C'	D/D'	E/E'	F/F'	G/G'
Equation 10.6	1.33246	6.65122	0.19232	0.00010	0.48949	−0.03741	−0.01120
Equation 10.7	−0.00399	0.23830	1.40929	0.05388	0.07567	−0.00050	−0.03525

ing at the model shown in Equation 10.6. This model was found to fit the data with an average relative error of 6.5% ($R^2 = 0.9804$).

The data were fitted to a second formula that is based on the total vanadium $[V_T]$ and total sulphate $[SO_4^{-2}]$ concentrations as shown in Equation 10.7. The average relative error was around 6.6% ($R^2 = 0.9842$). The calculated coefficients for both Equations 10.6 and 10.7 are shown in Table 10.8. The suitability of this model was confirmed by plotting the calculated conductivity values against the measured values as shown in Figure 10.14.

$$\eta = A + B \times [V_2(SO_4)_3]^{2.32} + C' \times [H_2SO_4]^{2.12} + D' \times T^{2.05} + E' \times [V_2(SO_4)_3]^{0.59}$$
$$\times [H_2SO_4]^{1.40} + F' \times [V_2(SO_4)_3]^{1.36} \times T^{1.83} + G'[H_2SO_4]^{1.90} \times T^{0.89}$$
(10.6)

$$\eta = A' + B' \times [V_T]^{3.13} + C' \times [SO_4^{-2}]_T^{0.61} + D' \times T^{0.73} + E' \times [V_T]^{0.74}$$
$$\times [SO_4^{2-}]_T^{2.05} + F' \times [V_T]^{2.02} \times T^{1.99} + G'[SO_4^{2-}]_T^{0.75} \times T^{0.89}$$
(10.7)

where η is the measured viscosity of the solution in cP (centi-Poise), $[V_2(SO_4)_3]$ and $[H_2SO_4]$ are the molar concentration (M) of vanadium (III) sulphate and sulphuric acid, respectively; $[V_T]$ and $[SO_4^{2-}]_T$ are the molar concentration (M) of total vanadium ions and sulphate ions in the solution, T is the temperature in °C and A, A', B, B', C, C', D, D', E, E', F and F' are empirical coefficients.

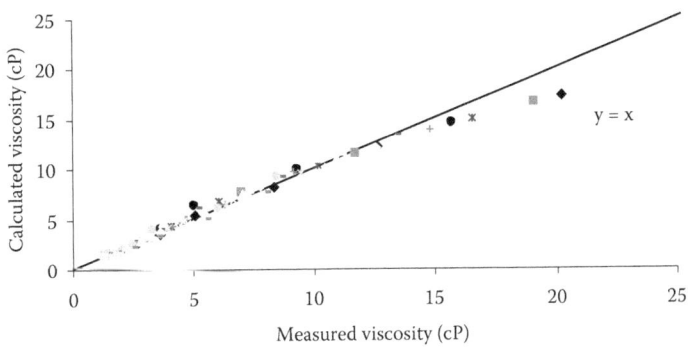

FIGURE 10.14
Correlation between measured viscosity values and those calculated from the empirical formula in Equation 10.7 and coefficients from Table 10.8.

The error in the calculated viscosity values is relatively high (average error = 6.6%, $R^2 = 0.9842$) compared to that in the density data (average error = 0.2%, $R^2 = 0.9947$). This could be due to the high degree of scattering in the data. The high error was reflected by the calculated values of the A and B coefficients obtained from the modified Jones–Dole formula. The same also applies on the variation in the activation energy values with the increase in the vanadium sulphate and the sulphuric acid concentrations as described earlier. It is also worth mentioning that the accuracy of the fitting is sensitive to the power of the different parameters. It was found by a trial-and-error approach that rounding the power of these coefficients to less than two digits results in increasing the error in the fitting by several fold.

10.2.3 Conductivity of Vanadium (III) Sulphate Solutions

The specific conductivity of vanadium (III) sulphate solutions was measured using a conductivity cell with a known cell constant value. The tested solutions used in this study were the same as those used in the density and viscosity studies. The variation in the conductivity of these solutions with the increase in the $V_2(SO_4)_3$ and sulphuric acid concentration and the solution's temperature was investigated in this study. The measured conductivity values are shown in Table 10.9.

TABLE 10.9

Measured Conductivity Values (mS cm^{-1}) for Vanadium Solutions at Different Acid Concentrations and Temperatures

[H$_2$SO$_4$] (M)	Temperature (°C)	[V$_2$(SO$_4$)$_4$] (M)						
		0	0.3	0.5	0.8	1.0	1.3	1.5
2.0	15	500	389	295	217	156	111	76
	20	540	424	324	241	173	122	79
	25	577	458	354	266	193	135	84
	30	611	489	381	289	211	149	92
	35	643	518	407	311	230	164	102
	40	672	545	433	334	250	180	113
1.5	15	429	342	275	198	138	90	50
	20	456	370	301	222	153	104	56
	25	483	398	325	247	169	118	63
	30	509	425	347	270	184	132	70
	35	535	449	366	291	199	144	78
	40	560	471	385	312	214	154	87
1.0	15	323	264	225	158	107	63	21
	20	339	284	244	179	119	78	29
	25	356	305	261	200	130	91	35
	30	374	325	277	220	141	103	41
	35	394	344	291	238	152	112	47
	40	415	361	303	256	163	118	52

All measurements were carried out in beakers of the same dimensions to minimise errors between measurements. Solutions were allowed to reach thermal equilibrium with the conductivity cell immersed in it to reduce errors due to temperature effects.

10.2.3.1 Effect of Vanadium (III) Sulphate Concentration

Figure 10.15 shows the measured specific conductivity values as a function of $V_2(SO_4)_3$ concentration in 2.0 M sulphuric acid solution at different temperatures. This figure shows a significant drop in the solution conductivity with the increase in vanadium (III) sulphate content of the solution. Similar behaviour was also observed for vanadium solutions in 1.5 and 1.0 M sulphuric acid solutions.

Two factors might be considered to explain the drop in conductivity of these solutions. Firstly, the increase in their viscosities due to the increase in their content of vanadium (III) sulphate as mentioned in the previous section. This increase results in a reduction in the mobility of ions, which is reflected by their lower ionic conductance.

Secondly, the drop in the conductivity might also be explained by changes in the chemical composition of the solution. The direction and extent of this effect depends on the behaviour of vanadium (III) sulphate in the aqueous solution. If the salt is assumed to behave as a strong electrolyte and to fully dissociate in the solution, it will increase the concentration of sulphate ions in the solution. Although the addition of vanadium (III) sulphate will result in an increase in the total ionic concentration in the solution, it will also shift

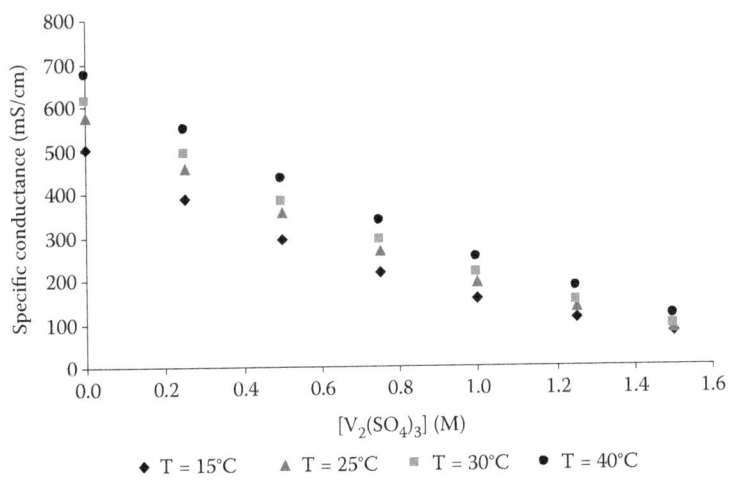

FIGURE 10.15
Measured conductivity values for vanadium (III) sulphate solutions at different temperatures. $[H_2SO_4] = 2.0$ M.

the bisulphate/sulphate equilibrium to the left due to the increase in the sulphate ion concentration. This shift will cause a reduction in the proton concentration according to Equation 10.8. The reduction in the proton concentration is expected to have a more significant effect due to the major role that the hydronium ions play in the ionic conductivity of aqueous solutions. Hence, the overall effect will be a drop in the solution conductance.

$$HSO_4^{-1} \rightleftharpoons SO_4^{-2} + H^+ \quad (10.8)$$

Several studies on the conductivity of concentrated electrolytes reported the drastic effects of the viscosity on the solution conductivity and a correction factor was suggested by many workers to account for this effect [20–23]. The correction factor is calculated as the ratio of the viscosity of the electrolyte (η_{sol}) to that of the solvent (η_o), which is always greater than 1 for concentrated solutions as shown in Equation 10.9 [20]. As discussed before, it is possible to consider the sulphuric acid–water mixture ($[V_2(SO_4)_4] = 0.0$ M) as the solvent.

$$\Lambda_{corr.} = \Lambda_{meas.} \eta_{sol.} / \eta_o \quad (10.9)$$

where $\Lambda_{meas.}$ is the measured conductivity value, $\Lambda_{corr.}$ is the corrected value and η_o and $\eta_{sol.}$ are the viscosity of the solvent and the solution, respectively.

Figure 10.16 shows the corrected specific conductance values for vanadium sulphate solution in 2.0 M sulphuric acid at different temperatures. The corrected values show a wave-like curve; however, the shape of the curve seems to differ between different temperatures. The degree of scattering in the data seems to be higher at the higher concentration range. This is most likely due to the abnormality in the viscosity behaviour especially at higher concentrations.

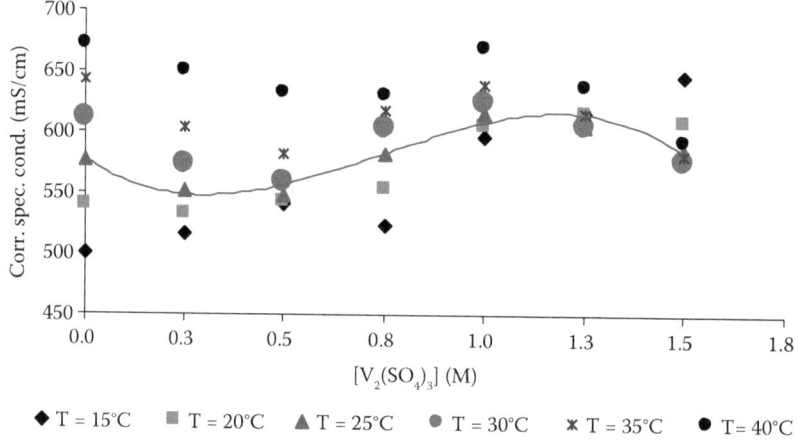

FIGURE 10.16
Conductivity values for vanadium (III) sulphate solutions at different temperatures after viscosity correction. [H_2SO_4] = 2.0 M.

Except for the curve at 15°C, the change in the corrected conductivity of the solutions follows the wave shape. A slight drop in the conductivity is initially observed at concentrations below 0.8 M, which is followed by an increase in the conductance to a maximum around 1.3 M. Similar behaviour was reported for a number of ternary systems MX–HX–H$_2$O, where MX is the salt and HX is the acid. Zinc sulphate–sulphuric acid–water mixture in particular was reported to show similar curves when the concentration of the salt was increased at constant acid concentration. The measured conductivity values were correlated to the concentration of different species in the solution. The calculation of different species concentrations was based on a set of equilibria with known equilibrium constant values. Unfortunately, it was not possible to carry out similar calculations on vanadium sulphate solutions due to the lack of values of equilibrium constants.

Nevertheless, it should be noted that the changes in the corrected value of conductivity with the increase in the V$_2$(SO$_4$)$_3$ concentration is small relative to that of the uncorrected values. The set of vanadium solutions in the 2.0 M acid at 25°C, for example, drops by less than 5% at 0.8 M and increases by about 7% at 1.3 M. These values are much less than the 85% drop in the uncorrected values. Hence, it is possible to assume that these variations might be within the 7% experimental error range and that the conductivity of sulphuric acid is not significantly affected by the addition of the V$_2$(SO$_4$)$_3$.

Another approach was considered to verify the behaviour of vanadium (III) sulphate in the sulphuric acid solution. In this approach, the density, viscosity and conductivity of several glycerol–sulphuric acid solutions in 2.0 M sulphuric acid were measured at 25°C as shown in Table 10.10. The variation of the conductivity values of these solutions with the increase in their viscosity was compared to that of vanadium (III) sulphate solutions in 2.0 M sulphuric acid at 25°C. The measured conductivity values for each of the solutions in the two sets were plotted against the corresponding viscosity values as shown in Figure 10.17.

Glycerol is a non-electrolyte that forms strong hydrogen bonds with water, causing a significant increase in its viscosity. It has been reported in many studies as an ideal additive to increase the viscosity of strong electrolytes—including sulphuric acid solutions—without affecting their ionic composition to test the validity of the Walden rule.

Figure 10.17 shows that the conductivity–viscosity curves for both vanadium (diamond markers) and glycerol solutions (triangular markers) are superimposed. This was further confirmed by fitting the data to an exponential equation (Equation 10.10), which is also plotted in Figure 10.17 (solid curve). Although the formula has no physical meaning, this figure shows that both sets of solutions are following the same conductivity–viscosity pattern.

$$C = A \times \eta^b \qquad (10.10)$$

where A and b are constants and their values are determined experimentally.

TABLE 10.10

Density, Viscosity and Conductivity Values of Glycerol–Sulphuric Acid Solutions

Temperature (°C)	%Glycerol (vol/vol)	Density (g cm^{-3})	Viscosity (cP)	Conductivity (mS cm^{-1})
15	0	1.1212	1.7	501.8
	10	1.1475	2.5	404.7
	20	1.1725	4.2	294.1
	30	1.1915	4.2	241.4
	40	1.2040	5.7	178.2
	50	1.2404	11.2	92.8
25	0	1.1156	1.5	577.0
	10	1.1418	1.9	454.4
	20	1.1669	2.5	445.6
	30	1.1854	4.3	272.2
	40	1.1996	4.2	216.2
	50	1.2446	8.5	116.5
40	0	1.1074	1.2	672.2
	10	1.1440	1.6	544.0
	20	1.1579	1.9	415.2
	30	1.1766	2.4	446.2
	40	1.1904	4.0	274.0
	50	1.2259	5.2	156.0

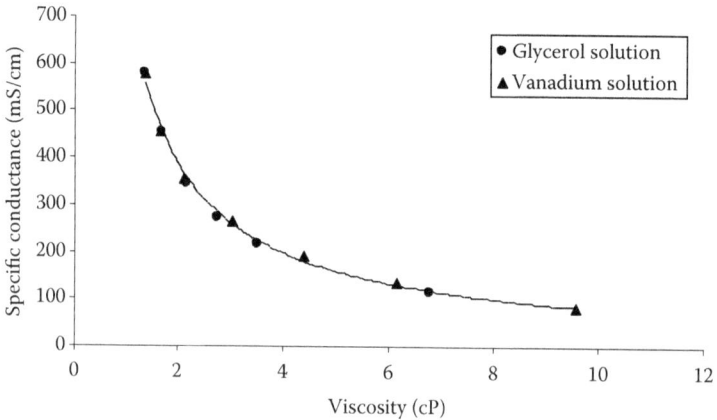

FIGURE 10.17

Conductivity–viscosity curve for vanadium and glycerol solutions in 2.0 M sulphuric acid at 25°C.

Physical Properties of Negative Half-Cell Electrolytes

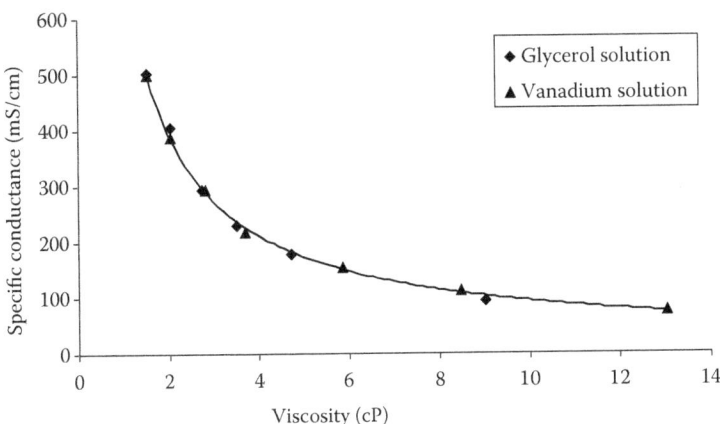

FIGURE 10.18
Conductivity–viscosity curve for vanadium and glycerol solutions in 2.0 M sulphuric acid at 15°C.

The conductivity–viscosity curves were also obtained for vanadium sulphate and glycerol solutions at 15°C and 40°C to confirm the validity of this assumption over the whole temperature range. Similar behaviour is observed as shown in Figures 10.18 and 10.19.

Hence, it could be assumed that vanadium (III) sulphate behaves mainly as a non-electrolyte in sulphuric acid solutions. It might undergo dissociation to a limited extent—as indicated by the slight variation in the conductivity

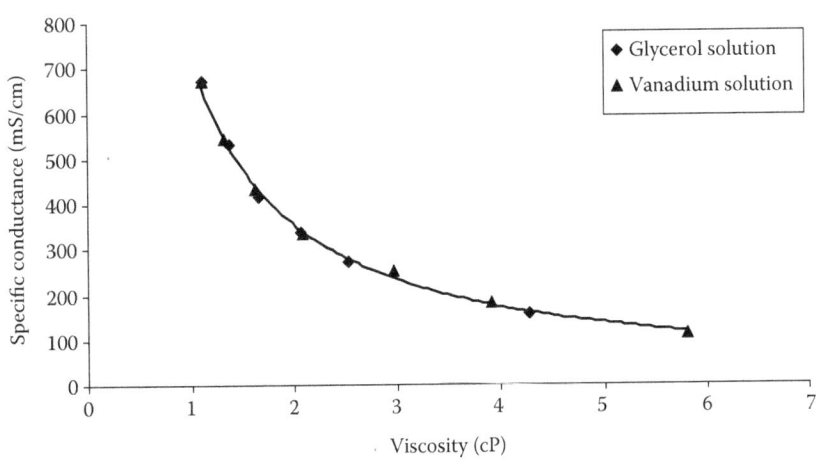

FIGURE 10.19
Conductivity–viscosity curve for vanadium and glycerol solutions in 2.0 M sulphuric acid at 40°C.

TABLE 10.11
Walden Product for Different Vanadium (III) Sulphate Solutions in 2.0 M Sulphuric Acid at Different Temperatures

[V$_2$(SO$_4$)$_4$] (M)	Temperature (°C)		
	15	25	45
0	766.2	794.0	742.9
0.3	790.8	759.5	718.2
0.5	827.8	755.4	698.0
0.8	801.2	802.8	697.0
1.0	912.8	849.1	749.4
1.3	944.7	844.2	705.4
1.5	991.1	807.6	654.4
Average	861.9	800.2	707.9
STD%	9.4	4.0	4.9

with the increase in the salt concentration—but most of the solute appears to remain in molecular form or as ion pairs in these concentrated solutions.

It is also possible to confirm the above conclusion by examining the applicability of the Walden rule to the experimental data. According to this rule, the product of the conductivity of an electrolyte and its viscosity should be constant (known as the Walden constant). The main condition for this rule to be applicable is that the number of ions in the electrolyte should remain constant.

Table 10.11 shows the Walden constant for both vanadium and glycerol solutions in 2.0 M sulphuric acid solution at 15°C, 25°C and 40°C. The average value and the standard deviation of these values are shown in the table as well.

10.2.3.2 Effect of Sulphuric Acid Concentration

The conductivity of 0–1.5 M V$_2$(SO$_4$)$_3$ solutions (equivalent to 0–3 M V(III) ions) in 1.0 and 1.5 M sulphuric acid was measured at different temperatures. The variation of the solution conductivity with the increase in the salt concentration in these solutions showed a similar behaviour to that observed for vanadium solutions in 2.0 M acid. Figure 10.20 shows the conductivity–vanadium (III) sulphate concentration curves for vanadium (III) sulphate solutions at 25°C.

Overall, a decrease in the solution conductivity is observed with an increase in the vanadium salt concentration. When comparing between solutions with different sulphuric acid concentrations, the difference in conductivity decreases as the vanadium (III) salt concentration is increased. The initial conductivity values ([V$_2$(SO$_4$)$_3$] = 0 M) reflect the difference in the concentration of the hydronium ions in the solutions; hence, higher conductivity

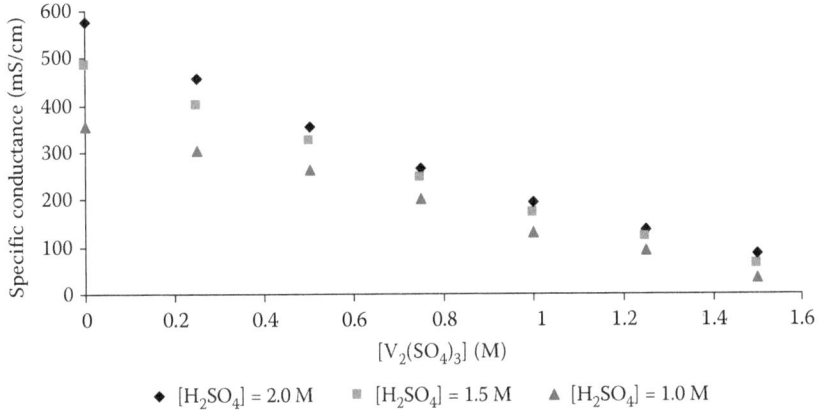

FIGURE 10.20
Variation in the conductivity of vanadium (III) sulphate solutions with the increase in the salt concentration in different acid solutions at 25°C.

is observed for the solution with 2.0 M acid while the lowest is for the solution containing 1.0 M acid. The decrease in the difference between the three sets indicates the dominant effect of the viscosity on solution conductivity.

Viscosity corrections were again applied to the measured conductivity values according to Equation 10.9. The corrected values were plotted against the vanadium (III) sulphate concentration as shown in Figure 10.21. This figure shows that compensating for the viscosity effect unmasks the differences in the conductivity values between the three sets. These curves support the argument that the differences in the conductivity of the solutions in each set are insignificant and lie within the limits of experimental errors.

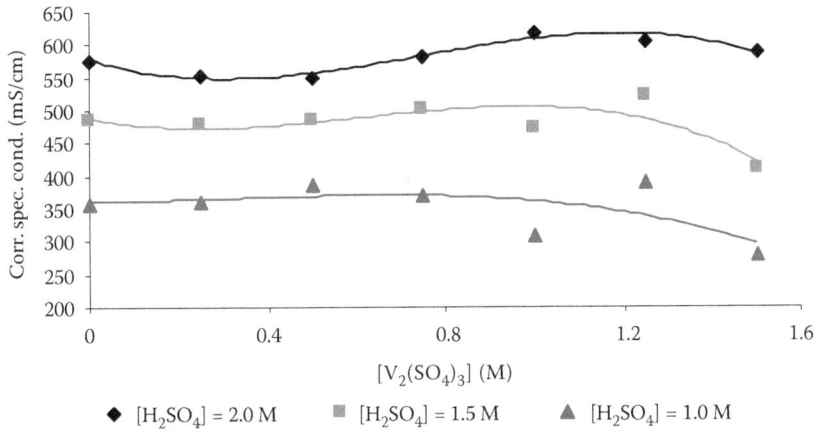

FIGURE 10.21
Variation in the corrected conductivity of vanadium (III) sulphate solutions with the increase in salt concentration in different acid solutions at 25°C.

10.2.3.3 Temperature Effects

Figure 10.22 shows a linear increase in the measured conductivity of vanadium (III) sulphate solution in 2.0 M sulphuric acid with the increase in temperature. The effect of the temperature (as indicated by the slope of the line) decreases with the increase in the vanadium (III) sulphate concentration.

Figure 10.23 shows the variation in the corrected conductivity values with the increase in the temperature. This behaviour reflects the compound nature of the temperature effect on the conductivity of the solutions. Increasing the temperature should result in an increase in the mobility of ions in the solution (due to the decrease in the viscosity of the solution) and, hence, higher ionic conductivity as observed for most of the solutions. Nevertheless, the 1.5 M vanadium (III) sulphate solution seems to show an initial decrease in conductivity as the temperature is increased. This behaviour might be attributed to a decrease in the hydrogen ion concentration caused by changes in the composition of the solution.

Temperature has a significant effect on the sulphate/bisulphate equilibrium and on the concentration of the hydronium ions in the solutions as well. However, the lack of information on the temperature effects on the formation of vanadium–sulphate ion pairs makes it difficult to explain the effect of the temperature on the solution conductivity.

10.2.3.4 Modelling the Conductivity Data

Attempts were made to find an empirical model similar to that obtained for the density and viscosity values. Such a model will allow calculation of the conductivity of vanadium solutions at different vanadium (III) and

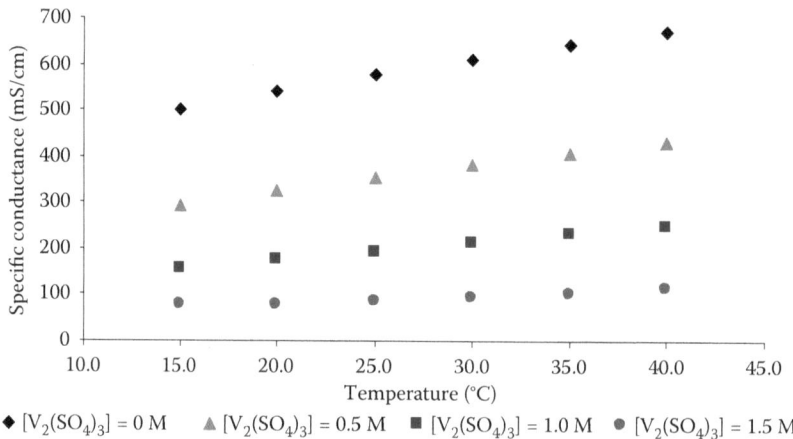

FIGURE 10.22
Conductivity of vanadium (III) sulphate solutions at different temperatures for different $V_2(SO_4)_3$ concentrations in 2.0 M sulphuric acid solution.

Physical Properties of Negative Half-Cell Electrolytes

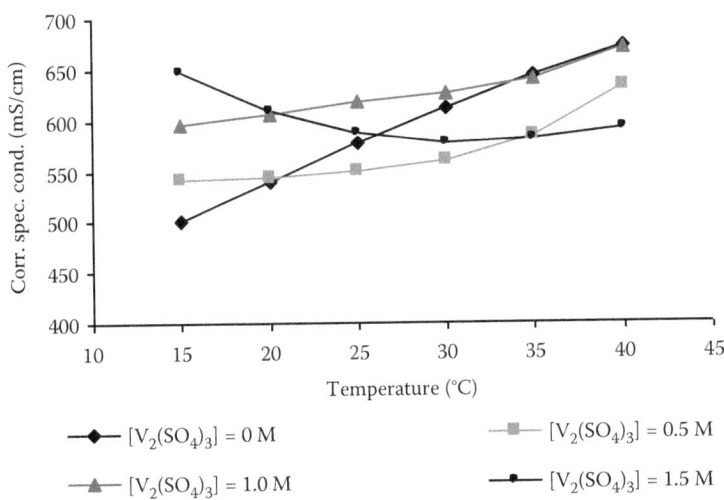

FIGURE 10.23
Corrected conductivity values of vanadium (III) sulphate solutions at different temperatures for different V$_2$(SO$_4$)$_3$ concentrations in 2.0 M sulphuric acid solution.

total sulphate concentrations and temperatures. Several models that were reported in the literature for similar systems were tested [10,11,13,24,25]. A model similar to those used for the viscosity data in Equations 10.6 and 10.7 was found to be suitable, as shown in Equations 10.11 and 10.12. The values of the calculated coefficients are shown in Table 10.12. The data were also fitted to a similar model using the total vanadium and total sulphate concentrations. The coefficients are shown in Table 10.12 as well.

Figure 10.24 shows the plot of the calculated values against the measured ones and indicates that the fitting is acceptable (average error = 9%, $R^2 = 0.9939$).

$$\kappa = A + B \times [V_2(SO_4)_4]^{1.06} + C \times [H_2SO_4]_T + D \times T + E \times [V_2(SO_4)_4]^{0.54}$$
$$\times [H_2SO_4]_T 1.91 + F \times [V_2(SO_4)_4]^{1.06} \times T^{1.87} + G[H_2SO_4]_T^{1.44} \times T^{1.29} \quad (10.11)$$

where κ is the solution's conductivity in mS cm^{-1}, [V$_2$(SO$_4$)$_4$] and [H$_2$SO$_4$]$_T$ are the concentration of vanadium (III) sulphate and free sulphuric acid,

TABLE 10.12
Values of the Empirical Coefficients for Equations 10.11 and 10.12

	A/A′	B/B′	C/C′	D/D′	E/E′	F/F′	G/G′
Equation 10.11	69.9544	−146.5425	182.8914	4.1145	−54.5444	−0.0687	0.2555
Equation 10.12	−484.752	−490.240	748.019	2.180	0.004415	0.000046	−0.000197

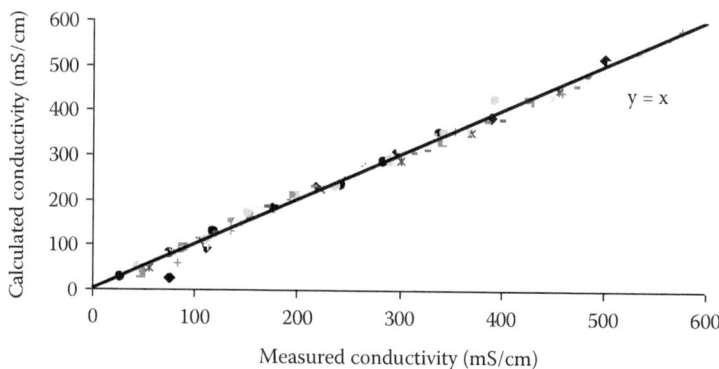

FIGURE 10.24
Correlation between measured conductivity values and those calculated from the empirical formula in Equation 10.12 and coefficients from Table 10.12.

respectively, T is the solution's temperature and A, B, C, D, E, F and G are empirical coefficients.

$$\kappa = A' + B' \times [V_T]^{0.80} + C' \times [SO_4^{-2}]_T^{0.44} + D' \times T^{1.21} + E' \times [V_T]^{4.24} \times [SO_4^{2}]_T^{4.14}$$
$$+ F' \times [V_T]^{2.74} \times T^{4.12} + G'[SO_4^{2}]_T^{2.51} \times T^{2.48}$$

(10.12)

where $[V_t]$ and $[SO_4^{-2}]_T$ are the total vanadium (III) ions and total sulphates in the solution, A', B', C', D', E', F' and G' are empirical coefficients and other parameters are as defined previously in Equation 10.11.

10.3 Physical Properties of Vanadium (II) Sulphate Solutions

Density, viscosity and conductivity of 0, 0.5, 1.0, 1.5 and 2.0 M vanadium (II) sulphate solutions in 2.0 M sulphuric acid solution were measured at 25°C. Measurements were conducted following the same procedure used for vanadium (III) solution measurements, but solutions were kept under a nitrogen blanket by maintaining a stream of nitrogen gas flowing above the solutions during the experiments or by using nitrogen bags. These efforts are believed to be less effective in the conductivity measurements compared to the density and viscosity measurements due the higher extent of exposure to the atmosphere. Hence, these results should be treated with caution.

The sensitivity of the solutions to oxidation by atmospheric oxygen limited the number of runs carried out since partial conversion of V(II) to V(III) was difficult to avoid. The data from many of the measurements had to be discarded due to partial oxidation of V(II) in some of the solutions. Attempts to maintain an inert atmosphere made the measurements more lengthy and troublesome. Hence, experiments were limited to 2.0 M sulphuric acid solutions. The temperature of the solution could not be varied as well. Increasing the solution temperature will increase the rate of solution oxidation, while decreasing the temperature reduces the stability of the solutions against precipitation, especially at high concentrations. This is a common problem in V(II) solution studies and quite often explains the lack of reproducibility of V(II) electrolyte measurements in the literature. Care was therefore taken in this study to use the data from solutions that clearly retained their distinctive violet V(II) colour during the measurement period.

Table 10.13 shows the measured density, viscosity and conductivity values for the 2.0 M vanadium (II) sulphate solution. These results show that while the density and viscosity values are less than those of vanadium (III) solutions, conductivity values are significantly higher. The difference between the properties of these solutions reflects the difference in the behaviour of vanadium (II) and vanadium (III) ions in sulphuric acid solutions.

The density of vanadium (II) sulphate solutions increases with the increase in the vanadium concentration. The extent of the increase is less than that observed for vanadium (III) sulphate solutions. Nevertheless, it should be noted that the difference between the density of vanadium (II) and vanadium (III) sulphate solutions is not significant compared to the differences in the viscosity and conductivity values. This reflects the insensitivity of the density measurements to the changes in the chemical compositions of the solutions compared to the other two methods.

Figure 10.25 shows the change in viscosity of vanadium (II) and vanadium (III) solutions with the increase in the vanadium ion concentration. This figure shows a distinct difference between the behaviour of the two solutions. This difference indicates a lesser degree of ion pairing in vanadium (II) sulphate solutions compared to vanadium (III) ones.

The differences in the behaviour of the two solutions is further confirmed by the change in the conductivity of each solution with the increase in the vanadium ion concentration as shown in Figure 10.26 This figure shows that

TABLE 10.13

Density, Viscosity and Conductivity of Vanadium (II) Sulphate Solutions in 2.0 M Sulphuric Acid at 25°C

[VSO$_4$] (M)	Density (g mL^{-1})	Viscosity (cP)	Conductivity (mS)
0.0	1.1072	1.4	565
1.0	1.2195	1.7	414
1.5	1.2765	2.0	350
2.0	1.3349	2.4	290

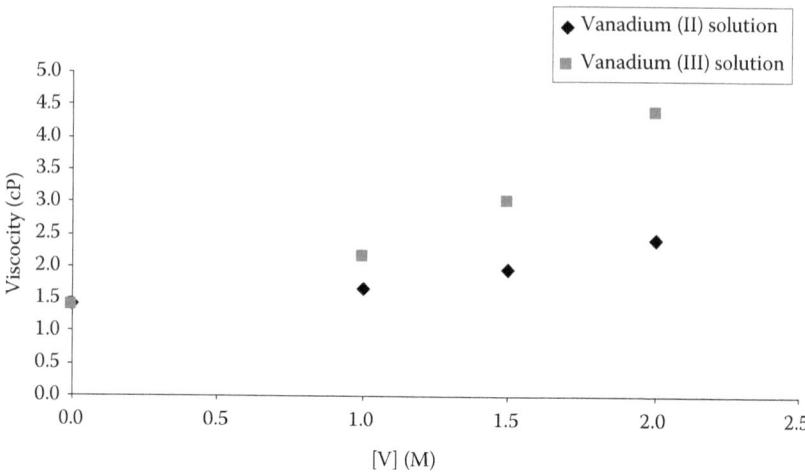

FIGURE 10.25
Viscosity of vanadium (II) and vanadium (III) sulphate solutions in 2.0 M sulphuric acid at 25°C.

vanadium (II) sulphate solutions are more conductive than vanadium (III) sulphate solutions at the same vanadium ion concentration. This behaviour confirms the significance of the viscosity effect on the conductivity of the solutions. The fact that the conductivity of vanadium (II) solution is higher than that of vanadium (III) solution despite the higher number of moles of sulphate ions in the latter further proves the significance of the viscosity effect.

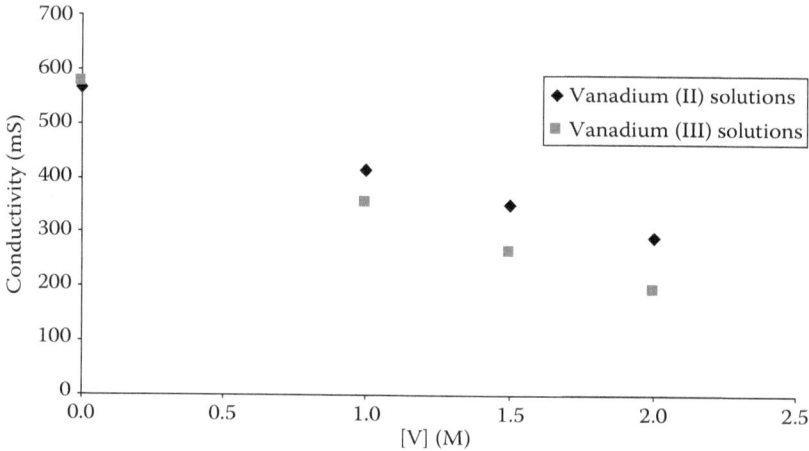

FIGURE 10.26
Conductivity of vanadium (II) and vanadium (III) sulphate solutions in 2.0 M sulphuric acid at 25°C.

10.4 Summary

The properties of the negative half-cell solutions are critical in the design and optimisation of the vanadium redox battery, especially for low-temperature operation where the solubilities of V(II) and V(III) sulphates will influence the maximum practical vanadium ion concentration in the VRB electrolyte. The selected composition of the VRB electrolyte will determine the solution properties such as density, viscosity and conductivity and these in turn are also influenced by temperature.

Accurate knowledge of solution density is important in calculating theoretical energy density values of flow battery electrolytes and for pump selection during battery system design. The results obtained in this study showed that the viscosity and conductivity of vanadium (III) sulphate solutions are more sensitive to changes in the solution's chemical composition compared to the density of the solution. The present studies showed that the density of vanadium (III) sulphate solutions is directly proportional to the concentration of vanadium (III) sulphate and to a lesser extent to the concentration of sulphuric acid. The density of the solution decreased slightly with the increase in the solution's temperature.

Viscosity data are important for pump selection and the calculation of pumping energy losses for different battery designs. The viscosity of vanadium (III) sulphate solutions increased exponentially with the increase in the salt concentration. The variation of the viscosity with vanadium (III) sulphate concentration was successfully fitted to a modified form of the Jones–Dole equation and the A and B coefficients were calculated under different conditions. The calculated values indicate strong solute–solute interaction and a structure-breaking effect. The viscosity of the solutions were also increasing with the increase in acid content but the extent of change was higher for solutions with higher content of vanadium (III) sulphate. Temperature increase results in a decrease in the solution viscosity. Activation energy of flow was calculated for different solutions and was found to increase with the increase in vanadium (III) sulphate and sulphuric acid contents.

Optimisation of electrolyte conductivity is important for reducing ohmic losses in the VRB and therefore increasing energy efficiency. This study showed that the conductivity of vanadium (III) sulphate solutions is inversely proportional to vanadium (III) sulphate and sulphuric acid in the solution, but directly proportional to the solution's temperature. These results showed that the conductivity of the solution is determined by its viscosity. Hence, factors that increase the viscosity of the solution result in a decrease in its conductivity.

The measured values of each of the physical properties (density, viscosity and conductivity) were fitted to an empirical formula that describes the property as a function of the vanadium (III) sulphate concentration, sulphuric acid concentration and temperature. The three properties showed good agreement between the measured and calculated values, indicating successful fitting.

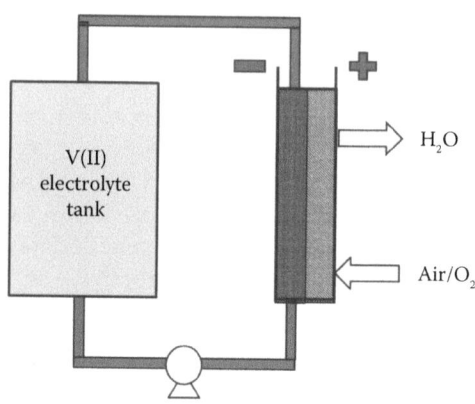

FIGURE 10.27
Schematic of the V/O$_2$ redox fuel cell.

In addition to their application in the VRFB, V(II)/V(II) electrolytes are also currently being developed for the vanadium–oxygen fuel cell system (V/O$_2$) that utilises an oxygen gas diffusion electrode for the positive half-cell reaction as illustrated in Figure 10.27 [26–30].

In this case, the cell employs a single electrolyte tank for the V(II)/V(III) redox couple solution, while the positive half-cell reactant is oxygen from the air. By eliminating the positive V(IV)/V(V) half-cell electrolyte, the specific energy of the cell is almost doubled for the same vanadium ion concentration; however, increased vanadium ion concentrations can more readily be used since the problem of the V(V) thermal precipitation reaction can be eliminated. Optimisation of the negative half-cell electrolyte composition becomes much easier therefore, so that a wider operating temperature range can also be achieved compared with the traditional all-vanadium redox flow battery. A detailed knowledge of the physical and chemical properties of the V(II) and V(III) solutions will therefore be of great value in the development of the V/O$_2$ redox fuel cell that shows great promise for electric vehicle applications [27].

References

1. M. Skyllas-Kazacos and R. Robins, All-Vanadium Redox Battery. U.S. Patent No. 4,786,567, 1986.
2. M. Skyllas-Kazacos, M.H. Chakrabarti, S.A. Hajimolana, F.S. Mjalli and D. Saleem, Progress in flow battery research and development, *J. Electrochem. Soc.*, 158, 2011, R55–R79.

3. M. Skyllas-Kazacos, G. Kazacos, G. Poon and H. Verseema, Recent advances with UNSW vanadium-based redox flow batteries, *Int. J. Energy Res. (Energy Storage Special Issue)*, 34, 2010, 182–189.
4. M. Rychcik and M. Skyllas-Kazacos, Characteristics of new all-vanadium redox flow battery, *J. Power Sources*, 22, 1988, 59–67.
5. R.L. Largent, M. Skyllas-Kazacos and J. Chieng, Improved PV system performance using vanadium batteries, In: *Proceedings of the IEEE 23rd Photovoltaic Specialists Conference*, Louisville, Kentucky, USA, May 1993.
6. Prudent Energy—Case study: VRB technology in Japan. http://www.pdenergy.com/pdfs/casestudy_japan.pdf, 2011
7. N. Tokudu, T. Kanno, T. Hara, T. Shigematsu, Y. Tsutsui, A. Ikeuchi, T. Itou and T. Kumamoto, Development of a redox flow battery system, *SEI Tech. Rev.*, 50, 2000, 88–94.
8. Z. Yang, J. Zhang, M.C.W. Kintner-Meyer, X. Lu, D. Choi, J.P. Lemmon and J. Liu, Electrochemical energy storage for green grid, *Chem. Rev.*, 111, 2011, 3577–3613.
9. P.K. Saho, D. Panda and S.K. Singh, *Indian Chem. Eng.*, XXXIII, 4, 1991, 77–79.
10. T.A. Torok and E. Berecz, Volumetric properties and electrolytic conductances of aqueous ternary mixtures of hydrogen chloride and some transition metal chlorides at 25°C, *J. Solution Chem.*, 18, 1989, 1117–1131.
11. T.A. Torok, J.A. Rard and D.G. Miller, Viscosities, electrolytic properties and volumetric properties of $HCl-MCl_x-H_2O$ as a function of temperature up to high molal ionic strengths, *Fluid Phase Equilib.*, 88, 1993, 263–275.
12. A.L. Horvath, *Handbook of Aqueous Electrolyte Solutions*, Ellis Horwood Limited, United Kingdom, 1985.
13. J. Hotlos and M. Jaskula, Densities and viscosities of $CuSO_4-H_2SO_4-H_2O$. solutions, *Hydrometallurgy*, 21, 1988, 1–7.
14. V.L. Pogrebnaya, N.P. Pronina and E.V. Kadzharova, *Zhurnal Prikladnoi Khimii*, 56, 1983, 2182–2185.
15. A. Mousa, *PhD Thesis*, University of New South Wales, Australia, 2002.
16. F. Rahman, *PhD Thesis*, University of New South Wales, Australia, 1998.
17. C.A. Angell, Free volume-entropy interpretation of the electrical conductance of aqueous electrolyte solutions in the concentration range 2–20N, *J. Phys. Chem.*, 70, 1966, 3988–3998.
18. C.A. Angell and R.D. Bressel, Fluidity and conductance in aqueous electrolyte solutions: An approach from the glassy state and the high concentration limit. I. $Ca(NO_3)_2$ solutions, *J. Phys. Chem.*, 76, 1972, 3244–3253.
19. A.J. Easteal and C.A. Angell, Glass-forming system $ZnCl_2$ + pyridinium chloride: A detailed low-temperature analogue of the $SiO_2 + Na_2O$ system, *J. Phys. Chem.*, 74, 1970 3987–3999.
20. A. Robinson and R.H. Stokes, *Electrolyte Solutions*, 3rd edn., London: Butterworth, 1970.
21. S.S. Islam, R.L. Gupta and K. Ismail, Extension of the Falkenhagen-Leist-Kelbg equation to the electrical conductance of concentrated aqueous electrolytes, *J. Chem. Eng. Data*, 36, 1991, 102–104.
22. M. Della Monica, Conductance equation for concentrated electrolyte solution, *Electrochim. Acta*, 29, 1984, 159–160.
23. M. Della Monica, A. Ceglie and A. Agostiano, Physico-chemical properties of NH_4I-formamide concentrated solutions, *Electrochim. Acta*, 29, 1984, 933–937.

24. M. Della Monica, A. Ceglie and A. Agostiano, Extension of the Falkenhagen equation to the conductivity of concentrated electrolyte solutions, *J. Phys. Chem.*, 88, 1984, 2124–2127.
25. J.T. Hinatsu, V.D. Tran and F.R. Foulkes, Electrical conductivities of aqueous $ZnSO_4$–H_2SO_4 solutions, *J. Appl. Electrochem.*, 22, 1992, 215–223.
26. E.M. Kartzmark, Densities, viscosities, and conductances of saturated solutions in the systems mercuric chloride-hydrogen chloride-water, mercuric chloride-potassium chloride-water, and mercuric chloride-indium chloride-water at 25°C, *J. Chem. Eng. Data*, 27, 1982, 38–41.
27. H. Haneko, N. Akira, N. Ken, S. Kanji and N. Masato, Redox Battery, U.S. Patent No. US5318865, 1994.
28. M. Skyllas-Kazacos and C. Menictas, Performance of vanadium-oxygen redox fuel cell, *J. Appl. Electrochem.*, 41, 2011, 1223–1232.
29. J. Noack, C. Cremers, K. Pinkwart and J. Tuebke, Air breathing vanadium/oxygen fuel cell, *218th ECS Meeting*, Las Vegas, Nevada, USA, October 10–15, 2010.
30. S.S. Hosseiny, M. Saakes and M. Wessling, A polyelectrolyte membrane-based vanadium/air redox flow cell. *Electrochem. Commun*, 13, 2011, 751–754.

11

pH Differential Power Sources with Electrochemical Neutralisation

Huanqiao Li, Chi-Ying Vanessa Li, Guo-Ming Weng and Kwong-Yu Chan

CONTENTS

11.1 Electrochemical Window of Power Sources with *pH*-Dependent Half-Cells 430
11.2 Examples of Electrochemical Neutralisation..432
11.3 Bipolar Membrane 438
11.4 Acid–Alkaline Hybrid Fuel Cells 442
11.5 Acid–Alkaline Microbial Fuel Cells 447
11.6 Acid–Alkaline Hybrid Fuel Cell Reactor for H_2O_2 Production 449
11.7 Hybrid PbO_2/MH_x Secondary Battery 450
11.8 MH-VRF Semi-Flow Battery...452
11.9 Alternative Separator of Acid and Alkaline Electrolyte 456
    11.9.1 CEM and AEM Combined with a Salt Compartment as the Separators for Acid–Alkaline Hybrid Cell....................456
    11.9.2 Gel Electrolyte 460
    11.9.3 Lithium Ceramic Super-Ionic Conductor (LISICON) Film as Acid–Alkaline Separator..462
    11.9.4 Membrane-Less Interface with Parallel Laminar Flows..........463
11.10 Summary and Outlook 464
Acknowledgements 465
References..465

The cell voltage of aqueous electrochemical power sources is limited to be within 1.229 V, the thermodynamic stability window of water. This cell voltage is significantly lower than the values of cells operating in organic electrolytes. For an aqueous electrode with a half-cell reaction that is *pH* sensitive, it is possible to have a higher cell voltage by coupling the positive electrode in acid with the negative electrode in an alkaline electrolyte. The electrolytes of different *pH* in these hybrid power sources are separated by a membrane to prevent bulk neutralisation and facilitate an ionic current. This chapter introduces the working principle and the development of various acid–alkaline hybrid electrochemical power sources over the past decades. A higher output voltage can be obtained in most of these acid–alkaline hybrid power sources, while in some fuel cells better water management is achieved. The shortcomings of these hybrid power sources, particularly in ionic interfaces, are discussed.

11.1 Electrochemical Window of Power Sources with pH-Dependent Half-Cells

In Figure 11.1, electrochemical power sources are coordinated by their specific energies (horizontally) and their energy densities (vertically) [1]. It can be seen that the energy densities are much lower in most aqueous electrolytes than those operating in non-aqueous solutions, resulted from the lower cell voltage limited by the electrochemical stability window of water. From thermodynamics, this window, ΔE is 1.229 V, is determined according to the Nernst equation

$$\Delta E = \frac{\Delta G}{-nF} \qquad (11.1)$$

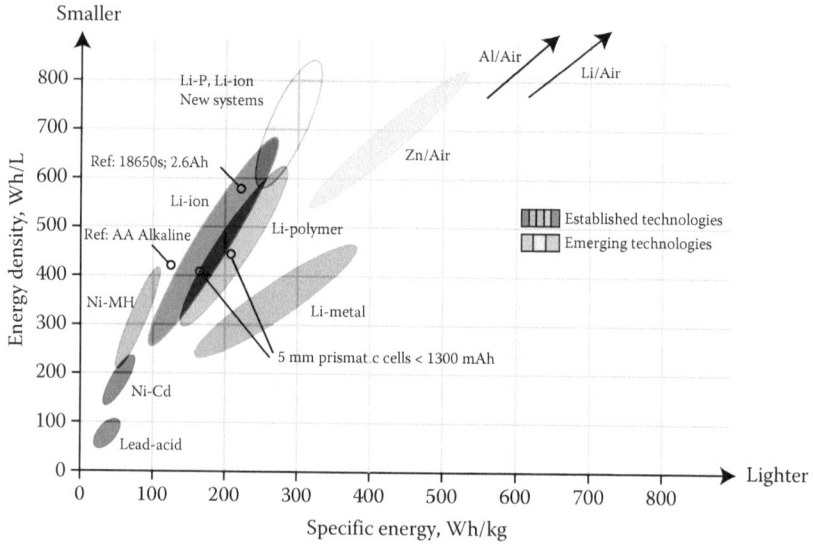

FIGURE 11.1
A comparison of energy density (volumetric energy density) and specific energy (gravimetric energy density) for different electrochemical power sources. (Reprinted from http://www.nexergy.com/battery-density.html, Accessed online 11 February 2010.)

where ΔE is the voltage at which water dissociated to hydrogen and oxygen, n is the number of electrons transferred in the redox reaction per mole of water, F is the Faraday constant, and ΔG is the corresponding free energy change of the electrochemical reaction.

The half-cell reactions of the electrodes, together with electrode potential dependence on pH are given as follows:

$$2H^+ + 2e^- \leftrightarrow H_2 \quad (11.2)$$

$$E_{H^+/H_2} = E^o_{H^+/H_2} + \frac{RT}{nF} \ln\left[\frac{a_{H^+}}{a_{H_2}}\right] = 0\text{ V} - 0.059\text{ V} \cdot (pH) \quad (11.3)$$

$$4H^+ + O_2 + 4e^- \leftrightarrow 2H_2O \quad (11.4)$$

$$E_{O_2/H_2O} = E^o_{O_2/H_2O} + \frac{RT}{nF} \ln\left[\frac{a_{O_2}}{(a_{H^+})^4(a_{H_2O})^2}\right] = 1.229\text{ V} - 0.059\text{ V} \cdot (pH) \quad (11.5)$$

The pH dependence can be illustrated in Figure 11.2, the Pourbaix diagram of water [2]. It can be seen that the thermodynamic stability window of water remains the same as the hydrogen and oxygen electrodes are shifted by the same amount when operated at the same pH.

Power sources containing aqueous electrolytes with an electromotive force (EMF) exceeding 1.229 V are not practical unless water decomposition is

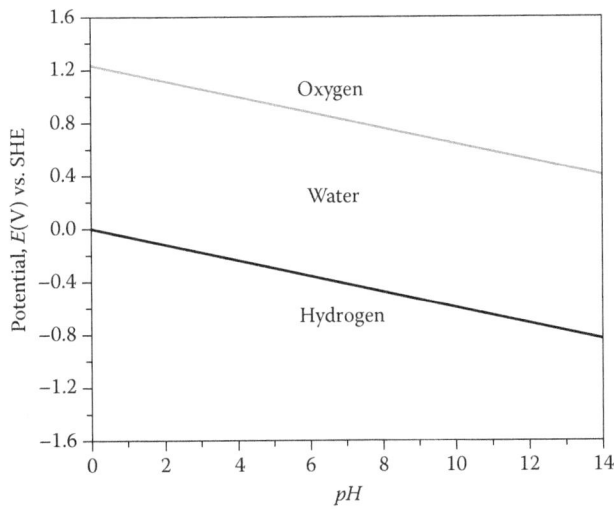

FIGURE 11.2
A Pourbaix diagram of water with curves generated by the Nernst equation (Equations 11.3 and 11.5).

prevented. Conventional batteries need to avoid oxygen or hydrogen evolution caused by overcharging. A few exceptional aqueous power sources[*] can have EMF above 1.229 V.

Pairing negative and positive electrodes in two electrolytes of different pH values can lead to a higher theoretical EMF of the cell. For example, the theoretical EMF of the hydrogen–oxygen fuel cell can be raised to 2.06 V, compared to 1.23 V in a single pH cell, by having the oxygen electrode located in acid ($pH = 0$) and hydrogen electrode in alkaline ($pH = 14$). To explore different combinations of negative and positive electrodes, the half-cell electrode reactions of some commonly used aqueous power sources are listed in Table 11.1. They are divided into pH-sensitive and pH-insensitive electrode reactions. As observed from Table 11.1 and Figure 11.3, coupling a pH-sensitive positive half-cell in acid solution with a pH-sensitive negative half-cell in alkaline solution can lead to a higher cell voltage. For example, the theoretical EMF of a hybrid cell constituted by a PbO_2-positive plate in 5 mol L^{-1} H_2SO_4 electrolyte and a MH_x negative plate in 10 mol L^{-1} KOH solution can operate at 2.80 V, which is comparable to the EMF of some non-aqueous batteries. The EMF of this acid–alkaline hybrid cell can be further increased by selecting a more positive half-cell and a more negative half-cell, like PbO_2 (5 mol L^{-1} H_2SO_4)/Zn (10 mol L^{-1} KOH) couple, where the theoretical EMF of this hybrid cell can be as high as 3.17 V [7].

11.2 Examples of Electrochemical Neutralisation

The higher voltage and additional energy of a hybrid acid–alkaline cell can be interpreted as voltage and energy from electrochemical neutralisation. The reaction of hydroxide ion and proton combining to water is

$$H_3O^+ + OH^- \leftrightarrow H_2O \qquad (11.6)$$

and has $\Delta G = -79$ kJ/mol, corresponding to a voltage of 0.828 V according to the Nernst equation. Therefore, a cell voltage of 0.828 V will be generated if the neutralisation of acid and base solution can be carried out electrochemically. In most situations, this neutralisation is carried out in the bulk

---

[*] In a lead-acid battery, lead is covered by a dense corrosion layer of electrically insulating but ionic conducting $PbSO_4$ film, across which there exist a steep potential gradient, which is the main reason that lead-negative plate exhibits a high hydrogen evolution overvoltage in H_2SO_4. Similarly, the high oxygen evolution overvoltage observed in nickel electrodes in alkaline electrolytes due to the presence of an electrically insulating but ion-conducting layer of $Ni(OH)_2$ on the surface in alkaline electrolyte is the primary reason of its higher working voltage. But if the charge voltage exceeds 2.4 V in lead-acid battery or 1.5 V in nickel metal hydride alkaline battery, 'gassing' results from water decomposition.

pH Differential Power Sources with Electrochemical Neutralisation

TABLE 11.1
Half-Cell Reactions of Common Aqueous Power Sources

Cell	Negative Electrode Reaction	Positive Electrode Reaction	Electrolyte	OCV
(a) pH Sensitive Power Source				
Alkaline				
Alkaline cell: (primary battery)	$Zn(s) + 2OH^-(aq) \rightarrow ZnO(s) + H_2O(l) + 2e^-$ $E^0 = 1.26$ V	$2MnO_2(s) + H_2O(l) + 2e^- \rightarrow Mn_2O_3(s) + 2OH^-(aq)$ $E^0 = 0.15$ V	Paste of 7 M KOH	1.43 V
Mercury button cell (primary battery)	$Zn(Hg) + 2OH^-(aq) \rightarrow ZnO(s) + H_2O(l) + 2e^-$ $E^0 = 1.26$ V	$HgO(s) + H_2O(l) + 2e^- \rightarrow Hg(l) + 2OH^-(aq)$ $E^0 = 0.098$ V	Paste of KOH	1.358 V
Silver button cell (primary battery)	$Zn(s) + 2OH^-(aq) \rightarrow ZnO(s) + H_2O(l) + 2e^-$ $E^0 = 1.26$ V	$Ag_2O(s) + H_2O(l) + 2e^- \rightarrow 2Ag(s) + 2OH^-(aq)$ $E^0 = 0.342$ V	Paste of KOH	1.602 V
Nickel–cadmium cell (NiCd) (secondary battery)	$Cd(Hg) + 2OH^-(aq) \rightarrow Cd(OH)_2 + 2e^-$ $E^0 = 0.809$ V	$NiO(OH)(s) + H_2O(l) + e^- \rightarrow Ni(OH)_2(s) + OH^-(aq)$ $E^0 = 0.45$ V	KOH	1.259 V
Alkaline fuel cell	$H_2(g) + 2OH^-(aq) \rightarrow 2H_2O(l) + 2e^-$ $E^0 = -0.828$ V	$O_2(g) + 2H_2O(l) + 4e^- \rightarrow 4OH^-(aq)$ $E^0 = 0.401$ V	KOH	1.229 V
Acid				
Lead-acid battery (or accumulator); 2.1 V (secondary battery)	$Pb(s) + HSO_4^-(aq) \rightarrow PbSO_4(s) + H^+(aq) + 2e^-$ $E^0 = 0.355$ V	$PbO_2(s) + 3H^+(aq) + HSO_4^-(aq) + 2e^- \rightarrow PbSO_4(s) + 2H_2O(l)$ $E^0 = 1.691$ V	4 M H_2SO_4	2.046 V
Proton-exchange membrane fuel cell	$H_2(g) \rightarrow 2H^+(aq) + 2e^-$ $E^0 = 0$ V	$O_2(g) + 4H^+(aq) + 4e^- \rightarrow 2H_2O(l)$ $E^0 = 1.229$ V	Proton-exchange membrane	1.229 V
Direct methanol fuel cell	$CH_3OH + H_2O \rightarrow CO_2 + 6H^+ + 6e^-$ $E^0 = 0.016$ V	$^3/_2 O_2 + 6H^+ + 6e^- \rightarrow 3H_2O$ $E^0 = 1.229$ V	Proton-exchange membrane	1.213 V
Soluble lead-acid flow battery	$Pb \rightarrow Pb^{2+} + 2e^-$ $E^0 = 0.13$ V	$Pb^{2+} + 2H_2O \rightarrow PbO_2 + 4H^+ + 2e$ $E^0 = 1.49$ V	CH_3SO_3H	1.62 V
H_2–Br_2 flow battery	$H_2 \rightarrow 2H^+ + 2e^-$ $E^0 = 0$ V	$Br_2 + 2e^- \rightarrow 2Br^-$ $E^0 = 1.09$ V	Proton-exchange membrane-HBr	1.09 V

continued

TABLE 11.1 (continued)
Half-Cell Reactions of Common Aqueous Power Sources

Cell	Anode (or Negative) Half-Cell	Cathode (or Positive Half-Cell)	Electrolyte	OCV
(b) Electrodes with Weak or No Sensitivity to pH				
Daniell cell (primary battery)	$Zn(s) \rightarrow Zn^{2+}(aq) + 2e^-$ $E^0 = 0.76$ V	$Cu^{2+}(aq) + 2e^- \rightarrow Cu(s)$ $E^0 = 0.34$ V	$ZnSO_4/CuSO_4$ (acid)	1.1 V
Leclanché or dry cell (primary battery)	$Zn(s) \rightarrow Zn^{2+}(aq) + 2e^-$ $E^0 = 0.76$ V	$NH_4^+(aq) + MnO_2(s) + e^- \rightarrow NH_3(aq) + MnO(OH)(s)$ $E^0 = 0.74$ V	Paste of $NH_4Cl + ZnCl_2$ (acid)	1.5 V
Vanadium poly-halide flow battery (1.3 V)	$V^{2+} \rightarrow V^{3+} + e^-$ $E^0 = 0.26$ V	$½Br_2 + e^- \rightarrow Br^-$ $E^0 = 1.09$ V	VCl_3-HCl/ NaBr-HCl (acid)	1.35 V
Bromine–polysulphide flow battery	$2S_2^{2-} \rightarrow S_4^{2-} + 2e^-$ $E^0 = 0.265$ V	$½Br_2 + e^- \rightarrow Br^-$ $E^0 = 1.09$ V	Na_2S_2/NaBr (acid)	1.355 V
Iron–chromium flow battery	$Cr^{2+} \rightarrow Cr^{3+} + e^-$ $E^0 = 0.42$ V	$Fe^{3+} + e^- \rightarrow Fe^{2+}$ $E^0 = 0.77$ V	HCl/HCl (acid)	1.19 V
All vanadium flow battery	$V^{2+} \rightarrow V^{3+} + e^-$ $E^0 = 0.26$ V	$VO_2^+ + 2H^+ + e^- \rightarrow VO^{2+} + H_2O$ $E^0 = 1.00$ V	H_2SO_4/H_2SO_4 (acid)	1.26 V
Zinc–bromine hybrid flow battery	$Zn(s) \rightarrow Zn^{2+}(aq) + 2e^-$ $E^0 = 0.76$ V	$Br_2 + 2e^- \rightarrow 2Br^-$ $E^0 = 1.09$ V	$ZnBr_2/ZnBr_2$ (acid)	1.85 V
Zinc–cerium hybrid flow battery	$Zn(s) \rightarrow Zn^{2+}(aq) + 2e^-$ $E^0 = 0.76$ V	$2Ce^{4+} + 2e^- \rightarrow 2Ce^{3+}$ $E^0 = 1.72$ V	CH_3SO_3H (acid)	2.48 V

pH Differential Power Sources with Electrochemical Neutralisation

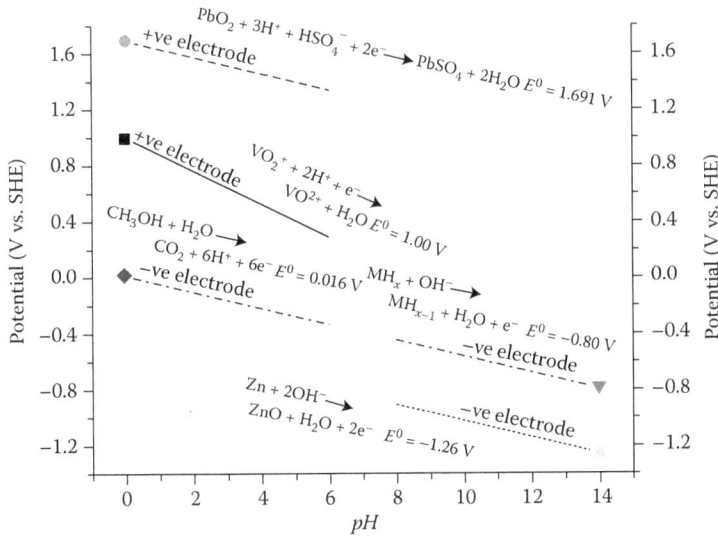

FIGURE 11.3
An electrode potential variation with electrolyte pH at standard conditions. (Constructed according to Verink, E. D. Simplified procedure for constructing Pourbaix diagrams. *Uhlig's Corrosion Handbook*, 2011: 111–124.)

solution, and energy released in the form of heat. Electrochemical neutralisation can only proceed if protons are consumed at the cathode and hydroxide ion reacted separately at the anode, accompanied by electron transfer via an external circuit. This conversion of neutralisation energy into electrical energy can be illustrated by several examples where there is no overall redox or faradic reactions.

Figure 11.4a shows two air electrodes immersed separately in alkaline and acid electrolytes. Oxygen reduction occurs in the acid electrode with consumption of protons, whereas oxygen is generated from hydroxide ion in the alkaline electrolyte. Oxygen is not consumed as a whole but transferred from one side to the other. In a way, this is an oxygen pump not needing a power supply but has additional power generated by electrochemical neutralisation. The electric circuit is completed by the transfer of electrons via an external circuit, driving a load. This is a fuel cell 'operating without a fuel'. A preliminary test shows (Figure 11.4b) that the proposed neutralisation cell has an open-circuit voltage up to 0.4 V and an operating voltage of around 0.12 V at a current density of 0.2 mA cm^{-2}. Thus, the power density (based on the active area of the electrode) of this cell is 240 mW m^{-2}. This value is much higher than those of an aerobic microbial fuel cell (MFC) (46.3 mW m^{-2}) [8] and other similar types of fuel cells [9,10].

A second example can be two electro-capacitors adsorbing protons and hydroxides separately in acid and alkaline, with the discharge process

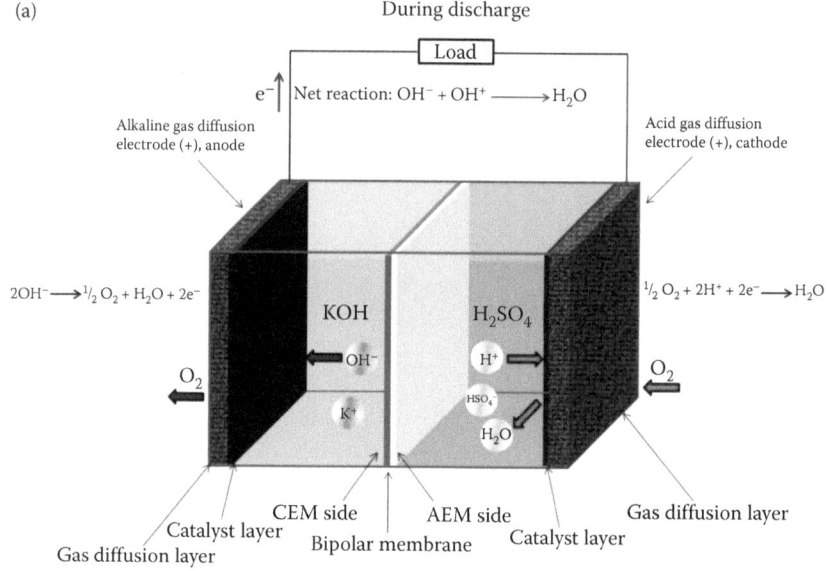

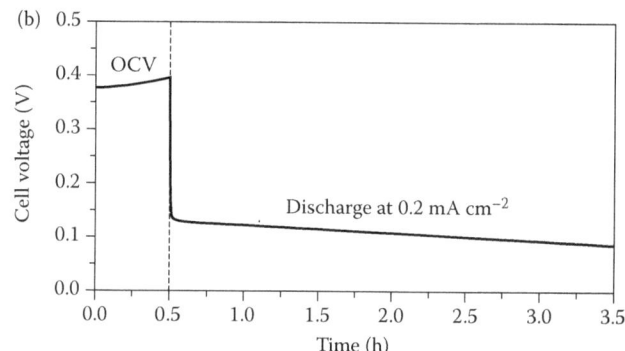

FIGURE 11.4
(a) An electrochemical neutralisation cell with oxygen transferred through two opposite electrochemical reactions at separate electrodes in alkaline and acid with a BPM. (b) Electrochemical performance of the proposed electrochemical neutralisation cell in Figure 11.4a. Acidic electrolyte is 3 mol L^{-1} H$_2$SO$_4$ and alkaline electrolyte is 3 mol L^{-1} KOH.

as illustrated in Figure 11.5a. In this setup, proton and hydroxide ions are adsorbed into porous electrodes and electrons passed via the external circuit. Ten milligram commercial Vulcan XC-72 carbon was used as active material for both positive and negative terminals (active area is 2 cm²), respectively. As shown in Figure 11.5b, the maximum voltage window of such a hybrid neutralised cell is up to 1.6 V, when using 1 mol L^{-1} H$_2$SO$_4$ and 2 mol L^{-1} KOH dual electrolytes. This voltage window is around 2 times larger than

pH Differential Power Sources with Electrochemical Neutralisation

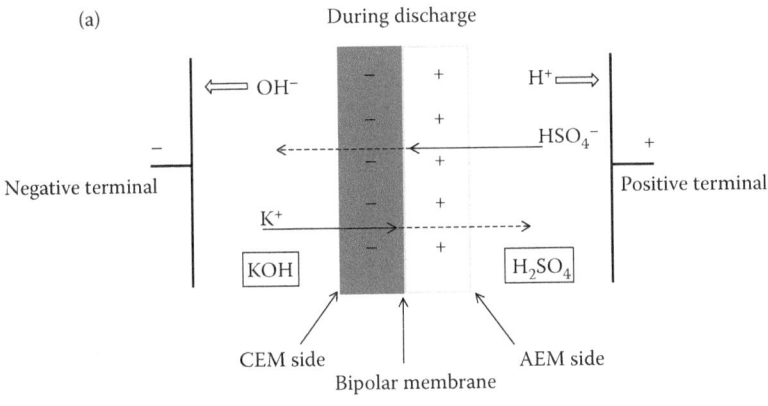

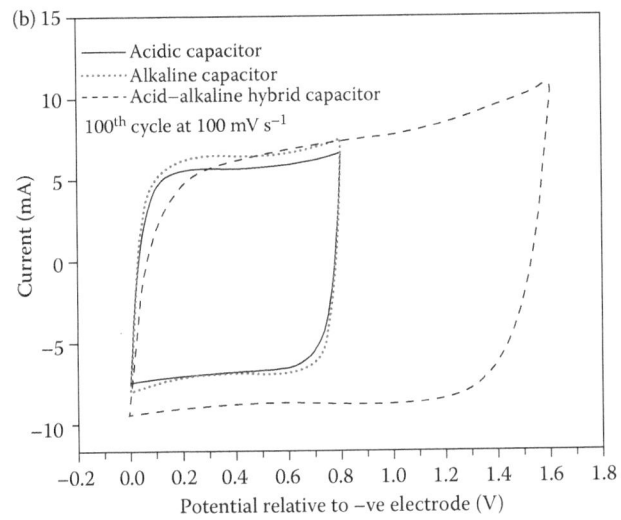

FIGURE 11.5
(a) The discharge process of an acid–alkaline hybrid neutralised electro-capacitor with a BPM as separator. (b) Typical cyclic voltammetry scanning of an acid–alkaline hybrid capacitor compared to corresponding single electrolyte acidic or alkaline capacitor.

a cell with a single electrolyte (0.8 V for acidic or alkaline capacitors) at a given scan rate, as shown in Figure 11.5b. In the charging process, salt ions move back toward the original chamber under electrical potential gradient and maintain ionic current and electroneutrality. The corresponding capacitance of acid–alkaline hybrid capacitors at a given scan rate of 10 mV s^{-1}) is 9.3 F g^{-1}, calculated by the integrated area of the cyclic voltammetry (CV) curves in Figure 11.5b and mass of the active material (20 mg for both electrodes); The corresponding specific energy for acid–alkaline hybrid capacitor is 3.31 Wh kg^{-1}, calculated using the equation, $E = \frac{1}{2}CV^2$.

Thus, this novel acid–alkaline hybrid capacitor has a larger voltage window and higher specific energy, when compared to that in a single electrolyte of either acid or alkaline. However, the maximum output power is still limited by the ionic interface/membrane that results in higher equivalent series resistance (ESR). Based on the bipolar membrane (BPM) we used, the ESR of acid–alkaline hybrid capacitor is around 4 times higher than that of either acidic or alkaline capacitor, in which no separator was used.

The application of electrochemical neutralisation has been considered by Walther [11] to utilise waste acids and bases for electricity. In the invention of Walther [11], the acid–base neutralisation voltage is superimposed to a Faraday cell of Ag/Ag_2O electrode. The theoretical cell voltage of the acid–alkaline hybrid battery can be 0.95 V, as shown by the half-cell reactions 11.7 and 11.8, and the additional acid–base neutralisation reaction 11.9.

$$Ag_2O + H_2O + 2e^- \xrightarrow{discharge} 2Ag + 2OH^- \text{ (at the cathode)}$$
$$E = 0.344 \text{ V} \qquad (11.7)$$

$$2Ag + 2Cl^- \xrightarrow{discharge} 2AgCl + 2e^- \text{ (at the anode)}$$
$$E = 0.222 \text{ V} \qquad (11.8)$$

$$2H^+ + 2OH^- \xrightarrow{discharge} 2H_2O \text{ (neutralisation)}$$
$$E = 0.828 \text{ V} \qquad (11.9)$$

The overall cell voltage is 0.828 V + (0.344 − 0.222) V = 0.950 V. A stack of 100 units of such Ag_2O–$AgCl$ acid–alkaline hybrid cell can have the output voltage as high as 40–80 V with an output power density of 30–60 kW m^{-2} [11].

11.3 Bipolar Membrane

The challenge of acid–alkaline hybrid power sources lies in how to separate the acid and alkaline electrolytes effectively while maintaining ion transport between these two electrolytes. As shown in Table 11.1, some conventional electrochemical power sources do operate in two chambers with two electrolytes separated by a membrane or an ionic interface, such as the Daniell cell and most flow batteries. In these well-known electrochemical cells with two electrolytes, the *pH* value of both anolyte and catholyte are almost the same and there usually exists a common species to function as a charge carrier between the anolyte and catholyte. Examples include SO_4^{2-} ion for Daniell battery, and H^+ ion for vanadium redox flow battery, in which there is a separator to avoid electrolyte mixing. An ion-exchange membrane is usually employed as a separator. Ion-exchange membranes have fixed charges on their backbones and, therefore, have the ability to transport selected ions [12–14]. Anion-exchange membranes (AEMs) which are permeable to anions and can be used in Daniell cell to allow SO_4^{2-} anion to pass through, though the original version of Daniell uses a porous ceramic separator.

A BPM consists of two polymer layers, one AEM with positive fixed charges permeable only to anions, the other cation-exchange membrane (CEM) with negative fixed charges permeable to cations. Between the AEM and CEM is a junction region where negative and positive fixed charges coexist. BPM has unique properties of current rectification and water splitting. The current rectification of a BPM is analogous to that of an inorganic P–N semiconductor junction. Figure 11.6 shows a typical I–V curve of a BPM with variable conductivity, which is asymmetric to voltage bias [15–23]. The I–V curve of a BPM can be divided into four regions. Region 1 is under forward bias ($V > 0$) where the orientation of the positive and negative fixed charge layers are opposite to the external field across the membrane, as shown in Figure 11.6, with the electric potential lower at the AEM/electrolyte interface. Protons, hydroxide ions and salt ions will transport towards the AEM/CEM junction. Ion accumulation in the BPM renders the Donnan exclusion less effective, and a small amount of accumulated ions can penetrate into the adjacent second barrier layer and eventually into the electrolyte on the other side [18,19]. In this region the resistance is low. Regions 2, 3 and 4 are under reverse bias

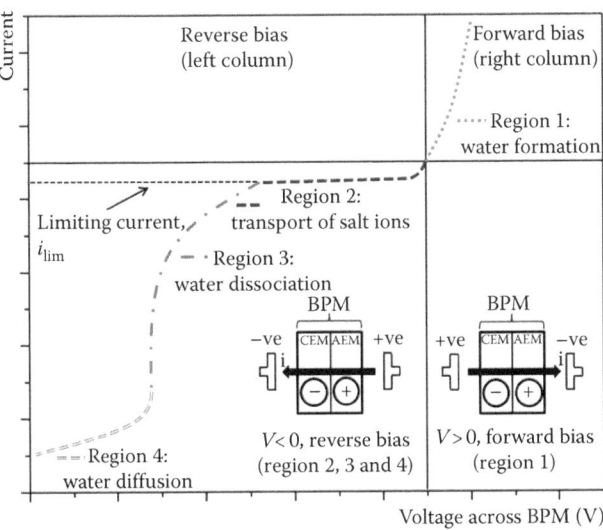

FIGURE 11.6
A typical I–V curve of a BPM over the full voltage range. (Redrawn based on A. Kemperman, *Handbook on Bipolar Membrane Technology*. Twente University Press, Enschede, 2000.)

($V < 0$) where the orientation of the positive and negative layers are in line with the external field across the membrane, as shown in Figure 11.6, with the electric potential higher at the AEM/electrolyte interface. Under reverse bias, ions adsorbed in the BPM will be attracted out of the bipolar junction. In Region 2, the I–V curve shows high resistance and corresponds to the situation that a narrow region of the BPM is almost devoid of mobile ions. Current is carried by ions leaving the BPM junction. In Region 3, water splitting can occur when reverse bias exceeds the thermodynamic threshold of −0.828 V. The current in this region is carried by H^+ and OH^-, the water-splitting products [24], as shown in Figure 11.7. The dissociation rate of water with a BPM is reported to be 106–107 times faster than in free solution, as enhanced by the second Wien effect in the high electric field region [24,25]. In Region 4, when the applied voltage is further increased, water transport into the BPM junction becomes limiting [26,27], resulting in an inflection point in the I–V curve with rapid rise in voltage.

Bipolar membranes (BPM) have found many applications in the last decades, such as acid and base production from corresponding salt, separation of ions with different valence, recovery or purification of acid and bases from the spent liquors, and production of organic acids [28]. Properties and performance of some commercial BPMs are tabulated in Table 11.2.

In summary, BPMs provide two important functions: (1) as a separator between the acid and alkaline solutions to avoid self-neutralisation; and (2) to provide various ion transport schemes in the four regions as described above according to bias voltage, and serve as a reactor for splitting H_2O into H^+ and OH^- ions.

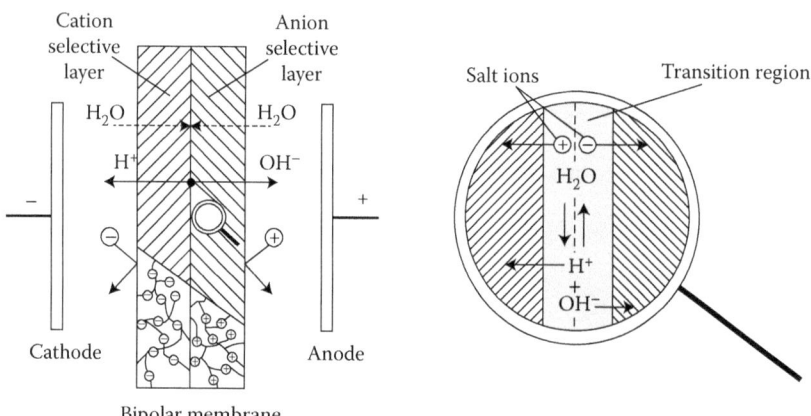

FIGURE 11.7
A schematic illustration of the function and structure of a BPM. The right-hand side illustrates the alternatives of salt ions separation or water-splitting reaction at the BPM junction region. (Reprinted from *Desalination* 90, H. Strathmann, Theoretical and practical aspects of preparing bipolar membranes, 1993, 303–323. Copyright 1993, with permission from Elsevier.)

TABLE 11.2
Characteristics of Bipolar Membrane at 25°C

Membrane		BP-1	AQ-6	WSI	MB-3	BPM
Manufacturer		Tokuyama Co., Japan	Aqualytics Inc., U.S.	WSI Technologies Inc., U.S.	Membrane Technology Centre, Russia	Membrane International Inc., U.S.
Structure and functional group	CEM layer	CM-1 membrane, sulphonic acid groups	Polystyrene and kraton G, sulphonic acid groups	Pall/Raipore R1010, sulphonic acid groups	Poly-ethylene, phosphoric acid groups	Poly-styrene/ divinylbenzene co-polymer, sulphonic acid groups
	AEM layer	Poly-sulphone, quaternary ammonium groups	Poly-styrene/vinyl benzoyl-chloride co-polymer, quaternary ammonium group	Pall/Raipore R1030, quaternary ammonium groups	Poly-ethylene, quaternary ammonium groups	Poly-styrene/divinyl benzene co-polymer, quaternary ammonium groups
Thickness (mm)		0.2–0.35	0.2	0.116	0.2–0.3	1
Limiting current density (mA/cm^2)		1.0	1.2	8.8	22	0.6
Electrical resistivity (Ω·cm)		8.5	10.2	1.8	5.3	CEM: <30 AEM: <40

Note: Information and data obtained from A. Kemperman, *Handbook on Bipolar Membrane Technology.* Twente University Press Enschede, 2000 and supplier's catalogues.

11.4 Acid–Alkaline Hybrid Fuel Cells

Application of the acid–alkaline dual electrolyte concept to hydrogen–oxygen fuel cell was reported by Cheng and Chan [29,30] and Abruña et al. [31]. The alkaline air electrode in Figure 11.4a was replaced by a hydrogen gas diffusion electrode in Cheng and Chan's acid–alkaline hybrid hydrogen fuel cell. Compared to a single pH hydrogen fuel cell, an increase of 0.4–0.5 V of operating voltage was observed, as shown in the polarisation curve of Figure 11.8.

The half-cell reactions in their fuel cell configuration are

$$\tfrac{1}{2}O_2(A) + 2H^+(A) + 2e^- \rightarrow H_2O\ (A)\ \text{(at the cathode)}$$

$$E^0 = 1.229\ \text{V} \quad (11.10)$$

$$H_2(B) + 2OH^-(B) \rightarrow 2H_2O\ (B) + 2e^-\ \text{(at the anode)}$$

$$E^0 = -0.828\ \text{V} \quad (11.11)$$

where A and B denote species in the acidic positive electrode and alkaline negative electrode compartments, respectively.

The overall reaction of the cell is

$$H_2(B) + \tfrac{1}{2}O_2(A) + 2OH^-(B) + 2H^+(A) \rightarrow H_2O(A) + 2H_2O(B)$$

$$E = 2.057\ \text{V} \quad (11.12)$$

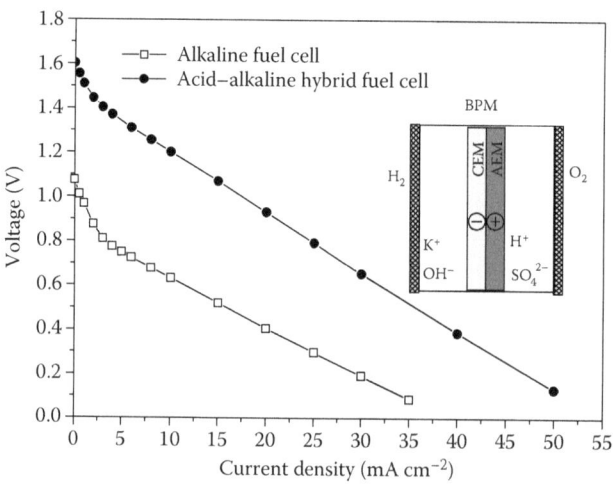

FIGURE 11.8
A polarization curve of the hydrogen-oxygen acid-alkaline fuel cell (setup as shown in inset) compared with a corresponding single electrolyte alkaline fuel cell.

with additional water formed from an overall acid–alkaline neutralisation that proceeds via half-reactions in different compartments:

$$2H^+(A) + 2OH^-(B) \rightarrow 2H_2O\,(B) \qquad (11.13)$$

To sustain electroneutrality and ionic current, different schemes of ion transport through the BPM are possible, according to Figure 11.6 (*I–V* curve). In Cheng and Chan's acid–alkali hybrid fuel cell, a higher operating voltage was observed in Figure 11.8, corresponding to the case of salt ion transport under forward bias (Region 1 of Figure 11.6) with one or more of the ion transfer reactions as follows:

$$2K^+(B) \rightarrow 2K^+(A) \qquad (11.14)$$

$$HSO_4^-(A) \rightarrow HSO_4^-(B) \rightarrow H^+ + SO_4^{2-}(B) \qquad (11.15a)$$

or

$$SO_4^{2-}(A) \rightarrow SO_4^{2-}(B) \qquad (11.15b)$$

The ratio of bisulphate to sulphate ions changes with *pH* and HSO_4^- originally in the acid chamber will dissociate after crossing the membrane to a high *pH* environment. SO_4^{2-} has a lower concentration in the acidic chamber, but may have higher selectivity under aligned electric field due to its charge.

The alternative ion transport via H$^+$ and OH$^-$ is less likely as they are present on the wrong side of the BPM, that is, H$^+$(A) and OH$^-$(B) are not favorable to enter the BPM junction. Water splitting does not occur, as the voltage bias is not in a favorable direction. Salt ions can carry current at a relatively low voltage under forward bias.

Instead of aiming at a higher operating voltage for hydrogen fuel cell, Kohl and coworkers configured the alkaline and acidic sides differently to achieve better water management. Their hybrid ion-exchange membrane H_2–O_2 fuel cells [32,33] have two different MEA configurations, as shown in Figure 11.9. The half-cell reactions and the corresponding *pH* environments, locations of water consumption and generation are summarised in Table 11.3. For the hybrid fuel cell with a high *pH* anode, Figure 11.9a, the corresponding electrode reactions are the same as Equations 11.10 and 11.11. However, the BPM is oriented under reverse voltage bias. Under moderate and high current operation, water-splitting reaction proceeds, corresponding to Region 3 of Figure 11.6 to sustain ion transport at the bipolar junction. This water-splitting reaction requires a voltage of 0.828 V, which is compensated by the same voltage created by operating the anode

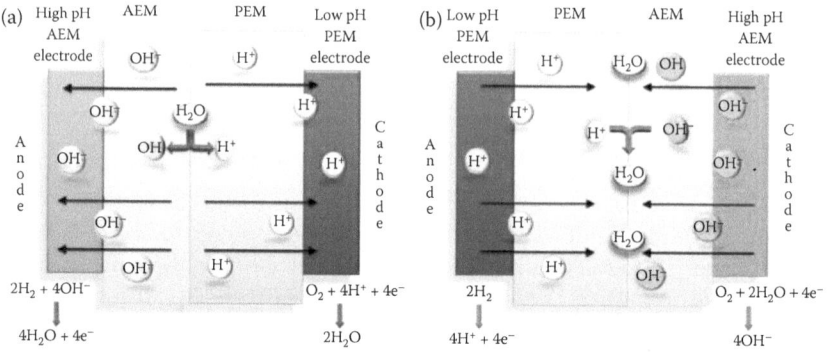

FIGURE 11.9
Two types of hybrid MEA cell configuration. (a) High *pH* anode versus low *pH* cathode, and the BPM is under reverse bias. (b) Low *pH* anode versus high *pH* cathode, and the BPM is under forward bias. (Reprinted with permission from M. Ünlü, J. Zhou and P.A. Kohl, Hybrid anion and proton exchange membrane fuel cells, *J. Phys. Chem. C* 113, 2009, 11416–11423. Copyright 2009, American Chemical Society.)

in alkaline and cathode in acid. The overall cell voltage is 0.4 V, the same as a single *pH* fuel.

Although no increase in voltage was observed, this BPM fuel cell has water generation distributed at the cathode, anode, and water consumed at the bipolar junction. This avoids the common problem of flooding at the air cathode where water is generated in a single *pH* fuel cell. In the other configuration of the hybrid fuel cell (Figure 11.9b), the *pH* environments are reversed for anode and cathode and the BPM is under forward bias. The thermodynamic cell voltage for both of these hybrid fuel cells are the same, since the overall hydrogen/oxygen reaction is the same in both cases. But the polarisation curves in Figure 11.10 shows that the hybrid MEA with a high *pH* cathode (dash curve) could provide much better performance, compared to a high *pH* anode (solid curve).

Enhanced performance was mainly due to favourable formation of water at the PEM/AEM junction to produce a self-hydrating effect [21]. At the same time, flooding is avoided at either gas diffusion-type anode or cathode. AEM/Nafion hybrid fuel cell with a self-humidifying hybrid anion–cation membrane fuel cell can be operated under totally dry conditions [34]. The self-humidifying hybrid fuel cell comprised both electrodes operating in high *pH* environment and one proton-exchange membrane, shown in Figure 11.11. In this hybrid fuel cell, two *pH* junctions are present between the proton conducting membrane and alkaline electrodes. The hybrid fuel cell shows better polarisation performance than the conventional proton-exchange membrane fuel cell (PEMFC), especially under low humidity conditions. Experimental results show that this hybrid cell can be operated steadily at 60°C with dry gas feeds at a current density of 100 mA cm^{-2} for several days. The increase in performance of the hybrid cells at low humidity is attributed to the superior water management configuration.

TABLE 11.3
Configurations of MEA Fuel Cells with Multiple pH Junctions Proposed by Kohl and Coworkers

Configure	Anode	BPM Junction	Cathode	BPM Bias	Water Generated	Water Consumed
Figure 11.9a High pH anode	H_2 (AEM) + $2OH^-$ → $2H_2O$ (AEM) + $2e^-$ $E = -0.8$ V	$2H_2O$ (BPM) → $2H^+$ (PEM) + $2OH^-$ (AEM)	½ O_2 (PEM) + $2H^+$ + $2e^-$ → H_2O (PEM) $E = 1.2$ V	Reverse bias	$2H_2O$ (AEM) + H_2O (PEM)	$2H_2O$ (BPM)
Figure 11.9b High pH cathode	H_2 (PEM) → $2H^+$ (PEM) + $2e^-$ $E = 0$ V	$2H^+$ (PEM) + $2OH^-$ (AEM) → $2H_2O$ (BPM)	½ O_2 (PEM) + H_2O (AEM) + $2e^-$ → $2OH^-$ (AEM) $E = 0.4$ V	Forward bias	$2H_2O$ (BPM)	H_2O (AEM)
Figure 11.10 High pH anode and cathode	H_2(AEM) + $2OH^-$ → $2H_2O$ (AEM) + $2e^-$ $E = -0.8$ V	$2H_2O$ (BPM) → $2H^+$ (PEM) + $2OH^-$ (AEM) $2H^+$ (PEM) + $2OH^-$ (AEM) → $2H_2O$ (BPM)	½ O_2 (PEM) + H_2O (AEM) + $2e^-$ → $2OH^-$ (AEM) $E = 0.4$ V	Forward bias	$2H_2O$ (Anode AEM) + $2H_2O$ (cathode BPM)	$2H_2O$ (Anode BPM) + H_2O (cathode BPM)

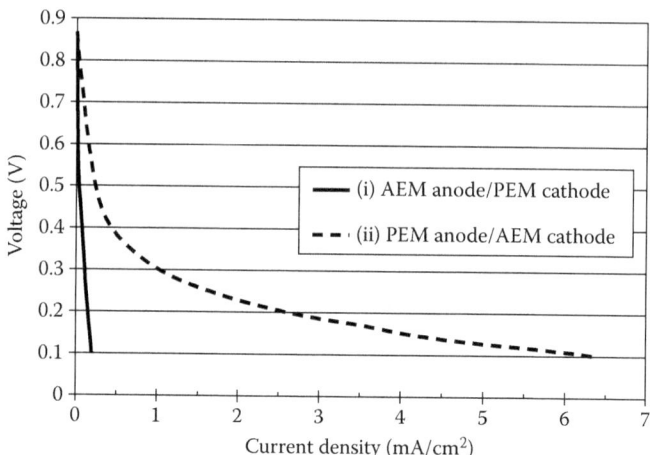

FIGURE 11.10
Polarisation curves of two hybrid MEAs. (Reprinted with permission from M. Ünlü, J. Zhou and P.A. Kohl, Hybrid anion and proton exchange membrane fuel cells, *J. Phys. Chem. C* 113, 2009, 11416–11423. Copyright 2009, American Chemical Society.)

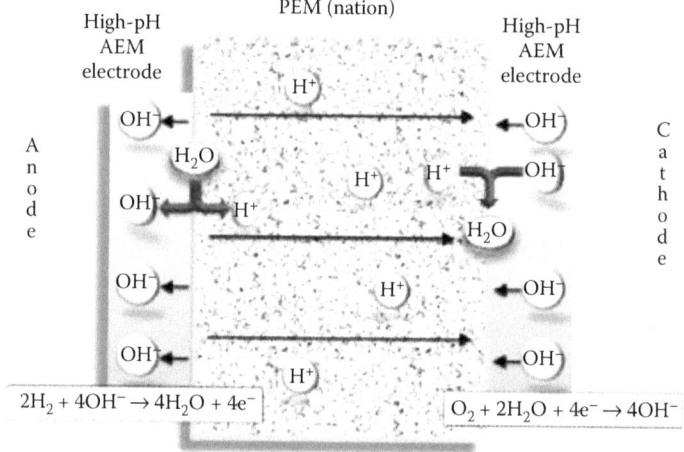

FIGURE 11.11
Self-humidifying hybrid fuel cell comprises both two electrodes operating in high *pH* environment and one proton exchange membrane. (Reprinted with permission from M. Ünlü, J. Zhou and P.A. Kohl, Hybrid polymer electrolyte fuel cells: Alkaline electrodes with proton conducting membrane, *Angew. Chem. Int. Ed.* 2010, 122, 1321–1323. Copyright 2010, John Wiley and Sons.)

Application of an acid–alkaline concept to direct alcohol fuel cells has also been reported. Cheng and Chan [29] reported operation of a 6-cell stack of dual *pH* electrolyte fuel cell using ethanol fuel. A BPM MB-3 supplied by Membrane Technology Centre (Russia) was used. Each anode chamber contained 4 mL 7 mol L^{-1} NaOH and 0.5 mol L^{-1} ethanol. Each cathode chamber

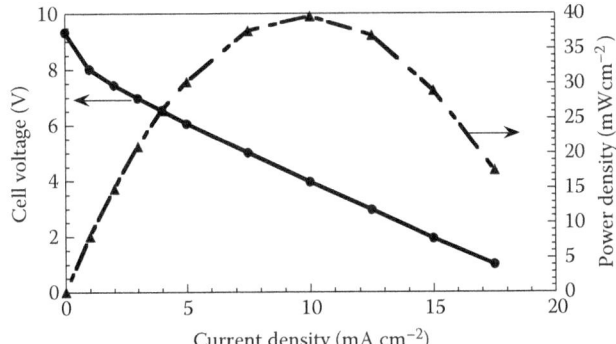

FIGURE 11.12
The performance of a direct ethanol fuel cell stack containing six cells each with acid–alkaline dual electrolytes. (Reprinted from S.A. Cheng and K.Y. Chan, High-voltage dual electrolyte electrochemical power sources, U.S. Patent 7, 344, 801, 2008.)

contained 4 mL 5 mol L^{-1} H$_2$SO$_4$. The active area of electrodes was 25 cm^2. The stack operated at room temperature using atmosphere pressure air. As shown in polarisation curve of the stack cell in Figure 11.12, an open-circuit voltage of over 9.3 V is obtained. This voltage is almost equivalent to that of a 10-cell stack normal fuel cell. At a current density of 10 mA cm^{-2}, the voltage of the stack cell is 4 V, which is higher than that of normal direct methanol fuel cell (DMFC) stack cell operating at 60–80°C using compressed air; and the power density can be around 40 mW cm^{-2} at a current density of 10 mA cm^{-2}.

To minimise shunt currents and leak currents that are commonly present in fuel cell stacks with parallel electrolyte feed, Kohl et al. [35] put forward a AEM–CEM hybrid bi-cell design with an AEM and PEM fuel cell in a series using a common methanol fuel tank. Such a bi-cell configuration was developed because the AEM cathode potential was essentially the same as the PEM anode potential. In addition to the higher voltage (theoretically, 2.4 V) and reduced volume by using a common fuel tank, self-humidification and easy water management are additional advantages of this AEM–PEM hybrid bi-cell DMFC stack.

11.5 Acid–Alkaline Microbial Fuel Cells

The MFC is a special type of fuel cell in which microorganisms produce electricity directly from biodegradable material [36]. Details of MFC are well described in Chapter 2. Usually, the MFC consists of two compartments: an anaerobic compartment with an anode and an aerobic compartment with a cathode. The overall reaction is the conversion of organic material and oxygen to carbon dioxide, water and electricity. The MFC performance is significantly limited by the air cathode, similar to the conventional low-temperature fuel cells. Aeration in the cathodic compartment and lowering the catholyte *pH* have been shown to increase cathode performance [37–39]. Some inexpensive alternatives to replace platinum are needed in MFC s [40], like Fe(II)- and cobalt-based cathodes [40]. Another alternative is biological catalysts combined with redox mediators, which are active at ambient temperatures and can be recycled [41], as well as applied in the form of enzymes [41,42] or microorganisms. The commonly used CEMs in MFC can transport protons and other cations, and cause a *pH* rise in the cathodic compartment, which induces extensive iron precipitation that will damage the membrane. Heijne et al. [43] explored the redox couple Fe^{3+}/Fe^{2+} as a cathodic electron mediator for oxygen reduction in a MFC with a BPM to keep Fe ions soluble without external acid dosage, as shown in Figure 11.13. The BPM was employed to maintain the *pH* differential, that is, low *pH* catholyte (<2.5) and high *pH* anolyte, and provide the cathodic compartment with protons and the anodic compartment with hydroxides as a result of water dissociation. The low *pH* catholyte can be maintained in this way without the need of acid dosage. This MFC demonstrated improved Coulombic efficiency (80–95%) and energy recovery (18–29%) in comparison to 40–75% [44,45] and 2–15%, respectively, for platinum-based cathode [46,47]) at similar power densities. The discharge performance of this

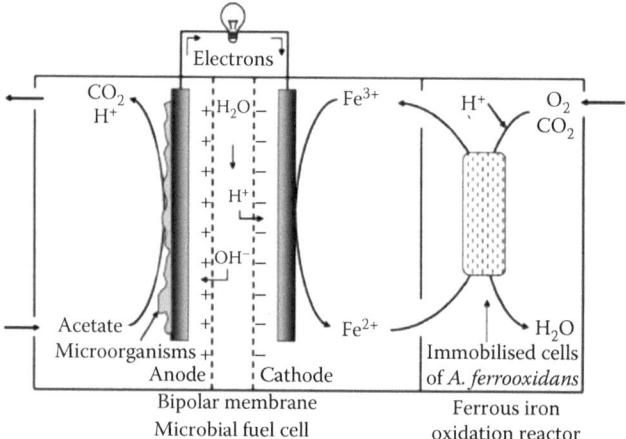

FIGURE 11.13
The working principle of a MFC with the iron-mediated cathode with a BPM. (Reprinted with permission from A.T. Heijne, H.V.M. Hamelers and C.J.N. Buisman, Microbial fuel cell operation with continuous biological ferrous iron oxidation of the catholyte, *Environ. Sci. Technol.* 41, 2007, 4130–4134. Copyright 2007, American Chemical Society.)

acid–alkaline hybrid MFC, however, is still far below the level for practical applications. Further research to improve the cell performance needs to be carried out in the future.

11.6 Acid–Alkaline Hybrid Fuel Cell Reactor for H_2O_2 Production

Most H_2O_2 is manufactured by the anthraquinone process in a multistep operation with high-energy consumption. The costly process and the toxicity of organic solvents make this preparation route unfavourable. The production of hydrogen peroxide through a fuel cell reactor is more economical, environmentally friendly and with co-generation of electricity [48,49]. Kallio et al. [50] proposed a fuel cell reactor with a BPM in forward bias for the production of H_2O_2. This configuration combines the advantages of faster H_2 oxidation kinetics in an acid environment, and faster O_2 reduction to H_2O_2 in alkaline, where H_2O_2 is more soluble and stable. The schematic structure of such a fuel-cell reactor is illustrated in Figure 11.14. At the anode, the feed humidified hydrogen is oxidised to H^+ in acid solution, and the resulting protons are transported toward the junction region of BPM via the CEM layer to react with HO_2^- ions, formed from oxygen reduction in an alkaline solution at the cathode. The electrode reactions during this process are listed as follows:

On the anode:

$$H_2(g) \xrightarrow{\text{discharge}} 2H^+ + 2e^- \quad (11.16)$$

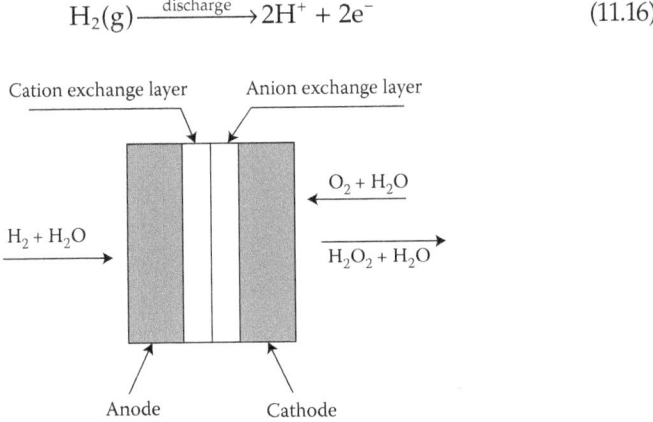

FIGURE 11.14
A schematic structure of a BPM-based fuel-cell reactors in forward bias for the production of H_2O_2. (Reprinted from *Electrochim. Acta* 51, E. Lobyntseva, T. Kallio and K. Kontturi, Bipolar membranes in forward bias region for fuel cell reactors, 1165–1171. Copyright 2006, with permission from Elsevier.)

On the cathode:

$$O_2(g) + e^- \xrightarrow{discharge} O_2^- \qquad (11.17)$$

$$O_2^- + H_2O \xrightarrow{discharge} HO_2^- + OH \qquad (11.18)$$

$$OH + e^- \xrightarrow{discharge} OH^- \qquad (11.19)$$

At the interface region:

$$HO_2^- + H^+ \xrightarrow{discharge} H_2O_2 \qquad (11.20)$$

$$OH^- + H^+ \xrightarrow{discharge} H_2O \qquad (11.21)$$

The overall reaction is

$$H_2(g) + O_2(g) \xrightarrow{discharge} H_2O_2 \qquad (11.22)$$

11.7 Hybrid PbO$_2$/MH$_x$ Secondary Battery

Although a higher operating voltage is evident in the acid–alkaline fuel cells described in [51–54] and Figure 11.8, the transport of salt ions from one compartment to another as in [55] and [56] are not sustainable. Replenishment or regeneration of acid and alkaline is required for sustained operation. Salt ion transport can be reversed in the opposite cycle of a rechargeable acid–alkaline power cell. The application of the acid–alkaline concept to rechargeable batteries was first demonstrated by Cheng and Chan [29,30]. They proposed a hybrid from two common aqueous rechargeable batteries: lead acid and the alkaline nickel metal hydride cells. The PbO$_2$–MH$_x$ hybrid battery couples a positive PbO$_2$ electrode in strong acid solution and a negative MH$_x$ electrode in strong alkaline solution [29,30] with pH dependence of electrode potentials presented in Table 11.1 and Figure 11.3.

A MB-3 BPM from Membrane Technology Centre, Russia, was employed as the separator between the acid and alkaline electrolytes chamber. The electrode reaction on positive electrode in acid chamber is

$$PbO_2(A) + HSO_4^-(A) + 3H^+(A) + 2e^- \underset{charge}{\overset{discharge}{\rightleftarrows}} PbSO_4(A) + 2H_2O(A)$$

$$E_+^0 = 1.695 \text{ V (vs. RHE)} \qquad (11.23)$$

with a pH dependence of

$$E_+ = E_+^0 - \left[2.303\frac{RT}{nF}\right]\left[\log\frac{1}{(a_{HSO_4^-})(a_{H^+})^3}\right] \quad (11.24)$$

Reaction in negative electrode in alkaline solution is

$$2MH_x(B) + 2OH^-(B) \underset{charge}{\overset{discharge}{\rightleftarrows}} 2MH_{x-1}(B) + 2H_2O(B) + 2e^-$$

$$E_-^0 = -0.80 \sim -0.85 V \text{ (vs. RHE)} \quad (11.25)$$

with a pH dependence expressed by

$$E_- = E_-^0 - \left[2.303\frac{RT}{nF}\right]\left[\log\frac{(a_{OH^-})}{(a_{H_2O})}\right] \quad (11.26)$$

and the overall electrode reaction is

$$PbO_2(A) + HSO_4^-(A) + 3H^+(A) + 2MH_x(B) + 2OH^-(B) \underset{charge}{\overset{discharge}{\rightleftarrows}}$$

$$PbSO_4(A) + 2H_2O(A) + 2MH_{x-1}(B) + 2H_2O(B)$$

$$E^0 = 2.495 \sim 2.545 V \quad (11.27)$$

with a cell voltage pH dependence as

$$\Delta E = \Delta E^0 - \left[2.303\frac{RT}{nF}\right]\left[\log\frac{1}{(a_{OH^-})^2(a_{HSO_4^-})(a_{H^+})^3}\right] \quad (11.28)$$

Here, (A) denotes the acid chamber and (B) the base chamber. For some BPM, like MB-3 and WSI, the membrane potential (ΔV) under an open-circuit condition is below 100 mV and does not affect the OCV significantly. By employing BPM MB-3 as the separator, Chan's group can obtain an OCV of 2.6 V for this PbO_2–MH_x hybrid battery with 3 mol L^{-1} H_2SO_4/5 mol L^{-1} KOH dual electrolytes [30]. Similar results can be obtained for another BPM from Membrane International Inc. with an OCV of 2.52 V in 1 mol L^{-1} H_2SO_4/2 mol L^{-1} KOH dual electrolytes. As shown in Figure 11.15, the operating discharge voltage of the hybrid battery is in the range of 2.0–2.4 V.

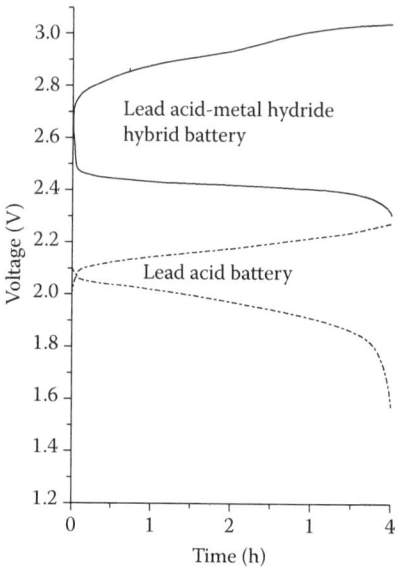

FIGURE 11.15
The charge and discharge curves of the metal-hydride/lead-acid cell (solid lines) compared to those of a single electrolyte lead-acid battery (dash dot dot). (Reprinted with permission from S.A. Cheng and K.Y. Chan, High-voltage dual electrolyte electrochemical power sources, *ECS Trans.* 25, 2010, 213–219. Copyright 2010, The Electrochemical Society.)

11.8 MH-VRF Semi-Flow Battery

Flow batteries are a promising electrochemical system for storing electricity intermittently generated from solar, wind and other renewable sources [57]. For this special type electrochemical system, energy is stored in the liquid phase, which can be transferred to a tank external to the electrochemical cell while the inert electrodes simply serve as current collector and do not participate in the electrochemical reaction. In other words, the power and energy ratings are independent of each other. This means the energy is determined by the quantity of the electro-active electrolytes, while the power is decided by the active area of the electrodes. The all-vanadium redox flow (VRF) battery [58] is one of the popular flow batteries, consists of two inert electrodes, where a positive electrode reaction of redox couple V^{4+}/V^{5+} (standard potential of 1.0 V) and a negative electrode reaction of redox V^{2+}/V^{3+} (standard potential of −0.26 V) take place. The two electro-active electrolytes are separated by a monopolar membrane. Details on how a VRF work can be found in Chapter 9, with electrochemistry described by

$$VO_2^+ + 2H^+ + e^- \underset{charge}{\overset{discharge}{\rightleftharpoons}} VO^{2+} + H_2O \text{ (at the positive electrode)}$$

$$E^0 = 1.00 \text{ V} \qquad (11.29)$$

$$V^{2+} \underset{charge}{\overset{discharge}{\rightleftharpoons}} V^{3+} + e^- \text{ (at the negative electrode)}$$

$$E^0 = -0.26 \text{ V} \qquad (11.30)$$

$$VO_2^+ + 2H^+ + V^{2+} \underset{charge}{\overset{discharge}{\rightleftharpoons}} VO^{2+} + H_2O + V^{3+} \text{ (overall reaction)}$$

$$\Delta E^0 = 1.26 \text{ V} \qquad (11.31)$$

A theoretical cell voltage of this VRF cell is 1.26 V which is inherently limited by the electrochemical window in aqueous electrolytes. The V^{2+} ions are found to be easily oxidised in air. It would be desirable to replace the negative component by another electrochemical reaction that has a potential lower than -0.26 V and at the same time increase the cell voltage.

Weng et al. [59,60] introduced a *pH* differential by coupling the positive V^{4+}/V^{5+} acidic compartment with a negative alkaline electrode/electrolyte of metal hydride electrode. The electrochemistry of their hybrid flow battery is described by the reactions:

$$VO_2^+ + 2H^+ + e^- \underset{charge}{\overset{discharge}{\rightleftharpoons}} VO^{2+} + H_2O \text{ (at the positive electrode)}$$

$$E^0 = 1.00 \text{ V} \qquad (11.32)$$

$$MH_x + OH^- \underset{charge}{\overset{discharge}{\rightleftharpoons}} MH_{x-1} + H_2O + e^- \text{ (at the negative electrode)}$$

$$E^0 = -0.80 \text{ V} \qquad (11.33)$$

The overall cell reaction is

$$VO_2^+ + 2H^+ + MH_x + OH^- \underset{charge}{\overset{discharge}{\rightleftharpoons}} VO^{2+} + 2H_2O + MH_{x-1}$$

$$\Delta E^0 = 1.80 \text{ V} \qquad (11.34)$$

The electrochemical neutralisation energy is stored in the electrolyte as in the case of redox couples in flow batteries. Mass flow of the electrolyte minimises membrane fouling and improves the mass transport. In principle, this vanadium–metal hydride (vanadium–MH) hybrid battery system can deliver an overall cell capacity of 110 mAh g^{-1}, cell voltage of 1.8 V, and specific energy of 200 Wh kg^{-1}, which are higher than those of conventional

VRF battery (48.4 mAh g^{-1}, 1.26 V and 60.5 Wh kg^{-1}, respectively) [59,60]. This hybrid battery prototype consists of a graphite felt positive electrode operating in a mixed solution of 0.128 mol L^{-1} VOSO$_4$ and 2 mol L^{-1} H$_2$SO$_4$, and a metal hydride negative electrode in 2 mol L^{-1} KOH aqueous solution. The two electrolytes of different *pH* are separated by a BPM. Detailed experimental information was reported in Ref. [60].

As shown in Figure 11.16a, the measured OCV of the vanadium–MH cell after fully charged is around 1.93 V, which is significantly higher than the theoretical OCVs of the VRF battery (1.26 V) and the NiMH$_x$ battery (1.25 V). At a given current, vanadium–MH semi-flow hybrid system shows a higher discharge voltage (1.70 V) than that of individual VRF (~1.2 V) with the same acidic electrolyte, and NiMH$_x$ (~1.28 V) batteries with the same alkaline electrolyte. At 1 mA cm^{-2}, a typical charging of the vanadium–MH system takes 4.8 h reaching the cutoff of 2.0 V and a discharge of 4.7 h to 1.6 V. This system also demonstrated high efficiencies in Coulomb (95%), voltage (88%) and energy (84%).

Figure 11.16b shows four voltage profiles V1–V4, which can be used to examine the breakdown of voltage losses in the vanadium–MH semi-flow system and performance of individual components. For the overall cell voltage, V4, the charge and discharge plateaus are stable over the typical 10 cycles at ca. 1.93, and 1.70 V, respectively. This is also attributed to the corresponding steady positive and negative electrode reactions, as shown in the curves of V1 and V2. The voltage across the BPM (V3), reflecting the mass transport through the ionic interface. After corrected for the difference in reference electrodes used, this voltage remains small and stable during charge and discharge, reflecting the steady and reversible ionic transport across the membrane. The present vanadium–MH prototype is positive-limited as the negative MH capacity is preset to be larger.

FIGURE 11.16 (continued)
(a) Typical charge/discharge curves at 1 mA cm^{-2} for NiMH$_x$, VRF, and hybrid semi-flow V-MH batteries. (b) Ten charge/discharge cycles of the V-MH battery. V1 is the potential difference between the positive graphite felt (GF) electrode and the Hg/Hg$_2$SO$_4$ reference: V1 = $\phi_{positive} - \phi_{Hg/Hg_2SO_4}$; V2 is the potential difference between the negative MH$_x$/MH$_{x-1}$ electrode and the Hg/HgO reference electrode: V2 = $\phi_{negative} - \phi_{Hg/HgO}$; V3 is the voltage across the BPM measured by the Hg/Hg$_2$SO$_4$ reference electrode and the Hg/HgO reference electrode: V3 = $\phi_{Hg/Hg_2SO_4} - \phi_{Hg/HgO}$; V4 is the cell voltage which is the potential difference between the GF electrode and MH$_x$/MH$_{x-1}$ electrode: V4 = $\phi_{positive} - \phi_{negative}$. (c) Theoretical cell voltage, cell capacity, and specific energy of selected pairing of positive and negative electrodes/electrolytes. (The theoretical cell capacity is calculated by C = C$_+$/(1 + C$_+$/C$_-$), where C$_+$ and C$_-$ represent the theoretical capacity of positive electrode/electrolyte material (width of bar at the top) and negative electrode material (width of bar at the bottom), respectively. C$_+$ and C$_-$ are scaled by Faraday's law to their individual active material. In VRF battery, VOSO$_4$ and V$_2$(SO$_4$)$_3$ are the active material for V^{4+}/V^{5+} and V^{2+}/V^{3+} redox couples, respectively. The specific energy is calculated as C× *Cell voltage*.) (Reprinted with permission from G.M. Weng, C.Y.V. Li and K.Y. Chan, High voltage vanadium-metal hydride rechargeable semi-flow battery, *J. Electrochem. Soc.* 160, 2013, A1384–A1389. Copyright 2013, The Electrochemical Society.)

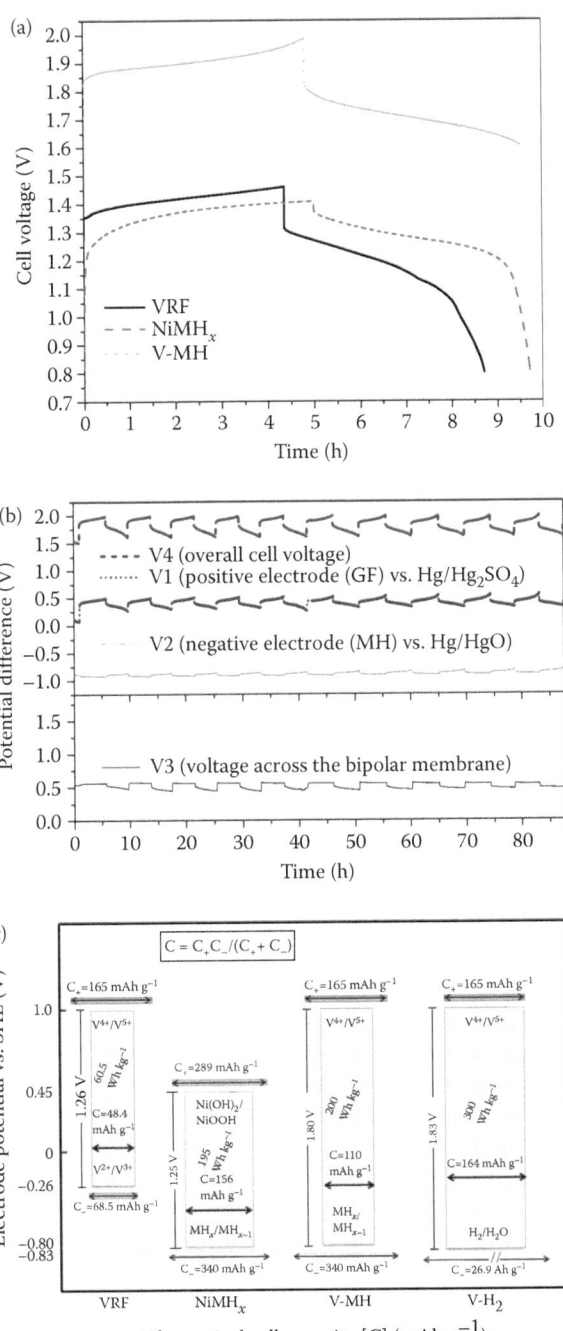

FIGURE 11.16 (continued)

The alkaline hydrogen electrode (–0.83 V vs. SHE), a close equivalent of metal hydride electrode (–0.8 vs. SHE) but with extremely high cell capacity and favourable reversible electrochemical kinetics, can be coupled with V^{4+}/V^{5+} as a full-flow *pH* differential battery. The electrochemistry of this vanadium–hydrogen full-flow battery is shown below.

The positive electrode reaction is

$$VO_2^+ + 2H^+ + e^- \underset{\text{charge}}{\overset{\text{discharge}}{\rightleftarrows}} VO^{2+} + H_2O$$

$$E^0 = 1.00 \text{ V} \qquad (11.35)$$

and the negative electrode reaction is

$$H_2 + 2OH^- \underset{\text{charge}}{\overset{\text{discharge}}{\rightleftarrows}} 2H_2O + 2e^-$$

$$E^0 = -0.83 \text{ V} \qquad (11.36)$$

whereas the overall reaction is

$$VO_2^+ + 2H^+ + \tfrac{1}{2}H_2 + OH^- \underset{\text{charge}}{\overset{\text{discharge}}{\rightleftarrows}} VO^{2+} + 2H_2O$$

$$\Delta E^0 = 1.83 \text{ V} \qquad (11.37)$$

Both the theoretical specific energy and power of vanadium–hydrogen flow battery are significantly higher than those of vanadium–MH semi-flow system and VRF, as illustrated in Figure 11.16c with the largest rectangle of theoretical specific energy (300 Wh kg^{-1}) due to the high voltage (1.83 V) and high cell capacity (164 mAh g^{-1}). In terms of costs, this hybrid system would not be more expensive than VRF, since hydrogen electrode in alkaline are matured systems.

11.9 Alternative Separator of Acid and Alkaline Electrolyte

11.9.1 CEM and AEM Combined with a Salt Compartment as the Separators for Acid–Alkaline Hybrid Cell

An alternative to the BPM configuration was reported by Li et al. [55] using a three-electrolyte setup. The three electrolytes are segregated by two monopolar membranes with the neutral solution in the middle between the CEM and AEM. This proposed three electrolyte acid–alkaline PbO_2/MH_x hybrid battery is shown in Figure 11.17a.

The ion transport coupled with electrode reactions are described below. The reactions in the acidic positive electrode (Equation 11.23) and the alkaline

pH Differential Power Sources with Electrochemical Neutralisation

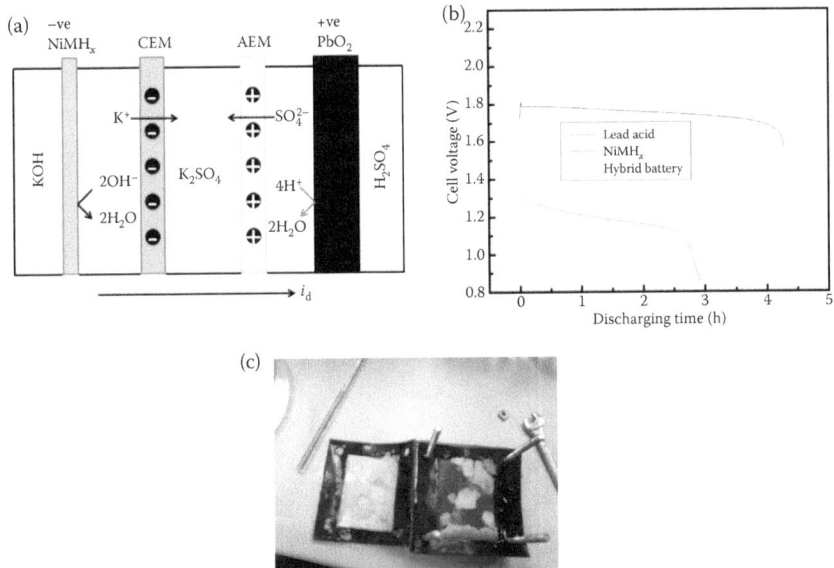

FIGURE 11.17
(a) A three-compartment cell arrangement of three-electrolyte PbO$_2$–MH$_x$ hybrid battery and the ion transport during discharge. (b) Three-compartment cell arrangement of three-electrolyte PbO$_2$–MH$_x$ hybrid battery and the ion transport during discharge. (Reprinted from *Electrochim. Acta* 56, H. Li et al., Three electrolyte high voltage acid–alkaline hybrid rechargeable battery, 2011, 9420–9425. Copyright 2006, with permission from Elsevier.) (c) Salting problem in three-electrolyte PbO$_2$–MH$_x$ hybrid battery with aqueous electrolyte solution.

negative electrode (Equation 11.24) are identical to that of the dual electrolyte battery (Section 11.7):

$$PbO_2(A) + HSO_4^-(A) + 3H^+(A) + 2\,e^- \underset{\text{charge}}{\overset{\text{discharge}}{\rightleftarrows}} PbSO_4(A) + 2\,H_2O(A) \tag{11.23}$$

$$2MH_x(B) + 2OH^-(B) \underset{\text{charge}}{\overset{\text{discharge}}{\rightleftarrows}} 2MH_{x-1}(B) + 2\,H_2O(B) + 2e^- \tag{11.24}$$

While the reaction in the neutral solution of the central compartment is

$$2K^+(B) + 2HSO_4^-(A) \underset{\text{charge}}{\overset{\text{discharge}}{\rightleftarrows}} 2K^+(C) + 2H^+(C) + 2SO_4^{2-}(C) \tag{11.38}$$

Equation 11.23 can be expressed with SO_4^{2-} replacing HSO_4^- in left. Since SO_4^{2-} is present at low concentration in acid, we do not express a corresponding equation.

This gives an overall cell reaction of

$$[PbO_2(A) + 3H^+(A) + HSO_4^-(A)] + 2HSO_4^-(A)$$

$$+ [2MH_x(B) + 2OH^-(B)] + 2K^+(B) \underset{\text{charge}}{\overset{\text{discharge}}{\rightleftarrows}} [PbSO_4(A) + 2H_2O(A)]$$

$$+ \left[2MH_{x-1}(B) + 2H_2O(B)\right] + [2K^+(C) + 2H^+(C) + 2SO_4^{2-}(C)] \quad (11.39)$$

Here (A), (B) and (C) mean the material in acid chamber, alkaline chamber or central salt chambers, respectively. Owing to the charge exclusion of CEM or AEM membranes and their orientation relative to the *pH* boundaries, only HSO_4^-/SO_4^{2-} and K^+ ions will be transported under charge and discharge to maintain charge balance and function as current carriers in the electrolytes.

During charging, excess $H^+(A)$ and $OH^-(B)$ ions are produced on the positive and negative electrode separately. The potassium and sulphate ions moved from the middle chamber to form more alkaline and acidic species in anode and cathode (Equation 11.38), respectively.

During discharge, $H^+(A)$ ion in the acid chamber and $OH^-(B)$ ion in the base chamber are consumed on the positive and negative electrode separately, as in reaction (Equations 11.23 and 11.24). The left over HSO_4^-/SO_4^{2-} ions in acid and alkaline chambers will transport into the middle compartment to maintain the charge balance of each chamber. Because K_2SO_4 can be easily dissociated into K^+ and SO_4^{2-} ions completely in water, the theoretical OCV of this hybrid cell should be 2.495 V under standard conditions. The three-electrolyte configuration has advantages of avoiding the more expensive BPM and an extra chamber to store salt, the by-product of electrochemical neutralisation, which can only be stored in either acid or a cathode chamber when a single BPM is used. The three-electrolyte configuration, however, is more bulky and lowers the specific energy and energy density of the acid–alkaline battery.

Experimental results show that an OCV of around 2.60 V can be obtained for such a three electrolyte acid–alkaline PbO_2/MH_x hybrid battery, which was significantly higher than the OCV of 2.05 V for lead acid battery and 1.35 V for $NiMH_x$ battery. Discharge curves shows that when using 1 mol L^{-1} H_2SO_4–0.2 mol L^{-1} K_2SO_4–2 mol L^{-1} KOH electrolytes, the hybrid cell was shown to operate with a voltage 20% higher than the conventional lead acid battery and 110% higher than nickel–metal hydride battery under at a 1/3 C discharge rate (Figure 11.17b). The distance between two membranes, and the concentration of the electrolytes, along with the type of salt solution in the middle compartment were found to be the main factors which could have significant impacts on cell performance [55].

For most ion-exchange membranes, the perm-selectivity for both AEM and CEM is only around 90% and, thus, there will be a high co-ion leakage across the membranes, which results in a severe salting problem for this three electrolyte hybrid battery (shown in Figure 11.17c). Since there is a large concentration gradient across the membrane between acid (or alkaline) chamber and middle salt chamber, large amounts of SO_4^{2-} and K^+ ions would diffuse into the middle chamber under concentration gradient and result in the super-saturation of salt. In addition, the low saturation concentration of K_2SO_4 in water (0.63 M) at room temperature means it will become super saturated easily. The salting problem can be overcome by using a suitable salt with a higher solubility, or apply CEM and AEM with a higher perm-selectivity.

A salt solution for electrochemical energy generation has been previously used in an interesting configuration of reverse electro-dialysis (RED). Conventional RED is the reverse of the better-known desalination method [61] and can be employed to harness renewable energy. The RED is essentially a concentration cell exploiting the concentration difference of seawater and fresh water. Figure 11.18 shows a stack of alternating cation and AEMs. The chemical potential difference between salt and fresh water could generate a voltage across each ion-exchange membrane. When the number of cells is high enough, the sum of cell-pair voltages could provide sufficient power. The RED is attractive conceptually for simultaneous desalination and energy generation. The technology depends on the availability of fresh water, which will become salty at the end of the RED process. The seawater will not be fully desalinated and the power output will decline with diminishing concentration differential between the two streams.

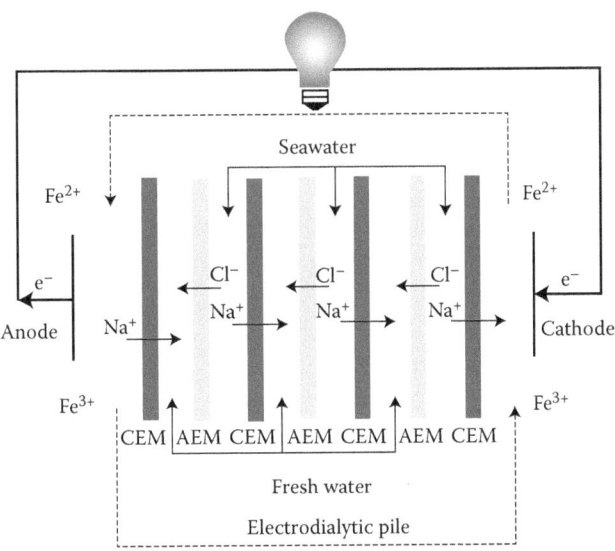

FIGURE 11.18
The principle of power generations of reversed electro-dialysis (RED). (Adapted from J.W. Post et al. Power generation by reversed electro dialysis (RED): Prevention of fouling. *Wetsus*, Centre of Excellence for Sustainable Water Technology, Leeuwarden.)

11.9.2 Gel Electrolyte

The same configuration of the hybrid PbO_2/MH_x secondary battery has been applied with gel electrolytes instead of aqueous electrolyte (cf. Sections 11.7 and 11.9.1) [56] to solve some issues on conventional batteries: maintenance cost and acid stratification for the lead-acid battery, high self-discharge and short life cycles for high current discharge for nickel–metal hydride battery. In addition, gel electrolytes offer advantages such as leak proof, maintenance-free and corrosion-free.

Alkaline gel electrolytes based on potassium-salt poly (acrylic acid) (PAAK), typically, PAAK–KOH–H_2O was used in metal hydride reactions to enhance the charge–discharge reactions and capacity retention [62]. A silica-based gel electrolyte was used in lead dioxide-positive electrode reaction to give superior discharge capacity and greatly reduced the rate of electrode corrosion [63]. The Na_2SO_4–PAAK gel electrolyte was employed in the middle to separate acid and alkaline chambers with an AEM and a CEM. Detailed information on preparation of gel electrolytes and setup is reported in Ref. [56].

It can be observed from Figure 11.19a that the three-gel–electrolyte hybrid cell had better discharge performance than that of individual lead-acid and $NiMH_x$ batteries in a typical cycle. The hybrid cell operated at a higher discharge plateau of around 2.27 V, which is much higher than that of individual lead-acid (1.87 V) and $NiMH_x$ (1.25 V) batteries with the same gel electrolyte concentration. Again, the boost in cell voltage is due to extra free energy obtained from the electrochemical neutralisation, which is proportional to the *pH* difference of the acid–alkaline system, as explained in Section 11.2.

Five charge and discharge cycles of a hybrid cell with aqueous and gel electrolytes at a current density of 0.83 mA cm^{-2} are compared in Figure 11.19b. Each hybrid system was charged for 5 h. As shown in Figure 11.19b, there is only a slight difference in charge and discharge behaviour for the hybrid batteries using gel and wet electrolytes. The discharge plateau for a wet electrolyte is around 2.3 V, which is slightly higher than that of a gel electrolyte (2.27 V) in the first cycle. However, it is apparent that the discharge plateau decreased over cycles (to 2.16 V) for the hybrid battery system with wet electrolytes. The 5% decrease in discharge capacity (to 100 mAh) and voltage for the system with wet electrolytes in the subsequent cycles could be caused by the accumulation of salt in the middle salt chamber and a faster electrode corrosion rate. In contrast, the discharge plateau and discharge capacity of the hybrid system with gel electrolytes maintained at 2.25 V and 106 mAh, respectively. The improvement could be attributed to the use of gel electrolyte which (i) reduces the problem of salting-out in central chamber in aqueous electrolyte system as the rigid polymer matrix restraints the growth of crystal salt, (ii) prevents acid stratification and (iii) greatly reduces the electrode corrosion rate.

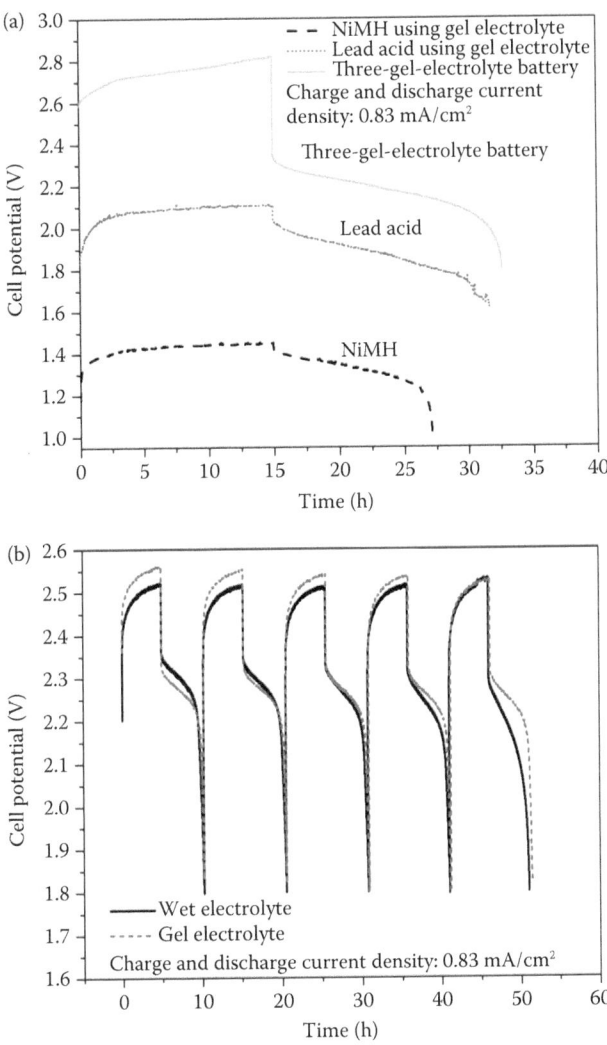

FIGURE 11.19
(a) The discharge performance of the PbO_2-MH_x hybrid battery in gel electrolyte. The distance between AEM and CEM are fixed at 1 cm and the electrolytes are 1 M H_2SO_4-silica gel, 1.28 M Na_2SO_4-PAAK gel and 3.24 M KOH-PAAK gel. The charge and discharge current density is 0.83 mA/cm². Individual performances of conventional lead-acid battery using 1 M H_2SO_4-silica gel electrolyte, and $NiMH_x$ battery using 3.24 M KOH-PAAK gel electrolyte are also included for comparison. (b) Five charge and discharge cycles of PbO_2-MH_x hybrid battery. The electrolytes are 1 M H_2SO_4-silica gel, 1.28 M Na_2SO_4-PAAK gel and 3.24 M KOH-PAAK gel electrolytes and 1 M H_2SO_4, 1.28 M Na_2SO_4 and 3.24 M KOH aqueous solutions, respectively. The inter-membrane distance is 1 cm, and the charge and discharge current density is 0.83 mA/cm². (Reprinted with permission from G.M. Weng, C.Y.V. Li and K.Y. Chan, *ECS Trans.* 41, 133–143. Copyright 2012, The Electrochemical Society.)

11.9.3 Lithium Ceramic Super-Ionic Conductor (LISICON) Film as Acid–Alkaline Separator

The direct borohydride fuel cell (DBFCs) usually operates in one single alkaline environment since the BH_4^- ion is unstable in neutral and acidic media. The theoretical OCV of a conventional DBFC in a single alkaline solution is 1.64 V. An acid–alkaline hybrid DBFCs can have oxidation of BH_4^- in alkaline and reduction of O_2 (or H_2O_2) in acid. The theoretical OCV of this acid–alkaline hybrid DBFCs can reach as high as 3.0 V. The reactions are

$$4H_2O_2 + 8H^+ + 8e^- \xrightarrow{discharge} 8H_2O \text{ (at the cathode)}$$
$$E = 1.77 \text{ V} \quad (11.40)$$

$$BH_4^- + 8OH^- \xrightarrow{discharge} BO_2^- + 6H_2O + 8e^- \text{ (at the anode)}$$
$$E = -1.24 \text{ V} \quad (11.41)$$

with the overall reaction:

$$BH_4^- + 4H_2O_2 \xrightarrow{discharge} BO_2^- + 6H_2O$$
$$\Delta E = 3.01 \text{ V} \quad (11.42)$$

Santos et al. [53] reported a hybrid DBFC having BH_4^- together with a Zn anode in 4 mol L^{-1} NaOH and H_2O_2 as the oxidant on a Pt cathode in 1 mol L^{-1} HCl. A Nafion 117 membrane was used as the separator. This hybrid DBFC had a high open-circuit voltage of 2.1 V, and a maximum power density of about 528 mW cm^{-2} can be obtained, but for less than 1 s. The operation's rapid decline after 6 h was mainly due to the bulk neutralisation from the crossover of H^+ through Nafion.

A ceramic lithium super-ionic conductor (LISICON) was used as the ionic interface for a similar alkaline–acid fuel cell by Zhou and coworker [54], as shown in Figure 11.20. The cell has two fuels, the zinc electrode and the borohydride solution. The lithium super-ionic conductor (LISICON) film is highly selective to Li^+ which crossover to the acid side during reaction for charge balancing, when OH^- is consumed in alkaline chamber and H^+ is consumed in the acid chamber. A very stable performance is obtained with the highly selective ceramic membrane. The OCV of this Li film direct borohydride fuel cell (DBFC) using peroxide oxidant remains at about 2.1 V after 24 h of testing. But the discharge performance of this hybrid fuel cell is greatly limited by the low ionic conductivity of the LISICON film. The maximum power density was only 1.54 mW cm^{-2} at a current density of 1.4 mA cm^{-2}. The application is therefore restricted to low current density, for example, in MFCs. A similar electrochemical system of an $Zn/KMnO_4$ aqueous cell was put forward with a 2.8 V operating cell voltage, approaching that of organic electrolyte-based batteries [64].

pH Differential Power Sources with Electrochemical Neutralisation

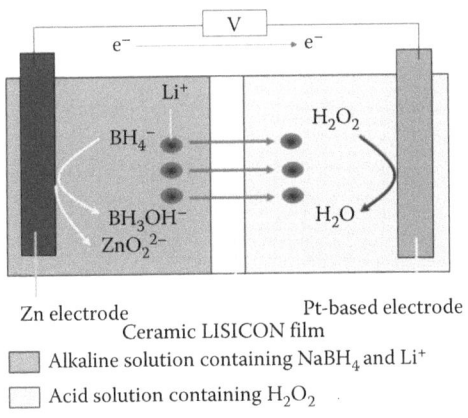

FIGURE 11.20
A schematic of an alkaline direct borohydride fuel cell coupled with acidic peroxide cathode and separated by a LISICON membrane. (From Y. Wang, P. He and H. Zhou, A novel direct borohydride fuel cell using an acid–alkaline hybrid electrolyte, *Energy Environ. Sci.* 3, 2010, 1515–1518. Reproduced by permission of The Royal Society of Chemistry.)

11.9.4 Membrane-Less Interface with Parallel Laminar Flows

Ferrigno et al. [65] reported the operation of a vanadium redox flow cell without using a membrane or separator. The mixing of anolyte and catholyte was prevented by maintaining parallel laminar flows of the electrolytes forced through a Y-shape entrance. The interface between the flow electrolytes plays the role of a separator across which ions diffuse and migrate.

The maximum operating voltage of a membrane-less all-vanadium redox flow battery was about 1.5 V, and varied accordingly with the state-of-charge, cell design and flow rate [65]. The same concept has been applied to dual electrolyte hydrogen–oxygen fuel cells by Ünlü et al. [33], and Cohen et al. [31], achieving an open circuit of 1.4 V. Membrane-less cells offer many advantages. Elimination of the membrane, which often is an expensive component, can reduce overall materials costs. No internal resistance of a membrane thus increases electrochemical efficiency and a higher operating voltage. Parallel laminar flow of two electrolytes can prevent convective mixing and minimise diffusion across the electrolyte/electrolyte interface caused by concentration gradients. However, migration of an ion under an electric field can still occur. Selective ion migration can be achieved by coherent alignment of the electric field with the gradients of ion species. This is similar to the case of a BPM where forward bias condition must be applied to prevent neutralisation between electrolytes on opposite sides of the interface.

Implementation of membrane-less cells, however, faces many inherent problems. Ancillary structures are needed to achieve the laminar flows, including pumps and special channel entrances. The operation is suitable to microfluidic designs with narrow channels and orientation independence

[66]. However, the pressure required for force convection increases with the decreasing channel size. The capacity of a pH change is also limited by the volume of acid and alkaline in the narrow channel. There are also critical regions in the operation that will have serious consequences when the laminar flow control is lost. An analysis of the operation of these membrane-less cells in microfluidics have been discussed by Chang et al. [67].

11.10 Summary and Outlook

The working voltage of aqueous electrolyte-based electrochemical power sources is limited by the thermodynamic window in water. Acid–alkaline dual electrolytes enable higher voltage of these power sources as a result of the internal pH difference. A wide range of devices from capacitors, batteries, fuel cells, flow batteries were demonstrated to perform with the advantageous features, but there are still unexplored devices and applications, for example, pairing different pH-sensitive half-cells listed in Table 11.1. Of these devices, adoption of the acid–alkaline dual electrolyte operation by flow batteries appears to be more promising, since flow batteries have similar operating characteristics of two electrolytes separated by a membrane. The flowing electrolyte can reduce the problems of non-electrochemical bulk neutralisation, fouling and salting.

The critical component to enable this type of power source is a separator preventing the mixing of acid and alkaline electrolyte, while allowing selective ion transport. BPM and its variation of two separate mono-polar membranes have been used to demonstrate feasibility of this technology. Depending on the bias (forward or reverse) across the membrane and the electrolytes adjacent to AEM and CEM, different ionic charge carriers are selected.

It should be noted that not all the BPM can be useful in this hybrid acid–alkaline cell. The particular BPM chosen for this acid–alkaline hybrid cell should possess the following characteristics: (1) adequate durability, mechanical strength, chemical stability, and resistance to strong bases and acids; (2) high perm-selectivity in the concentrated electrolyte; (3) low electrical resistance; and (4) thermal stability of the functional groups and structure of the membrane during thermal expansion and increased swelling.

The low current density of this acid-alkaline hybrid electrochemical system still restricts its further development. The key limiting factor is the ionic interface. Therefore, this technique will be widely applied when there is a specific functional membrane, like selective and improved ionic-conduction.

One scheme of preventing non-electrochemical self-neutralisation of protons and hydroxide ions is to exclude transport of another type of ion. The idea is demonstrated with a lithium ceramic super-ionic conductor (LISICON) film-based acid–alkaline hybrid DBPFC. A much higher cell voltage was demonstrated, but its discharging performance is greatly limited by the low ionic conductivity of the LISICON film with a maximum power density of only 1.54 mW cm^{-2}. This is the same order of magnitude of present Li ion batteries using LISICON. The membrane-less electrolyte/electrolyte interface under laminar flow control appears to be promising in microfluidic devices and offers the possibility of higher operating currents due to the reduced ionic resistance.

With the latest interests on sodium ion, potassium ion and other types of batteries, it is optimistic to anticipate development of new types of ion-selective interfaces that will function as barrier to *pH* gradient. In addition, there is room for optimising operation parameters using present interfaces for the various acid–alkaline devices.

Acknowledgements

The authors acknowledge the financial support for this project from Research Grants Council of Hong Kong (GRF HKU 700210P), Hong Kong Innovation Technology Fund (ITS/290/12), Small Project Funding (201209176091), University of Hong Kong SRT on Clean Energy, The University Development Fund on the Initiative of Clean Energy and Environment.

References

1. Nexergy, Energy density comparison, Accessed online 11 Feb. 2010, http://www.nexergy.com/battery-density.htm.
2. M. Pourbaix, *Atlas of Electrochemical Equilibria in Aqueous Solutions*, 2nd edn. National Association of Corrosion Engineers, Houston, 1974.
3. C. Wessells, R. Ruffo, R.A. Huggins and Y. Cui, Investigations of the electrochemical stability of aqueous electrolytes for lithium battery applications, *Electrochem. Solid-State Lett.* 13, 2010, A59–A61.
4. R.A. Huggins, Cause of the memory effect in nickel electrodes, *J. Power Sources* 165, 2007, 640–645.
5. T. Lee, Hydrogen over potential on pure metals in alkaline solution, *J. Electrochem. Soc.* 118, 1971, 1278–1282.
6. J. Tafel, On the polarization during cathodic hydrogen evolution, *Z. Phys. Chem.* 50, 1905, 641.
7. D.Y. Turaev, Use of ion-exchange membranes in chemical power cells, *Russ. J. Appl. Chem.* 78, 2005, 1615–1619.
8. O. Schaetzle, F. Barrière and U. Schröder, An improved microbial fuel cell with laccase as the oxygen reduction catalyst, *Energy Environ. Sci.* 2, 2009, 96–99.

9. A.M. Dreizler and E. Roduner, A fuel cell that runs on water and air, *Energy Environ. Sci.* 3, 2010, 761–764.
10. I. Ieropoulos, J. Greenman and C. Melhuish, Urine utilisation by microbial fuel cells; energy fuel for the future, *Phys. Chem. Chem. Phys.* 14, 2012, 94–98.
11. J.F. Walther, Process for production of electrical energy from the neutralization of acid and base in a bipolar membrane cell, *U.S. Patent* 4, 311, 771, 1982.
12. A. Yaroslavtsev and V. Nikonenko, Ion-exchange membrane materials: Properties, modification, and practical application, *Nanotechnol. Russ.* 4, 2009, 137–159.
13. R. Nagarale, G. Gohil and V.K. Shahi, Recent developments on ion-exchange membranes and electro-membrane processes, *Adv. Colloid Interface Sci.* 119, 2006, 97–130.
14. T. Xu, Ion exchange membranes: State of their development and perspective, *J. Membr. Sci.* 263, 2005, 1–29.
15. A. Kemperman, *Handbook on Bipolar Membrane Technology*. Twente University Press, Enschede, 2000.
16. R. Lacey, Energy by reverse electrodialysis, *Ocean Eng.* 7, 1980, 1–47.
17. F.L. Ramp, Secondary batteries powered by forced ionisation, *Nature* 278, 1979, 335–337.
18. J.H. Balster, *Membrane Module and Process Development for Monopolar and Bipolar Membrane Electrodialysis*. University of Twente, 2006.
19. F. Wilhelm, Bipolar membrane electrodialysis, PhD thesis, University of Twente, 2001.
20. S. Mafe and P. Ramirez, Electrochemical characterization of polymer ion–exchange bipolar membranes, *Acta Polym.* 48, 1997, 234–250.
21. J.J. Krol, *Monopolar and Bipolar Ion Exchange Membranes*. Universiteit Twente, Enschede, 1997.
22. H. Hurwitz and R. Dibiani, Investigation of electrical properties of bipolar membranes at steady state and with transient methods, *Electrochim. Acta* 47, 2001, 759–773.
23. H. Strathmann, J. Krol, H.J. Rapp and G. Eigenberger, Limiting current density and water dissociation in bipolar membranes, *J. Membr. Sci.* 125, 1997, 123–142.
24. H. Strathmann, H.J. Rapp, B. Bauer and C. Bell, Theoretical and practical aspects of preparing bipolar membranes, *Desalination* 90, 1993, 303–323.
25. R. Simons, Water splitting in ion exchange membranes, *Electrochim. Acta* 30, 1985, 275–282.
26. T. Aritomi, T. Van den Boomgaard and H. Strathmann, Current–voltage curve of a bipolar membrane at high current density, *Desalination* 104, 1996, 13–18.
27. J. Krol, M. Jansink, M. Wessling and H. Strathmann, Behaviour of bipolar membranes at high current density: Water diffusion limitation, *Sep. Purif. Technol.* 14, 1998, 41–52.
28. T. Xu, Development of bipolar membrane-based processes, *Desalination* 140, 2001, 247–258.
29. S.A. Cheng and K.Y. Chan, High-voltage dual electrolyte electrochemical power sources, *U.S. Patent* 7, 344, 801, 2008.
30. S.A. Cheng and K.Y. Chan, High-voltage dual electrolyte electrochemical power sources, *ECS Trans.* 25, 2010, 213–219.
31. J.L. Cohen, D.J. Volpe, D.A. Westly, A. Pechenik and H.D. Abruña, A dual electrolyte H_2/O_2 planar membraneless microchannel fuel cell system with open circuit potentials in excess of 1.4 V, *Langmuir* 21, 2005, 3544–3550.

32. M. Ünlü, J. Zhou and P.A. Kohl, Study of alkaline electrodes for hybrid polymer electrolyte fuel cells, *J. Electrochem. Soc.* 157, 2010, B1391–B1396.
33. M. Ünlü, J. Zhou and P.A. Kohl, Hybrid anion and proton exchange membrane fuel cells, *J. Phys. Chem. C* 113, 2009, 11416–11423.
34. M. Ünlü, J. Zhou and P.A. Kohl, Hybrid polymer electrolyte fuel cells: alkaline electrodes with proton conducting membrane, *Angew. Chem. Int. Ed.* 122, 2010, 1321–1323.
35. H. Kim, M. Ünlü, J. Zhou, I. Anestis-Richard and P.A. Kohl, Anionic–cationic bi-cell design for direct methanol fuel cell stack, *J. Power Sources* 195, 2010, 7289–7294.
36. S.K. Chaudhuri and D.R. Lovley, Electricity generation by direct oxidation of glucose in mediatorless microbial fuel cells, *Nat. Biotechnol.* 21, 2003, 1229–1232.
37. G.C. Gil, I.S. Chang, B.H. Kim, M. Kim, J.K. Jang, H.S. Park and H.J. Kim, Operational parameters affecting the performance of a mediator-less microbial fuel cell, *Biosens. Bioelectron.* 18, 2003, 327–334.
38. J.K. Jang, I.S. Chang, K.H. Kang, H. Moon, K.S. Cho and B.H. Kim, Construction and operation of a novel mediator- and membrane-less microbial fuel cell, *Process Biochem.* 39, 2004, 1007–1012.
39. S. Oh, B. Min and B.E. Logan, Cathode performance as a factor in electricity generation in microbial fuel cells, *Environ. Sci. Technol.* 38, 2004, 4900–4904.
40. F. Zhao, F. Harnisch, U. Schröder, F. Scholz, P. Bogdanoff and I. Herrmann, Application of pyrolysed iron (II) phthalocyanine and CoTMPP based oxygen reduction catalysts as cathode materials in microbial fuel cells, *Electrochem. Commun.* 7, 2005, 1405–1410.
41. G.T.R. Palmore and H.H. Kim, Electro-enzymatic reduction of dioxygen to water in the cathode compartment of a biofuel cell, *J. Electroanal. Chem.* 464, 1999, 110–117.
42. S. Topcagic and S.D. Minteer, Development of a membraneless ethanol/oxygen biofuel cell, *Electrochim. Acta* 51, 2006, 2168–2172.
43. A.T. Heijne, H.V.M. Hamelers and C.J.N. Buisman, Microbial fuel cell operation with continuous biological ferrous iron oxidation of the catholyte, *Environ. Sci. Technol.* 41, 2007, 4130–4134.
44. K. Rabaey, P. Clauwaert, P. Aelterman and W. Verstraete, Tubular microbial fuel cells for efficient electricity generation, *Environ. Sci. Technol.* 39, 2005, 8077–8082.
45. H. Liu, S. Cheng and B.E. Logan, Production of electricity from acetate or butyrate using a single-chamber microbial fuel cell, *Environ. Sci. Technol.* 39, 2005, 658–662.
46. H. Liu, S. Cheng and B.E. Logan, Power generation in fed-batch microbial fuel cells as a function of ionic strength, temperature, and reactor configuration, *Environ. Sci. Technol.* 39, 2005, 5488–5493.
47. K. Rabaey, W. Ossieur, M. Verhaege and W. Verstraete, Continuous microbial fuel cells convert carbohydrates to electricity, *Water Sci. Technol.* 52, 2005, 515–523.
48. I. Yamanaka, T. Onisawa, T. Hashimoto and T. Murayama, A fuel–cell reactor for the direct synthesis of hydrogen peroxide alkaline solutions from H_2 and O_2, *ChemSusChem* 4, 2011, 494–501.
49. I. Yamanaka, T. Onizawa, S. Takenaka and K. Otsuka, Direct and continuous production of hydrogen peroxide with 93% selectivity using a fuel–cell system, *Angew. Chem. Int. Ed.* 42, 2003, 3653–3655.
50. E. Lobyntseva, T. Kallio and K. Kontturi, Bipolar membranes in forward bias region for fuel cell reactors, *Electrochim. Acta* 51, 2006, 1165–1171.

51. U.B. Demirci, Direct borohydride fuel cell: Main issues met by the membrane–electrodes-assembly and potential solutions, *J. Power Sources* 172, 2007, 676–687.
52. B. Liu and Z. Li, Current status and progress of direct borohydride fuel cell technology development, *J. Power Sources* 187, 2009, 291–297.
53. D. Santos and C. Sequeira, Zinc anode for direct borohydride fuel cells, *J. Electrochem. Soc.* 157, 2010, B13–B19.
54. Y. Wang, P. He and H. Zhou, A novel direct borohydride fuel cell using an acid–alkaline hybrid electrolyte, *Energy Environ. Sci.* 3, 2010, 1515–1518.
55. H. Li, G.M. Weng, C.Y.V. Li and K.Y. Chan, Three electrolyte high voltage acid–alkaline hybrid rechargeable battery, *Electrochim. Acta* 56, 2011, 9420–9425.
56. G.M. Weng, C.Y.V. Li and K.Y. Chan, Lead acid-NiMH hybrid battery system using gel electrolyte, *ECS Trans.* 41, 2012, 133–143.
57. C. Ponce de Leon, A. Frías-Ferrer, J. González-García, D. Szánto and F.C. Walsh, Redox flow cells for energy conversion, *J. Power Sources* 160, 2006, 716–732.
58. M. Skyllas–Kazacos and F. Grossmith, Efficient vanadium redox flow cell, *J. Electrochem. Soc.* 134, 1987, 2950–2953.
59. G.M. Weng, C.Y.V. Li and K.Y. Chan, Study of the electrochemical behavior of high voltage vanadium-metal hydride hybrid semi-flow battery, *ECS Trans.* 53, 2013, 39–50.
60. G.M. Weng, C.Y.V. Li and K.Y. Chan, High voltage vanadium-metal hydride rechargeable semi-flow battery, *J. Electrochem. Soc.* 160, 2013, A1384–A1389.
61. J.W. Post, H.V.M. Hamelers, G.J.W. Euverink, S.J. Metz and C.J.N. Buisman, Power generation by reversed electro dialysis (RED): Prevention of fouling. *Wetsus*, Centre of Excellence for Sustainable Water Technology, Leeuwarden.
62. C. Iwakura, S. Nohara, N. Furukawa and H. Inoue, The possible use of polymer gel electrolytes in nickel/metal hydride battery, *Solid State Ionics* 148, 2002, 487–492.
63. S. Martha, B. Hariprakash, S. Gaffoor and A. Shukla, Performance characteristics of a gelled-electrolyte valve-regulated lead-acid battery, *Bull. Mater. Sci.* 26, 2003, 465–469.
64. L. Chen, Z. Guo, Y. Xia and Y. Wang, High-voltage aqueous battery approaching 3 V using an acidic–alkaline double electrolyte, *Chem. Commun.* 49, 2013, 2204–2206.
65. R. Ferrigno, A.D. Stroock, T.D. Clark, M. Mayer and G.M. Whitesides, Membraneless vanadium redox fuel cell using laminar flow, *J. Am. Chem. Soc.* 124, 2002, 12930–12931.
66. J. Xuan, M.K. Leung, D.Y. Leung and H. Wang, Towards orientation-independent performance of membraneless microfluidic fuel cell: Understanding the gravity effects, *Appl. Energy* 90, 2012, 80–86.
67. M.H. Chang, F. Chen and N.S. Fang, Analysis of membraneless fuel cell using laminar flow in a Y-shaped microchannel, *J. Power Sources* 159, 2006, 810–816.

Index

A

AB_2 alloy, 266, 269
AB_5 alloy, 266, 269
ab initio, 196, 204
Absorbent glass mat (AGM), 325
AC, see Activated carbon (AC); Alternating current (AC)
Accumulator, see Rechargeable battery
Acetonitrile (AN), 177
Acid and alkaline electrolyte
  CEM and AEM, 456
  ion-exchange membranes, 458
  PbO_2–MH_x hybrid battery and ion transport, 457
  RED, 459
Acid–alkaline hybrid fuel cells, 442
  BPM-based fuel-cell reactors, 449
  direct ethanol fuel cell stack, 447
  H_2O_2 production, 449
  half-cell reactions, 443
  hybrid MEA cell configuration, 444, 445
  hydrogen–oxygen fuel cell, 442
  polarization curve, 442, 446
  self-humidifying hybrid fuel cell, 446
Acid–alkaline microbial fuel cells
  compartments, 447, 448
  redox mediators, 448
  working principle, 448
Acid stratification, 460
Acidity, 23, 305
Activated carbon (AC), 89
  anode, 325
  charge and discharge reactions, 326
  coulombic efficiency, 333
  microporous, 164
  with nickel oxide, 7
  12 V lead–carbon HUC, 329
Activation barrier, 11, 204, 229, 238, 240, 241, 243, 333
Activation energy, 104, 204, 405, 412, 425

Activation losses, 75–77
Activation polarisation, see Activation losses
Adenosine triphosphate (ATP), 59
AEM, see Anion-exchange membranes (AEM)
Aggregation, 152, 180
AGM, see Absorbent glass mat (AGM)
Air cathode, 75, 88, 93, 96, 98, 444
Air-cathode microbial fuel cells, 96–98
Alkali-metal-based oxides, 8
All-solid-state thin-film lithium ion battery, 206; see also Layer-structured cathode materials
  BE-ESM technique, 211, 213
  binding energies, 209
  cross-sectional layout, 210
  layered cathode for, 206–209
  $LiMnO_2$ XRD patterns, 208
  $LiNi_{1-x}Co_xO_2$ XRD patterns, 207
  nanoscale mapping of Li^+ diffusion, 209–214
  oscillatory surface displacement, 212
  Q-factor mapping, 212, 214
  resonance amplitude, 214
Alloying, 133, 269
Alternating current (AC), 322
Amine scrubbing, 6
Ammonia, 29, 66, 99
AN, see Acetonitrile (AN)
Anaerobic microbial fuel cell, 67, 100, 109, 110, 447
Anion, 14, 32, 138, 357, 439
Anion-exchange membranes (AEM), 357, 439
  membrane selectivity and ion conductivity, 360
  morphology changes, 360
  NF membrane, 358, 359
  PAN-H, 358
  properties and performance, 362–365
  quaternised groups, 357, 358

Anion-exchange (*Continued*)
   vanadium ions and protons, 361
   V/H selectivity, 359
Anode materials, 89, 286; *see also* Cathode materials; Lithium–ion batteries (LIBs)
   carbonaceous materials, 89
   characteristics comparison, 90
   electrode materials, 91
   graphite fiber brush, 90
   lithium metal, 286–288
   modifications and improvements, 92
   performance, 84
   Si–C composites and inter-metallic phases, 313
Anodes, advanced, 129; *see also* Cathodes, advanced; Lithium batteries
   carbon matrix, 133
   characteristics and properties, 131
   electrochemical process, 131
   expansion–contraction process, 130
   LTO, 129
   nanocomposite approach, 132
   Sn–C nanocomposite, 132
Anodic electron transfer mechanism, 68; *see also* Exo-cellular electron transfer
   direct cell-surface electron transfer, 68–69
   direct electron transfer, 70
   electron transfer, 69, 70–72
   requirements, 68
Anolyte, 20, 27, 349, 372, 375, 438, 448
Anthraquinone-1, 6-disulphonic acid (AQDS), 102
Anthrone, 42, 43
Anti-site defect, 243, 244
Aqueous air-cathode MFCs, 98
Aqueous electrolyte, 147; *see also* Non-aqueous electrolyte
Arrhenius plot, 406
Artificial leaves, 41
Artificial photosynthesis, 9
   electrochemical procedures, 11
   methanol economy, 10
   multi-stage reaction, 9
   reactions, 10
ATP, *see* Adenosine triphosphate (ATP)

B

Bacterial batteries, 106, 107
   anode and cathode compartments, 107
   cell voltage and power of cell, 108
Bacterial metabolic losses, 77
Band-excitation electrochemical strain microscopy (BE-ESM), 209, 210
   feature, 212
   high-resolution imaging capability, 211
   resonance amplitude and Q-factor, 214
   variations, 213
Basicity, 305, 369
Batch reactors, 25, 31
Battery
   all-solid-state thin film Li-ion battery, 206–214
   iron-chrome flow battery, 349
   lead acid battery, 257–261
   lead oxide/metal hydride, 259
   LiPBs, 135
   lithium ion battery, 206–214
   lithium-air battery, 145–156
   lithium-sulfur battery, 142–145
   Ni-Cd batteries, 262–264
   $NiMH_x$ battery, 266–270
   PbO_2/MH_x hybrid battery, 458
   (PbO_2/MH_x) secondary battery, 258, 259
   polysulfide-bromine flow battery, 434
   potassium ion battery, 269
   sodium ion battery, 247
   three electrolyte acid alkaline, 456, 458
   $V-MH_x$ hybrid battery, 453
   VFB, 349
   VRLA, 257, 260
   zinc-bromine flow battery, 349, 434
BE-ESM, *see* Band-excitation electrochemical strain microscopy (BE-ESM)
BEAMRs, *see* Bioelectrochemically assisted microbial reactors (BEAMRs)
BET, *see* Brunauer–Emmett–Teller (BET)

Index

BET surface area, 170, 177, 275, 302
Bi-functional Pt/Au catalyst, 150
Bimetallic catalysis, 17–18
Bimetallic electrode, 42
Bio-cathode microbial fuel cells, 98–99
Biocatalysed electrolysis system, 104
Biocathode, 85, 92
Bioelectrochemically assisted microbial reactors (BEAMRs), 104
Bioelectrochemistry, 58, 104
Biofilm, 64, 67, 70, 73, 107
Bipolar membrane (BPM), 438
    acid and alkaline electrolytes, 438
    AEM, 439
    applications, 440
    characteristics of, 441
    function and structure, 440
    I–V curve, 439
    junction, 440
Bipolar plate, 350, 351, 377
Bisulfate, *see* Bisulphate
Bisulphate, 414, 443
BMIMBF$_4$, *see* 1-butyl-3-methylimidazoliumtetrafluoborate (BMIMBF$_4$)
Bode diagram, 263
Bond-valence method (BV method), 229
Bond-valence sums (BVS), 244
Bonding energy, 203
Born–Oppenheimer approximation, 225
BPM, *see* Bipolar membrane (BPM)
Bravais lattice, 193
Brunauer–Emmett–Teller (BET), 170
Buffer solution, 27, 31, 86
Bulk neutralisation, 429, 462, 464
1-butyl-3-methylimidazoliumtetrafluoborate (BMIMBF$_4$), 32
BV method, *see* Bond-valence method (BV method)
BVS, *see* Bond-valence sums (BVS)

C

C-type cytochrome, 68
C$_{18}$-TMS, *see* n-octadecyl trimethoxysilane (C$_{18}$-TMS)
Calcination, 167, 279
Capacitance, 179, 329, 334

Capacitors, 322
    EDLC, 323
    electrochemical, 342
    parallel-plate electrolytic, 325
Capacity decay, 142, 269, 355
Capacity fade, 272, 276, 313
Capacity retention, 152, 184, 279, 374, 460
Carbon
    black, 24, 37, 152
    cloth, 38, 84, 95, 302
    CNT, 37
    composite, 133, 156, 285
    felt, 89, 91, 366, 370
    monolith, 307, 308, 309, 310
    nanofiber, 306
    nanostructure, 179
    xerogel catalyst, 40
Carbon-based materials, 366–367
Carbonaceous materials, 89; *see also* Anode materials
    GNS, 310, 311, 312
    graphitic carbon, 288
    lithium de-intercalation, 289, 309
    lithium storage in, 311
    natural graphite, 290–292
Carbon cotton (CC), 305
Carbon dioxide (CO_2), 3, 5, 7, 19
Carbon dioxide capture and storage (CSS), 5
    alkali-metal-based oxides, 8
    amine scrubbing, 6
    calcium oxide, 7
    membrane processes, 6, 7
    MOFs, 8
Carbon dioxide, electroreduction
    artificial photosynthesis, 9–11
    cathode materials, 21
    cell separators, 27–28
    copper, 21–24
    cost-effective, 11
    CSS, 5–8
    electrochemical reactors, 25–27
    emission, 2–5
    ethanol and 2-propanol synthesis, 37
    experimental methodology for, 20
    formic acid synthesis, 29–31
    future directions, 39–43
    GDEs, 24
    high-purity electrolytes, 25

Carbon dioxide, electroreduction (*Continued*)
  hydrocarbon synthesis, 37–39
  imidazolonium cations, 16, 17
  kinetics of, 14–15
  mediated electrocatalytic reduction, 15–18
  methanol synthesis, 34–37
  organic carbonates preparation, 31–33
  in organic media, 18–20
  polar interaction of, 19
  products and standard reduction potentials, 12
  synthesis of fuels from, 8–9
  thermodynamics of, 12–14
  water and organic solvents, 18
Carbon dioxide emission, 2
  automotive applications, 3
  burning fossil fuels, 2
  into chemicals, 5
  greenhouse effect, 2
  increase in atmospheric, 3
  reduction, 4
Carbon nanotube (CNT), 37, 310
Carbonization, 167, 293
Carbonyl sulphide (COS), 20
Catalyst, 34, 93, 153, 436
  catalyst-bed, 266
  support, 165
Catalyst support in fuel cell, 180
  chronoamperogram, 182, 183
  SEM image, 181
  structural parameters for HMC, 182
  TEM image, 180, 181
Cathode materials, 91, 271; *see also* Lithium–ion batteries (LIBs)
  aerobic biocathodes, 92
  air-cathode, 93
  design, 224
  galvanostatic profiles, 275
  lattice parameters, 277
  LIB cathode materials, 272
  $LiCoO_2$ and $LiTiS_2$ crystal structures, 273, 274
  $LiFePO_4$, 272, 273, 284
  $LiMn_2O_4$, 282, 283
  Li–Mn–O system, 280, 282
  $LiNi_{1/3}Co_{1/3}Mn_{1/3}O_2$, 278
  $LiNiO_2$, 276
  $Li_xMn_yO_z$, 279
  Mn^{3+}, 281
  nanostructured materials, 271
  non-Pt catalysts, 93
  olivine, 284
  PEDOT-coated nanorods, 286
  performance, 85–86
  Pt catalyst, 91
  spray-drying method, 274
Cathodes, advanced, 138; *see also* Electrolytes, advanced
  characteristics and properties for, 141
  intrinsic resistance, 139
  lithium iron phosphate, 138, 140
  thin surface carbon coating, 140
Cathodic electron mediator, 448
Catholyte, 20, 66, 372, 463
Cation, 136, 440, 449
Cation exchange membrane (CEM), 27, 31, 98, 355, 439
Cation mixing, 195, 197, 198, 205, 272
CC, *see* Carbon cotton (CC)
Cell polarization, 182
Cell separators, 27–28
Cell stacks, 99, 351, 375, 377, 379, 447
Cell yield, 81
CEM, *see* Cation exchange membrane (CEM)
Chalcogenides, 272, 273
Charge balance, 458
Charge plateau, 202
Charge transfer resistance, 333
Chemical energy, 29, 58, 123, 256, 348
Chemical oxygen demand (COD), 80
Chemical stability, 89, 315, 352, 357, 464
Chemical vapor deposition (CVD), 164
Chronoamperogram, 182, 183
CMK-3 carbon, 168
  CV plots, 171
  cycling and rate performance, 175
  galvanostatic plots, 172, 173
  nyquist plots, 174
  parameters, 174
CNT, *see* Carbon nanotube (CNT)
Co-electrolysis, 4–5
Co-precipitation method, 199, 205, 206
COD, *see* Chemical oxygen demand (COD)

Index

Coke-based anode, 288
Columbic efficiency, 133, 172, 173
Combustion method, 206
Competing mechanisms in CO_2 electroreduction, 15
Composites, 27, 132, 144, 312, 313
Computational methods, 224, 237
Concentrated photovoltaic field (CPV field), 380
Concentration gradient, 459, 463
Concentration losses, 77–78
Concentration polarisation, *see* Concentration losses
Conductivity-viscosity curve, 415, 416, 417
Constant-power discharge data, 333–335
Continuous flow modes, 99
Convective mixing, 463
Conventional layer-structured materials, 193; *see also* Integrated layer-structured materials
   $LiCoO_2$, 193–195
   $LiNi_{1/2}Mn_{1/2}O_2$ and $LiNi_{1/3}Co_{1/3}Mn_{1/3}O_2$, 196–200
   $LiNiO_2$ and $LiMnO_2$, 195–196
Conversion electrode, 142
Convex hull construction, 228, 246
Copper, 21
   approaches, 23–24
   bronze electrodes, 23
   catalytic effects, 21
   ethylene, 22
Copper cathode, 14, 20, 21–22, 32, 37
Core-shell, 199, 200
Coronene, 303, 304
Correlation factor, 213
Corrosion, 16, 106, 259, 266, 460
Corrosion potential, 106, 107
COS, *see* Carbonyl sulphide (COS)
Cost reduction, 95
Coulombic efficiency, 67, 80, 99, 333, 378, 448
CPV field, *see* Concentrated photovoltaic field (CPV field)
Crosslink, 356, 357
Crystal structure, 193, 230–232, 241, 276, 284

CSS, *see* Carbon dioxide capture and storage (CSS)
Cubic spinel structure, 196
Current carrier, 458
Current density, 15, 28, 67, 177, 179, 356, 378, 462
Current rectification, 439
Current-voltage curve (I-V curve), 76
Cut-off voltage, 263, 279, 329, 333
CV, *see* Cyclic voltammetry (CV)
CVD, *see* Chemical vapor deposition (CVD)
Cycle life, 131, 274, 286, 296, 304
Cyclic carbonates, 32
Cyclic voltammetry (CV), 32, 107, 171, 370, 437
Cycling performance, 148, 173, 184, 192, 196, 200, 309
Cycling stability, 133, 314

D

Dalian Institute of Physics, Chinese Academy of Sciences (DICP), 380, 382
Daniell cell, 434, 438, 439
DBFC, *see* Direct borohydride fuel cell (DBFC)
DC, *see* Direct current (DC)
De-alloying, 130, 133
Deintercalation, 140, 193, 213
Delithiation
   crystal structure and volume change, 230–232
   electronic structure and redox reaction, 232–236
Delivery rate, 322
Dendrite, 125, 126
Density functional theory (DFT), 13, 204, 226
Density of state (DOS), 234, 235
Dentritic growth, 125
Depth of discharge (DOD), 127
Desalination, 459
Desulfuromonas species, 66, 67, 101
DFT, *see* Density functional theory (DFT)
Dibromobenzenes, 42

DICP, see Dalian Institute of Physics, Chinese Academy of Sciences (DICP)
Diffusion, 24, 97, 204–205, 209, 210, 228–299, 238–244
Diffusion coefficient, 195, 213, 273
Diffusion layer, 24, 98, 148
Diffusion pathway, 229, 238, 240, 243
Diffusion theory, 86
Dimension stability, 356
Dimensionally stable electrode (DSA), 361
Dimerisation process, 15
Dimethyl carbonate (DMC), 19, 32, 33, 171, 290
Direct borohydride fuel cell (DBFC), 462, 463
Direct cell-surface electron transfer, 68–69; see also Exo-cellular electron transfer
Direct current (DC), 322
Direct electron transfer, 70; see aslo Exo-cellular electron transfer
Direct methanol fuel cell (DMFC), 433, 447
Discharge plateau, 155, 204, 460
Dissociation, 35, 417, 440, 448
Dissolved oxygen (DO), 58, 98
DMC, see Dimethyl carbonate (DMC)
DMFC, see Direct methanol fuel cell (DMFC)
DO, see Dissolved oxygen (DO)
DOD, see Depth of discharge (DOD)
Donnan exclusion, 439
DOS, see Density of state (DOS)
DSA, see Dimensionally stable electrode (DSA)
Dual electrolyte, 442, 451, 457, 463
Dual-chamber reactor, 79, 96, 98, 100
Durability, 84, 89, 183, 315, 375

E

E-TEK, 180, 182
EC, see Electrochemical capacitor (EC); Ethylene carbonate (EC)
EC–DMC, see Ethylene carbonate–dimethyl carbonate (EC–DMC)
EDLC, see Electrical double layer capacitor (EDLC)
EE, see Energy efficiency (EE)
EES, see Electrical energy storage (EES)
Efficiency
  Columbic efficiency, 133, 172, 173
  energy efficiency, 15, 30, 81, 343, 377–378, 425
  voltage efficiency, 142, 352, 378
EIS, see Electrochemical impedance spectroscopy (EIS)
Electric conductivity, 84, 89
Electric current, 256, 322
Electric vehicle (EV), 121, 122, 124, 174, 192, 342
Electrical conductivity, 73, 165, 281, 366, 369; see also Electric conductivity
Electrical double layer capacitor (EDLC), 176, 323
  electrochemical characterisation, 177
  FE-SEM image, 176
  galvanostatic charge–discharge, 177
  high rate capability, 178
  HMC capacitance retention, 179–180
  nitrogen adsorption–desorption isotherm, 176, 177
Electrical energy, 322
Electrical energy storage (EES), 347
  applications, 348
  FBS, 349
Electrical field potential gradient, 378
Electrical potential gradient, 437
Electricity, 322
Electro-capacitors, 435, 436
Electrocarboxylation, 42, 43
Electrocatalysis, 15, 16
Electrocatalyst, 10, 32, 37, 368
Electrochemical
  activation, 16, 32, 42, 182, 202
  degradation, 212, 213
  performance, 184, 283, 290, 292, 305
  power source, 429, 430, 438
  reaction pathways, 203
  strain, 212, 213
Electrochemical capacitor (EC), 164; see also Electrical double layer capacitor (EDLC); Electro-capacitor

Index 475

Electrochemical impedance
    spectroscopy (EIS), 107, 174
Electrochemical manganese dioxide
    (EMD), 278
Electrochemical neutralisation, 432
    acid–alkaline hybrid capacitor, 438
    application of, 438
    discharge process, 437
    electro-capacitors, 435, 436
    oxygen reduction, 435, 436
Electrochemical reactors, 25
    batch and flow reactors, 25
    for CO_2 reduction, 25
    modes, 25–26
    PEC cell, 26
Electrode corrosion, 460
Electrode/electrolyte interface, 16, 172, 175, 323, 369
Electrode materials, 89, 91, 255, 361; see also Vanadium flow batteries (VFBs)
    anode materials, 89–91, 92
    batteries, 256–257
    carbon paper coating, 371, 372
    carbon-based materials, 366–367
    cathode materials, 91–93
    cubic nanocatalysts, 370, 371
    electrolytes for lithium batteries, 313–315
    graphene edge plane or carbon atoms, 368
    lead–acid batteries, 257–261
    modification, 367–372
    N-MPC, 369, 370
    nickel-based batteries, 261–270
    non-Pt catalysts, 93
    oxidation methods, 369
Electrode modification, 370
Electrode potential, 21, 179, 431, 435, 450
Electrolytes, 372, 376; see also Gel-type electrolyte (GPE)
    active species, 372
    aqueous electrolyte, 147
    capacity retention profiles, 374
    chemistry, 87–88
    composition, 94, 396
    conductivity, 333, 425
    discharge curves and cycle response, 150, 152
    durability of system, 375
    gel electrolyte, 460, 461
    glyme electrolytes, 149
    graphene, 152
    instability, 148
    issue, 149
    Li–air batteries, 149
    liquid electrolyte, 206, 373
    non-aqueous electrolyte, 148
    OER catalysts, 150, 151
    organic electrolyte, 25, 184, 429
    organic liquid electrolyte, 128, 143
    performance of VFBs, 373
    polymer electrolyte, 24, 126
    solid electrolyte, 155, 206
    TEGDME, 149
    viscosity, 414
Electrolytes, advanced, 133; see also Lithium batteries
    characteristics and properties for, 138
    GPEs, 135
    ILs, 136, 137
    imidazolium-based IL, 136
    ionic conductivity, 137
    issues, 133
    LiPB, 135
    PEO-based membranes, 134
    polyethylene oxide formula, 134
Electromobility, 123
Electromotive force (EMF), 164, 431
Electron acceptor, 57, 59, 66, 83, 103
Electron donor, 42, 61, 100
Electron shuttle, 70, 71, 72
Electron transfer, 70; see also Exo-cellular electron transfer
    endogenous redox mediators, 71
    exogenous redox mediators, 70–71
    reduced metabolic products, 72
Electroneutrality, 58, 437, 443
Electronic conductivity, 139, 142, 224, 269, 286
Electronic Hamiltonian operator, 225
Electronic interaction, 198
Electronic structure, 232, 234, 244
Electroreduction in organic media, 2, 13
EMD, see Electrochemical manganese dioxide (EMD)
EMF, see Electromotive force (EMF)
Empirical coefficient, 407, 408, 411, 421

Empirical model for Vanadium (III) sulphate solution
 conductivity effect, 420–422
 for density, 399–401
 for viscosity, 410–412
Endogenous redox mediators, 71
Energy density
 gravimetric energy density, 270, 322
 volumetric energy density, 132, 270, 322
Energy efficiency (EE), 81, 343, 353, 377, 425
Equilibrium, 179, 258, 414, 420
Equilibrium constant, 415
Equilibrium intercalation voltage, 228
Equivalent series resistance (ESR), 330, 438
Et_4NBF_4, see Tetrafluoroborate (Et_4NBF_4)
Ethanol synthesis, 37, 42
Ethylene, 12, 22
Ethylene carbonate (EC), 171, 290, 314
Ethylene carbonate–dimethyl carbonate (EC–DMC), 127
EV, see Electric vehicle (EV)
Exchange-correlation energy, 227
Exo-cellular electron transfer, 59
 bacterial membrane respiration, 60
 biological standard potentials, 61
 electron transfer mechanism, 60
 energy flux in MFC, 61, 62
 MFC process, 61
Exoelectrogen, 57, 61, 63, 71, 100
Exogenous redox mediators, 68, 70–71

F

Face seal, 379, 380
FAD, see Flavin adenine dinucleotide (FAD)
Faraday's law, 454
FBS, see Flow battery system (FBS)
Fe-Cotetramethoxyphenylporphyrin (FeCoTMPP), 102
FePc, see Iron phthalocyanine (FePc)
Ferricyanide ($K_3[Fe(CN)_6]$), 92
Filter-press reactor, 15
Firmicutes, 66
First-principles approach, 224
 average voltage, 246

calculations, 224–227
crystal structure and volume change, 230–232
density of, 235
electronic structure and redox reaction, 232–236
integrated spin density, 233
lattice parameters, 232
Li diffusion activation barriers, 248
Li rechargeable battery, 223
Li/Na ion diffusion, 228–229
lithium intercalation voltages, 236–238
lithium ion diffusion, 238–244
lithium/sodium intercalation voltage, 228
$Na_4Fe_3(PO_4)_2(P_2O_7)$ crystal structure, 245
for new cathode materials, 244–248
polyanionic cathode materials, 229
sodium vacancy activation barriers, 247
3D sodium diffusion, 247
Flavin adenine dinucleotide (FAD), 59
Flooding, 444
Flow battery system (FBS), 349; see also Vanadium redox flow battery (VRFB)
 iron-chrome flow battery, 349
 polysulfide-bromine flow battery, 434
 VFB, 349
 zinc-bromine flow battery, 349, 434
Flow field, 378
Flow pattern, 375, 377, 378
Flow reactors, 25, 31
Formate, 2, 15, 21, 29–31, 72
Formic acid synthesis, 29; see also Methanol synthesis
 GDEs, 30
 hydrogen, 29
 ruthenium phosphine systems, 30
 tin-based GDEs, 31
Forward bias, 443, 445, 449
Fourier-filtered image, 196
Free energy, 10, 13, 60, 228, 460
Fuel cell
 acid alkaline fuel cell, 442, 450
 anaerobic microbial fuel cell, 67, 100, 109, 110, 447

Index 477

DBFC, 462, 463
DMFC, 433, 447
low temperature fuel cell, 165
MFC, 58, 82, 435
PEMFC, 165, 184
PMFC, 102–104
SMFC, 100–102
V/O$_2$ fuel cell, 396
Functional group, 104, 355, 368, 369, 464

G

Galvanostatic cycling, 173, 212
Gas chromatography (GC), 28
Gas diffusion electrode (GDE), 2, 24, 426, 442
Gas diffusion membrane (GDM), 28
GCE, *see* Glassy carbon electrode (GCE)
GDE, *see* Gas diffusion electrode (GDE)
Gel electrolyte, 460–461
Gel-type electrolyte (GPE), 135
Generalised gradient approximation (GGA), 227; *see also* Born–Oppenheimer approximation
GGA + U, 229, 234, 236
Generalized gradient approximation density functional theory (GGA DFT), 204
Geobacter species, 101
Geobacter sulfurreducens, 63, 65, 69, 70
Geobacteraceae, 67
GGA, *see* Generalised gradient approximation (GGA)
GGA DFT, *see* Generalized gradient approximation density functional theory (GGA DFT)
Gibbs free energy, 61, 228
Glass transition concentration, 403
Glassy carbon electrode (GCE), 41, 367
Global warming, 2, 6, 56, 121
Glycerol, 415, 417
GNS, *see* Graphene nanosheets (GNS)
GO, *see* Graphene oxide (GO)
GPE, *see* Gel-type electrolyte (GPE)
Grain boundary, 211, 212, 213
Grain size, 211

Gram negative bacteria, 64, 66
Graphene
 GNS, 310, 312
 layer, 289, 369
 plane, 295
 sheet, 305, 310, 312
Graphene nanosheets (GNS), 310, 312
Graphene oxide (GO), 366
Graphite, 89, 92, 129, 366, 370
Graphite felt, 89, 92, 366, 370
Graphite fiber brush, 89, 90, 92
Graphitic carbon, 165, 288, 290, 295
Graphitization, 301
Gravimetric energy density, 270, 322, 430
Green Freedom concept, 39
Greenhouse gas, 3, 9, 163
Grid corrosion, 260
Grid support, 348
Ground state energy, 226, 227

H

Half-cell, 15, 396, 426, 431, 434, 443
Hamiltonian operator
 electronic, 225, 226
 nuclear, 225, 226
Hard carbon materials, 288, 293, 298
Hard carbon spherules (HCS), 144, 145
Hard lead sulphate, 259
Hartree term, 227
Hartree–Fock method, 226
H/C ratio, *see* Hydrogen-to-carbon ratio (H/C ratio)
HCS, *see* Hard carbon spherules (HCS)
Heavy metal, 100
HEPES, *see* 4-(2-hydroxyethyl)-1-piperazineethanesulphonic acid (HEPES)
HEV, *see* Hybrid electric vehicle (HEV)
Hexacyanoferrate (K$_3$[Fe(CN)$_6$]), 85
Hexa phenyl benzene (HPB), 298, 299
Hexagonal, 30, 168, 211, 230, 288
HF acid, *see* Hydrofluoric acid (HF acid)
Hierarchical nanostructure, 164, 170
High-energy conversion rate, 56
Highest occupied molecular orbital (HOMO), 11, 258
Hohenberg-Kohn theorem, 226

Hollow mesoporous carbon (HMC), 165; see also Electrical double layer capacitor (EDLC)
  catalyst support in fuel cell, 180
  chronoamperogram, 182, 183
  SEM image, 181
  structural features, 166–168
  structural parameters for HMC, 182
  TEM image, 180, 181
HOMO, see Highest occupied molecular orbital (HOMO)
Homopolymer, 134
HPB, see Hexa phenyl benzene (HPB)
HUC, see Hybrid ultracapacitor (HUC)
Hybrid electric vehicle (HEV), 121, 163, 270, 324
Hybrid PbO_2/MHx secondary battery, 450
  charge and discharge curves, 452
  electrode reaction, 450
  pH dependence, 451
Hybrid ultracapacitor (HUC), 323
Hydration sphere, 410
Hydrocarbon synthesis, 37
  carbon dioxide reduction to, 37
  copper electrodes, 38
  energy balance, 39
Hydrofluoric acid (HF acid), 199
Hydrogen
  evolution, 12, 15, 22, 266, 370
  gas diffusion electrode, 442
  hydrogen-oxygen recombinant catalyst, 266
  storage, 29, 164, 269
Hydrogen peroxide (H_2O_2), 449
Hydrogen-to-carbon ratio (H/C ratio), 297
Hydrolysed polyacronitrile (PAN-H), 358
Hydronium ion, 42, 414, 418, 420
Hydrophilic, 354
  carbon black, 24
  clusters, 353
  interaction of gelators, 315
  terminal sulphonic acid groups, 352
Hydrophobic, 354
  carbon black, 24
  interaction of gelators, 315

4-(2-hydroxyethyl)-1-piperazineethanesulphonic acid (HEPES), 87
  capacitance data, 334
  constant-current discharge, 331, 332
  constant-power discharge data, 333–335, 336
  Coulombic efficiency, 333
  energy density data, 337–339
  impedance data, 335–337
  internal resistance vs. potential resistance, 337
  leakage current, 339–341
  parallel resistance data, 339–341
  performance curves, 333
  performance data, 330–333
  power density data, 337–339
  pulse-cycling data, 341–342
  response time data, 338
  self-discharge, 339–341
Hysteresis, 142, 176, 180

I

I-V curve, see Current-voltage curve (I-V curve)
ICE, see Internal combustion engine (ICE)
ILs, see Ionic liquids (ILs)
Imidazole, 16, 136
Impedance, 107, 175, 355, 370
Impregnation, 166, 180, 370, 371
Inoculum source, 64, 66, 83
Inorganic redox couples, 17
In situ propellant production plant, 40
Integrated composite, 201
Integrated layer-structured materials
  delithiation process, 203, 204
  electrochemical reaction pathways, 203
  lithium-rich, 200
  LNMO nanoparticle, 201, 202
  oxygen release, 202
Intercalation, 228, 236, 272
  voltage, 228, 236, 245
Interfacial resistance, 147
Intermetallic phases, 313
Internal combustion engine (ICE), 343
Internal resistance, 79, 86, 94, 104, 335, 337

Index

Internal strain, 195
International Union of Pure and Applied Chemistry (IUPAC), 170
Inverted micelles, 352
Ion clusters, 352, 354
Ion conductivity, 100, 352, 360, 361
Ion crossover, 353
Ion diffusion, 228–229, 238–244
Ion diffusivity, 228–229
Ion exchange membrane, 30, 350, 439, 443, 459
Ion selectivity, 352, 353, 357, 360
Ion transport, 125, 134, 367, 440, 443, 456
Ion-exchange method, 197, 205
Ion-pairing, 410, 423
Ionic conductance, 413
Ionic conductivity, 137, 149, 414, 462; see aslo Ion conductivity
Ionic diffusion, 212
Ionic interface, 430, 438, 464
Ionic liquids (ILs), 136, 313
Ionic strength, 83, 88, 94
Iron phthalocyanine (FePc), 92
Iron-chrome flow battery, 349
Iron-reducing bacteria, 64
Irreversible capacity, 292
IUPAC, see International Union of Pure and Applied Chemistry (IUPAC)

J

Jahn-Teller distortion effect (JT distortion effect), 196, 197, 272
Jones-Doles equation, 407, 409

K

Kinetic energy, 226, 227, 405
Kinetic reversibility, 371
Kinetic stabilisation, 127
Kinetics of CO_2 electroreduction, 14–15
Kohn-Sham equation, 227

L

Large scale energy storage, 348, 349, 396
Lattice distortion, 194
Layer-structured cathode materials, 191
  conventional, 193–200
  energy-storage devices, 191
  integrated, 200–204
  $LiCoO_2$ realisation, 192
  Li diffusion, 204–205
  lithium migration paths, 205
  synthesis and characterizations, 205–206
LDA, see Local density approximation (LDA)
Lead–acid batteries, 257, 324; see also Nickel-based batteries
  corrosion and sulphation, 259
  electrode reactions, 258
  galvanostatic discharge profiles, 261
  nanostructural design, 260
  SEM micrographs, 261
  SLA batteries, 259
  thermodynamic stability, 257
  types of, 257
Lead–carbon hybrid ultracapacitor (Lead–carbon HUC), 325
  applications, 342–345
  cost-effective approach, 324–325
  electrical energy, 322
  electrochemical capacitors, 323
  equivalent circuit for capacitors, 328
  lead–acid battery, 324
  with LED lights, 344
  operating principle, 325–329
Lead oxide/metal hydride secondary battery (PbO_2/MH_x secondary battery), 258, 259
Leakage current, 339–341
Li rechargeable battery, 223, 224
Li–air battery, see Lithium–air battery (Li–air battery)
$Li_4Ti_5O_{12}$/LTO, see Lithium titanium oxide ($Li_4Ti_5O_{12}$/LTO)
LIBs, see Lithium–ion batteries (LIBs)
$LiCF_3SO_3$, see Lithium trifluoromethanesulfonate ($LiCF_3SO_3$)
$LiCoO_2$, see Lithium cobalt oxide ($LiCoO_2$)
$LiFePO_4$, see Lithium iron phosphate ($LiFePO_4$)

Lightning, 322
LiMBO$_3$, see Lithium metal borates (LiMBO$_3$)
LiMn$_2$O$_4$, 282
   discharge curves for, 283
   magnifications for, 284
   SEM micrograph, 283, 284
Li/Na ion diffusion, 228–229
Line seal, 379, 380
Linear model, 399
Linear regression function, 407
LiNiO$_2$, see Lithium nickel oxide (LiNiO$_2$)
Linking species, 68
LiPBs, see Lithium ion polymer batteries (LiPBs)
LiPF$_6$, see Lithium hexafluorophosphate (LiPF$_6$)
Liquid electrolyte, 25, 128, 206, 373
Li/S, see Lithium–sulphur battery (Li/S)
LISICON, see Lithium Ceramic Super-Ionic Conductor (LISICON)
LiTFSI, see Lithium N,N-bis(trifluoromethanesulfonyl) imide (LiTFSI)
Lithiation, 171, 197, 290, 313
Lithium batteries, 121, 313
   all-solid battery, 127
   climate change phenomenon, 122
   electric transportation, 128
   electrolytes for, 313, 314
   galvanostatic lithium intercalation, 314
   HEVs and EV, 122, 124
   ICE vehicles replacement, 122
   and lithium-ion battery systems, 125
   lithium metal, 125
   mass and volume energy density, 123
   PEO–LiX polymer membranes, 126
   rechargeable batteries, 123
   strategies, 315
Lithium Ceramic Super-Ionic Conductor (LISICON), 462, 463
Lithium cobalt oxide (LiCoO$_2$), 193
   advantages, 193
   from cycled cathode, 195
   delithiation process, 194
   lattice parameters variation, 194
   LiCoO$_2$–graphite-based system, 223
   superstructural, 193

Lithium hexafluorophosphate (LiPF$_6$), 31, 127, 314
Lithium intercalation voltages, 236
   average voltages, 236, 237
   LiMPO$_4$, 238
   Li$_x$MBO$_3$, 237
Lithium ion diffusion, 238
   activation barriers and threshold values, 242
   BV mismatch map, 239
   intrinsic properties, 244
   in LiMnBO$_3$ and LiCoBO$_3$, 243
   Li modulation effect, 241–242
   NEB calculation, 241
   trajectories and activation barriers, 240, 242
Lithium ion polymer batteries (LiPBs), 135
Lithium ion storage capacity, 164, 173, 224
Lithium iron phosphate (LiFePO$_4$), 138
   cycle life data and rate capability for, 286
   electrochemical process of, 139
   electronic conductivity data for carbon-coated, 286
   TEM micrographs and HRTEM images, 285
Lithium manganese oxide (LiMnO$_2$), 195
   Ni-substituted, 272
   TEM image of nanostructured, 196
Lithium metal, 286
   anode issue, 155–156
   discharge profiles, 287
   LiFePO$_4$, 287
   rechargeable non-aqueous batteries, 286
Lithium metal borates (LiMBO$_3$), 229
   crystal structure and volume change, 230–232
   density of, 235
   electronic structure and redox reaction, 232–236
   integrated spin density, 233
   lattice parameters, 232
   lithium intercalation voltages, 236–238
   lithium ion diffusion, 238–244
Lithium modulation, 241, 243

Index

Lithium nickel oxide (LiNiO$_2$), 195, 196
Lithium *N,N*-bis(trifluoromethanesulfonyl) imide (LiTFSI), 136, 137
Lithium titanium oxide (Li$_4$Ti$_5$O$_{12}$/LTO), 129, 130
Lithium trifluoromethanesulfonate (LiCF$_3$SO$_3$), 134, 149, 314
Lithium–air battery (Li–air battery), 145; *see also* Super energy density batteries
  aqueous electrolyte, 147
  catalysts for, 153–154
  electrolyte issue, 149–154
  lithium metal anode issue, 155–156
  non-aqueous electrolyte, 148
  oxygen reduction process, 145, 148
Lithium–ion batteries (LIBs), 125, 128, 163, 257, 270, 271; *see also* Electrode materials; Nickel-based batteries
  CMK-3 carbon, 168
  CV plots, 171
  cycling and rate performance, 175
  discharge capacity, 173
  electrode material in, 168
  galvanostatic cycling behaviour, 173
  galvanostatic plots, 172
  Li$^+$ insertion/extraction reactions, 171
  nitrogen adsorption–desorption isotherm, 170
  nyquist plots, 174
  parameters, 174
  properties, 170
  SEM and TEM, 169
Lithium–sulphur battery (Li/S), 142
  charge–discharge voltage profiles, 144
  cycling response, 146
  discharge voltage profile, 143
  HCS–S, 144, 145
  low electronic conductivity, 142
  solubility issue, 143
Load leveling, 381, 383
Local density approximation (LDA), 227
Long-chain hydrocarbons, 2, 26

Lowest unoccupied molecular orbital (LUMO), 11, 258
Low temperature fuel cell, 165

M

Macropores, 170, 175
Magic-angle spinning nuclear magnetic resonance (MAS NMR), 203
MAS NMR, *see* Magic-angle spinning nuclear magnetic resonance (MAS NMR)
Mass spectrometry (MS), 28
Mass transfer of CO_2, 25, 28, 31; *see also* Mass transport
MCMB, *see* Mesocarbon micro-beads (MCMB)
MEA, *see* Membrane electrode assembly (MEA)
MEC, *see* Microbial electrolysis cell (MEC)
Mechanical stability, 147
Mechanical strength, 89, 375, 464
Mediated electrocatalytic reduction
  bimetallic catalysis, 17–18
  of CO_2, 15–16
  electrocatalytic effect of pyridinium salts, 16
  imidazol-catalysed electrochemical reduction, 17
  imidazolonium cations, 16
  inorganic redox couples, 17
  trialkylborane additives, 16, 17
Membrane
  AEM, 439
  BPM, 438–441
  CEM, 27, 31, 98, 355, 439
  fouling, 453
  ion exchange, 28, 350, 439
  monopolar, 452, 456
  NF membrane, 358
  non-fluorinated, 355–357
  PEM, 86
  perfluorinated, 351–355
  processes, 6, 7
  selectivity, 353, 356
  solid-state, 143
  solvent-free-lithium conducting, 134
  surface-area-to-system volume, 87

Membrane electrode assembly (MEA), 443
  cell configuration, 444, 445
Membrane-less interface, 463–464
Membranes separators, 351; *see also* Vanadium flow batteries (VFBs)
  anion exchange membranes, 357–361
  non-fluorinated membranes, 355–357
  perfluorinated membranes, 351–355
Memory effect, 263
MEMS/NEMS, *see* Micro/nano-electromechanical system (MEMS/NEMS)
MES, *see* 2-[N-morpholino] ethane sulphonate (MES)
Mesocarbon micro-beads (MCMB), 292
Mesophase pitch, 308
Mesopores, 167, 168, 170, 178
Mesoporous carbon
  cathode, 144
  ordered, 168
Mesoporous shell, 175, 165 168
Metal hydride electrode (MH_x electrode), 269
Metal-organic framework (MOF), 8
Methane, 9, 22, 38
  analysis for CO_2 reduction, 13
Methanol, 9
  CO_2-saturated, 20
  cold, 20
  economy, 10
  pyridinium ions, 34–37
  synthesis of, 34
MFC, *see* Microbial fuel cell (MFC)
MH-VRF semi-flow battery, 452
  cell reaction, 453
  charge/discharge curves, 455
  electrochemical system, 452, 456
  voltage profiles, 454
MH_x electrode, *see* Metal hydride electrode (MH_x electrode)
Microbattery, 206
Microbial community, 83
  analysis, 67
  molecular analysis, 100, 101
Microbial electrochemistry, 73
  anode performance, 84
  cathode performance, 85–86
  electrolyte chemistry, 87–88
  MFC affecting factors, 83
  power generation, 78–82
  proton transfer efficiency, 86–87
  voltage losses in MFCs, 73–78
Microbial electrolysis cell (MEC), 104
  H_2 production rate, 106
  two-chamber, 105
Microbial fuel cell (MFC), 56, 435
  Microbial–electrochemical conversion devices
  air cathode MFCs, 96–98
  anode materials, 89–91, 92
  anodic electron transfer mechanism, 68–72
  bio-cathode, 98–99
  biocatalysis, 59
  bioelectrochemical system, 58
  cathode materials, 91–93
  cell stacks, 99
  configuration, 97
  continuous flow modes, 99
  current and power generation, 64
  electricity-producing microorganism, 63
  electricity producing, 96
  electrode materials, 89, 91
  electron transfer pathway, 57
  exo-cellular electron transfer, 59–62
  high-energy conversion rate, 56
  microorganism, 57
  microorganism communities, 62–67
  non-Pt catalysts, 93
  phylogenetic tree, 65
  PMFCs, 102–104
  pollutants removing, 99–100
  recent advances, 72–73
  scale-up, 93–95
  SMFCs, 100–102
  types, 95
  working principle, 58
Microbial sensor, 106
Microbial–electrochemical conversion devices
  bacterial batteries, 106–108
  MECs, 104–106
  microbial sensors, 106

Index 483

Microfluidics, 464
Microfracture, 195
Microgrid, 384, 385
Micro/nano-electromechanical system (MEMS/NEMS), 206
Microorganism, 57
   current and power generation, 64
   electricity-producing, 63
   in MFCs, 62
   phylogenetic tree, 65
Micropores, 164, 168, 170, 175
Microporous carbon, 164
Microporous materials, 164
Misch metal alloys, 266
Mobility, lithium ion, 238, 273
MOF, *see* Metal-organic framework (MOF)
Molar volume, 399
Monoclinic symmetry, 194, 206
Monolithic macroporous carbon, 307
Monopolar membrane, 452, 456
2-[N-morpholino] ethane sulphonate (MES), 87
MS, *see* Mass spectrometry (MS)
Multi-walled carbon nanotube (MWCNT), 367
MWCNT, *see* Multi-walled carbon nanotube (MWCNT)

N

Na rechargable batteries, 224, 244
N-methyl-2-pyrrolidone (NMP), 171
N-MPC, *see* Nitrogen-doped mesoporous carbon materials (N-MPC)
n-octadecyl trimethoxysilane (C_{18}-TMS), 166
$Na_4Fe_3(PO_4)_2(P_2O_7)$, 245
   first-principle, 248
   sodium vacancy activation barriers, 247
   3D sodium diffusion path in, 247
$Na_4M_3(PO_4)_2(P_2O_7)$, 244
Nafion®, 352
Nafion proton, 86
Nano silicon, 313
Nano-domain structure, 196
Nanocasting method, 165, 166

Nanocomposite, 132, 311
Nanofiltration membranes (NF membranes), 358
Nanostructured carbon, 164
Nanowire, 58
NAS, *see* Norit A Supra (NAS)
NASICON, 147
National Oceanic and Atmospheric Administration (NOAA), 3
Natural graphite, 290; *see also* Synthetic graphite
   electrochemical performance, 292
   galvanostatic lithium intercalation, 293
   lithium intercalation, 291
   physical parameters, 291
   utilisation, 290
NCM, *see* $LiNi_{1/3}Co_{1/3}Mn_{1/3}O_2$
NEB method, *see* Nudged elastic band method (NEB method)
Negative electrode, 256
Negative plate, *see* Negative electrode
Nernst equation, 85, 430
NF membranes, *see* Nanofiltration membranes (NF membranes)
Ni-Cd batteries, *see* Nickel-cadmium batteries (Ni-Cd batteries)
Nickel cobalt manganese oxide ($LiNi_{1/3}Co_{1/3}Mn_{1/3}O_2$), 196, 278
   cycling performance, 200
   electrochemical properties and structural stability, 197
   performance of macroporous, 200
   rate capability tests on, 199
   superstructural layer, 198
Nickel manganese oxide ($LiNi_{1/2}Mn_{1/2}O_2$), 196, 197, 205
Nickel-based batteries, 261
   Ni–Cd batteries, 262–264
   Ni–Fe batteries, 264–266
   Ni–MH batteries, 266–270
Nickel-cadmium batteries (Ni-Cd batteries), 262
   Bode diagram, 263
   discharge profiles of, 264
   electrochemical reactions, 262
   memory effect, 263
   operating principle, 262

Nickel-iron batteries, 264–266; *see also* Sealed Ni-Fe batteries
Nickel–iron batteries (Ni–Fe batteries), 264; *see also* Sealed Ni-Fe batteries
   cell reactions, 264
   charge–discharge reactions, 264, 265
   float-charge applications, 266
   galvanostatic discharge profiles, 265
Nickel–metal hydride batteries (Ni–MH batteries), 266, 454, 458, 460
   cell voltage and internal pressure, 268
   electrochemical reactions, 268
   hydrogen absorption–desorption, 267
   magnesium-based alloys, 267
   metallurgical processes, 269
   misch metal alloys, 266
   specific energy of, 270
Ni–Fe batteries, *see* Nickel–iron batteries (Ni–Fe batteries)
NiMH$_x$ battery, *see* Nickel–metal hydride batteries (Ni–MH batteries)
Nitrate
   chemical oxidiser, 58
   electron acceptor, 57, 61
   LiNiO$_2$ preparation, 207
Nitrogen-doped mesoporous carbon materials (N-MPC), 369
Nitrogenisation treatment, 368
NMP, *see* N-methyl-2-pyrrolidone (NMP)
NMR spectroscopy, *see* Nuclear magnetic resonance spectroscopy (NMR spectroscopy)
N,N-bis(trifluoromethanesulphonyl) imide (TFSI), 136
NOAA, *see* National Oceanic and Atmospheric Administration (NOAA)
Non-aqueous electrolyte, 148
Non-fluorinated membranes, 355; *see also* Perfluorinated membranes
   CEM, 355
   SDPEEK, 356, 357
   SPEEK membranes, 356
   sulphonated poly(arylenethioether ketone), 355
Non-graphitisible carbon fibres, 301
Non-Pt catalysts, 93
Norit A Supra (NAS), 176
Nuclear Hamiltonian operator, 225
Nuclear magnetic resonance spectroscopy (NMR spectroscopy), 28
Nucleophilically stable, 149
Nudged elastic band method (NEB method), 229
Nyquist plot, 174, 175

O

OCP, *see* Open circuit potential (OCP)
Octahedral site, 193, 198
OER, *see* Oxygen evolution reaction (OER)
Off-grid, 384
OGS, *see* Oxidised graphene sheets (OGS)
Ohmic loss, 75
Ohmic polarisation, *see* Ohmic loss
Olivine structure, 276
OM, *see* Outer membrane (OM)
Open circuit potential (OCP), 77, 228
Operation temperature, 67
Organic carbamates, 16
Organic carbonates, 31
   carbonyl equivalents, 31
   dimethyl carbonate, 32, 33
   lithium ion batteries, 31
Organic electrolyte
   cell voltage, 429
   CO$_2$ reduction, 25
Organic liquid electrolyte, 128, 143
Organosilica, 165
ORR, *see* Oxygen reduction reaction (ORR)
Oscillatory surface displacement, 212
Outer membrane (OM), 68
Overheating, 128, 143
Overpotential
   anode material, 84
   carbon dioxide reduction, 34
   electrocatalytic activity evaluation, 16
   electrodes, 75

Index 485

Ovshinsky's concept of disorder, 266
Oxidation treatment, 368
Oxidised graphene sheets (OGS), 311
Oxygen evolution reaction (OER), 150
Oxygen pump, 435
Oxygen reduction reaction (ORR), 150, 165

P

PA, *see* Polyacrylic acid (PA)
PAAK, *see* Potassium-salt poly (acrylic acid) (PAAK)
PAH, *see* Polycyclic aromatic hydrocarbon (PAH)
PAN, *see* Poly(acrylo nitrile) (PAN)
PAN-H, *see* Hydrolysed polyacronitrile (PAN-H)
PANI, *see* Polyaniline (PANI)
Parallel laminar flow, 463–464
Parallel resistance, 339–341
Passivation, 171, 361
Pauli exclusion principle, 226
PAW, *see* Projector-augmented wave (PAW)
PBE functional method, *see* Perdew-Burke-Ernzerhof functional method (PBE functional method)
PbO_2/MH_x secondary battery, *see* Lead oxide/metal hydride secondary battery (PbO_2/MH_x secondary battery)
PC, *see* Propylene carbonate (PC)
PCP pincers, 18
Peak shaving, 380, 383
PEC, *see* Photoelectrocatalytic cell (PEC)
PEEK, *see* Polyether ether ketone (PEEK)
PEI, *see* Polyethylenimine (PEI)
Pelotomaculum thermopropionicum bacteria, 70
PEM, *see* Proton exchange membrane (PEM)
PEMFC, *see* Proton exchange membrane fuel cell (PEMFC)
PEO, *see* Poly(ethylene oxide) (PEO)
PEPs, *see* Power energy plants (PEPs)

Perdew-Burke-Ernzerhof functional method (PBE functional method), 227
Perfluorinated membranes, 351
 characteristics, 351, 352
 cluster-network model, 352
 fluorinated polymers, 354
 inorganic–organic composite membranes, 353
 Nafion membrane, 352
 PVDF, 353
Perm-selectivity, 458, 464
Permanganate ($KMnO_4$), 85, 92
Permeability
 vanadium, 357
 V/H ratio, 360
Permeation
 high ion selectivity, 352
 membranes, 6
3,4,9,10-peryleneterta-3,4,9,10-carboxylic di-anhydride (PTCDA), 297
PES membranes, *see* Polyether sulphone membranes (PES membranes)
Petroleum pitch, 292
PEV, *see* Plug-in electric vehicle (PEV)
Phase separation, 354
Phase transformation, 196, 202
pH dependence, 431, 451
pH differential power sources, 429
 electrochemical power sources, 430
 electrode potential variation, 435
 half-cell reactions, 433–434
 Nernst equation, 430
 Pourbaix diagram of water, 431
pH insensitive electrode reaction, 432
pH sensitive electrode reaction, 429, 432
Photoelectrocatalytic cell (PEC), 26
Piperazine-*N*,*N*-bis [2-ethanesulphonate] (PIPES), 87
Plant-microbial fuel cell (PMFC), 96, 102
 advantage, 104
 aquatic plants, 103
 electricity production, 103
Platinum (Pt), 94, 104, 448
Plug-in electric vehicle (PEV), 163
PMFC, *see* Plant-microbial fuel cell (PMFC)

PMMA, *see* Poly(methyl methacrylate) (PMMA)
Polarisation
　activation, 75–77
　concentration, 77–78
　curve, 442, 444, 446
　effect, 199
　H_2-fueled fuel cell, 182
　ohmic, 75
Poly(acrylo nitrile) (PAN), 135, 303
Poly(ethylene oxide) (PEO), 125
Poly(methyl methacrylate) (PMMA), 307
Poly(vinylidene fluoride), *see* Polyvinylidene (PVDF)
Poly (vinyl pyrrolidone) (PVP), 360
Polyacrylic acid (PA), 374
Polyaniline (PANI), 84
Polyanionic cathode materials, 229
　crystal structure and volume change, 230–232
　density, 235
　electronic structure and redox reaction, 232–236
　integrated spin density, 233
　lattice parameters, 232
　lithium intercalation voltages, 236–238
　lithium ion diffusion, 238–244
Polyanionic materials, 224
Polycrystalline, 30, 67
Polycyclic aromatic hydrocarbon (PAH), 298
Polyether ether ketone (PEEK), 355
Polyether sulphone membranes (PES membranes), 360
Polyethylenimine (PEI), 8, 354
Polymer electrolyte, 24, 124, 149
Polypropylene separator, 171
Polypyrole, 354
Polysulfide-bromine flow battery, 349, 434
Polysulphone-2-amide-benzimidazole (PSf-ABIm), 353
Polytetrafluoroethylene (PTFE), 24
Polyvinyl chloride (PVC), 291
Polyvinylidene (PVDF), 135, 171, 353
Pore size distribution (PSD), 168
Porous carbon, 292
　as anode material, 172
　HMC, 166–168

Positive electrode, 256
Positive plate, *see* Positive electrode
Potassium ion battery, 269, 465
Potassium-salt poly (acrylic acid) (PAAK), 460
Pourbaix diagram, 431
Power density, 79, 335
Power energy plants (PEPs), 163
Power generation, 78
　COD removal, 81–82
　Coulombic efficiency, 80
　maximum cell voltage, 78
　MFC energy efficiency, 81
　power density, 79
Power output
　fuel cell characteristics, 79
　MFC energy, 78
　seawater, 459
Precipitation inhibitor, 374
Precursor
　carbon, 166
　hydroxide, 205
　impregnation method, 180
　organic, 303
Projector-augmented wave (PAW), 227
Propanol, 37
Propylene carbonate (PC), 290, 314
Proteobacteria, *see* Iron-reducing bacteria
Proton conductivity, 352, 361
Proton exchange membrane (PEM), 86
Proton exchange membrane fuel cell (PEMFC), 165, 184, 444
Proton transfer efficiency, 86–87
PSD, *see* Pore size distribution (PSD)
PSf-ABIm, *see* Polysulphone-2-amide-benzimidazole (PSf-ABIm)
Pt catalyst, 91
PTCDA, *see* 3,4,9,10-peryleneterta-3,10-carboxylic di-anhydride (PTCDA)
PTFE, *see* Polytetrafluoroethylene (PTFE)
Pulse-cycling data, 341–342
Pulverisation, 130, 269
PVC, *see* Polyvinyl chloride (PVC)
PVDF, *see* Polyvinylidene (PVDF)
PVP, *see* Poly (vinyl pyrrolidone) (PVP)

Index

Pyridinium electrocatalyst, 16
Pyridinium ions, 34
   carbon dioxide, 34
   formaldehyde, 36
   formic acid to methanol, 36
   inner-sphere-type electron transfer, 35
   step-by-step reduction, 37
Pyrochlore, 152
Pyrolysis, 296, 303

Q

Q-factor mapping, 212
Quaternization, 357, 358

R

Rate capability, 174
Rate performance
   cathode materials, 204–205
   from Nyquist plots, 175
Re-oxidation process, 150
Reaction overpotential, 92, 93, 369
Rechargeable battery, 256
Rectisol process, 20
Recyclability, 323
RED, *see* Reverse electro-dialysis (RED)
Redox couple, 17, 202, 279
Redox mediator, 84
   endogenous, 71
   exogenous, 70–71
Redox potential, 68, 77, 236–237
Redox reaction, 10, 75
   with delithiation, 232–236
   V^{3+}/V^{2+} redox, 372
Reduced metabolic products, 72
Reed manna grass (*Glyceria maxima*), 103
Regenerative braking, 324, 343
Regression coefficient, 399, 400
Relative viscosity, 407
Repellent effect, 357
Resonance
   amplitude, 211, 213
   frequency, 211, 212
   quality factor, 211
Resorcinol–formaldehyde resin (RF resin), 307
Response time, 323

Reverse bias, 440
Reverse electro-dialysis (RED), 459
Reversibility, 174, 367, 371
Reversible capacity, 193, 196, 203
Rhombohedral, 193, 288
Ruthenium (Ru), 29, 30, 152
Ruthenium phosphine systems, 30

S

S-Radel, 356
SAC, *see* Super active carbon (SAC)
Sacrificial anode, 32, 43
Salting, 458, 459
Saturation, 20, 407
Scale-up, 93
   cost reduction, 95
   electrode spacing reducing, 94–95
   electrolyte composition, 94
   on mL-scale MFCs, 93
Scanning electron microscopy (SEM), 167
Scanning probe microscopy (SPM), 209
Scanning transmission electron microscopy (STEM), 201
Schrödinger equation
   equation, 226
   time-dependent Schrödinger, 225
   time-independent Schrödinger, 225
SCMS, *see* Solid core–mesoporous shell (SCMS)
SDLF, *see* Self-discharge energy loss factor (SDLF)
SDPEEK, 356, 357
Sealed lead–acid batteries (SLA batteries), 257
Sealed Ni-Fe batteries, 266
Second order polynomial, 399
Second Wien effect, 440
Sediment microbial fuel cell (SMFC), 96, 100
   electricity generation, 101
   electrode modification, 102
   Geobacteraceae, 101
   molecular analysis, 100–101
SEI, *see* Solid-electrolyte interphase (SEI)
Selective transfer, 361
Self-discharge, 339–341

Self-discharge energy loss factor (SDLF), 339
Self-humidifying, 444
SEM, see Scanning electron microscopy (SEM)
Separator
　cell, 27–28
　membranes, 351–361
　proton transfer, 95
Shewanella oneidensis, 107
SHMP, see Sodium hexametaphosphate (SHMP)
Short-chain hydrocarbons, 2
Shunt current, 378, 379
SI, see Substrate-integrated (SI)
Silica gel, 330, 333, 341
Silicon-carbon composite, 155
Single cell, 182, 327, 349, 351, 356
Single-chamber reactor, 80, 96, 105
SLA batteries, see Sealed lead–acid batteries (SLA batteries)
Slater determinant, 226
SLI, see Starting–lighting–ignition (SLI)
SMFC, see Sediment microbial fuel cell (SMFC)
SOC, see State of charge (SOC)
Sodium hexametaphosphate (SHMP), 374
Sodium ion battery, 247; see also Na rechargable batteries
Soft carbon, 292
Sol-gel method, 199, 281, 353
Solar photovoltaic, 343
Solid electrolyte, 155, 206
Solid polymer electrolyte (SPE), 27
Solid solution, 197, 201
Solid state insertion/deinsertion, 289
Solid core–mesoporous shell (SCMS), 165
Solid-electrolyte interphase (SEI), 127, 171, 199, 292, 314
Solid-state membrane, 143
Solubility
　CO_2 in water, 18
　electrolyte, 375
Solute-solute attraction, 403
Solution
　density, 399

　viscosity, 404
Solvent-free-lithium conducting membrane, 134, 143
SPE, see Solid polymer electrolyte (SPE)
Specific capacitance, 178, 179
Specific capacity
　$LiMBO_3$ and $LiMPO_4$, 238
　for $LiNiO_2$, 276
　LTO, 130
Specific conductance, 414
Specific energy
　electrolyte cell, 426
　Ni–MH batteries, 270
SPEEK, see Sulphonated poly(ether ether ketone) (SPEEK)
Spin
　functions, 226
　moment, 232, 233
　state, 232
Spinel structure, 195, 279
　cubic spinel structure, 196
　tetragonal spinel structure, 196
SPM, see Scanning probe microscopy (SPM)
Stack, 375
　cell stack construction, 375
　flow field, 378
　flow pattern of electrolyte, 375, 377
　power output and energy efficiency, 377, 378
　sealing of stack, 379, 380
　shunt current, 378, 379
Staging, 289
Stainless steel, 31, 92, 95
Starting–lighting–ignition (SLI), 260
Starved-electrolyte, 266
State of charge (SOC), 373
STEM, see Scanning transmission electron microscopy (STEM)
Stiffness, 212
Structure-breaking effect, 410, 425
Substrate-integrated (SI), 324
　lead-dioxide, 324
　$SI–PbO_2$, 324
Sulfate, see Sulphate
Sulfation, see Sulphation
Sulfide, see Sulphide
Sulfonated polyethylene, 355
Sulfuric acid; see also Sulphuric acid

Index

Sulphate, 57, 61
Sulphation, 259
Sulphide, 101, 142
Sulphonated poly(ether ether ketone) (SPEEK), 353
Sulphuric acid
　conductivity effect, 418–419
　density effect, 399
　viscosity effect, 403–404
Super active carbon (SAC), 371
Supercapacitor, 163, 322
Super energy density batteries, 141; *see also* Lithium batteries
　insertion reactions, 142
　lithium–sulphur battery, 142–145
Superlattice, 198
Support electrolyte, 17, 28, 33
Surface roughness, 211, 212
Swelling, 28, 353
Syn gas, 4
Synthetic graphite, 292; *see also* Carbonaceous materials
　aromatic hydrocarbons pyrolysis, 296
　carbon fibres, 300
　carbon monolith, 309
　carbon nanofibrils, 305, 306
　carbon sphere, 308
　CC, 305
　CNFs, 306
　corrolenulene structure, 298, 299
　cycle life data for, 296
　cyclic voltammograms, 299
　graphitic carbons, 295
　hard carbons, 293, 298
　HPB, 299, 300
　HRTEM images, 301
　lithium intercalation, 294, 301, 307
　MCMB carbons, 294
　monolithic macroporous carbon, 307
　non-graphitisible carbon fibres, 301
　PAN, 303
　physical, chemical and electrochemical properties, 296
　propene pyrolised carbon fibre, 302
　PTCDA, 297
　soft carbon, 292
　3DOM, 308, 309

T

TEGDME, *see* Tetra (ethylene) glycol dimethyl ether (TEGDME)
TEM, *see* Transmission electron microscopy (TEM)
Temperature effect
　conductivity effect, 420
　density effect, 399
　viscosity effect, 405–407
Template
　carbon monoliths preparation, 308
　carbon sphere with and without, 308
　SBA-15 silica, 168
　SCMS silica as, 165
Tetra (ethylene) glycol dimethyl ether (TEGDME), 149
Tetraethylorthosilicate (TEOS), 166, 353
Tetrafluoroborate (Et_4NBF_4), 177
Tetragonal spinel structure, 196
Tetrahedral site, 241
Tetrahedral site hop (TSH), 204
Tetraphenylporphyrin catalyst, 41
TFSI, *see* N,N-bis(trifluoromethanesulphonyl) imide (TFSI)
Thermal precipitation, 396
Thermal stability, 28, 197, 224, 373
Thermodynamic instability, 246
Thermodynamics of CO_2 electroreduction, 12–14
Thermodynamic stability, 257
Thermodynamic window; *see also* Thermodynamic stability
Three electrolyte acid alkaline PbO_2/MH_x hybrid battery, 456, 458
Three-dimensionally interconnected nanostructures, 184
Three-dimensional macroporous ordered carbon monoliths (3DOM), 307
3D sodium diffusion, 247
Time-dependent Schrödinger equation, 225
Time-independent Schrödinger equation, 225
Titanium, 324
Topography, 211, 212
Transition metals (TMs), 224
Transition state theory, 228

Transmission electron microscopy (TEM), 194
Trialkylborane additives, 16
Triangular phase diagram, 280
Trickle-bed reactor, 30
TSH, *see* Tetrahedral site hop (TSH)
12 V/kF-range lead–carbon HUCs, 329; *see also* Lead–carbon hybrid ultracapacitor (Lead–carbon HUC)
Two-phase electrochemical process, 130, 138, 141

U

Ultracapacitor, 322, 323, 345
Uninterrupted power supply (UPS), 266, 348
Universal functionals, 227
Up-flow biofilter circuit (UBFC), 109

V

V-MH$_x$ hybrid battery, *see* Vanadium-metal hydride hybrid battery (V-MH$_x$ hybrid battery)
Valence, 100, 196, 206, 349, 369
Valve-regulated lead–acid batteries (VRLA batteries), 257, 260
Vanadium (II) sulphate solution, 396, 422; *see also* Vanadium redox flow battery (VRFB)
　change in viscosity, 423
　conductivity, 423, 424
　density, 423, 424
　viscosity, 423, 424
Vanadium (III) sulphate solution, 396
　Arrhenius plot, 406
　concentration. effect, 397–399
　conductivity, 412–422, 424
　conductivity–viscosity curve, 416, 417
　density, 397, 398–401
　parameters and regression coefficients, 399, 400
　V$_2$(SO$_4$)$_3$ density values, 401
　V$_2$(SO$_4$)$_3$ viscosity values, 402
　viscosity, 401–412, 424
　viscosity–acid concentration, 405

Vanadium flow batteries (VFBs), 347, 349
　application and demonstration, 380–384
　assembly, 351
　challenges and prospective, 384–387
　charge and discharge process, 349
　cost map of Rongke Power, 387
　DICP, 382
　EES, 347
　electrolytes, 372–375
　FBS, 349
　5 MW/10 MWh VFB system, 383
　megawatt-class power generation, 382
　microgrid assembled in SEI, 385
　MW/6 MWh VFB system, 381
　peak shaving and load leveling, 383
　principle, 350
　stacks, 375–380
　2 kW/10 kWh VFB system, 385
　200 kW/800 kWh VFB system, 386
Vanadium redox flow battery (VRFB), 396, 397
Vanadium sulfate, 403, 417, 420
Vanadium to proton permeability (V/H) ratio, 360
Vanadium-metal hydride hybrid battery (V-MH$_x$ hybrid battery), 453
Vanadium/oxygen fuel cell (V/O$_2$ fuel cell), 396
VASP, *see* Vienna abinitio simulation package (VASP)
VE, *see* Voltage efficiency (VE)
VFBs, *see* Vanadium flow batteries (VFBs)
Vienna abinitio simulation package (VASP), 227
Viscosity-concentration curve, 403
Viscosity activation energy, 405, 425
V/O$_2$ fuel cell, *see* Vanadium/oxygen fuel cell (V/O$_2$ fuel cell)
Voltage, 73, 256, 322
　activation losses, 75–77
　bacterial metabolic losses, 77
　bias, 439, 443
　concentration losses, 77–78
　cut-off voltage, 263, 279, 329
　depression, 263
　efficiency, 352
　equilibrium intercalation voltage, 228

losses, 73
ohmic losses, 75
profile, 155, 156, 228, 454
standard electrode potentials, 74
Voltage efficiency (VE), 352
Volume
expansion, 269
stress, 133
Volumetric energy density, 132, 270, 322
VRFB, *see* Vanadium redox flow battery (VRFB)
VRLA batteries, *see* Valve-regulated lead–acid batteries (VRLA batteries)
Vulcan XC-72, 38, 165, 436

W

Walden constant, 418
Walden rule, 415, 418
Wastewater treatment, 81
aeration treatment, 109
MFCs producing electricity, 96–100
sludge treatment, 109
Water management, 430, 444, 447
Water splitting, 440, 443
Wavefunction, 225, 226
Wind power, 380, 384
Wind turbine, 343, 344
Working electrode, 24, 38

X

X-ray adsorption spectroscopy (XAS), 196
X-ray diffraction (XRD), 206, 260
X-ray photoelectron spectroscopy (XPS), 206

Z

Zinc-bromine flow battery, 349
Zview code, 175